NATO ASI Series

Advanced Science Institutes Series

A series presenting the results of activities sponsored by the NATO Science Committee, which aims at the dissemination of advanced scientific and technological knowledge, with a view to strengthening links between scientific communities.

The Series is published by an international board of publishers in conjunction with the NATO Scientific Affairs Division

A Life Sciences B Physics	Plenum Publishing Corporation London and New York
C Mathematical and Physical Sciences D Behavioural and Social Sciences E Applied Sciences	Kluwer Academic Publishers Dordrecht, Boston and London
F Computer and Systems Sciences G Ecological Sciences H Cell Biology I Global Environmental Change	Springer-Verlag Berlin Heidelberg New York London Paris Tokyo Hong Kong Barcelona Budapest

PARTNERSHIP SUB-SERIES

1. Disarmament Technologies	Kluwer Academic Publishers
2. Environment	Springer-Verlag/Kluwer Acad. Publishers
3. High Technology	Kluwer Academic Publishers
4. Science and Technology Policy	Kluwer Academic Publishers
5. Computer Networking	Kluwer Academic Publishers

The Partnership Sub-Series incorporates activities undertaken in collaboration with NATO's Cooperation Partners, the countries of the CIS and Central and Eastern Europe, in Priority Areas of concern to those countries.

NATO-PCO DATABASE

The electronic index to the NATO ASI Series provides full bibliographical references (with keywords and/or abstracts) to about 50000 contributions from international scientists published in all sections of the NATO ASI Series. Access to the NATO-PCO DATABASE is possible via a CD-ROM "NATO Science & Technology Disk" with user-friendly retrieval software in English, French and German (© WTV GmbH and DATAWARE Technologies Inc. 1992).

The CD-ROM can be ordered through any member of the Board of Publishers or through NATO-PCO, Overijse, Belgium.

Series G: Ecological Sciences, Vol. 42

Springer
Berlin
Heidelberg
New York
Barcelona
Budapest
Hong Kong
London
Milan
Paris
Singapore
Tokyo

Desulfurization of Hot Coal Gas

Edited by

Aysel T. Atimtay

Middle East Technical University
Environmental Engineering Department
06531 Ankara, TURKEY

Douglas P. Harrison

Louisiana State University
Department of Chemical Engineering
Baton Rouge, LA 70803, USA

With 166 Figures

Springer

Proceedings of the NATO Advanced Study Institute on Desulfurization of Hot Coal Gas with Regenerable Metal Oxide Sorbents: New Developments, held in Kusadasi, Turkey, July 1996

Cataloging-in-Publication Data applied for

Die Deutsche Bibliothek - CIP-Einheitsaufnahme

Desulfurization of hot coal gas / ed. by Aysel T. Atimtay. - Berlin ; Heidelberg ; New York ; Barcelona ; Budapest ; Hong Kong ; London ; Milan ; Paris ; Singapore ; Tokyo : Springer, 1998
(NATO ASI series : Ser. G, Ecological sciences ; Vol. 42)
ISBN 3-540-64726-0

ISSN 0258-1256
ISBN 3-540-64726-0 Springer-Verlag Berlin Heidelberg New York

Printed in Germany

Typesetting: Camera ready by authors/editors
Printed on acid-free paper
SPIN 10531508 31/3136 - 5 4 3 2 1 0

Preface

Optimal development of high efficiency advanced coal-fired power systems such as the Integrated Gasification Combined Cycle (IGCC) process requires that contaminants, including sulfur compounds (primarily H_2S), be removed at high temperature. Energy conversion efficiencies as high as 50 % may be possible in the future with improvements in gas turbines and implementation of high temperature desulfurization technology. Current emphasis on the control of greenhouse gas emissions provides added incentive for increasing efficiency. Otherwise, coal, the world's most abundant fossil fuel, may be largely replaced by lower carbon-content fossil fuels.

Therefore, the NATO Advanced Study Institute on the subject of **"Desulfurization of Hot Coal Gas with Regenerable Metal Oxide Sorbents"** was quite timely. Leading scientists and engineers from different countries met in Kusadasi, Aydin, Türkiye to discuss the current status and to identify the future research needs in this area.

Eleven lecturers covered a broad range of related topics such as an examination of world energy resources, gasification technology, and IGCC process development in addition to the primary topic of hot coal gas desulfurization. This book includes the original contributions of these specialists, each of whom deserves special thanks. Without their efforts the ASI and this proceedings would not have been possible.

In addition to the lecturers, special thanks are expressed to the Organizing Committee and the participants of this ASI for making the institute so stimulating and successful. The ASI was generously supported by the Scientific and Environmental Affairs Division of NATO and their contribution is gratefully acknowledged.

Finally, we thank all those people who contributed to the organizational aspects of the ASI, in particular to Valör Tourism, Drs. Serpil Kayin, Gökhan Ölmez, Süleyman Kaytakoglu and two graduate students of the Environmental Engineering Department of the Middle East Technical University, Ms.Canan Derinöz and Ms. Perin Ayse Cengiz.

Douglas P. Harrison
Aysel T. Atimtay

Contents

World Energy Resources, Their Use and the Environment

P W Sage

Coal Technology Development Division, British Coal Corporation,
PO Box 199, Stoke Orchard, Cheltenham, GL52 4ZG, UK

Abstract. Energy demand continues to increase with population growth and economic development. Analysis suggests that non-renewable energy will not be resource limited in the medium term but that patterns of energy use may alter. Environmental impacts of different fuels are overviewed with special reference to coal. Technologies to lessen the environmental impact of coal use are presented.

1 Historical Energy Demand

Increasing world population and economic development is resulting in an ever growing global primary energy demand. These factors together with advances in technology influence the patterns of primary energy use.

The World Energy Council/IIASA[1] describe developing global energy needs. Before the industrial revolution burning traditional renewables such as wood or tallow candles was the principal conversion of chemical energy into heat and light. Also natural energy flows were harnessed by the use of watermills and windmills. During this period energy consumption was typically less than 0.5 toe per capita per year[2].

The 'steam age' which took off early in the 19th century following the development of the coal-fired steam engine resulted in radical changes in energy end use. Using the steam cycle fossil energy could be converted into work. Early applications included pumping water from coal-mines and driving machinery in the first factories. Because coal was easily transported provision of energy could be site dependent and relatively high power intensities achieved. Later mobile steam engines were used in railway locomotives and ships. This spawned a transport revolution. Energy consumption of 2 toe per capita per year was typical during the steam age.

Coal's share of global primary energy demand increased from about 10% in the early 19th century to approaching 50% by 1890 and reaching a peak of over 60% by the start of World War 1. The traditional renewable share fell from about 90% to 30% over the same period.

In the last 100 years or so the introduction of electricity and the development of the internal combustion engine has produced further radical changes in energy end use patterns. The use of traditional renewables has continued to decline and now accounts for only about 10% of global primary energy demand. Coal's share has declined steadily and now accounts for just under 30% of total demand. Oil and gas on the other hand now have significant market share. Both took a negligible share in 1900. Oil's share peaked at over 40% in about 1970 and has

NATO ASI Series, Vol. G 42
Desulfurization of Hot Coal Gas
Edited by A. T. Atimtay and D. P. Harrison

since declined to just over 30%. Natural gas' share of the market has increased steadily this century and now accounts for about 20% of total global primary energy demand.

Whilst patterns of primary energy shares have changed dramatically during the last two centuries, total demand has increased relentlessly because of population growth coupled with economic development and technological advancement. World population has risen from just over 1 billion in 1850 to 5.3 billion in 1990[3]. Increase has not been linear with a doubling occurring over the period 1950 to 1990. Prior to 1850 total global energy demand associated with a population of c.1 billion was less than 0.5 Gtoe per annum. Demand is now some twenty-fold this value at about 9 Gtoe per annum. The table below[4] shows consumptions for commercially traded fuels for 1969 and 1994. The difference between the 1994 total in the table of 8 Gtoe and the 9 Gtoe referred to above is due to traditional renewables which are unreliably documented in terms of consumption statistics but are thought to account currently for about 10% of global primary energy demand.

Table 1.1 Global Fuel Consumption 1969 and 1994[4]

	1969 Gtoe	1994 Gtoe
Oil	2.1	3.2
Gas	0.9	1.8
Coal	1.5	2.2
Nuclear	Negligible	0.6
Hydroelectric	0.1	0.2
Total	4.6	8.0

An increase in global primary energy demand of over 70% has occurred in the last 25 years. This has been characterised by a steady rise from 1969-1973 and a plateau in demand from 1973-1975 following the first oil crisis and associated price shock. Demand then increased through to the second oil crisis of 1979. This precipitated a slight drop in demand during the period 1979-1982. A fairly rapid increase ensued during the boom of the mid to late 1980s followed by a plateau in the early 1990s caused by global recession and restructuring in the former USSR and former Eastern Bloc Countries. Demand is now again showing an increase.

2 Energy Resources and Future Energy Demand

Whilst historical information can be reported with some confidence future global energy demand can not be predicted accurately. It is certain however that in the foreseeable future demand will increase because of both growth in population and continued economic development within the less mature economies.

Population is clearly a major factor. The World Bank scenario[3] predicts a global population growth to 10.6 billion by 2060 which will stabilise at about 12 billion just post 2100. This compares with projections ranging from 9 billion (IIASA low) to 19 billion (UN high) in 2100.

Reference 1 presents three scenarios for energy use for the year 2050 and 2100. Some important numbers and assumptions for these scenarios are given in Table 2 below:

Table 2.1 Future Energy Scenarios (Adapted from Ref 1)

	Case		
	A High Growth	B Middle Course	C Ecologically Driven
Population in 10^9			
Year 2050	10.1	10.1	10.1
Year 2100	11.7	11.7	11.7
Global World Product in 10^{12} US (1990) $			
Year 2050	100	75	75
Year 2100	300	200	220
Primary energy demand Gtoe			
Year 2050	25	20	14
Year 2100	45	35	21
Resource availability			
Fossil	High	Medium	Low
Non-fossil	High	Medium	High
CO_2 emission constraint	No	No	Yes
Carbon emissions, GtC			
Year 2050	9-15	10	5
Year 2100	7-22	14	2
Environmental taxes	No	No	Yes

The population projections used are those of the World Bank[3]. Values for global world product represent three to five-fold increases on the current value by the year 2050 and ten to fifteen-fold increases by the year 2100. These figures mean that per capita income in most of the currently developing countries will be greater than that of developed countries today. Primary energy demand will increase by between 150 and 250% by 2050 and between 200 and 500% by 2100. Demand for energy services will increase by a significantly larger amount with the differences being met by technological advancement and efficiency improvement.

Case A is based on high rates of economic growth (c.2% p.a in OECD countries and 4% p.a in the developing countries) and technological progress. It includes three sub-scenarios on energy supply. One is dominated by oil and gas, one by a massive return to coal towards the end of the 21st century and a third where rapid nuclear and renewable technological advancement leads to an economic phase out of fossil fuels.

Case B incorporates more modest estimates of economic growth and technological development. It results in heavy reliance on fossil fuels without major changes in energy use patterns in the short to medium term. In the longer term economic factors rather than resource limitations (see later) will cause a move away from fossil fuel energy sources.

The 'ecologically driven' Case C is based on concerted international cooperation on environmental protection and international equity. A raft of policies would be developed to encourage careful and efficient energy use. These would include policies to reduce global carbon emissions to one third of today's level. Within Case C two options for nuclear energy are considered. First there is development of a new generation of nuclear reactors that find widespread acceptability. Alternatively nuclear power proves to be a transient technology and is phased out by the end of the 21st century.

The WEC/IIASA study suggest massive increases in future global energy demand for all three cases. The patterns for meeting these demands vary but in the short to medium term usage of fossil fuels will continue to rise. It is therefore pertinent to consider global availability of non-renewable energy. Table 2.2 below has been adapted from Reference 1.

Table 2.2 Global Fossil and Nuclear Energy Reserves, Resources, and Occurrences in Gtoe[1]

	Consumption		Reserves	Resources	Additional Occurrences
	1850-1990	1990			
Fossil:					
Oil					
Conventional	90	3.2	150	145	
Unconventional	-	-	193	332	1900
Natural Gas					
Conventional	41	1.7	141	279	400
Unconventional	-	-	192	258	18700
Hydrates	-	-	-	-	
Coal	125	2.2	606	2794	3000
Nuclear:					
Uranium (thermal)	17	0.5	57	203	150
Uranium (FBRs)	-	-	3390	12150	8900

In the above table 'reserves' are those occurrences that are known, and are recoverable with present technologies. 'Resources' are those occurrences, in addition to reserves, with less certain geological assurance, or lacking present economic feasibility or both. The sum of reserves and resources is known as the 'resource base'. This includes all potentially recoverable coal, conventional oil and natural gas, unconventional oil resources (such as oil shale, tar sands and heavy crude), and unconventional natural gas resources (such as gas in Devonian shale, tight sand formations, gas pressurised aquifers, and coal seams). 'Additional occurrences' are not included in the resource base of future, potentially recoverable resources. They include methane hydrates and natural uranium dissolved in sea water. Economic factors together with technological and scientific advances mean that there is a steady transfer from 'resources' to 'reserves'. Similarly the 'resource base' is increasing with new discoveries.

The National Academies Policy Advisory Group[5] have considered how long reserves and resources will last at current usage rates. Their analysis based on WEC data[6] is summarised in the following table.

Table 2.3 Lifetime of Reserves and Resources at Current Usage Rates

	Lifetime in Years	
	Reserves	Resources
Hard Coal	200	} 1250
Lignite	300	
Oil	40	230
Natural Gas	56	115
Uranium	40	200 or 10000*

*Assumes successful development and deployment of fast breeder reactor technology.

It is apparent from Tables 2.2 and 2.3 that in the absence of fast breeder reactors both reserves and resources are dominated by coal. At current usage rates reserves of all non-renewables would last a little over 100 years. Inclusion of resources extends the timescale to some 600 years. However if fast breeder technology developed then the dominant fuel would be uranium which would account for some 60% of the energy recoverable from non-renewable sources (cf 30% from hard and soft coal). Without fast breeder technology thermal nuclear technology would account for just 3% of energy recoverable.

Renewable energy sources currently contribute about 15% to World Energy use. Renewables can be split into two categories, viz, traditional and new. Traditional sources include fuel wood, crop residues and hydroelectric. New sources include solar, wind, geothermal and modern biomass (eg, miscanthus, coppicing, esterification of rape seed oil to give 'biodiesel' and ethanol produced from sugar cane products). The exploitation of renewables is very capital intensive and it is unlikely that there will be any significant increase in their market share in the foreseeable future.

In broad terms global energy demand will continue to increase and although fossil reserves are finite it is unlikely that there will be a major shift in fuel use patterns in the next hundred years or so.

3 Environmental Impacts of Energy Use

All forms of energy use both renewable and non-renewable impact to a greater or lesser extent on the environment. This section will give an overview of such impacts and then look at those issues pertinent to fossil fuel use in more detail.

Environmental impacts can either be as a result of accidents or an inevitable consequence/byproduct of the energy cycle itself. Accidents by their nature tend to be well documented. During the period 1969-1986[5] nearly 4000 people have been reported killed in dam failures worldwide. This just exceeds the 3600 killed in mining disasters and far exceeds the 2000 killed in oil fires and 1500 killed in gas explosions. As technology advances and safety requirements and consciousness increases then fatality rates decrease. For example 476 miners were killed in the UK in 1950 compared with 18 in 1989/90 which represents a reduction of 90% per tonne of coal mined. Not all disasters are occupational though with Britain's worst and most infamous disaster being the slippage of the Aberfan coal tip with over one hundred casualties.

Accidents during the production and transport of oil and gas have been well documented. During the period 1975-1990, the UK oil and gas activities on the continental shelf had a fatality rate in the range 0.1 to 1.0 per 1000 employees. This was sadly marred by the Piper Alpha disaster of 1988 which claimed over 100 lives. Oil spillages during transportation can lead to major environmental damage. Notable examples are the Torrey Canyon, the Amoco Cadiz, the Exxon Valdez and the Sea Empress. Depending on location and sensitivity of the environment damage can be quantified in terms of losses of flora and fauna or in some instances economic impacts because of loss of tourist revenue.

Operations within the nuclear industry always remain under close scrutiny. Mosey[7] has listed failure related to reactor internals, core assemblies and steam generators. Well documented disasters within the nuclear industry include Windscale in 1957, Three Mile Island in 1979 and Chernobyl in 1986. At Windscale the accident related to a plutonium producing reactor of primitive design - all reactors of this type have now been shut down. The accident at Three Mile Island destroyed the reactor but the containment systems functioned correctly and radioactive release was small. Public concern following the Three Mile Island incident caused additional safety circuits to be installed elsewhere. The destruction of the reactor at Chernobyl in 1986 is clearly a disaster of epic proportion. There were over thirty immediate deaths and some 200 cases of radiation sickness. Vast tracts of land close to Chernobyl are highly contaminated. Indeed contamination still causes problems remote from the site, eg, caesium contamination of upland areas in the UK still affects grazing sheep. It is impossible to quantify the final impact of a disaster such as Chernobyl or to determine the timescale over which Chernobyl will impact.

Renewable sources of energy are not without their impacts on the environment. Hydroelectric and tidal schemes necessitate the construction of dams and barriers. These either flood land areas or contribute to the silting of estuaries and thereby alter the local ecosystems. Windmills and solar energy systems are visually intrusive and the former can be noisy. Modern biomass practices to produce fuel as a 'crop' inevitably means growing something that would not otherwise be there. This will impact on local soil conditions, drainage, etc. Judgements have to be taken concerning any damage to the environment and benefits to be accrued.

4 Fossil Energy Use - Historical Perspective

Pollution from man's activities has been recognised for many centuries. Early cave and hut dwellers would have noticed smoke from their fires filling their homes before finding an escape. The need for controls has been recognised for almost as long. In the UK for example an Environmental Committee was set up in 1285, the use of sea coal was banned in 1306 and an early offender hanged in 1307. During the last few decades concerns have grown from local air pollution issues to aspects of regional and global scale such as acidification and anthropogenic greenhouse gas contribution. The Earth Summit of 1992 was the focus for global environmental issues.

Until the middle of this century high ground level concentration of smoke and sulphur dioxide from uncontrolled fossil fuel was, if not completely accepted, regarded as the norm. Local pollution arose from intense use of coal and oil for domestic and industrial purposes and increasingly of late oil use in the transport sector. The impact of emissions is sometimes exacerbated by local topographical and meteorological conditions, eg, the now infamous Los Angeles photochemical smogs. Single events can precipitate changes in public perception and acception of adverse environmental conditions. An example of this is the London smog of 1952 for which several thousand extra deaths were attributed. This resulted in legislation forcing householders to use either 'authorised' smokeless fuel or an 'exempted' appliance designed to burn fuel smokelessly. Limits were also placed on particulate solids emissions from industrial boilers and these limits were met by the use of appropriate gas cleaning devices including mechanical arrestors and electrostatic precipitators. The UK legislation at that time did not stipulate emission limits for gaseous combustion products. Instead minimum chimney heights were specified to ensure that ground level concentrations were kept to an acceptably low value. Similar types of legislation were deployed elsewhere and whilst this approach was effective in improving the local situation it did not address potential regional impacts.

In the 1970s and 1980s 'acid rain' came onto the agenda. Acidification of surface waters in parts of northern Europe and north America was attributed to acid gas emissions from fossil fuel use. Damage to forests, particularly in Central Europe, was also cited as a further impact of acid rain. Several major research initiatives including the National Acidic Precipitation Assessment Programme[8], the Surface Waters Acidification Programme in Scandinavia[9] and the Loch Fleet Project in South West Scotland[10] have been undertaken in an effort to understand the science involved. Originally it was thought that the link between acid rain and terrestrial and aquatic eco systems was quite straightforward. However the scientific findings are not conclusive and we now know that many factors are involved. For example if trees are otherwise healthy acid rain will not normally harm them. Only if trees are stressed from drought or disease or there is an imbalance of trace elements will acid rain have any direct effect. Surface waters have acidified in marginal areas only. The acidification would have occurred anyway in due course as a result of natural weathering although additional acid inputs from man's activities may have accelerated the process.

Inevitably there has been intense political pressure to do something about acid rain and environmental legislation now features strongly. Regulatory systems are well advanced in most developed countries and in many developing ones. Examples include the US Code of Federal Regulations for Combustion Plant and the more recent Clean Air Act Amendments, the TA Luft Regulations in Germany, various European Directives and the UK Environmental

Protection Act. Specifically the European Community Directive on the Combatting of Air Pollution from Industrial Plant[11] or 'Framework' Directive specified process categories for control and introduced the BATNEEC concept (best available technology not entailing excessive cost). Later the Large Combustion Plant Directive[12] (LCP) specified emission limit values for all +50 MWe input new plant and national emissions reduction targets for existing plant. The form of the Directive was driven by more stringent legislation already in place in Germany. In the UK the Environmental Protection Act (EPA) of 1990 has provided a framework for a major revision of UK pollution control legislation and also the vehicle by which European Directives are adopted. The UK EPA has introduced strong legislation into the UK by which to implement an advanced system of Integrated Pollution Control (IPC). The BATNEEC Concept embraced provides a mechanism of continuous tightening of requirements. As increasingly more stringent BATNEEC is recognised the EPA requires its implementation and the reduced releases into the environment which go with it. The UK EPA is strong enough to match any other pollution control legislation in the world.

In the late 1980s carbon dioxide emissions and the threat of global warming overtook acid rain as a major international environmental concern. Parallels can be drawn between the acid rain and global warming debates, not least the intense political pressure to embark on expensive courses of action in advance of the underlying science being fully understood. Assembling the science of climate change is the task of the Intergovernmental Panel on Climate Change (IPCC) Working Group 1. The reports of this Working Group[13],[14],[15] indicate that there remain considerable uncertainties regarding predictions of temperature increases, the timescale over which they may appear and whether or not they will lead to significant problems for the human race. Nonetheless the latest IPCC report says "The Balance of Evidence Suggests a Discernible Human Influence on Global Climate". Following the Earth Summit in 1992 the World is taking climate change issues seriously. The United Nations Framework Convention on Climate Change is now come into force having been ratified by more than fifty countries. This commits developed country parties to draw up programmes of measures aimed at returning emissions of carbon dioxide and other greenhouse gases to their 1990 levels by the year 2000.

5 Fossil Fuel Use - Principal Environmental Impacts

The major environmental concerns of fossil fuel relate to the various extraction processes and major releases during combustion. Within the very broad scope of this paper I shall concentrate on environmental impacts of fuel winning with particular emphasis on coal and releases of acid and greenhouse gases and particulate solids from fuel use.

Coal has been mined for several centuries with scant regard for any damage caused to the environment. When reserves have been exhausted at a particular site the operator has generally walked away leaving dereliction and a host of associated problems. Fortunately this is no longer the case in most developed countries; nor can it be allowed to happen in the developing countries as fossil fuel helps to meet the world's increased future energy demands. Coal can be mined with minimal environmental impact so long as recognised good practice is followed. Before any new mining operation is undertaken a full environmental impact statement should be produced to enable strict operational guidelines to be drawn up. Items to be considered should include minimising spoil extraction, provision of adequate waste water treatment, site screening and reclamation, transport requirements and social impacts. New mine sites can be designed to be rehabilitated to a state equivalent to and sometimes better than that prior to development. Surface and ground water should be protected by secure handling and disposal of toxic and acid materials. Excess overburden spoils should be disposed of safely or ideally utilised in some other development. Top soil removed from mining areas should be stockpiled and subsequently replaced. Where these practices have been done post mining land uses have included pasture, agriculture and the creation of wildlife habitats.

A potential global impact of oil, gas and coal production is the release of methane during the operation. Methane is a greenhouse gas and isotopic data indicates that about 100 Mta methane is fossil derived. Specific methane emissions are extremely low from surface mining operations and are not considered further. Releases can be higher from deep mining and practices are now available to drain and utilise this methane. Indeed for very gassy coal seams this makes strong economic sense irrespective of the potential environmental benefits. Closer attention should be paid to minimising methane release during oil and gas production. Wherever possible by-product methane should be utilised but failing this flaring is preferable to venting from a greenhouse view point. In the past methane losses from natural gas transmission systems have been a major problem. There have been suggestions that transmission losses in the former USSR alone could be in excess of 30 Mta or about 30% of the total attributable to fossil sources. New gas transmission systems can operate with minimal losses. Recent improvements in methane management during fossil fuel production may be the reason that there has been a slow-down in the build-up of atmospheric methane concentrations.

The principal regional challenge to fossil fuel is the emission of pollutants including particulate solids, sulphur dioxide and oxides of nitrogen.

In the past particulate solids have been an issue in areas of intense fossil fuel use because of their soiling effect on buildings and their impact on health especially in combination with sulphur dioxide. As indicated earlier the 'tall stacks' policy coupled with limits on particulate emissions has tackled this problem. Technologies are readily available and easily retrofitted to remove in excess of 99% of particulates from exhaust gases from stationary combustion plant. There is therefore no reason for particulate emissions to be a future concern from such plant even if there is as projected massive increase in use particularly in the developing countries. Particulate emissions from the transport sector, especially diesels, is an issue that still needs to be properly addressed. Modern diesel engines are low smoke and hydrocarbon emitters but improved systems can not be easily or economically fitted to existing units. Thus in many urban areas poor air quality due principally to transport emissions is still a problem and represents a priority area for improvement.

Sulphur dioxide and nitrogen oxides from fossil fuel use are implicated in acid rain as discussed earlier.

During combustion almost all of the sulphur content of the fuel is converted t sulphur dioxide (plus a small amount of sulphur trioxide). Thus the final release of sulphur dioxide is directly related to fuel sulphur content. Emission limit values for the European LCP Directive are fairly representative of new plant performance standards elsewhere in the world. In summary the smaller plant have to meet limits of 2000 and 1700 mg/m^3 for solid fuel and oil respectively reducing to 400 mg/m^3 for larger plant (>500 MW). In general the limit for gaseous fuels is 35 mg/m^3 throughout. As a general guide, for every 1% by weight of sulphur in coal or oil there will be, on combustion, an SO_2 emission of about 2000 mg/m^3 (expressed on a dry basis at 6% O_2 v/v) or 1700 mg/m^3 (expressed on a dry basis at 3% O_2 v/v) for coal and oil respectively. Thus for example about 90% sulphur dioxide removal would be required to enable a new 500 MW plant to meet the requirements of the Directive when firing a 2% sulphur coal. Most existing stationary combustion plant was not designed with sulphur dioxide control in mind but technologies to reduce sulphur dioxide emissions are available and subject to space requirements can usually be retrofitted. The scope for reducing sulphur dioxide emissions from existing plant and the intrinsic design capability of new and emerging technologies to limit sulphur dioxide release is mentioned later.

Release of nitrogen oxides on combustion is not solely a function of the nitrogen content of the fuel but related to the combustion regime itself. During combustion competing 'NO_x' formation and destruction reactions take place. Generally speaking 'NO_x' emissions are enhanced by high combustion temperatures and excess oxygen levels. Uncontrolled emissions of nitrogen oxides can be as high as 2000 mg/m^3 for some categories of stationary plant which necessitates about 70% reduction to meet the European LCP Directive limit of 650 mg/m^3 for solid fuels (450 and 350 for liquid and gaseous fuels respectively). Already new plant performance standards are emerging which would require approaching 90% nitrogen oxides abatement for the example given above. Options for reducing nitrogen oxides emissions from existing stationary plant and the capabilities of new and emerging technologies are covered later.

Internal combustion enquiries are very high specific nitrogen oxides emitters. The use of catalytic converters and lean burn techniques can reduce significantly specific emissions and deployment of these approaches should be further encouraged as vehicle use continues to expand apace. Further consideration should also be given to full fuel cycle issues for the transport sector and areas for attention should include development of public transport infrastructure and the use of electric vehicles (zero emission at the point of use).

Concerns about global warming pose a major challenge to fossil fuel use. Currently about 6GT per annum carbon is released from fossil fuel use worldwide. If the projections given in Table 2.1 are accurate carbon emissions will increase significantly for two of the three sub-scenarios considered. Only the ecologically driven case predicts a decrease in carbon emissions. The likely impact of sustained increase in atmospheric carbon dioxide concentrations is being assessed by the IPCC. Although the consequences are not known with any certainty action is already being taken by the developed countries to limit emissions of greenhouse gases. New and emerging technologies that will be discussed later do achieve higher efficiencies which means that specific carbon emissions per unit of electricity produced are lower. Assessment of the options to remove carbon dioxide from flue gases is currently being undertaken and the status of this is also presented later.

6 Reducing Acid Gas Emissions

6.1 Sulphur Dioxide

Options for reducing sulphur dioxide emissions include switching to low sulphur fuels and removal of sulphur from the fuel prior, during or post combustion.

Fuel switching can be used locally eg, to meet sulphur emissions budgets but is not an overall solution given the increasing global future fossil fuel demand and the limited availability of sufficiently low sulphur fuel.

The scope for sulphur removal prior to combustion is limited. Current coal cleaning practices separate much of the pyritic sulphur but do not remove that sulphur intimately bound within the coal structure. There are opportunities to ensure that the developing countries adopt best practice in terms of coal cleaning. Hydrodesulphurisation of oil is technically possible but very expensive compared to downstream sulphur removal options.

In furnace techniques[16] are available to capture SO_2 released during combustion either within the burning zoned itself or immediately afterwards. These techniques rely on the addition of a dry sorbent usually limestone on dolomite to capture the SO_2 in solid form as calcium sulphate. Such materials react with SO_2 at temperatures between 800-1200°C so the specific success of this approach depends on the temperature of the combustion regime itself. Current fossil fuel use for power generation almost exclusively uses suspension firing where either pulverised solid fuel or oil is burnt in large combustion chambers with individual boilers up to 600 MWe. In these systems flame temperatures of typically 1600°C mean that sulphur capture by feeding sorbent with the fuel is ineffective. Instead sorbent can be added to the furnace above the burners where temperatures are less than 1200°C or alternatively at the periphery of 'low NO_x' burners sometimes known as Limestone Injection Multistage Burner (LIMB). These techniques can achieve up to 50% sulphur abatement at a sorbent:sulphur molar ratio of 2:1. Variations of stoker firing are extensively used in the industrial boiler sector and injection of sorbent above the fuel bed or above the burners in an oil-fired plant can achieve sulphur captures of up to about 60% at a sorbent: sulphur molar ratio of 2:1. The sorbent is entrained in the combustion gases and reacts as they cool, eventually passing out of the furnace and removed in the downstream gas cleaning equipment. Drawbacks of both 'suspension' and stoker firing in relation to sorbent injection in-furnace include increased risk of deposition and fouling, higher dust burdens for collection and curtailed utilisation options for the ash because of its altered composition.

Fluidised bed combustion (FBC) is a relatively new technology which is now gaining extensive usage in the industrial and power generation sectors. FBC is characterised by its relatively low operating temperatures, typically 850-950°C and its capability to burn a wide range of fuels. The combustion conditions prevailing with FBC systems are generally within the optimum range for sulphur capture and sulphur removal efficiencies of up to 90% can be achieved at a sorbent:sulphur ratio of 2:1. FBC Systems are now being deployed as components of advanced coal fired power generation systems. The environmental merits of such systems are overviewed later.

A wide range of technologies generically are known as flue gas desulphurisation (FGD) are available to remove sulphur dioxide from flue gases post combustion. New source performance standards and emission limit values for existing plant mean that FGD is already extensively deployed. The barriers to more widespread use of FGD include tensions between the generators and the coal producers. The former bear extra investment and operational costs and seek lower prices from the latter who is almost certainly already operating at very low margins. The cost of FGD is site specific but in round terms it adds about 0.5p/kWh to existing power plant and somewhat less if part of a new installation. Nonetheless FGD is available and is capable of achieving >90% sulphur dioxide removal. Care must however be taken to ensure that use of FGD does not transfer environmental impacts from one medium to another eg, in terms of limestone quarrying requirements or disposal of increased quantities of residues. Some systems produce useful byproducts such as gypsum, elemental sulphur and sulphuric acid. This can offset other environmental impacts, eg, gypsum from FGD can reduce gypsum quarrying for wallboard manufacture. Regenerable FGD systems are also available which significantly reduce net sorbent requirement although these systems are generally more expensive than the 'once through' type. As an alternative to FGD several 'in duct' sorbent injection and reactivation techniques are available. These tend to be lower cost and efficiency than FGD and choice in a retrofit situation will be dictated by a multitude of factors.

The choice of sulphur control system is dictated by its suitability for use with a particular combustion technology, its technical performance in terms of meeting environmental emission targets and of course its cost. Large capital cost systems, such as semi dry and wet FGD systems are more suited to larger applications, where 'economies of scale' can be made[17]. For example a spray dry system operating on a 100 MWt plant is estimated to remove sulphur at a seventh of the cost of a similar system operating on a 10 MWt plant. For lower capital cost systems, such as dry injection desulphurisation, specific abatement costs are again less at the larger scale. But, the difference between large and small plant is not as significant since a major portion of the abatement cost will be the sorbent which makes a fixed contribution and is essentially independent of plant size. Sulphur dioxide abatement costs for a range of control techniques are presented in Table 6.1. These costs should be regarded as a guide rather than in absolute terms. The basis for the estimates is a 300 MWe coal-fired plant burning a 1.5% sulphur coal at a load factor of 65%. It should be noted that the techniques differ in their ability to control sulphur so environmental targets may preclude some options from certain applications.

Table 6.1 Estimate Costs of SO_2 Control Systems[17]

Process	SO_2 Abatement (% Fuel Sulphur)	Cost ($/tonne S removed)
Dry Sorbent Injection		
· In-Furnace	50	300-500
· In-Duct	50	500-750
· Sodium In-duct	70	500-600
Semi Dry Systems		
· Spray Dry	90	400-560
· In-duct Reactivation	80	530-800
Wet Scrubbing Systems		
· Limestone-Gypsum	>90	400-600

6.2 Oxides of Nitrogen

Since the formation of nitrogen oxides in combustion systems is largely dependent on the firing method rather than the nitrogen content of the fuel, fuel switching from a higher to a lower nitrogen content fuel, or fuel cleaning to remove nitrogen, could at best play only a minor role in reducing emissions. These are not considered to be viable options. In practice there are two approaches for controlling nitrogen oxides emissions, namely, modification to the combustion system and treatment of the flue gases. The usual approach is to concentrate first on the combustion system because this is generally recognised as the lower cost option. Then, if further reductions are necessary, the flue gases can be treated but this is expensive both in terms of capital and operating costs.

For 'suspension' firing the general principle for reducing nitrogen oxides is to change the distribution of fuel and air so that fuel rich conditions favouring nitrogen oxides destruction are present. This also lowers the temperature and retards thermal nitrogen oxides formation. This combustion 'staging' can be achieved by removing burners from service for use as overfire air ports or the incorporation of overfire air ports in new boiler designs. Staging can also be achieved within the burner and most manufacturers now offer staged or low NO_x burners (LNBs). These techniques are known as air staging. Alternatively fuel staging or reburning can be used. This can be done either within individual burners or within the furnace itself. Reburning technology has now reached the demonstration stage with assessments underway in both the US and in Europe. Indications are that NO_x reductions of around 40-50% are possible using natural gas as the reburning fuel.

Techniques involving combustion modifications to suspension firing are relatively cheap, but there are potential disadvantages. In general, systems that operate in the staged mode run the risk of promoting slagging and corrosion on the furnace walls because of the presence of reducing gases. Furthermore combustion efficiency may be reduced due to inadequate burnout of fuel.

For stoker fired plant the fuel bed on the grate itself promotes natural staging of combustion. Nitrogen oxides emissions are therefore relatively low and would normally comfortably meet the new plant requirements of the European Commission LCP Directive.

The nitrogen oxides emission from FBC is dependent on the variant of the technology. Circulating FBCs are inherently staged so nitrogen oxide emissions are low. Emissions from bubbling FBCs are higher and air staging has been used to reduce nitrogen oxides emissions.

Whilst combustion modification is usually the preferred and cheaper approach for nitrogen oxides control, flue gas treatment may be required for abatements of over 50%. Flue gas treatment generally involves injection of ammonia or ammonium based compounds into the flue gas with the aim of reducing nitrogen oxides to nitrogen. These reactions proceed at temperatures of 950-1050°C, or at 300-400°C in the presence of a catalyst. Flue gas treatment is expensive and can give rise to operational problems including contamination of flyash by ammonia due to slippage and the formation of small quantities of ammonium bisulphite, a sticky compound, that can clog boiler components.

It is generally advisable to maintain flue gas NH_3 concentrations below 5 ppm but this tends to restrict the amount of reducing agent that can be added and hence the maximum achievable NO_x reduction.

Typically NO_x reductions of 30-70% can be achieved with ammonia slippage rates of 5 ppm.

The selection of the NO_x control method is dependent on site specific factors such as prevailing emission legislation, plant size and type, new or retrofit application, etc. Table 6.2[18] presents estimated costs for NO_x control systems. The basis for these estimates is a 300 MWe pulverised coal-fired plant operating at a 65% load factor. Again as per the costs presented for sulphur earlier these should be regarded as a guide rather than in absolute terms. As an initial measure, most utility scale plant use some form of combustion modification technique. Generally the retrofitting of low NO_x burners possibly combined with overfire air. Currently about 185 GWe of coal-fired capacity worldwide uses combustion modification techniques for NO_x reduction. Where there is the need for larger NO_x reductions SCR is most commonly applied with about 40 GWe coal-fired capacity with SCR currently operational in Europe and Japan.

Table 6.2 Estimated Costs of NO_x Control Systems[18]

Process	NO_x Abatement (% uncontrolled emissions)	Cost (£/tonne NO_2 abated)
Combustion Modifications		
· Low excess air	20	20-60
· Low NO_x burners	50	180-300
· Reburning	50	90-350
Flue Gas Treatment		
· SNCR	70	470-600
· SCR	80	600-760

7 Control of Particulate Solids Emissions

Both solid fuels and to a lesser extent heavy fuel oils contain mineral matter which largely constitutes any ash product on combustion. These ash residues from combustion need to be removed from the following locations, namely:

i. where the fuel is burnt (either as fused or unfused ash); for example from the grate of a stoker fired boiler.
ii. boiler and furnace flueways (as gritty deposits); removal frequently being achieved by on-line sootblowing.
iii. from the flue gases by the use of particulate removal equipment, leaving very small quantities of fine particles that are emitted from the chimney.

Four general classes of equipment are available for the removal of particulate solids from exhaust gases, namely: cyclones, electrostatic precipitators, bagfilters and wet scrubbers. As for sulphur dioxide and nitrogen oxides control the choice of particulate control equipment is dictated by a range of factors. These include the type of combustion system, the target emission requirement and of course cost.

Cyclones are inertial mechanical collectors with no moving parts. They are best suited for the removal of coarse particles and are therefore particularly suited for use with fixed and travelling grate stoker fired boilers which give an inherently low elutriation of relatively coarse particles in the flue gas stream. In some circumstances cyclones may also be adequate for use with small fluidised bed fired boilers burning washed, graded coal. Cyclones may be either of single or multicell design. High efficiencies are achieved with single cell devices but these incur a penalty in terms of their size, particularly their height. Multicell units are more compact but can suffer from irregular flow distribution through individual cells which limits their performance. This is some scope to enhance the performance of cyclones at modest additional cost by the use of a base purge or blowdown[19]. Here a draw off of gas from the disengagement hopper area of a single cell cyclone or from below the tubeplate of a multicyclone helps to stabilise the gas vortex and prevent particle re-entrainment. The small gas stream drawn off (perhaps 5% of the inlet gas flow) is heavily solids laden and subsequently cleaned by a small bagfilter. Success has also been achieved with cyclones of novel design[20] to overcome size limitations of high efficiency single cell units. These devices incorporate vortex collector pockets (VCPs) attached to the main cyclone body. A contra rotating vortex established in the VCP allows transfer of dust from the main cyclone with no net fluid transfer. Designs have been demonstrated with several VCPs which have achieved different particle cut sizes. This technology therefore has the potential for process applications where particle classification is required.

Electrostatic precipitators (ESPs) and bagfilter are typically between four and six times the cost of cyclones but they can achieve much high collection efficiencies. They are usually the only technological options for large combustion plants where legislation demands low emissions.

ESPs can be designed to meet particulate solids emission levels of as low as 30 $mg.m^{-3}$. However to achieve these extremely low emissions low gas velocities and additional fields have to be used. The performance of electrostatic precipitators can be affected by upstream control of sulphur and nitrogen oxides. Addition of sorbents to retain sulphur as a solid sulphate adds to the particulate burden for collection and reduces the sulphur trioxide content which impacts adversely on particle resistivity with respect to particle collectability by the ESP. Combustion manipulation to reduce nitrogen oxides formation can increase the carbon content of the ash residues. This again adversely impacts on particle collectability. Increase in moisture content of the gas stream or decrease in temperature on the other hand enhances particle collectability in an ESP.

Bagfilters are essentially barrier filters and when operating correctly can achieve almost complete particulate solids removal. The hostile environment in combustion flue gases however demands the use of exotic and expensive filter media such as acid resistant Nomex, glassfibre or ptfe. Again there are implications if upstream sulphur control is being undertaken because of the risk of damage by 'concreting' of the bags from temperature excursions below dewpoint in the presence of ash, calcium sulphite/sulphate and unspent sorbent mixtures.

Wet scrubbers can be very efficient particulate solid removal devices although they are not deployed widely on combustion plant because they have the additional complications of a water circuit with pumps, settling tanks and pH control. Also to achieve high efficiency removal a large operational pressure drop is required which often places an unacceptably large parasitic energy loss on the process. The use of wet scrubbers is favoured in process applications where expertise is available for effluent control.

A suite of particulate control technologies are available which can enable solid fuel and heavy oil fired combustion plant to meet very stringent limits on emissions. As indicated earlier the final choice of control equipment will be dictated by cost, boiler type, fuel type, legislative requirement, new or retrofit situation, etc.

8 Minimising Greenhouse Gas Emissions

Greenhouse gases of relevance to fossil fuels are methane, nitrous oxide and carbon dioxide. Methane release associated with fuel winning has previously been discussed. Methane emissions from fossil fuel combustion are generally very low and so long as good combustion practice is followed methane from combustion should not be a concern. Fossil fuel combustion can give rise to trace releases of nitrous oxide but these emissions are very small compared with those from natural sources[13][14]. Carbon dioxide emission from combustion is by far the most important greenhouse gas issue for fossil fuel use.

Carbon dioxide release is directly related to the carbon content of the fuel. Thus there is some limited scope in the short term to increase use of lower carbon fuels such as natural gas. However given the projected growth in global fossil fuel demand and the fact that coal represents about 75% of global recoverable fossil fuel reserves then increased use of natural gas is not a long term solution.

It is technically possible to remove carbon dioxide from combustion exhaust gases and then to transport the carbon dioxide to a disposal site from which it will not escape into the atmosphere. Previously such approaches have been discounted because of their costs but recent concerns about the threat of global warming have prompted a reappraisal, eg, the assessment work currently being carried out by the IEA Greenhouse Gas R&D Programme. This assessment[21] shows that the application of carbon dioxide capture and disposal technologies to large stationary sources is a viable method for reducing carbon dioxide emissions from the combustion of fossil fuel. Preliminary conclusions of this work indicate increased power generation costs of between 30 to 70%. These costs being associated both with the removal technology itself and the parasitic energy demands of the process. Further R&D offers the potential for reducing these costs for example with the development of membrane separation as described by Bower[22]. Storage/disposal of carbon dioxide in the deep ocean, in aquifers and in depleted oil or gas reservoirs can be achieved at relatively low cost but further work is needed to establish the environmental credibility of each disposal option.

9 New Technologies for Power Generation

The overall aims of new technologies are to improve efficiency, reduce costs and to minimise environmental impacts.

Conventional pulverised fuel power plant fitted with FGD can achieve a net efficiency of about 38% (LHV basis). In practice however much of the plant currently deployed is operating at much lower efficiencies down to 20-25% in some cases. Integrated Gasification Combined Cycles, Circulating Fluidised Bed Combustion and Pressurised Fluidised Bed Combustion systems can now achieve well over 40% and natural gas combined cycle plant over 50%. Emerging technologies such as the Air Blown Gasification Cycle (formerly known as the British Coal Topping Cycle) will achieve about 47%. With improved materials performance at high temperatures together with other developments, eg, fuel cells, etc, the promise of overall conversion efficiencies in excess of 55% could become a reality. Another paper at this meeting[23] overviews Clean Coal Technologies and presents a status report on the Air Blown Gasification Cycle.

The much higher efficiencies achievable by these new technologies mean that in the absence of any controls the specific emissions of sulphur dioxide and carbon dioxide per unit of electricity generated will be correspondingly lower. In reality the situation is much better because the new systems feature either fluidised beds or total gasification as components. As previously stated sorbent injection into FBC systems can remove up to 90% of sulphur. Conventional chemical scrubbing of gasification products prior to final combustion can achieve 99% removal of sulphur. With respect to nitrogen oxides the advanced technologies tend to be intrinsically low NO_x emitters achieving levels of typically 200 mg/m^3 which is less than one third of the LCP limit value for new plant. Should the need ever arise to move towards 'zero' emissions of acid gases from advanced fossil fuel plant then the opportunity still remains to fit post combustion treatment technologies. This would of course add to generating costs and it is difficult to see how it could be justified on environmental grounds. Strategies to remove carbon dioxide from flue gases could if required be deployed with new fossil fuel technologies.

10 The Way Forward

On the basis of current energy demand projections it is inconceivable that there will be anything but a substantial growth in global fossil fuel use in the foreseeable future. Such an increase is seen as a major threat to the environment because principally of concern about global warming. A key part of any precautionary strategy must therefore be the development of advanced fossil fuel technologies and their subsequent deployment in those areas of rapidly increasing demand. If the developed world adopts policies which encourage utilities to back out of fossil fuel use in general and coal in particular, it will mean that the adoption and spread of advanced higher efficiency technologies will be slower and more uncertain. Instead the developing world will use dated technologies offering perhaps 25-30% overall efficiency and rudimentary or no control at all for particulate solids and acid gases. They will also produce considerably more carbon dioxide than they need to. Thus if we seriously want to minimise carbon dioxide and other emissions production in the future and to manage responsibly global fossil fuel reserves then the development and deployment of advanced fossil fuel systems with higher efficiency must be part of our policy.

11 Conclusions

1. Energy demand is dependent on population, economic development and technological advancement. Global primary energy demand has increased about twenty-fold over the last 150 years. Estimates predict up to a further five-fold increase by the year 2100.

2. Fossil fuels will continue to make a major contribution to meeting global primary energy demand for the foreseeable future.

3. The impact of fossil fuel winning can be considerably minimised by the use of modern methods and strict environmental practices.

4. Controls are available to enable existing technologies to meet new plant performance standards when utilising fossil fuels.

5. Advanced higher efficiency technologies are emerging and it is important that political barriers do not hinder their use in the developed world as this will delay deployment in developing areas where the major growth in energy demand is anticipated.

6. It is technically possible to remove and dispose of carbon dioxide from combustion flue gases should a future need be demonstrated.

7. Technologies to reduce emissions from the transport sector are available and their deployment should be encouraged.

Acknowledgements

The author acknowledges the assistance given by colleagues in the preparation of this paper. The views expressed are those of the author and do not necessarily represent those of British Coal Corporation.

References

1. 'Global Energy Perspectives to 2050 and Beyond', WEC/IIASA Report. 1995.
2. Smil, V. 1994. 'Energy in World History', Westview Press, Boulder, USA.
3. Bos E, Vu MT, Leven A and Bulatao RA, 1992. World Population Projections 1992-1993, John Hopkins University Press, Baltimore, USA.
4. BP Statistical Review of World Energy, 1995.
5. 'Energy and the Environment in the 21st Century'. National Academies Policy Advisory Group, July 1995.
6. 'Energy for Tomorrow's World', WEC, Kogan Page, London 1993.
7. Mosey D 'Reactor Accidents', Nuc. Eng. Int. Special Publication, 1990.
8. NAPAP 'Acid Deposition: State of Science and Technology', Volume I-IV, National Acidic Precipitation Assessment Program, Washington DC, 1990.
9. Mason BJ 'The Surface Waters Acidification Programme', Cambridge University Press, 1990.
10. Howells G and Dalziel TRK (editors) 'Restoring Acid Waters: Loch Fleet 1984-1990' Elsevier Applied Science, 1992.
11. EEC 'Council Directive on the Combatting of Air Pollution from Industrial Plant' (84/360/EEC), 1984.
12. EEC 'Council Directive on the Limitation of Emissions of Certain Pollutants into the Air from Large Combustion Plants' (88/609/EEC), 1988.

13. Houghton JT, Jenkins GJ and Ephraums JJ (editors), 'Climate Change: The IPCC Scientific Assessment', Cambridge University Press, 1990.
14. Houghton JT, Callander BA and Varney SK (editors), 'Climate Change 1992: The Supplementary Report to the IPCC Scientific Assessment', Cambridge University Press, 1992.
15. IPCC Working Group One: Second Assessment Report, Published June 1996.
16. Ford NWJ and Sage PW. 'Review of Sorbent Injection Processes for Low Cost Sulphur Dioxide Control'. Proc. Instn. Mech. Eng. Vol 210, pp183-190, 1996.
17. Cooke MJ and Pragnell RJ, 'Costs of Flue Gas Desulphurisation on Industrial Coal-Fired Plant'. Paper to costs of Flue Gas Desulphurisation Conference, London UK, 19 September 1990.
18. Eskinazi D, Cichanowicz JE, Linak WP and Hall RG, 'Stationary Combustion NO_x Control: A Summary of the 1989 Symposium', NAPCA, (39), 8, p1131-1138.
19. Sage PW and Wright MA. 'The Use of Gas Bleeds to Enhance Cyclone Performance'. Filtration and Separation Jan/Feb 1986.
20. Biffin M, Syred N and Sage PW. 'Enhanced Collection Efficiency for Cyclone Dust Separators'. Chem. Eng. Res. Des. 62, July 1984.
21. Ormerod WG, Webster IC, Audus H, Riemer PWF and Saroff L. 'International Perspectives and Preliminary Results from Studies of Carbon Dioxide Sequestration'. Paper to 'Global Climate Change: Science Policy and Mitigation Strategies'. April 5-8 1994, Pheonix, Arizona, USA.
22. Bower CJ, Goldthorpe SH and Fynes G. 'Low CO_2 Power Generation Options and their Environmental Impact'. 37th ASME International Gas Turbine and Aeroengine Congress, June 1992, Cologne, Germany.
23. Sage PW and Mills SJ. 'Overview of Clean Coal Technologies and Current Status of the Air Blown Gasification Cycle'. Paper to NATO:ASI 'Desulphurisation of Hot Coal Gas with Regenerable Metal Oxide Sorbents: New Developments'. July 7-19 1996, Kusadasi, Turkey.

Overview of Clean Coal Technologies and Current Status of the Air Blown Gasification Cycle

P.W. Sage and S.J. Mills

Coal Technology Development Division, British Coal Corporation,
PO Box 199, Stoke Orchard, Cheltenham, GL52 4ZG, UK

Abstract. Historical trends in energy use and future projections indicate that global energy demand will continue to increase. Coal is expected to maintain a significant share of this demand for the foreseeable future. However, coal's continued acceptance as a primary fuel will depend on its ability to deliver energy cheaply and with low environmental impacts. To this end a suite of Clean Coal Technologies are emerging and these are overviewed in this paper. A more detailed description is given of the Air Blown Gasification Cycle which is being developed by the UK lead Clean Coal Power Generation Group.

1 Introduction

Global future energy demand is reviewed elsewhere[1]. Estimates suggest up to a five-fold increase in demand compared with current levels by the year 2100. It is generally accepted that fossil fuels, of which coal is by far the most abundant, will continue to meet the bulk of future energy demand for the next hundred years or so. Fossil fuel use is not expected to be resource limited over this timescale.

Increasing public awareness and political pressures demand that energy is supplied with due regard for the environment. To meet this demand Clean Coal Technologies (CCTs) are being developed which offer improved efficiency and lower environmental impacts compared with those technologies currently in use. The following sections briefly overview competing CCTs and describe in more detail the Air Blown Gasification Cycle (ABGC) which is being developed by the UK lead Clean Coal Power Generation Group (CCPGG).

2 Overview of Emerging Clean Coal Technologies

As economies develop and mature there tends to be a transition from direct end use of coal, eg, for heating and cooking to large scale primary use to generate electricity. The vast majority of coal use in developed countries is at large power stations with a small and diminishing percentage being used within the domestic and commercial environments.

Pulverised fuel power stations are the standard technology used worldwide for electricity generation. Finely pulverised fuel usually coal (typically 70% $<75\ \mu m$) is burnt

NATO ASI Series, Vol. G 42
Desulfurization of Hot Coal Gas
Edited by A.T. Atimtay and D. P. Harrison

in a high temperature flame and power is produced by a condensing steam turbogenerator. Efficiency is maximised in larger units (>100 MWe) by the use of a reheat cycle. Over the years the efficiency of pulverised fuel stations has increased by the use of ever larger individual units (up to 900 MWe) and the gradual improvement of steam conditions. State of the art conventional pulverised fuel power stations can now achieve LHV efficiencies of about 38% even when flue gas desulphurisation equipment is fitted. This compares very favourably with the approximately 30% efficiency achieved by early variants of the technology.

CCTs are now emerging that will challenge the dominance of conventional pulverised coal firing. These technologies include advanced pulverised fuel, variants of fluidised bed combustion, integrated gasification combined cycle and hybrid systems. Often features of one technology can benefit another, for example use of advanced steam conditions, gas cleaning developments, new materials of construction and advances in turbine technology. In the medium term we can expect perhaps two or three technologies to emerge as winners which will capitalise on such developments and will be best suited to local conditions, eg, fuel type, ash characteristics, etc.

Emerging CCTs for electricity generation are now briefly summarised.

2.1 Advanced Pulverised Fuel Technology

The basic technology is the same as for conventional pulverised fuel with a finely ground fuel usually coal being burnt in suspension in a high temperature flame. Combustion manipulation techniques are employed to restrict nitrogen oxides formation and sulphur dioxide and particulate solids emissions are controlled by conventional post combustion technologies. Depending on legislative requirements additional flue gas treatment may be required to reduce further final nitrogen oxide emissions. Aspects of pulverised fuel technology are being developed to overcome the limitations of the traditional design of plant. These include the use of advanced steam conditions (increased temperature and pressure), made possible by the use of improved materials of construction and the use of double reheat steam turbines. Using these techniques LHV efficiencies of 43-45% have been achieved although site specific factors, for example exceptional scope for low temperature heat recovery, contribute towards these high efficiency figures. A typical flow diagram for pulverised fuel technology is shown in Figure 1.

There are several advanced pulverised fuel plants now in operation. Avedore belonging to SK Power of Denmark is one such example, producing 250 MWe and supplying steam to a district heating scheme. Other examples in Denmark include Studstrup, Fynvaerket and Esbjorg. The efficiency improvements achieved by these plants have been obtained using 250 bar, 560°C steam conditions, improved fuel pulverisation, greater low temperature heat recovery through the use of oversized air heaters, use of state-of-the-art steam turbines, and newly developed technological products in the boiler and turbine regions. Two Danish plants currently under construction will utilise 290 bar, 580°C steam conditions which would suggest a further two percentage points efficiency improvement.

The potential for advanced steam cycles has been reviewed[2]. Steam conditions of 320 bar, 620°C with double reheat cycles are anticipated by the year 2005. These steam conditions increase to over 350 bar and over 700°C by the year 2015 (a potential LHV efficiency of ~52%). Bottle-necks are identified that may prevent these goals being achieved. These include high temperature corrosion and magnetite growth. Exotic superalloy materials will need to be developed for turbine rotors, thick walled sections and superheaters. Such materials, even if available, will inevitably have cost implications. As indicated earlier many of these technological advances could be deployed in other competing CCTs.

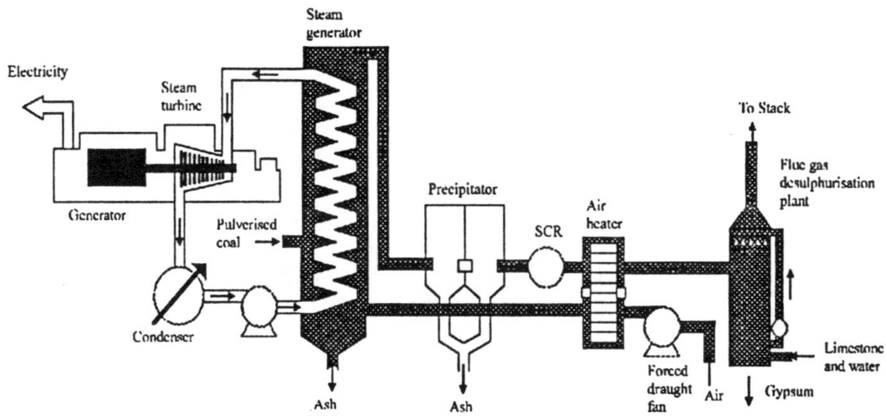

Fig.1. Flow diagram for pulverised fuel technology

2.2 Circulating Fluidised Bed Combustion (CFBC) Power Plants

In the CFBC system, coal or other fuels are burnt in a bed of solids, fluidised by a high velocity air stream. The off-gases carry entrained solids out of the bed that are separated either in a high efficiency cyclone or an impingement separator, and are returned to the base of the bed via a non-mechanical valve, such as a J or L valve. Usually, heat is extracted from the combustor and from a waste heat boiler which cools the combustion gases before final clean-up in an electrostatic precipitator or bag filter. Superheated steam is raised in the boiler system and used to drive a conventional condensing steam turbine which generates power.

CFBC offers a viable and commercially-proven alternative to conventional pf technology, and is ideally suited to plants in the range 30- ~300 MWe. If plants are smaller, capital costs can be disproportionately high. If larger, economies of scale are unlikely to apply although efforts to scale-up designs to ~400 MWe are under way by at least one manufacturer[3]. If increased capacity is required, it is normal practice to use multiple units, resulting in a higher degree of flexibility.

Control of sulphur emissions from CFBC may be achieved through the addition of limestone or dolomite to the bed. This in-bed sulphur capture obviates the necessity of an FGD unit. NO_x emissions are inherently low although where even lower levels are required, selective catalytic reduction (SCR) or selective non-catalytic reduction (SNCR) systems can be added. Such systems are in use on several CFBC power plants including the Tampella-based Norkopings plant in Sweden, the York County plant in Pennsylvania, and the 180 MW Barbers Point station in Hawaii, which utilises two Pyropower CFBC units. A typical CFBC flow diagram is shown in Figure 2.

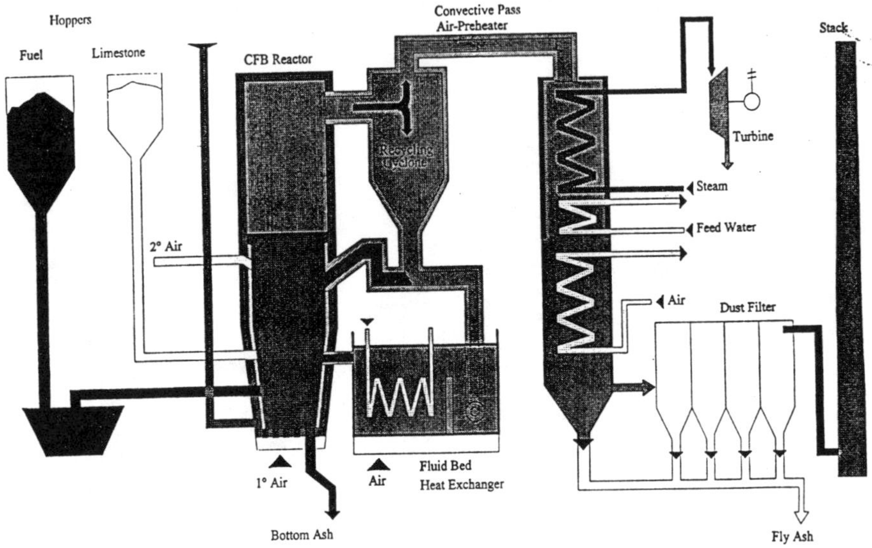

Fig.2 CFBC flow diagram

Recent developments have included the introduction of the CYMIC CFBC, manufactured by Tampella. Here, a water cooled cyclone is situated **inside** the furnace. The design offers improved thermal transfer and a reduction in overall plant size[4].

CFBC units are inherently fuel-flexible and plants throughout the world operate on a wide range of fuels and mixtures of fuels. Apart from operation on coal, there are plants that operate successfully on bituminous and anthracitic coal wastes, petcoke and refinery wastes, biomass fuels, lignites, oil shale, flyash, coke and coal washery wastes. Often, a blend of fuels is used.

CFBC offers a viable alternative to pulverised fuel firing particularly for smaller plant or where variable/low quality fuel is available. CFBC stands to benefit from future advances in steam cycle technology but by the same token is ultimately limited by the steam cycle Carnot efficiency.

Currently over 200 CFBC units are in operation worldwide.

2.3 Pressurised Fluidised Bed Combustion (PFBC)

In PFBC, coal is burnt in a bubbling or circulating fluidised bed under pressure. The combustion takes place at a higher intensity because of the increased partial pressure of oxygen, resulting in a reduced plant size (compared to pf plant) for a given output. The hot pressurised gases produced can be cleaned and used to drive a gas turbine. A steam turbine is also used, driven by steam generated from the remainder of the plant. Figure 3 shows a typical PFBC flow diagram.

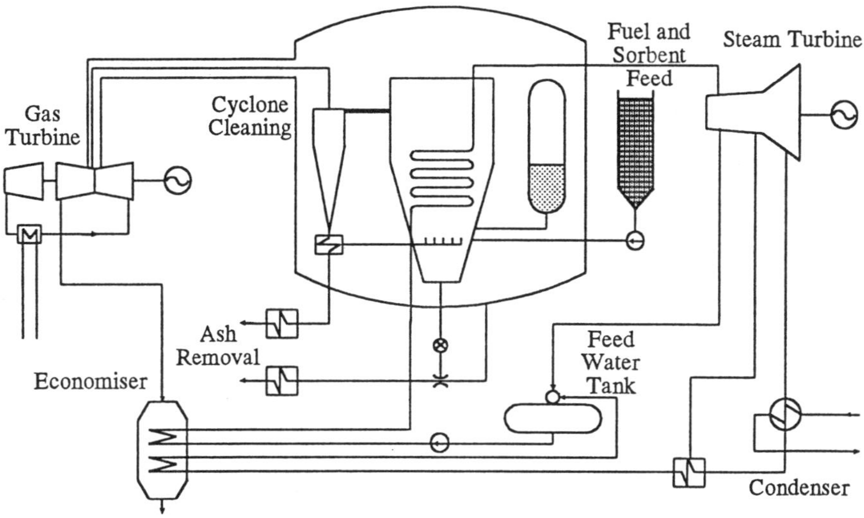

Fig.3 PFBC Flow Diagram

Currently, all commercial-scale PFBC plants rely on bubbling beds, typically operating at 12 bar and 850°C. Air for combustion is taken from the compressor stage of the turbine and fed to the bed. Off-gases from the PFBC are passed through cyclones and then on to the gas turbine inlet. The turbine exhaust gases are fed to a waste heat recovery section, followed by particulate removal and final discharge. Heat from the recovery section and combustion heat exchangers is used to raise superheated steam for a reheat cycle[5].

Sulphur oxide emissions are controlled by feeding crushed limestone or dolomite, either dry or as a paste, to the bed. Coal is usually fed in the form of a 70% solids coal-water mixture although one major plant (Escatron in Spain) utilises a dry coal feed. NO_x control is generally by ammonia injection with SCR. Currently, cyclones are used for dust removal from the pressurised gas stream although not all of the particulates are removed. Consequently, the gas turbine is a suitably modified 'ruggedised' unit, capable of withstanding the dust levels present. At one PFBC plant (Wakamatsu in Japan) a hot gas filter developed by Asahi Glass is under evaluation[6].

Cycle efficiency is typically in the 41-42% (LHV) range although there are variations between the four commercial plants; once fully operational, the Wakamatsu plant efficiency is expected to be the highest.

ABB Carbon are the main supplier in the field of PFBC, having built the four commercial plants:

- **Vaertan, Sweden.** This was the world's first commercial plant, commencing operations in 1991. The plant uses two ABB P200 PFBC bubbling bed units, each of 200 MW. It operates in cogeneration mode and is rated at 135 MWe + 224 MWth, supplying heat to a district heating scheme. Once an availability of 80% had been demonstrated, the plant was officially accepted by Stockholm Energy in 1994. Emission levels are low, the plant firing imported Polish coal[7].

- **Tidd, Brilliant, Ohio.** This was a repowering exercise carried out for Ohio Power Company. A vintage coal-fired boiler was replaced with a 70 MW ABB P200 unit, the balance of the equipment being reused. Pittsburgh high sulphur coal and dolomite was fed to the bed. The programme included a hot gas clean-up demonstration, one seventh of the plant's flow being directed via a slipstream through a hot gas filter test facility. The plant was shut down in 1995 having completed successfully its test programme[8].

- **Escatron, Spain.** This was another repowering exercise, a single ABB P200 unit replacing an old coal-fired boiler. The existing Westinghouse steam turbine etc was retained. The feedstock for the plant is black lignite with a sulphur content of up to ~7%. This is fed in dry form. Emissions have fallen well within specified limits and the site power output has been increased from 62 to 80 MWe. Around 20k hours of operation have so far been achieved.

- **Wakamatsu, Kita Kyushu City, Japan.** This is a further repowering exercise, a single ABB P200 unit being used alongside the existing Mitsubishi steam turbine. As part of a three year demonstration phase, electricity production started in 1995. Low sulphur Australian coal is fed, with limestone, both in paste form, to the bed. Hot gas clean-up is being evaluated through the use of an Asahi Glass ceramic filter and a high performance bag filter. NO_x control is by means of ammonia injection and selective catalytic reduction[9].

Whilst the four plants referred to above have inevitably suffered some teething problems the technology is now accepted as commercially proven. A number of other plants are now under construction or consideration.

These include the Trebovice Station in the Czech Republic, the Osaki plant of Chugoko Electric in Japan, the Hokkaido Electric Power plant at Tomatoazumi, and the 360 MWe Kyushsu Electric plant at Karita, Japan. The latter will consist of the first P800 unit supplied by ABB. This is a larger bubbling bed unit, based on the design of the existing P200 module.

Although bubbling bed PFBC remains the only commercially proven form of the technology, an alternative design (circulating PFBC) is being developed by Ahlstrom and Deutsche Babcock. It is suggested that these designs will have advantages in that they will have lower capital and operating costs, simpler operation, lower sorbent consumption, and better load following capabilities.

On the basis of evidence to date PFBC appears to have a future for both new station construction and for repowering applications.

2.4 Integrated Gasification Combined Cycle (IGCC)

In IGCC systems, coal or other fuel is fed to a gasifier operating at elevated pressure, where it reacts with oxygen or air and often steam, to produce a raw fuel gas. The latter is subsequently cleaned to remove particulates and other gaseous pollutants prior to firing in a gas turbine. Hot gas from the gas turbines exhaust is passed through a waste heat recovery system in order to raise part or all of the steam to drive a conventional steam turbine. Electricity is thus produced from both the gas and steam turbines. A typical IGCC flow diagram is shown in Figure 4.

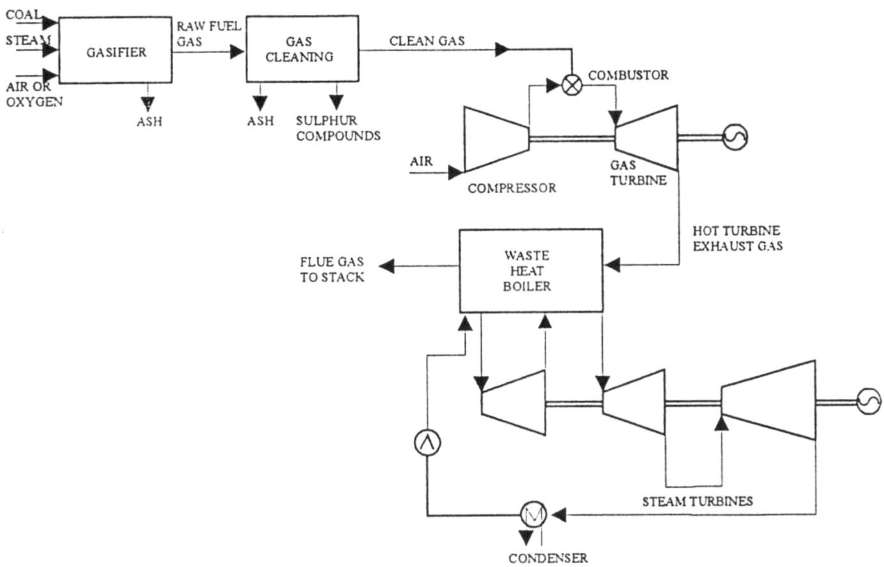

Fig.4 IGCC Flow Diagram

IGCC systems involve the gasification of the coal/fuel using sub-stoichiometric amounts of air or oxygen, to produce a fuel gas comprised primarily of carbon monoxide and hydrogen. Gasifier design varies widely although it falls into three main categories.

- **Fixed Bed Gasifiers** - coal is fed onto the top of an almost stationary bed through which steam and oxidant rise. Maximum temperature is at the base of the bed, the product gas emerging from the top at a relatively low temperature. As a consequence, these systems usually produce large quantities of oils and tars. The presence of these complicates heat recovery processes and makes gas clean-up more difficult. In-bed sulphur control methods cannot be used, a separate sulphur recovery system being necessary. Fixed bed gasifiers are sensitive to high levels of fines and require their coal feed in lump form. Residence times tend to be lengthy although load following can be quite rapid. Examples of fixed bed gasifiers include Lurgi and British Gas/Lurgi.

- **Entrained Flow Gasifiers** - here, the coal passes rapidly through the reaction zone while entrained in oxygen, steam and product gases. These gasifiers operate at higher temperatures and consequently, generate high alkali salt levels in the fuel gas as well as a sticky ash. The gas requires cooling in order to remove alkali salts (detrimental to downstream components and gas turbines), and for sulphur removal. The fuel gas needs extensive heat removal via waste heat recovery prior to scrubbing - this is associated with a cycle efficiency penalty. In addition, because of the low carbon inventory in the gasifier, variations in coal feed and properties may make control difficult. Entrained flow gasifiers are now commercially proven and offer high carbon utilisation and fuel flexibility. Commercial scale entrained flow systems developed by Texaco, Shell, Destec, CRIEPI and Krupp-Koppers are currently in use or under construction (eg, Texaco - Tampa Electric Polk County, Florida; Shell - Buggenum; Destec - Wabash River; CRIEPI - Nakoso Power Plant; and Krupp-Koppers Prenflow - Puertollano).

- **Fluidised Bed Gasifiers** - in a fluidised bed system, the reacting solids are kept in turbulent motion by a rapid updraught of air or oxygen and steam. These operate within a relatively narrow temperature range, usually just below the ash fusion point to avoid bed defluidisation and the formation of tars. Residence times are long enough to obviate the need for fine grinding of the coal feed. In order to achieve an appropriate operating temperature and a suitable level of conversion, oxygen blowing may be used, although several designs operate successfully with air blowing; the choice between oxygen or air-blowing can have a significant impact on costs and performance. Alkali concentrations in the fuel gas are lower than for entrained flow systems. Particularly where air blowing is used, conversion can be lower than for other systems, therefore, residual char may be utilised in a separate stage, eg, combustion in a CFBC. Systems under development or in use include High Temperature Winkler (HTW) - Rheinbraun, Berrenrath; Tampella/U-Gas - Toms Creek; KRW - Pinon Pine; and CTDD.

Recent years have seen the steady progress of a number of major IGCC projects and the demise of several other important initiatives, such is the site-specific nature of many of these. What has become apparent is that on a worldwide basis, there has been an upsurge in interest in gasification-based power plants although not all of these have been focused on coal; alternative fuels such as biomass fuels and refinery wastes are being utilised in a number of cases. In this context, a number of substantial IGCC plants are currently under construction with proposals/negotiations for a further tranche in coming years. It is predicted that the majority of refineries will eventually make use of IGCC-based systems in order to dispose of wastes. However, it is anticipated that the proliferation of such plant will make a major contribution to the maturity of gasification technology and will ultimately bring down the unit costs of coal-based systems. This will also have spin-off benefits for other competing CCTs.

2.5 Hybrid Combined Cycle Systems

Hybrid cycles were originally conceived as a means of increasing the maximum cycle temperature (of ~900°C) usually associated with PFBC systems, to the much higher temperatures (~1250°-1300°C) utilised by modern gas turbines, allowing for greater overall plant efficiency.

Such systems usually combine both gasification and combustion technologies. During the first stage, a proportion of the coal is gasified to produce a hot product gas suitable for firing in a high temperature gas turbine. Full carbon conversion is not necessary as long as sufficient fuel gas of the appropriate characteristics is produced to allow for correct turbine operating conditions. In the second stage, the unconverted carbon, in the form of char, is burnt in a separate combustor in order to raise steam for a conventional steam cycle. A major advantage of a separate combustion stage is that it allows for the adoption of advanced steam conditions (increased steam pressure and temperature), making supercritical conditions feasible.

A variety of systems is currently under development in the USA, Europe and Japan. In the US, Foster-Wheeler is leading a US DOE-sponsored programme to develop an advanced PFBC-based process which incorporates a pressurised fluidised bed carboniser and a pressurised circulating fluidised bed combustor, the latter used to burn the char generated. Westinghouse are developing a topping combustor to burn the fuel gas from the carboniser in the hot vitiated air from the combustor.

Also in the USA, a further system has been developed by Ahlstrom, known as the Pyroflow Advanced Pressurised CFB System. Here, plans revolved around a second generation PFBC system, based on a CFB carboniser and CFB combustor, both operated under pressure. Currently, the future for the project remains unclear as proposals for a 140 MW plant were rejected.

In Germany, RWE Energie AG and Rheinbraun AG have developed a combined cycle based on an integrated HTW brown coal gasifier known as the KoBra Project. It was intended to build the 320 MWe plant at the Goldenbert power station site, near Cologne, the plant comprising two air-blown gasifiers and gas trains; an overall efficiency of ~46% was suggested. This project is currently on hold.

Japan is also undertaking development of a hybrid topping cycle system. This is based on a pressurised fluidised bed gasifier linked to a PFBC. Their proposed gasifier is air blown and will retain sulphur within the gasifier using a novel upper bed arrangement. Japanese Government Endorsement is currently awaited for a 5 MWth pilot unit which is scheduled for design and build by 1998 and operation during the period 1998-2000.

A hybrid system known as the Air Blown Gasification Cycle is also being developed in the UK. This is described in more detail in the following section.

3 History and Current Status of the Air Blown Gasification Cycle

As outlined in Section 2, clean coal power generation systems based on gasification combined cycle processes are being developed because they promise higher thermal efficiencies, lower cost of electricity and improved environmental performance compared with existing technologies. The Air Blown Gasification Cycle (ABGC), formerly known as the British Coal Topping Cycle, is an example of such a gasification combined cycle process. The ABGC is a hybrid partial gasification cycle based on a novel, air blown pressurised fluidised bed gasifier (PFBG) coupled to a circulating fluidised bed combustor (CFBC) to burn the residual char from the PFBG. Fuel gas from the PFBG is cleaned and burnt to produce a gas at high temperature and pressure which is expanded through a gas turbine. The gas turbine exhaust gas is used to raise steam and augment the steam cycle associated with the CFBC.

British Coal first proposed the ABGC concept in the 1970s and undertook initial development within the Grimethorpe Topping Cycle Project. In early 1993 the UK Department of Trade and Industry (DTI) through its Executive Agency, ETSU, published a report recommending the further development of the ABGC through to a commercial process. To achieve this goal an industry led group known as the Clean Coal Power Generation Group (CCPGG) was set up. This group comprised GEC ALSTHOM, through its subsidiaries European Gas Turbines Ltd, Stein Industrie SA and EVT GmbH; PowerGen plc; Mitsui Babcock Energy Ltd and the Coal Technology Development Division (CTDD) of the British Coal Corporation. The aims of the CCPGG are to undertake research programmes to develop the components of the system and to produce a design specification for a 90 MWe prototype integrated plant by early 1996. In addition the CCPGG is seeking to develop other market opportunities for ABGC components. Component development consists of both the establishment of new components, such as the PFBG and the hot gas clean up (HGCU) system, and specific development of already established components, such as the raw gas cooler, heat recovery steam generator (HRSG) and gas turbine.

There follows a brief historical perspective on the ABGC development and associated component development activities. The ABGC process is described and the expected performance of both prototype and commercial plants are presented.

The ABGC, formerly known as the British Coal Topping Cycle, is based on the partial gasification of coal in an air blown spouted bed at pressures of typically 20-25 bar and temperatures of about 1000°C. About 70-80% of the coal is converted to a low calorific value gas which is cooled to about 400°C and then cleaned using ceramic barrier filter technology. Concentrations of gaseous contaminants are minimised using appropriate sorbents and catalysts. This fuel gas is burnt to produce a gas at high temperature and pressure which is expanded through a gas turbine, and the turbine exhaust gas is used to raise steam in a waste heat recovery unit. A sorbent, such as limestone, is fed with coal into the gasifier to trap as sulphide most of the sulphur which would otherwise have exited with the fuel gas. Solid residues from the gasifier, comprising mineral matter, char and sulphided sorbent residue occur as two main streams, viz: as a coarse solid removed from the base of the gasifier and as fines material elutriated from the gasifier and trapped in the cyclone and ceramic filter units. Both solid residue streams are depressurised, cooled and burnt in a CFBC boiler, thereby raising steam which is used to generate further power in a steam turbine. The ABGC is shown diagrammatically in Figure 5.

3.1 Historical Perspective of ABGC Development

British Coal has a long standing association with the development of Clean Coal Power Generation Technology. This can be traced back to the early research and proving of pressurised fluidised bed combustion (PFBC) at the Coal Utilisation Research Laboratories (formerly BCURA) at Leatherhead, UK from the late 1960s through to the early 1980s. Larger scale work was undertaken at the Grimethorpe PFBC during several operational phases from the late 1970s to the early 1990s[10], culminating in the Grimethorpe Topping Cycle Project (GTCP) between 1990 and 1993. This joint programme[11] was the initial development of the Topping Cycle concept first proposed by British Coal in the 1970s. Participants were British Coal, PowerGen and GEC ALSTHOM (through its subsidiaries European Gas Turbines (EGT), Stein Industrie and EVT) with significant funding from the UK DTI.

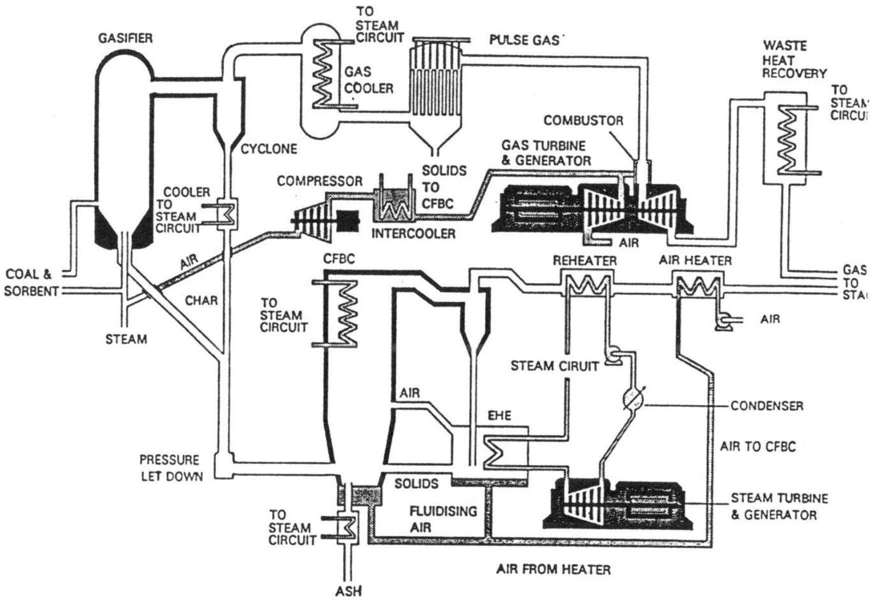

Fig.5. The Air Blown Gasification Cycle

Objectives of the GTCP were:

- to appraise the feasibility of operating a commercial gas turbine on coal-derived fuel gas at elevated temperature.
- to assess the viability of further development and exploitation of Topping Cycle technology.

The PFBC at Grimethorpe was operated in conjunction with a gas turbine in 1991/2. The aim was to characterise the hot gas paths and to collect data on corrosion performance of materials. Combustor exhaust gases were cleaned by a high temperature high pressure ceramic element filter owned by EPRI. The gas could then be split with the bulk passing to an EGT TB5000 gas turbine with an entry temperature raised to 950°C by the addition and combustion of propane. The remainder went to a high temperature sidestream where the gas temperature was raised to about 1240°C before entering a cascade of eight static aerofoils.

The overall conclusions of the GTCP were:

- acceptable blade lifetimes will be achieved with gas turbines for utility ABGC plants utilising 400°C fuel gas cooling

- the viability of blade lifetime where 600°C cooling is utilised is less certain because of predictions of alkali salt depositions at that temperature.

Studies at Grimethorpe were supported by pilot plant, materials and analytical programmes at British Coal's Coal Research Establishment (CRE) at Cheltenham. Coal Technology Development Division (CTDD) is one of the successor companies to CRE following the privatisation of the UK Coal Industry in 1994. CRE studies centred on researching several aspects of the ABGC including gasifier and hot gas clean-up development. A pressurised fluidised bed gasifier (PFBG) was constructed at Cheltenham by the substantial modification of an existing PFBC. This plant was commissioned in 1990 and modified to include a gas utilisation facility (GUF) in 1993[12]. Details of the operations at Cheltenham as part of the ABGC component development strategy appear later in this paper.

In late 1991 the UK DTI set up a Working Party led by EGT to examine the potential of the ABGC for commercialisation. The report of this Working Party[13] concluded that the ABGC appeared to offer performance and electricity cost benefits over more developed, competing gasification systems. However it also stated the need for significant development of both the system and its components. A strategy was produced identifying the need to progress through component development to a demonstration size plant of about 75 MWe size before proceeding to a commercial unit. The gasifier was identified as the main area of technical risk.

As part of the GTCP an appraisal of ABGC technology was undertaken in parallel. Key findings of this appraisal presented elsewhere[14] confirm the main conclusions of the DTI Working Party, ie, assuming successful development of the main components that a commercial plant based on the ABGC can offer significant efficiency and cost advantages compared with conventional coal-fired plant, and improved environmental performance.

Following the publication of the UK DTI Working Party Report an industry led group known as the Clean Coal Power Generation Group (CCPGG) came together to realise the commercial potential of the ABGC technology.

CCPGG membership currently comprises EGT (leader), Stein Industrie and EVT (all GEC ALSTHOM Subsidiary Companies), PowerGen plc, Mitsui Babcock Energy Ltd,

CTDD British Coal, and ETSU who represent the UK DTI which is providing substantial governmental funding to the programme.

3.2 ABGC Component Development

CCPGG members are undertaking several projects on ABGC component development. These studies will input to the ABGC prototype integrated plant design which will be available in 1996; they will also identify exploitation options for these components in non-ABGC applications. The current status of component development is given below:

3.2.1 The Gasifier

Gasifier process development studies have been undertaken exclusively by CTDD at Cheltenham. Process designs for both 15 tonne/h and 30 tonne/h gasifiers have been produced by CTDD and converted into engineering designs by Mitsui Babcock Energy Ltd as part of the prototype plant specification.

Practical studies at CTDD have utilised a range of gasifiers and cold models with support from fundamental laboratory studies and mathematical modelling. The aim of this work is to specify confidently, design and operating parameters for a full size gasifier. Most of the rig work at CTDD centres on the 0.5 t/h pilot plant. This rig consists of a 10.8m high refractory lined pressure vessel with an internal diameter of 0.32m, which expands to 0.46m at 6m above the base flange. The gasifier can operate at pressures up to 20 bar, temperatures of up to 1050°C and with fluidised bed heights up to 6m. The fluidising media are air and steam and these gases are injected in combination through the central spout of the reactor and through nozzles in the wall of the conical base section. Crushed coal and limestone are injected with air into the reactor via the central spout. In the gasifier, the coal is gasified producing fuel gas, char and ash. Cyclones remove the elutriated solids from the fuel gas prior to cooling and pressure let down. The gas is then either burnt in an enclosed flare prior to being exhausted to atmosphere or routed to the GUF which is described later.

The CTDD PFBG test facility was commissioned in 1990. Since then it has operated for over 4500 hours of which over 3000 hours has been at 12 bar or above. Performance targets determined at the outset for the gasifier were:

- a coal conversion efficiency in the range 70 to 80% (dry-ash-free basis)
- a minimum outlet fuel gas calorific value of 3.6 MJm^{-3} (wet, net)
- a sulphur retention greater than 90% for the overall cycle with a calcium to sulphur molar ratio of 2.1 for a typical UK coal.

All of these targets have been met or exceeded for a wide range of fuels including coals, lignite, coal/sewage sludge mix and sewage sludge alone. Coals tested include a range of UK run of mine, power station and low ash coals, and coals from Columbia, South Africa, Australia and the USA.

During gasifier test runs the effects of the main operating parameters on process performance have been studied. These parameters have included: coal type, limestone type and addition rate, pressure, bed temperature, fluidising velocity, coal to air mass ratio and steam to air mass ratio. A fuller description of the CTDD gasification process and the effect of operating parameters on process performance is given elsewhere[15]. Typical CTDD PFBG test facility performance at selected operating conditions is presented in Table 1.

Table 1. Typical CTDD gasifier performance at selected operating conditions

Operating Conditions	
Pressure, bar	12
Fluidising velocity, ms^{-1}	0.8
Bed height, m	4
Coal to air mass ratio	0.38:1
Ca:S molar ratio	2:1
Process Performance	
Gas composition, % by volume	
CO	17
H_2	17
CO_2	12
CH_4	1.5
N_2	47.5
H_2O	5
Gas calorific value MJm^{-3}	4.5
Coal Conversion efficiency, %	77
Sulphur retention efficiency, %	93

3.2.2 The CFBC Boiler/Steam Generation System

The design of the raw gas cooler, HRSG and CFBC for the ABGC prototype plant will be based largely on proven technology within GEC ALSTHOM. However there will need to be some specific thermal design work for the various components which will be based on information currently being gained at commercial gas turbine combined cycle and CFBC installations. In addition char combustion programmes are being undertaken by both CTDD and Stein Industrie to prove the CFBC technology on the low CV mixed char/fines feedstock and to assess the extent of burn out and oxidation of sulphided sorbent from the gasifier. Preliminary experimental work has been performed in a

0.12 MWt CFBC test rig at the University of British Columbia (UBC) in Canada[16]. Combustion efficiencies of up to 99% were achieved and sulphur retentions of 96% without additional limestone being added. However, small amounts of calcium sulphide remained in the final ash which had not been fully oxidised to sulphate. It is considered that the conditions for maximising sulphide to sulphate oxidation were not optimised in this testwork, for example original sorbent size and solids residence times within the CFBC. Further testwork will be undertaken at CTDD a CFBC of 0.1 MWt output which has just been commissioned.

3.2.3 Hot Gas Clean Up

Hot gas clean up (HGCU) studies in support of the ABGC prototype plant development to date are being undertaken almost exclusively by CTDD. An overview of various HGCU techniques is presented elsewhere in this book[17]. The CTDD studies build on the earlier Grimethorpe work that demonstrated the efficiency of systems using ceramic filter elements. A range of HGCU test rigs are used at CTDD. Their description and detailed work programme is given elsewhere[18].

Findings from the CTDD test programmes will form the design basis for a hot gas filter for the prototype ABGC plant. Negotiations are currently underway with a potential industrial collaborator who will provide the engineering design for the ABGC prototype hot gas filter.

In parallel to particulate removal work a suite of projects is underway at CTDD to investigate the reduction/removal of contaminant gases from the fuel gas stream prior to combustion in the gas turbine. These studies centre on H_2S, ammonia and HCl removal and are aimed towards achieving an acceptably low final emission of acid gases to atmosphere from the gas turbine.

The methodology being assessed for sulphur polishing is to contact the H_2S with a metal oxide thus producing a metal sulphide which is then regenerated using an appropriate oxidant to produce a concentrated sulphur dioxide stream. Various approaches are being investigated including the use of fixed, moving and fluidised bed reactors. Ammonia removal is being assessed using a range of proprietary ammonia reduction catalysts; this testwork is still at a very early stage and there are no firm conclusions. Hydrogen chloride will be reduced by contacting with a suitable sorbent. A fuller description of gaseous contaminant removal, especially H_2S removal using metal oxides, is presented separately at this meeting[19].

The developmental status of gaseous contaminant removal for the ABGC is less advanced than that of particulate solids removal. Discussions are underway with a potential industrial partner to design a sidestream hot gas cleansing facility for the ABGC prototype plant in advance of full stream gaseous contaminant cleansing for a commercial ABGC plant.

3.2.4 Gas Turbine Combustion

The gas turbine for the ABGC prototype plant will be based on the proven EGT Frame 6B machine. However it will be necessary to ensure that the gas turbine will burn the gasifier product gas at good efficiency and with low contaminant emissions. Major combustion test programmes are underway both at CTDD and at EGT using both 5 inch and 10 inch combustor test rigs. CTDD studies form part of the GUF operation (see Figure 5). This arrangement will enable the combustor to be fired either on cleaned fuel gas from the CTDD PFBG facility or synthetic gas from an adjacent SYNGAS trailer.

3.2.5 Cycle Performance Studies

Steady state and dynamic performance of both prototype and commercial ABGC plants need to be modelled in order to give confidence in the plant's abilities. British Coal, PowerGen and CRE Group Ltd (the other successor company to the former Coal Research Establishment) have well developed steady state and dynamic process modelling packages for the ABGC; these are being further refined to accept new information from the component development programmes.

3.3 Performance and Costs of the ABGC

A number of studies have been undertaken to address the relative efficiencies and economics on a common basis from a range of advanced power generation systems. Ultimately the customer is only interested in the final cost of the product so the cost of electricity is the key issue. A recent study[13] shows the ABGC has a 7% cost of electricity advantage over a typical IGCC plant. Another study[14] shows that the ABGC has a 12% cost of electricity advantage over advanced pulverised coal fired plants. These results are summarised in Table 2.

Studies cannot be directly compared because different baseline criteria were used for each; in particular the steam conditions considered were different. The results clearly show the benefits of the ABGC technology over other competing technologies for power generation.

A significant feature of the ABGC which sets it apart from many of its competitors is that it is air blown. This factor is a contributor to the overall economics of the ABGC as presented in Table 2. The relative merits of oxygen and air blown gasification systems have recently been reported[20] and conclude that air blown systems are the best choice for power generation applications. Oxygen blown gasifiers initially developed for synthesis gas production can of course be used for IGCC but they are not necessarily the best choice. Separating air into oxygen and nitrogen, only to mix them back together again at the gas turbine, is a fundamental thermodynamic inefficiency of oxygen blown IGCC.

Table 2. A comparison of the cost of electricity from power generation cycles

Reference 13	ABGC	IGCC
Efficiency (LHV %)	46.5	44.1
Relative Capital Cost (per kW)	1	1.11
Relative Cost of Electricity	1	1.07

Reference 14	ABGC	Ultra Supercritical PF+FGD+SCR
Efficiency (LHV %)	47.8	44.7
Relative Capital Cost (per kW)	1	1.23
Relative Cost of Electricity	1	1.14

Environmental emissions predicted for commercial ABGC plant, on the basis of research undertaken to date and the successful completion of developments in hand, will at least match those of other competing clean coal technologies. Predicted CO_2 emissions at <700 g kWh^{-1} generated for ABGC are lower than IGCC at 750 g kWh^{1} generated. With full fuel gas cleansing ABGC should achieve overall SO_2 and NO_x emissions of <0.19 and <0.4 g kWh^{-1} generated respectively. These figures are similar to those predicted for commercial IGCC plant.

4 Conclusions

Several CCTs are currently being developed which offer significant advantages in terms of efficiency, environmental performance and costs compared with conventional pulverised fuel systems. One CCT known as the Air Blown Gasification Cycle is now at an advanced stage of development. Individual components of this system are being proven and a design specification for an ABGC prototype plant of c.90 MWe size is now available.

Recent costing studies have shown that at the commercial scale the ABGC will produce electricity more cheaply than either advanced pulverised fuel or IGCC. Commercial ABGC will also match the very low environmental emissions predicted from other CCTs.

Acknowledgements

The authors gratefully acknowledge the financial support of the European Union, the European Coal and Steel Community, the UK DTI and CCPGG Industrial Partners towards the work described. Most of the development work related to the ABGC has been undertaken at CTDD and the contributions of colleagues at CTDD is gratefully acknowledged. Views expressed are those of the authors and not necessarily those of the British Coal Corporation or the supporting organisations.

References

1. Sage PW. 'World Energy Resources, their Use and the Environment'. Paper to NATO:ASI 'Desulphurisation of Hot Gas with Regenerable Metal Oxide Sorbents: New Developments'. July 7-19, 1996, Kusadasi, Turkey.
2. Kjaer S. 'Advanced Steam Cycles'. Paper to TOTeM 10 'Advanced Power Generation', 28-30 September 1994, Veldhoven, Netherlands.
3. Maitland JE. 'Today's Scaled-up CFBs meet Commercial Cogen Demands'. Private Power Executive Nov/Dec 1994.
4. 'CYMIC. Ready for Worldwide Sales', Tampella Power Lines, Winter-Spring 1995.
5. 'Combined Cycle with Pressurised Fluidised Bed Combustion'. Unipede Report 02001 Ren 9405, 1994.
6. Fugita M. 'Start-up of the Wakamatsu PFBC Plant', Japan PFBC Workshop, 1994.
7. Nordin S and Dahe A. 'Experience for the First Year of Commercial Operation of the Vaertan PFBC Plant'. VGB Kraftwerkstechnik, 73, 1993.
8. 'Tidd concludes four year PFBC test programme', Clean Coal Today (US DOE) Summer 1996.
9. Goto H. 'Operation Experience for the 71 MW Wakamatsu PFBC Demonstration Plant'. 13th FBC Conference, Orlando, USA, May 1995.
10. Dawes SG 'Reflections on the History and Achievements of the Grimethorpe PFBC Experimental Facility from Conception to Demolition 1970-1993'. Institute of Energy, Combustion and Emissions Control Conference, Cardiff, September 1993.
11. Clark RK, Arnold MStJ, Fackrell JE, Mordecai M and Dawes SG. 'The Grimethorpe Topping Cycle Project'. Proc. 5th Intl. FBC Conference, Adam Hilger for IOP Publishing, pp353-362, London, 1991.
12. Sage PW. 'Development of the British Coal Air Blown Gasifier'. Paper to ETSU Coal Contractors' Meeting, Nottingham, UK, November 1994.
13. Soothill C and Brown D. 'The Department of Trade and Industry's Topping Cycle Working Party Final Report'. Contractor Report R011, ETSU for DTI, 1993.
14. Dawes SG, Mordecai M, Brown D and Burnard GK. 'The Air Blown Gasification Cycle'. Proceedings of the 13th Intl Conf on FBC, Orlando, FA, May 1995.
15. Welford GB and Marshall AR. 'The Pressurised Gasifier in the Air Blown Gasification Cycle'. 3rd International Conference on Combustion Technologies for a Clean Environment, Lisbon, Portugal, 1995.
16. Sage PW, Welford GB, Brereton C and Julien S. 'Development Issues for the Char Combustor Component of an Integrated Partial Gasification Combined Cycle System'. Proceedings of the 13th Intl Conf on FBC, Orlando, FA, May 1995.
17. Fantom IR, Cahill P and Sage PW. 'Hot Gas Cleaning - An Overview'. Paper to NATO:ASI 'Desulphurisation of Hot Coal Gas with Regenerable Metal Oxide Sorbents: New Developments'. July 7-19, 1996, Kusadasi, Turkey.

18. Cahill P, Dutton MC, Tustin M, Rasmussen G and Sage PW. 'Development of Hot Gas Filtration for Air Blown Gasification Plant'. 12th Pittsburgh Coal Conference, September 1995.
19. Fantom IR and Radford S. 'Experience of Regenerable Sulphur Sorbents in Real Fuel Gas Systems'. Paper to NATO:ASI 'Desulphurisation of Hot Coal Gas with Regenerable Metal Oxide Sorbents: New Developments'. July 7-19, 1996, Kusadasi, Turkey.
20. Dawes SG, Sage PW, Welford GB and Davison JE. 'Air Blown Gasification for Power Generation'. IChemE Conference 'Coal Gasification - An Alternative to Natural Gas', London, Nov 1995.

Asssessment of Coal Gasification Processes - Relevance to Sorbent Development

R.E. Ayala

GE Corporate Research and Development, One Research Circle, Schenectady, New York, 12309 USA

Abstract. Coal is one of the most abundant fossil fuel sources in the world with available reserves for several hundred years worth of consumption. Use of coal for electricity generation is expected to increase at least for the next thirty years. The increase in coal consumption is dependent on the development of advanced processes that are more energy efficient and compliant with current environmental regulations.
Integrating the gasification with an existing power generating cycle, such as the gas (combustion), steam, or fuel-cell cycles, provides a means of achieving this goal. The objective of this paper is to rewiew coal gasification technology, highlight the most recent developments, define gas cleaning needs at high temperature, and attempt to put in perspective the overall process needs for the development of regenerable metal-oxide sorbents for high-temperature desulfurization.

1 Introduction

The development of regenerable sorbents for high-temperature desulfurization requires an understanding of the process constraints that govern the operation and optimization of power generation systems. In power generating cycles, the ultimate objective is to produce cheap electricity at a high overall efficiency for the system, i.e., the coal-to-busbar efficiency. Integrating the gasification with an existing power generating cycle, such as the (gas) combustion, steam, or fuel-cell cycles, provides a means of achieving this goal.

The first step in the coal-to-electricity conversion is the preparation and gasification of coal. Development of more advanced and efficient gasification processes affects the allowable window of operation for other technologies involved in the power cycle, such as the fuel cleanup, the power generating island, and the waste disposal system. Regenerable sorbents for hot gas cleanup (HGCU), as part of the fuel cleanup technology, are of course affected by the operation of the gasification process. The choice of one sorbent over another is greatly determined by the coal gas temperature, pressure of operation, and gas composition, just to name a few. Consequently, the selection and optimization of the overall power generating system is influenced by the gasification conditions and the quality of its product, i.e., the coal fuel gas.

It is the objective of this paper to review current gasification processes and technology, highlight the most recent developments, define gas cleaning needs at high temperature, and attempt to put in perspective the overall process needs for the development of regenerable metal-oxide sorbents for high-temperature desulfurization.

NATO ASI Series, Vol. G 42
Desulfurization of Hot Coal Gas
Edited by A.T. Atimtay and D. P. Harrison

2 Fossil Fuel Sources and World Energy Demand

According to Holt [1], coal and lignite account for approximately 3400 gigatons of oil equivalents (Gtoe), or 76%, of the total recoverable fossil fuel reserves in the world, which is estimated at 4400 Gtoe. In contrast, natural gas is only 220 Gtoe or 5% of the total recoverable fossil fuel reserves. Unconventional oils (including heavy crudes, natural bitumen, and oil shale) account for 14% of the total, while conventional oil represents an additional 5%. From these figures, it is clear that coal and lignites account for a large percentage of total recoverable fossil fuel reserves in the world and, as such, will have an important role in the future of energy usage and power generation worldwide.

Using population projections from the World Bank, the World Energy Council [2,3], has calculated that the annual global fuel use will approximately double between 1990 and the year 2020, from 8.8 Gtoe to 11.3-17.2 Gtoe, depending on the predictions of population and economic growth taken. Table 2.1 compares the world total electricity generation (TEG) expressed as TeraWatt-hours (TWh) and world population for 1990 and the projections for the period 1990-2020. Most of the world total energy generation occurs in the highly industrialized regions of North America and Western Europe. On a relative scale, most of the expected growth rate in electricity generation will occur in the Middle East-North Africa region, South Asia, and Latin America at nearly 5%.The rest of the regions will have a moderate or a small growth rate. However, the absolute growth rate expressed as additional TWh produced, rather than percentages, is still the highest for the industrialized regions. These results are in contrast with the world population statistics per region: most of the population growth will occur in regions other than North America and Europe (western, central, and eastern), where the projected electricity generation growth is lower, making the TEG growth rate per capita the lowest. Wolk and Holt [4] have pointed out that there is a proven strong positive correlation between electric usage per capita and gross domestic product (GDP) per capita.

The comparison between recoverable fossil fuel reserves and energy consumption indicates that the supply of fossil fuel, coal in particular, could fulfill the needs of today's world population for hundreds of years. Given that 69% of the natural gas reserves are concentrated between the former USSR (37%) and the Middle East countries (32%), many countries with domestic coal reserves (e.g., China, Australia, Germany, South Africa, India, Poland, Hungary, and Colombia) are looking to exploit and develop alternate energy sources for power generation, such as their own domestic coal. Other countries having both sources, like the US, are looking to reach a reasonable balance between the different fossil fuels. Overall, energy utilization is a dynamic process driven by economic as well as environmental factors.

Table 2.1 Projections of TEG and world population growth (after Holt [1]and Wolk[4])

Region	1990 TEG, TWh	TEG Growth Rate, (1990-2020), %	1990 World Population, Millions	World Population Growth Rate (1990-2020), %
North America	3534.7	1.0	275.9	18.3
Latin America	624.6	4.7	448.1	59.9
Western Europe	2427.3	1.5	454.1	7.7
Central Eastern Europe	372.8	1.7	100.2	10.8
Newly Independ. States of USSR	1722.1	1.1	288.6	19.2
Middle East and North Africa	273.4	5.0	271.0	100.5
Sub-Saharan Africa	220.5	3.8	501.6	138.3
Pacific (including CPA*)	1997.7	3.4	1806.9	34.4
South Asia	318.6	4.7	1146.0	69.1
World	11490.9	2.3	5292.4	52.9

* CPA = Central Planned Asia

3 Coal Utilization Processes

Coal is one of the most abundant fossil fuel sources on earth with reserves available for several hundred years worth of consumption. The most important use of coal in the economically developed countries is for electricity generation. Several technologies can be used to extract the energy from coal and convert it into useful electricity (Figure 1). They are categorized into combustion, liquefaction, and gasification. Combustion processes rely on the total oxidation of coal to primarily carbon dioxide and steam with the subsequent generation of electricity via steam production in a waste heat boiler and steam expansion in a steam turbine. Atmospheric combustion and liquefaction are considered conventional coal-conversion technologies, as opposed to pressurized combustion and gasification cycles which are considered advanced coal conversion technologies.

Atmospheric combustion has been used commercially for years and includes configurations such as fluidized-bed combustion, pulverized-coal (PC) combustion with flue gas desulfurization (FGD)/deNO$_x$, and PC/slurry combustion with physical/chemical cleaning. Efficiencies of atmospheric combustion processes are low by today's standards and will not be discussed further in this paper.

Pressurized combustion and gasification cycles are considered the advanced coal-based power generation technologies because they promise higher efficiencies for production of power, have been made commercially available more recently, or a combination of both. Pressurized combustion, as the name implies, performs the coal combustion at high pressures and is more efficient than the atmospheric case. Pressurized combustion has three major modalities: pressurized pulverized coal (PPC) or direct coal-fired turbine (DCFT) cycles, pressurized fluidized-bed combustion (PFBC) and hybrid (topping) cycles, and the magneto-hydrodynamic (MHD) cycle. A review of the

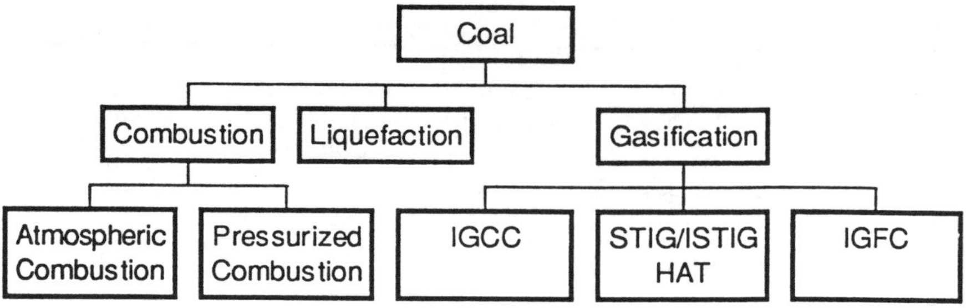

Figure 1. Technologies for the conversion of coal to electricity

and hybrid (topping) cycles, and the magneto-hydrodynamic (MHD) cycle. A review of the advanced pressurized combustion technology is given elsewhere in these proceedings [5].

In the gasification area, three technologies can be integrated to the gasification (Figure 1): gas turbine cycles (i.e., STIG, ISTIG, HAT), the gas turbine/steam turbine combined cycle (i.e., IGCC), and fuel cells (i.e., IGFC). In the gasification, the coal is only partially combusted in the gasifier to produce a fuel rich in CO and H_2 under high pressure and temperature.

After gas cleaning, the high pressure and high temperature gaseous fuel is sent to drive a gas turbine or a fuel cell. Besides generating electricity in a gas turbine as a simple cycle, the residual heat from the combustion gases can be used to generate steam in a heat recovery steam generator (HRSG) (Figure 2). If the steam is injected to the gas turbine, in addition to the fuel, the cycle is called steam-injected gas turbine (STIG). Compared to a simple gas cycle, the injected steam, with its higher specific heat capacity and mass flow, contributes to a large increase in efficiency and power output of the gas turbine. If an intercooling step between two stages of the gas turbine compressor is used, the modification of the STIG cycle is called intercooled steam-injected gas turbine (ISTIG) cycle. A cycle recently proposed by EPRI for development is called the humid air turbine (HAT) cycle, which is a variation of the ISTIG cycle. Here, the air is humidified to a water content of 20-40% and heated to around 538 °C prior to being introduced in the gas turbine. ISTIG and HAT offer the potential for improvements in efficiency and lower capital costs, but the higher risk, because of being in the early stages of development, may hinder its rapid evolution towards a commercial system.

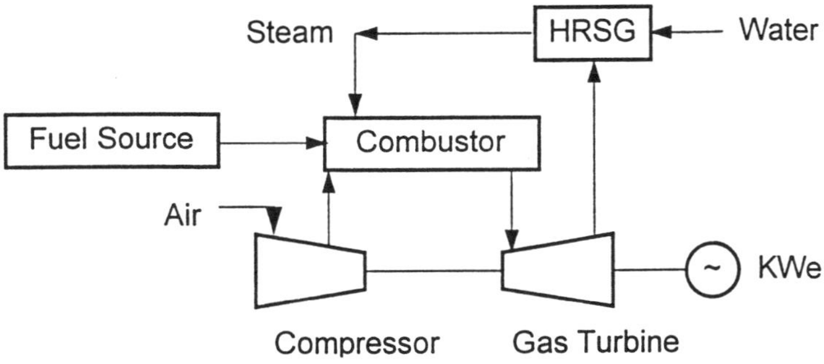

Figure 2. Schematic diagram of a STIG cycle

When the steam generated from heat exchanging an HRSG with the gas turbine exhaust gases is used to drive a steam turbine for generation of additional power, the process is called the integrated gasification-combined cycle (IGCC). In the IGCC system, a waste heat boiler or HRSG and a steam turbine are added to the gasification and gas turbine components to generate steam and further extract energy from the hot combustion gases in the combined gas turbine/steam turbine cycle. The IGCC stands as a configuration having the capability of achieving higher efficiencies and minimizing the waste products and emissions simultaneously. IGCC differs from STIG/ISTIG cycles in that a steam turbine is used in addition to the gas turbine. The combined cycle increases

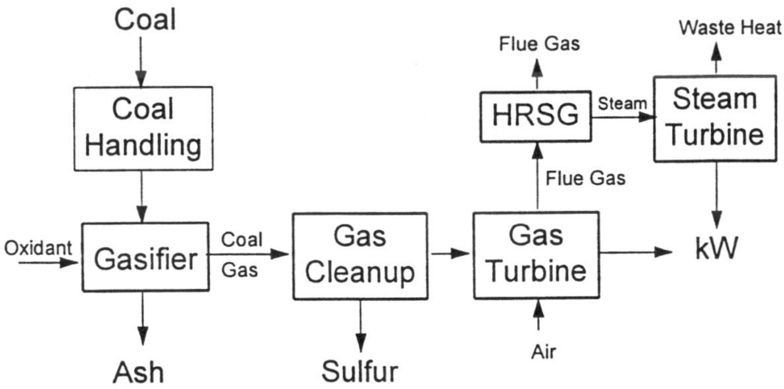

Figure 3. Schematic diagram of the IGCC system

the overall efficiency of the system. The IGCC cycle is described in great detail in the accompanying article "Application of IGCC Technology to Power Generation" by the same author [6].

The gasification can also be linked to a fuel cell in the integrated gasification fuel cell (IGFC) cycle, where electricity is generated via electrolytic oxidation of the fuel gas (primarily H_2 and CO) (Figure 4). The fuel cell cycle gives a very high thermodynamic efficiency, and the exothermic nature of the reactions allows to recuperate some of the wast heat to drive a bottoming gas turbine or a steam turbine. The degree of development of fuel cells is less than that of IGCC systems and fuel cell-based systems have only been built and demonstrated below 100 MW. Two types of fuel cells have received a higher degree of development: one is the molten carbonate fuel cell (MCFC) the other is the solid oxide fuel cell (SOFC). Fuel cells have more stringent requirements for gas cleaning than gas turbines, thus a sulfur guard is generally required in addition to hot gas cleanup for MCFC operation. Reactivities of catalytic electrodes in MCFC's are quite

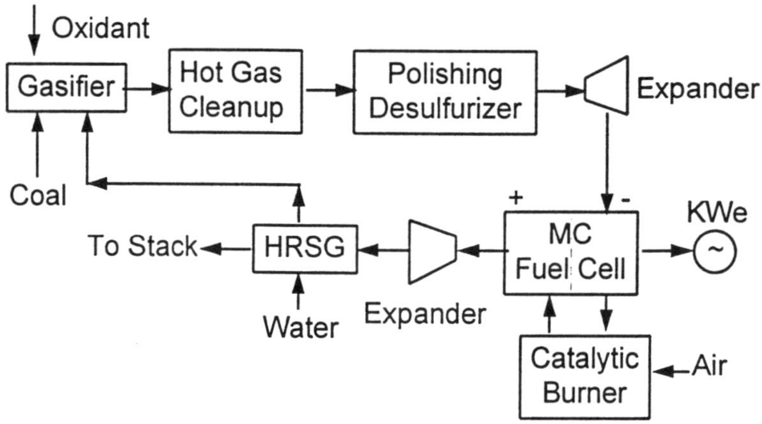

Figure 4. Schematic diagram of an IGFC cycle using a molten carbonate fuel cell (MCFC)

sensitive functions of several impurities in coal gasifier effluent, the most serious problem species being H_2S, COS and HCl. Total sulfur species must be less than 1 ppmv and HCl must be below 0.2-10 ppmv [7].

PFBC, gasification, gas cleanup, gas turbines, and fuel cells are all technologies which have shown significant improvement and promise growth potential. Furthermore, gas turbines can be designed to handle both natural gas and coal-derived gaseous fuels.

4 Coal Properties and Gasifiers

The performance of the gasification system is a key factor affecting the overall power generating efficiency of any power cycle. It is important, therefore, to ensure that the gasification and gas cleanup islands convert the coal into a clean fuel with the greatest possible efficiency.

Coal and its properties have been studied for a long time, and it is well known that coal is one of the most diverse solid substances available. Gorbaty and Ouchi [8] have compiled recent studies of coal organic constituents (macerals) and inorganic structures. Analytical methods for coal and coal products (including analyses for physical properties, proximate and ultimate analyses, trace elements, and coal-derived liquids) have been compiled by Karr [9]. A good review of coal properties and their relation to gasification is given also by Takematsu and Maude [10]. Among the properties of coal, the following are considered of importance for the quality of gasification:

- **Reactivity and volatile matter**. Highly reactive coals are high in volatile matter and desirable because it is easy to produce a fuel gas. However, friability, which also occurs in the same low-rank coals coals such as lignites, may result in compaction during gasification. These coals may not be as suitable for gasification configurations that rely on the flow of gases through a packed, fixed, or moving-bed configuration because channelling and uneven flow (bypassing) of gases result in difficulties for temperature control during gasification. Bricketting may then be a necessary preparation step prior to gasification.
- **Caking and swelling**. Bituminous coals tend to swell due to their high volatile content, and agglomeration may present problems in moving- and fluidized-bed gasifiers.
- **Fixed carbon content**. It is the remainder of coal matter after devolatilization. Usually, the residual char must be reactive to accelerate gasification.
- **Ash properties**. Ash is the solid non-gasifiable residue from coal and must be removed from the gasifier as a liquid slag or dry solid ash. Ash fusion temperature of the coal determines the applicability for operation under slagging or non-slagging conditions in a gasifier. Low-fusion temperature is desired for slagging gasifiers. Those ashes with a high-fusion temperature may require the addition of fluxing agents, e.g., limestone, to promote ash removal from the gasifier, or else, the gasifier must be operated in a dry-ash mode.
- **Particle size distribution**. Operation of coal particles with the flowing gas within the gasifier is highly sensitive to coal particle size due to solid/gas flow considerations. For example, fluidization of coal particles requires a relatively narrow size distribution and even gas velocity distribution across the gasifier. Permeation of gas through a fixed/moving-bed gasifier is sensitive to fine particle carryover and gas or solid maldistribution in the bed. Both moving-bed and fluidized-bed gasifiers are more sensitive than entrained-flow gasifiers, which operate with pulverized coal.

Gasification is a very complex technology, and the conversion rate of dry coal or slurry to gaseous fuel involves multi-phase transformations, mass transfer, chemical kinetics, thermodynamic equilibria, and many other complexities that occur simultaneously and are beyond the scope of this paper. For further information, the reader is referred to more extensive sources on gasification technology [10]. Four types of gasifiers are usually recognized as having demonstrated higher performance in large scale systems (Figure 5):

- **Moving-bed gasifiers**. These are also called fixed-bed gasifiers and the coal feed moves in a countercurrent direction to that of the gaseous reactants (e.g., air or steam) and fuel. Depending on the method used for ash removal in moving-bed gasifiers, operation is in either dry ash or slagging ash mode. The coal is fed in 3-30 mm particle size. The internal bed temperatures run between 800 °C and 2000 °C and the outlet gas temperatures are in the 400 °C to 800 °C range. These outlet gas temperatures are compatible with existing zinc-based desulfurization sorbents under development. Because of the relatively lower temperatures of the outlet gas, however, condensable tars or oils are produced that will deposit onto the sorbent surfaces and may decrease sorbent desulfurization performance. Typical moving-bed gasifiers are the Lurgi dry ash gasifier originally developed in Germany, the British Gas/Lurgi (BGL) slagger gasifier tested in Scotland, and the (rotary) Kiln Gas gasifier developed in the USA.

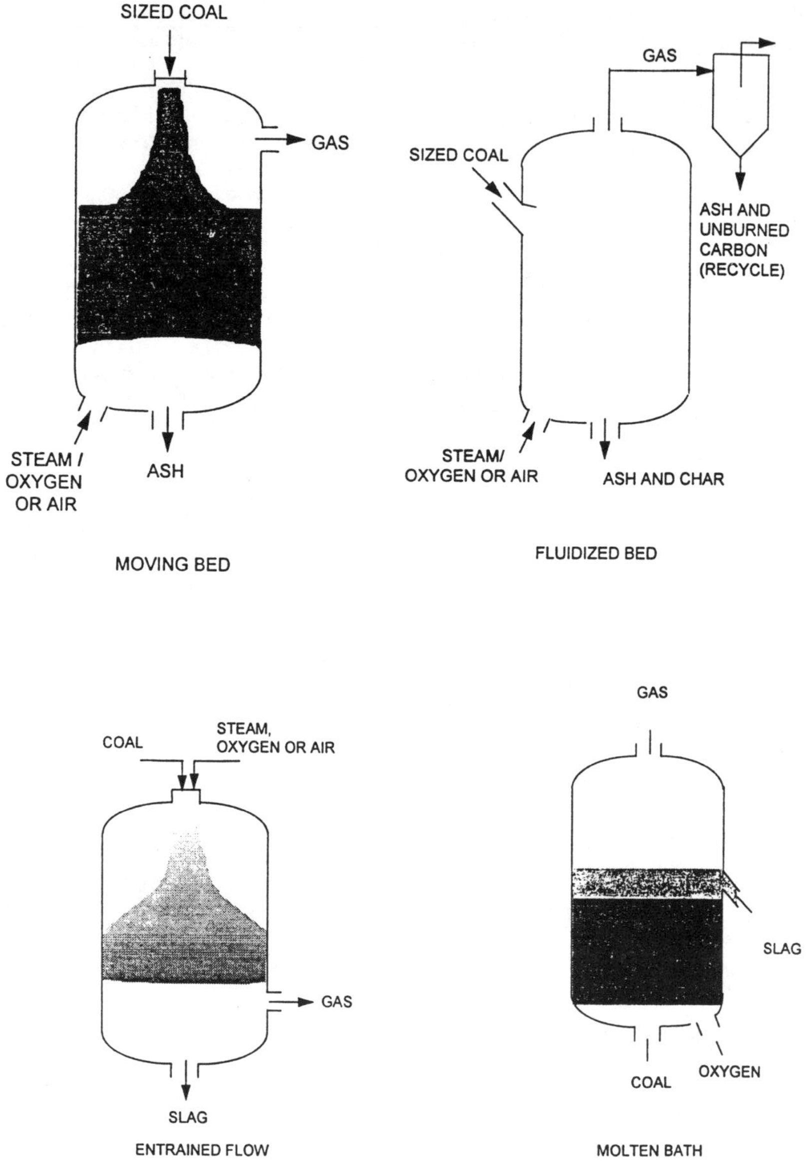

Figure 5. Gasifier types: moving bed, fluidized bed, entrained flow, and molten bath

- **Fluidized-bed gasifiers**. These are subcategorized into non-agglomerating (H-T Winkler) and agglomerating (U-Gas, Kellog-Rust-Westinghouse (KRW)) types. Both use dry feed of coal in the 1-5 mm size. Because lignites are very reactive, they are very suitable for fluidization in this type of gasifier. The gas temperatures are in the range 800-1100 °C and operate in the range of 1 to 25 bar. One definite advantage of fluidized-bed gasifiers over other types is that in-bed sulfur removal is possible by addition of limestone or dolomite into the gasifier. For high-ash fusion coals, mixing of a fluxing agent to the coal may be necessary to achieved the desired fluidization characteristics. Flux type and size need to be carefully selected to maintain proper solids flow inside the gasifier, particularly when in-bed sulfur removal is simultaneously carried out by addition of a calcium-based sorbent. Outlet gas cooling is necessary prior to external desulfurization with most hot gas cleanup regenerable sorbents.
- **Entrained-flow gasifiers**. These can use either a coal slurry feed (Texaco, DOW) or a dry feed (Shell, PRENFLO, NEDO). Coal is typically 0.1 mm in size. Internal gasifier temperatures are in the 1200-2000 °C, and consequently, gases need additional cooling before use with most hot gas cleanup sorbents.
- **Molten bath gasifiers**. These exhibit some of the characteristics of the three major types of gasifiers described above, making them a "hybrid" type and, strictly speaking, cannot be assigned to a single category. The molten bath gasifier concept is a development from the steel converter technology and operates on a dry feed system (Saarberg-Otto, MIP, ATGAS). Coal is fed into the molten bath, either iron or fused salts of sodium carbonate, together with limestone and oxygen. Ash and sulfur are chemically bonded to the ash. Typical conditions are temperatures near 1500 °C and pressures from 1 to 3 bar [11]. Gasification temperatures, usually above 1600 °C, are generally higher than those for entrained flow systems. No information is available on the suitability of the coal gas with existing regenerable metal oxide sorbents.

Table 4.1 shows the comparison of preferred coal feed for various gasifiers [1]. High ash coals tend to give difficulties to most gasifiers. Also, coals having a high ash fusion temperature require the use of flux (typically limestone) to produce acceptable operation of the ash removal system in the gasifiers.

Table 4.1. Preferred coal feed for various gasifiers (after Holt[1])

	Texaco	Destec	Shell, Prenflo	BGL	HT Winkler	KRW, U-Gas*
Bituminous	√	√	√	√	some	√
Subbitum./ Lignite		√	√	some	√	√
High Sulfur	√	√	√	√	√	
Low Sulfur	√	√	√	√	√	√
High Ash					<—— OK	——>
High ash fusion temp.	<——	Needs flux	——————>			

* Others include British Gas, Kellogg Transport, Foster-Wheeler

Besides the characteristics of coal, the next most important item in the operation of a gasifier is the choice of air or oxygen as an oxidant. The choice of oxidant will determine to a large extent, temperature, pressure, flow rates, and gas compositions for most suitable operation. The issue of oxidant selection has been discussed over the years and whether one or the other choice is more efficient and cost effective than the other is still the subject of debate. The answer fluctuates depending on the relative merit that it is given to the economic versus the process characteristics of each as well as the end use of the fuel desired. As stated by Swanekamp [12], EPRI sponsored in the late 1970's detailed technical and economic analyses of both types of oxidants under various configurations. The analyses showed essentially no economic differences between the two concepts; however, according to Swanekamp, it appears that oxygen-blown gasifiers have a certain non-economic advantages over the air-blown case. For the purpose of this paper, it is more appropriate at this point to enumerate the differences in operation, rather than providing absolute statements on overall economic outcome of one configuration versus the other since degree of system integration and operation may affect the end results.

Some of the advantages of oxygen-blown gasification include [12]:

- Gasification capacity is higher for a given-size gasifier. The output capacity of a given-size gasifier is cut in half when it is air-blown and, hence, two air-blown gasifiers are needed for the same throughput.
- The cold-gas efficiency, defined as heating value of the cooled and purified fuel gas divided by the heating value of the gasifier feed, is higher for an oxygen-blown gasifier
- Gas cleanup costs are lower for an oxygen-blown system based on the lower volume of gas to be processed.
- There is better flexibility if cogeneration and trigeneration of synthesis gas are considered as future applications. In addition, there is a choice of integration levels with the air separation unit, which produces the oxygen needed for gasification, and gives nitrogen for process use.

In comparison, Takematsu and Maude [10] list some of the advantages of an air-blown gasifier:

- The efficiency of the air-blown gasifier was 2% above that for an oxygen-blown gasifier in case studies performed in plants of comparable size, rather than of comparable gasifier size.
- Based on figures by Schwartz [13], costs for a 240 t/h system were 7% lower for the air-blown system, and the higher cost of the combined cycle section of the plant for the oxygen-blown case, by about 10%, was due to the added ccomplexity necessary to accomodate steam injection into the gas turbine combustor, which was required for NO_x control.
- No steam injection is required for the air-blown case because the large quantity of N_2 in the low-BTU gas acts as a diluent and helps prevent the combustion temperature from reaching high enough levels to promote fixation of nitrogen as NO_x. It is worth pointing that nitrogen from the air separation plant is also available for temperature control so that the differential should be less with newer systems.
- Part of the attractiveness of air-blown gasifiers is a result of their efficiency, lower cost to build, and simpler design compared to oxygen-blown systems.

In general, moving-bed gasifiers are suitable for cases where separation of the gasification and power generation block is desirable, fluidized-bed gasifiers are desirable where for highly reactive types of coals and for in-bed desulfurization, and entrained-bed gasifiers are more suitable for plants where a high degree of plant integration is necessary.

Table 4.2 Gasifier coal gas compositions examples

	Moving (Fixed) Bed[1], %	Fluidized Bed[2], %	Entrained Flow[3], %
H_2O	30	7	12
CO	10	23	44
H_2	15	14	32
CO_2	10	6	10
CH_4	3	2	0.2
N_2	34	47	1
Other	1	1	0.8

1 High-steam air-blown Lurgi gasifier
2 Air-blown U-Gas gasifier
3 Oxygen-blown Texaco gasifier

Typical coal gas compositions for the three major types of gasifiers (moving beds, fluidized beds, and entrained flow) are shown in Table 4.2. The oxygen-blown entrained flow gasifier has the highest concentrations of H_2 and CO, which increase the reducing power of the gas towards the metal oxides in the desulfurization sorbent. Overall, decreasing the reducing power of the gas is beneficial for metal oxides, but not necessarily desirable for power generation since the higher enthalpy of combustion for CO and H_2 (compared to CO_2 and H_2O) makes them more desirable. In this sense, the air-blown gasifier gas composition is more amenable for hot gas desulfurization than the oxygen-blown case in view of the lower temperature and reducing power of the gas, but the heating value per standard unit volume is lower for an air-blown gasifier gas. Although the steam content does affect to a certain extent the equilibrium values of H_2S in the clean gas, for the purpose of desulfurizing at a target level of 98-99%, the differences are very small between the two cases.

5 Integration of Gasification and HGCU Sorbents

When integrating gas cleanup systems with the gasification cycle it is necessary to match conditions of the gasifier exit coal gas with the operating conditions of downstream equipment (i.e., gas cleanup systems and gas turbine fuel inlet conditions). With respect to gas cleanup systems for IGCC operation, two temperature regions have been reported

[14]: the low-temperature region below 300 °C and the high-temperature region above that.

In the low-temperature cleanup (LTCU) range, also called cold gas cleanup (CGCU), the gas coming from the gasifier is typically cleaned of particulate matter and sulfur species such as H_2S, COS, and CS_2, at temperatures below 300 °C. Generally, CGCU systems operate around 20-100 °C, with some that include cryogenic stages that require sub-zero temperatures. CGCU systems are proven technologies and commercially available. The main limitation is that they are energy intensive, and are parasitic processes to the high cycle efficiencies desired from IGCC or any other system.

In the high-temperature cleanup (HTCU) range, also called hot gas cleanup (HGCU), the gas from the gasifier is cleaned of particulate matter, sulfur species, and other contaminants at temperatures above 400 °C. HGCU technologies are at various stages of development or in the early stages of commercialization. High-temperature desulfurization is typically accomplished using throwaway calcium-based sorbents or regenerable mixed-metal oxide sorbents. Cost, removal efficiency, and operability determine the advantages of one type of sorbent over the other. The performance of regenerable mixed-metal oxide sorbents for HGCU desulfurization will be solely discussed in this paper.

The term "warm" temperature region has been proposed also to describe gas cleanup above 200 °C and below 500-550 °C, where some degree of gas cooling is still necessary, and to reserve the term hot gas cleanup strictly for those cases above 550 °C where the coal gas exiting the gasifier requires very little or no cooling at all. Since the distinction between CGCU and HGCU is almost synonymous with being a commercial or developmental process, a third descriptive term to differentiate the temperature regimes has not been necessary to date.

The performance of desulfurization of mixed-metal oxides sorbents depends on several properties of the incoming coal gas:

- Pressure and temperature of gas
- Coal gas composition (reducing power)
- Minor and trace contaminants (type and concentration)
- Particulate matter in the gas (carbon and fly ash)

Higher pressures of the coal fuel gas result in higher molar concentrations of H_2S (i.e., mol/l) and therefore increase the rate of sulfur removal. For sorbents used in fixed/moving bed applications, the rate of sulfur capture by the individual pellets is mostly determined by pore diffusion resistances through the pellet and, at lower temperatures, by the intrinsic rate of reaction between the H_2S and the solid oxides. For agglomerated sorbents in fluid-bed or transport reactor applications, the pore diffusion term is not as predominant, and intrinsic kinetic rates of reaction have a larger role in the measured desulfurization activity of the sorbent.

Coal fuel gas temperature is also an important parameter in the determination of the optimum sorbent to use for desulfurization. As mentioned before, at lower temperatures the intrinsic rate of reaction determines the global rate of desulfurization. For instance, desulfurization using zinc oxide sorbents is generally pore diffusion limited at temperatures above 538 °C, while it is intrinsic rate limited below 482 °C. In the intermediate temperature range, both pore diffusion and intrinsitc reaction rate may determine the global rate of reaction. Other sorbents have comparable rates of desulfurization to zinc-based sorbents at different temperatures, some of them require much higher temperatures, such as cerium oxide, and some others, such as iron- and copper-based sorbents, require the same or lower temperatures than zinc. In addition, the regeneration of the sorbent must be initiated at temperatures similar to or lower than the desulfurization temperature so that there is no need for external heating of the sorbent in order to initiate the regeneration.

Coal gas composition is also very important in the selection of metal oxide sorbents for desulfurization. The presence of H_2 and CO in the coal gas determines the reducing power of the gas, i.e., the propensity of the gas to reduce the metal oxides to a lower oxidation state. For instance, the zinc component in zinc oxide and zinc titanate sorbents is reduced to elemental zinc at temperatures above 650 °C in the presence of H_2 and CO from coal gas, with the resulting complication that zinc metal has a higher vapor pressure at those conditions and is lost from the sorbent matrix. There is no problem below 650 °C at the same reducing gas compositions. Other metal oxides also exhibit similar behavior. Common oxides of copper, tin, and iron will be reduced to a lower oxidation state or even metallized at higher H_2 and CO concentrations. Oxygen-blown gasifiers tend to have hydrogen contents above 30% H_2 while air-blown gasifier gas compositions are typically 10-20%.

Of secondary importance are the minor and trace contaminants as well as particulate matter in the coal gas. Halogens, alkali metals, tars, and heavy metals may affect the desulfurization performance by reacting with the desulfurizing metal oxides of the sorbent or interacting with inert matrix components. HCl is notorious to attack metal oxides producing metal chlorides, which typically have vapor pressures higher than their respective oxides, thus volatilizing to the gas phase at higher temperatures. Air-blown gasifier coal gases operating at lower temperatures with higher steam content tend to present a less severe environment fo desulfurization with metal oxide sorbents, but they have higher levels of condensable tars. Tar condensation on the sorbent may reduce somewhat the available surface area of the sorbent for desulfurization. Again, it depends on the sorbent component characteristics to define whether the tars really represent a problem for desulfurization or are advantageous for initiation of regeneration at lower temperatures by ignition of the C-H components of tars.

6 Summary

Coal is one of the most abundant fossil fuel sources in the world with available reserves for several hundred years worth of consumption. Use of coal for electricity generation is expected to increase at least for the next thirty years. The increase in coal consumption is dependent on the development of advanced processes that are more energy efficient and compliant with current environmental regulations.

Gasification processes integrated with advanced coal-based power generation technologies, such as fuel cells and IGCC are at the forefront of technology development. IGCC is the most cost effective and power efficient cycle under commercialization to date. Many different combinations of gasification technology, gas cleanup, and power generation integration are being implemented throughout the world, adding a great deal of flexibility to the process to match a local utility's particular needs for power generation. A comparison of the basic types of gasifiers has shown that the choice of a particular gasifier configuration is dependent on the specific fields of application by the local utility. In general, moving-bed gasifiers are suitable for cases where separation of the gasification and power generation block is desirable, fluidized-bed gasifiers are desirable where for highly reactive types of coals and for in-bed desulfurization, and entrained-bed gasifiers are more suitable for plants where a high degree of plant integration is necessary.

Regenerable metal oxide sorbents for hot gas cleanup (HGCU) can be matched to operate with coal gasification systems. The selection of sorbent candidates for HGCU is highly dependent on the gasifier type and operation and therefore, the choice of cleanup system (particulate, sulfur, halogens, etc.) has to be made in the scheme of the overall IGCC configuration. Type of coal, gasification conditions (temperature, pressure, and gas composition), and hot gas cleanup conditions determine the desired characteristics of the regenerable metal oxide to be used.

Proper integration of HGCU sorbent operation in the power generation cycle increases the overall efficiency (above that of CGCU systems), and consequently, HGCU sorbents must be included in the integration analysis of IGCC and other high-efficiency power generation cycles.

7 Acknowledgments

The author acknowledges A. Furman, D. Najewicz, and S.G. Kimura at GE Corporate Research and Development for reviewing the manuscript and encouraging participation in this NATO/ASI symposium. Special thanks to D. Todd, GE Power Generation, for providing sources of information and references on the combined-cycle efficiency, status of commercialization of IGCC technology in power generation, and operation of IGCC components. Also, the help of R. Gupta, Research Triangle Institute, in providing some reference information is also acknowledged. Finally, my appreciation to the staff of the U.S. Dept. of Energy, Morgantown Energy Technology Center/Gasification and Gas Cleanup Division for sponsoring the development and demonstration of GE hot gas cleanup and IGCC technology in the U.S.

References

1 Holt, N. 1995. World Market: A Survey of Opportunities for Advanced Coal-Fired Systems. In *Proceedings of the Coal-Fired Power Systems 95 – Advances in IGCC and PFBC Review Meeting,* p. 5. DOE/METC-95/1018. NTIS/DE95009732. Springfield, Va.: National Technical Information Service.
2 Global Energy Perspectives to 2050 and Beyond. 1995. World Energy Council (WEC)IIASA Report.
3 Energy for Tomorrow's World - The Realities, the Real Options and the Agenda for Achievement, World Energy Council.
4 Wolk, H.R., and N.A. Holt. 1994. Overview of Global Utility Market. In Proceedings of the *Coal-Fired Power Systems 94 – Advances in IGCC and PFBC Review Meeting,* p. 27. DOE/METC-94/1008. NTIS/DE94012252. Springfield, Va.: National Technical Information Service.
5 Sage, P.W. 1996. Overview of Clean Coal Technologies and Current Status of the Air Blown Gasification Cycle. In *Proceedings of the NATO/Advanced Study Institute Conference - Desulfurization of Hot Coal Gas with Regenerable Metal Oxide Sorbents: New Developments*. Kusadasi, Turkey, July 7-19 1996.
6 Ayala, R.E. Application of IGCC Technology to Power Generation. In *Proceedings of the NATO/Advanced Study Institute Conference - Desulfurization of Hot Coal Gas with Regenerable Metal Oxide Sorbents: New Developments*. Kusadasi, Turkey, July 7-19 1996.
7 Ham, D. A. Gelb, G. Lord, and G. Simons. 1984. Hot-Gas Cleanup for Molten Carbonate Fuel Cells-Dechlorination and Soot Formation. Final Report to the U.S. DOE. DOE/MC/16242-1554. NTIS/DE84003072. Springfield, Va.: National Technical Information Service.
8 Gorbaty, M.L. and Ouchi, K. 1981. Coal Structure. Adv. Chem. Ser. 192. American Chemical Society, Washington, D.C.

9 Karr, Jr. C. Analytical Methods for Coal and Coal Products. 1978. Academic Press, New York.

10 Takematsu, T. and Maude, C. 1991. Coal Gasification for IGCC Power Generation. IEACR/37. IEA Coal Research, London, 1991.

11 Mueller, R. And J. Karg. 1987. Assessment of Coal Gasification Systems for Combined Cycle Power Plants. Paper presented at the Joint ASME/IEEE Power Generation Conference, Miami Beach, Florida, October 4-8, 1987.

12 Swanekamp, R. 1996. Startup of large-scale projects casts spotlight on IGCC. Power, pp. 59-68, June 1996.

13 Schwartz, C.W. 1982. Combined cycle coal gasification. *Energy Progress* 2 (4), pp 207-212 (1982).

14 Thambimuthu, K.1993. Gas Cleaning for Advanced Coal-Based Power Generation. IEACR/53. IEA Coal Research, London, 1993.

Effect of Process Conditions on Thermodynamics of Gasification

A.E. van Diepen and J.A. Moulijn

Department of Chemical Engineering, CPT-STM, Delft University of Technology, Julianalaan 136, 2628 BL Delft, The Netherlands

Abstract. Coal gasification involves the conversion of coal with steam (allothermic gasification), or with steam and oxygen (autothermic gasification). In this paper, the effect of temperature and pressure on the thermodynamics of gasification are discussed. The most important products of gasification are carbon monoxide, hydrogen, carbon dioxide, and methane. With increasing temperature, the concentrations of hydrogen and carbon monoxide in the product gas increase, while both carbon dioxide and methane go through a maximum. A higher pressure negatively influences the hydrogen and carbon monoxide production. Lowering of the feed temperature very much increases the heat required in allothermic gasification, and the oxygen to steam ratio in autothermic gasification. A comparison is made between gasification of pure carbon and CH, which gives a more accurate representation of coal. Naturally, CH yields a product gas with higher hydrogen content. Finally, the thermodynamic equilibrium composition is related to typical operating ranges of industrial gasifiers. This gives an indication of optimal operating conditions for specific applications, although an absolute measure of performance cannot be based on these data.

1 Introduction

Although coal is by far the most abundant fossil fuel available world-wide, it contributes only 34 % to the energy consumption [1]. However, with the limited crude oil and natural gas resources, coal will have to play a more important role in the future. Since the 1973 oil crisis, increasing efforts have been put into coal research.

In principle, any fuel or other chemical produced from oil or natural gas can also be produced from coal. The advantage of the use of coal is that it is cheap compared to oil and gas. However, the handling of solids is more difficult than the handling of gases and liquids. Therefore, central conversion of coal into gas by means of coal gasification is a logical step, which also has environmental advantages over a highly dispersed use of coal [2]. Fig. 1 summarizes the coal gasification process and its applications. Coal gasification involves the conversion of coal in a reactive gaseous atmosphere. Usually a mixture of coal, steam, and oxygen is fed to the gasifier. Their feed ratio depends among others on the reactivity of the coal, and the type of gasifier.

NATO ASI Series, Vol. G 42
Desulfurization of Hot Coal Gas
Edited by A.T. Atimtay and D. P. Harrison

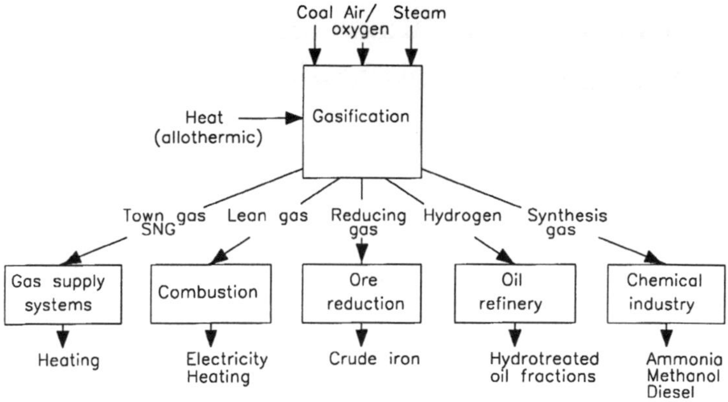

Fig. 1. Process and applications of coal gasification (adapted from Ref. [1])

When coal is gasified in air/steam mixtures low calorific gas is produced ('lean gas') due to the presence of nitrogen. The gas is usually applied for heating or power generation. In electrical power generation, coal gasification combined with the use of gas turbines offers two important advantages over the classical pulverised coal fired boiler and steam turbine combination, viz. higher thermal efficiency and minimal pollution. Gasification using steam and pure oxygen yields an undiluted gas (synthesis gas or syngas) containing mainly carbon monoxide and hydrogen, which can be used as a raw material for the chemical industry. Examples are Fischer-Tropsch type reactions to produce hydrocarbons and alcohols, methanol production, and the production of hydrogen via the water gas shift reaction. Hydrogen is used in a wide variety of chemical processes, such as ammonia production, and the reduction of ores. An application of increasing importance is the use of hydrogen in hydroprocessing operations in oil refineries. Coal gas is also applied in gas supply systems ('Substitute Natural Gas', SNG). Most applications are not unique for coal: in many cases any fossil fuel can be used.

2 Basic Processes and Reactions

Coal is a complex mixture of organic and mineral compounds. When it is heated to temperatures above 650 K, various thermal decomposition reactions occur. Usually, the organic matter melts, while gases are evolved (H_2, CH_4, aromatics, etc.). Upon further heating the organic mass is transformed into porous graphite-type material, called 'char'. Fig. 2 illustrates this complex process, called pyrolysis. When reactive gases are present in the gas phase, several following reactions occur on extended exposure to high temperatures.

During coal gasification, coal is subjected to reaction with a gasifying agent, e.g. steam or a mixture of steam and oxygen, at high temperature (1000 - 2000 K), and the organic matter is completely converted to gas (H_2, CO, etc.).

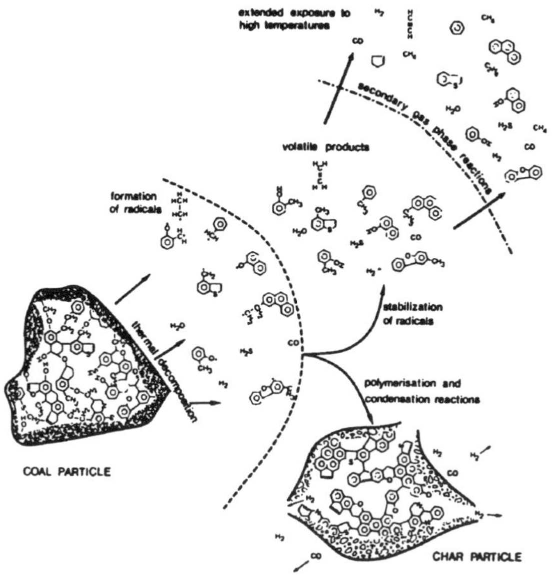

Fig. 2. Pyrolysis network [3]

Only the mineral matter (ash) remains. Depending on the temperature of the gasifier the mineral matter will be released as a liquid or a solid. When the mineral matter is released as a liquid, the gasifier is called a 'slagging' gasifier. Of course, during the gasification process, pyrolysis or devolatilization always occurs, either as a pre-reaction or simultaneously with the gasification reactions. The reactions occurring during carbon gasification are indicated in Table 1. The classification of the reactions is somewhat arbitrary, because the products of the primary reactions always can participate in following reactions with each other, the gasifying agent or unconverted carbon. Note that all reactions can be derived from the set of independent fundamental gasification reactions (1), (5), (6), and (7).

The most important reaction is the gasification of carbon by steam (1). The products of this reaction, hydrogen and carbon monoxide, can react further with steam in the water gas shift reaction (2) or with each other to form methane and carbon dioxide (3), (4). Reaction (1) is endothermic, while the following reactions are exothermic. At high temperature the overall steam gasification process is endothermic because reaction (1) is the dominant reaction. At low temperature the exothermic reactions become more important so that the process becomes nearly thermoneutral.

For the production of methane as a substitute for natural gas (SNG), hydrogasification, i.e. the direct reaction of hydrogen with coal (5) is very important. This exothermic reaction is favoured by low temperature. Low temperature however, lowers the reaction rate. Therefore, hydrogasification is carried out in the presence of a catalyst in order to maintain sufficiently high reaction rates.

Table 1. Gasification reactions

Reaction				ΔH_r^o (800 K) in kJ/mol
Steam gasification				
(1)	$C + H_2O$	$\leftrightarrow$	$CO + H_2$	135.6
(2)	$CO + H_2O$	$\leftrightarrow$	$CO_2 + H_2$	- 36.9
(3)	$CO + 3\ H_2$	$\leftrightarrow$	$CH_4 + H_2O$	- 222.9
(4)	$2\ CO + 2\ H_2$	$\leftrightarrow$	$CH_4 + CO_2$	- 259.8
Hydrogasification				
(5)	$C + 2\ H_2$	$\leftrightarrow$	CH_4	- 87.3
Carbon dioxide gasification				
(6)	$C + CO_2$	$\leftrightarrow$	$2\ CO$	172.5
Combustion and partial oxidation				
(7)	$2\ C + O_2$	$\rightarrow$	$2\ CO$	- 222.0
(8)	$C + O_2$	$\rightarrow$	CO_2	- 394.2
(9)	$2\ CO + O_2$	$\rightarrow$	$2\ CO_2$	- 566.4
(10)	$CH_4 + 2\ O_2$	$\rightarrow$	$CO_2 + 2\ H_2O$	- 799.8
(11)	$2\ CH_4 + O_2$	$\rightarrow$	$2\ CO + 4\ H_2$	- 47.4

Carbon dioxide gasification (6) is not applied as such, but the reaction occurs during steam gasification and during partial oxidation of carbon with oxygen. The overall partial oxidation reaction (7), by which carbon monoxide is formed, is the sum of complete carbon combustion (8) and reaction (6). At low temperature, or when an excess of oxygen is added, the product merely is carbon dioxide.

3 Thermodynamics

Thermodynamic equilibrium calculations help to determine the optimal operating conditions for the production of a particular product gas from coal. From the previous section it is clear that coal gasification involves a great number of reactions, many of which are equilibrium reactions. The presence of solid carbon in the reaction system further complicates the situation.

The relationship between the equilibrium constant and the gas composition for the most important gasification reaction, $C + H_2O \leftrightarrow H_2 + CO$, can be expressed as (the reference pressure has been omitted for convenience):

$$K_{eq} = \frac{p_{H_2} p_{CO}}{p_{H_2O}\ a_c} = \frac{x_{H_2} x_{CO}}{x_{H_2O}} \frac{p}{a_c} \tag{1}$$

where: x_i = mol fraction of component i (-)
p = total pressure of the reaction system (bar)
p_i = partial pressure of component i (bar)
a_c = "activity" of a particular carbon compared to graphite (for graphite $a_c = 1$)

The equilibrium constants for the other reactions can be expressed similarly.

The coal char, which results from exposing the coal to high temperatures, resembles a highly disordered structure of graphite-like domains [2]. Therefore, equilibrium gas compositions over a wide range of conditions relevant to coal gasification may be calculated, employing a value for a_c of unity.

The equilibrium constant is a function of temperature and reacting molecules and can easily be calculated from tabulated chemical data such as the JANAF tables [4]:

$$RT \ln K_{eq} = -\Delta G_r^0(T) = -\sum_i \nu_i \Delta G_{f_i}^0(T) \tag{2}$$

where: ΔG^o_r (T, p) = standard Gibbs energy change of reaction (kJ/mol) at temperature T and pressure p
ΔG^o_{fi} (T, p) = standard Gibbs energy of formation of component i (kJ/mol) at temperature T and pressure p
ν_i = stoichiometric coefficient of component i

In order to calculate equilibrium compositions at specified temperature and pressure, and initial feed composition, two strategies can be applied, once the equilibrium constants of all reactions, or the standard Gibbs energy of formation of all species involved are known:

1. Equilibrium constant method
2. Minimization of total Gibbs energy, G^t

One criterion for equilibrium is that the total Gibbs energy must decrease during an irreversible process such as chemical reaction (dG^t (T, p) < 0), and that at equilibrium there is no change (dG^t (T, p) $= 0$). At that point the equilibrium composition is reached, and reaction in any direction would result in an increase of the Gibbs energy. Method 1 is based on this criterion. An alternative way of stating the equilibrium criterion is that the equilibrium state at constant temperature and pressure is that state for which the total Gibbs energy is a minimum with respect to all possible changes. This criterion is the basis for method 2.

Method 1 requires the determination of a set of independent reactions (e.g. reactions (1), (5), (6), and (7)), with their associated equilibrium constants, and solving the equilibrium constant equations (1) simultaneously, subject to the material balance constraints [5]. This method is quite elaborate if multiple reactions take place.

For method 2, it is not important which reactions are involved. We only need to specify the reactants and probable products present, and then the problem is to find the number of moles of all chemical species which minimizes the total Gibbs energy, subject to the material balance constraints. The solution is based on the method of Lagrange's undetermined multipliers [5]. More information on both methods can be found in reference [5]. Method 2 is the easiest to apply, particularly for multiple reaction systems, and we have chosen this method to calculate the equilibrium compositions in coal gasification.

3.1 Equilibrium Compositions for Individual Reactions

Fig. 3 presents the equilibrium mol fraction of methane in the product gas of coal hydrogasification (5) as a function of temperature and pressure. The formation of methane from coal thermodynamically requires a low temperature. At high pressure, higher temperatures can be applied to achieve high conversion, because pressure increase shifts the equilibrium to the side of the least molecules in the gas phase.

Fig. 4 shows the mol fraction of CO as a function of temperature and pressure for the carbon dioxide gasification reaction (6). High temperature and low pressure result in the highest equilibrium CO yield, and consequently, lowest CO_2 content of the gas. If oxygen is added instead of carbon dioxide, with a O_2/C ratio of 0.5, the same result is obtained. In that case, the oxygen reacts with carbon to CO_2 first (8). With increasing temperature, CO_2 will react with the remaining C to a greater extent and form CO. If more oxygen is added ($O_2/C \geq 1$), the carbon will be completely combusted to yield only CO_2, even at high temperature.

The main gasification reaction, i.e. the reaction of steam with coal (1), cannot be considered separately because, mainly at low temperatures, also CO_2 and CH_4 are observed in the gas mixture. CO_2 is formed by the water gas shift reaction (2), while CH_4 is formed in the methanation reactions (3) and (4).

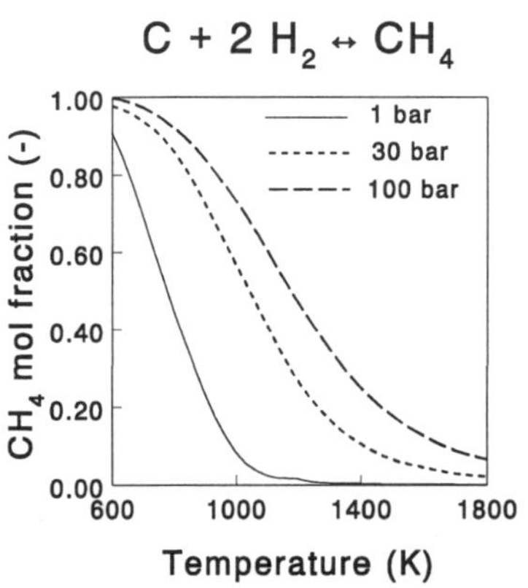

Fig. 3. Equilibrium mol fraction of methane in hydrogasification as a function of temperature and pressure

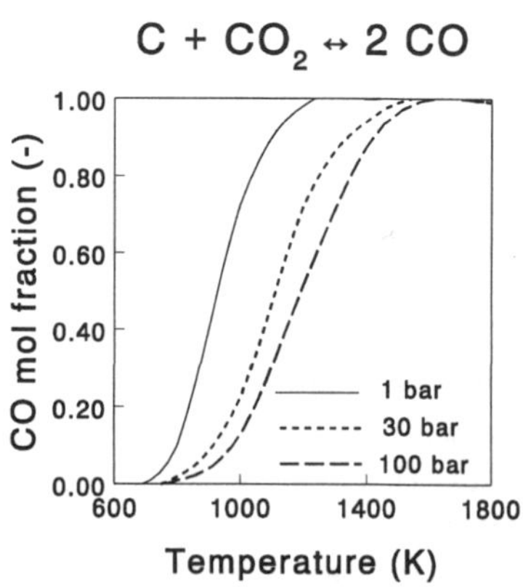

Fig. 4. Equilibrium mol fraction of carbon monoxide in carbon dioxide gasification as a function of temperature and pressure

3.2 Equilibrium Composition in Steam Gasification

Under practical gasification conditions usually the following species exist in the gas phase: H_2O, CO, CO_2, H_2, and CH_4. The equilibrium composition of the C-H-O system can be expressed in terms of temperature, pressure, and H/O ratio. Starting with pure H_2O and C (H/O = 2), Fig. 5 is obtained. In fact, coal is not entirely composed of carbon. The most prominent other constituent on a molar basis is hydrogen. The molar C/H ratio in coal may vary from 1 to 4. So, in order to check the influence of the C/H ratio, the equilibrium composition of the C-H-O system has also been determined with the assumption that coal can be represented as CH. For convenience, in the calculations it was assumed that CH is decomposed to C and H_2. The results for this system (H/O = 3) are shown in Fig. 6.

Effect of temperature. With increasing temperature, the CO and H_2 concentrations increase due to the increasing importance of the endothermic H_2O-gasification reaction. Accordingly, H_2O decreases with temperature. Both CO_2 and CH_4 go through a maximum (The maximum of CH_4 at 1 bar occurs at slightly below 600 K.). This can be explained as follows: At low temperature any CO and H_2 formed by gasification react to CH_4 and CO_2 in the very exothermic reactions (3) and (4). With increasing temperature, the equilibria of reactions (3) and (4) shift to the left.

Effect of pressure. At 30 bar, up to high temperature the production of CO and H_2 is lower than at a pressure of 1 bar due to the fact that both components are predominantly formed in reactions increasing the number of molecules present (reactions (1), (3), (4), and (6)). Thus, low pressure is favourable for these reactions. A higher pressure increases the CH_4 content at low temperatures on the basis of a similar argument (decrease in number of molecules). Both the CH_4 and CO_2 maxima shift to higher temperature with increasing pressure, because the H_2 and CO forming reactions become significant only at higher temperature.

Effect of H/O ratio. The same trends apply with respect to temperature and pressure for C and CH. However, when coal is represented as CH, for every mol of coal reacted, an additional 0.5 mol of H_2 is formed. At high temperature, where CO and H_2 are the primary products, the H_2/CO ratio shifts from 1 to 1.5. Also at low temperatures the increased amount of hydrogen causes a higher fraction of H-containing compounds. Actual coal gasification will yield a gas with a composition in between those shown in Figs. 5 and 6 but directed more towards the latter situation.

The trends in Figs. 5 and 6 are summarized in Table 2. The arrows indicate increase (↑), decrease (↓), or the occurrence of a maximum (↑↓). The trends indicate the temperature and pressure required to produce a gas with the right composition for a particular application. The production of synthesis gas (CO and H_2) requires high temperature and low pressure. For the production of methane on the other hand, a relatively low temperature and high pressure is optimal.

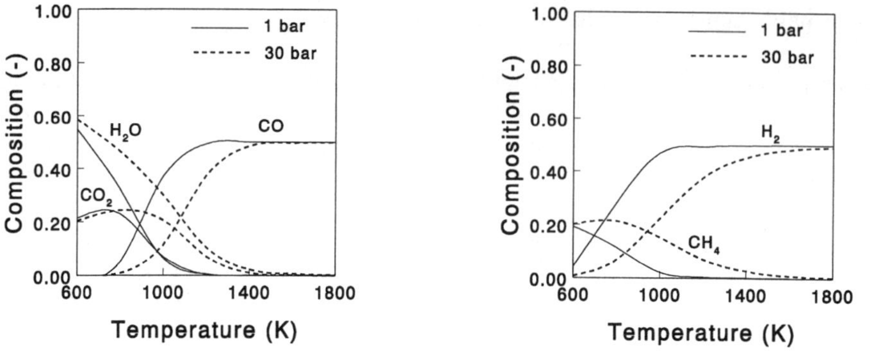

Fig. 5. Equilibrium composition in steam gasification of C (H/O = 2) as a function of temperature and pressure

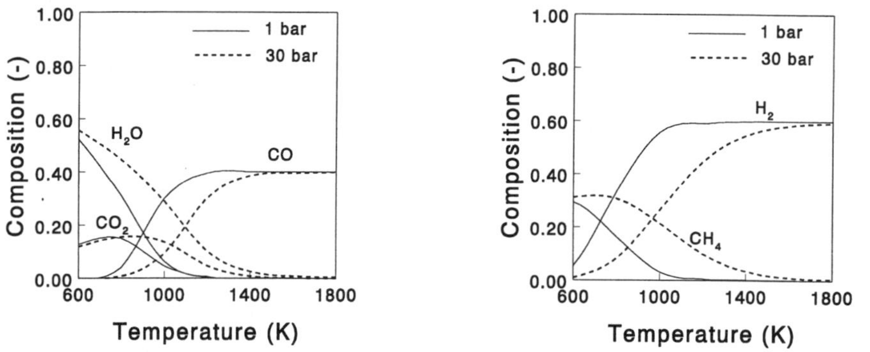

Fig. 6. Equilibrium composition in steam gasification of CH (H/O = 3) as a function of temperature and pressure

Table 2. Equilibrium trends for the C-H-O system (adapted from Ref. [2])

Mol fraction	Temperature ↑	Pressure ↑	H/O ratio ↑
H_2O	↓	↑	↓
H_2	↑	↓	↑
CO	↑	↓	↓
CO_2	↑↓	shift[1]	↓
CH_4	↑↓	↑, shift[1]	↑

1 maximum shifts to higher T

The preceding discussion and Table 2 are based on allothermic coal gasification, i.e. gasification without addition of oxygen. However, the trends apply equally well to the autothermic systems, which are discussed in Sections 3.3 and 4.

3.3 Heat Effects

During coal gasification several processes take place that involve enthalpy changes:

- Heating of reactants (coal, steam, oxygen) to reaction temperature. The associated enthalpy change for each component can be expressed as:

$$\Delta H = \int_{T_{in}}^{T_{gas}} C_p dT \tag{3}$$

  in which ΔH is the molar enthalpy change when the reactant is heated from T_{in} to T_{gas}, the inlet temperature of the reactant and the outlet temperature of the product gas, respectively. Correlations for the heat capacity, C_p for coal are given by Kirov [6]. These include devolatilization phenomena at low temperatures, < 773 K. Correlations for the C_p of gases can be found in for instance references [4] and [7].
- High temperature devolatilization and carbonization reactions, leading to char formation.
- Enthalpy changes associated with the gasification reactions.

The heat effects associated with the gasification reactions are much larger than those of the devolatilization and carbonization reactions [8, 9, 10, 11]. Therefore, and because the latter two processes involve a complex mixture of reactions, only gasification reactions will be considered in the determination of the heat effects involved in coal gasification. It will become clear from the subsequent discussion that the heat effects related to enthalpy changes of the reactants are not negligible.

The main gasification reaction, i.e. reaction of coal with steam is rather endothermic. In gasifiers the heat necessary for these reactions and for heating up to reaction temperature may be supplied externally (allothermic reaction system) or internally by addition of oxygen to the gasifier (autothermic reaction system). The latter option is applied in most industrial gasifiers. Both allothermic and autothermic gasification are discussed below. Two situations will be addressed:

- Reactants are at gasification temperature when they enter the reactor:
  minimum enthalpy change = heat of reaction at reaction temperature.
- Reactants are at 298 K (common temperature to determine thermodynamic properties) when they enter the reactor:
  'maximum' enthalpy change = heat of reaction + sensible heat of gas.

The word 'maximum' is placed between quotation marks, because factors, which are unaccounted for in the calculations, also play a role. These factors include heat losses through the reactor walls, drying of the coal feed (some coals contain a lot of moisture), heat for melting of the ash, heat lost as sensible heat of the ash, etc.

3.3.1 Heat Input for Allothermic Coal Gasification

Fig. 7 shows the enthalpy changes associated with the allothermic carbon gasification of Fig. 5, starting with pure H_2O and C, as a function of temperature and pressure (The results for CH are nearly identical and are therefore not shown.). The minimum enthalpy change (shown left), expressed as kJ per mol of H_2O converted, equals the total change in reaction enthalpy of all the reactions involved. Thus, at e.g. 1600 K, 135 kJ is shown to be required per mol H_2O converted, and this is the reaction enthalpy of reaction (1), the dominant reaction at high temperature. The energy input required to overcome the enthalpy changes increases with temperature, because at higher temperature the endothermic gasification reactions predominate, whereas at lower temperature the system is nearly thermoneutral owing to the exothermic methane forming reactions and the water gas shift reaction. Table 3 shows the near thermoneutrality of the reaction system at 800 K.

Table 3. Allothermic gasification of C with steam, enthalpy change at 800 K

No.	Reaction			ΔH_r (800 K, kJ/mol)
(1)	$2\,C + 2\,H_2O$	$\leftrightarrow$	$2\,CO + 2\,H_2$	271.2
(2)	$CO + H_2O$	$\leftrightarrow$	$CO_2 + H_2$	- 36.9
(3)	$CO + 3\,H_2$	$\leftrightarrow$	$CH_4 + H_2O$	- 222.9
Overall	$2\,C + 2\,H_2O$	$\leftrightarrow$	$CH_4 + CO_2$	11.4

An increase in pressure results in lower enthalpy changes, because the production of H_2 and CO by the endothermic reactions (Fig. 5) decreases.

The 'maximum' enthalpy change per mol of H_2O converted, shown on the right in Fig. 7, is considerably higher than the minimum values. The contribution of the enthalpy changes associated with raising the temperature of the reactants from 298 K to reaction temperature increases strongly with temperature. ΔH_r^o stays rather constant at high temperature, but the reactants have to be heated to increasingly higher temperatures, resulting in large enthalpy changes.

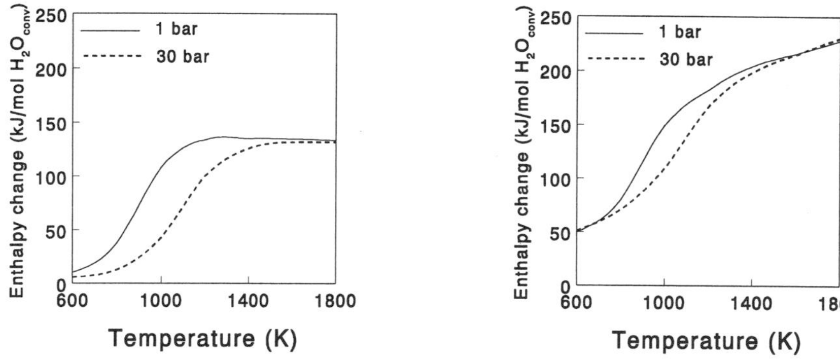

Fig. 7. Enthalpy changes associated with the steam gasification of coal (C, Fig. 5) as a function of temperature and pressure; left: minimum; right: 'maximum' (see text for explanation)

In practice, the energy consumption will be in between minimum and 'maximum' requirements, because steam is usually introduced in the reactor at about 500-600 K. However, drying of the coal feed, heat losses from the gasifier, both through the walls and as the sensible heat of the ash leaving the gasifier, also have to be taken into account.

3.3.2 Autothermic Coal Gasification

In industrial practice, the energy is usually supplied by partial oxidation of part of the coal feed in the reactor: autothermic gasification. Addition of oxygen changes the equilibrium composition of the gas. For instance, at high temperature, where CO and H_2 are the primary products, the minimum amount of oxygen required per mol steam can be derived theoretically by considering the most important reactions:

(1)	$C + H_2O$	$\leftrightarrow$	$CO + H_2$	ΔH_r^o (1200 K) = 135.9 kJ/mol
(7)	$C + 0.5\ O_2$	$\rightarrow$	CO	ΔH_r^o (1200 K) = - 113.3 kJ/mol

Based on the reaction enthalpies, reaction (7) has to take place about 1.2 times as much as reaction 1 for the system to be autothermic. So the minimum amount of oxygen required per mol of H_2O reacted is approximately 0.6 mol, provided that the feed enters the reactor at reaction temperature, and no heat losses occur. The resulting H_2/CO ratio, becomes $1/(1 + 1.2) = 0.45$, while in the allothermic process equal amounts of H_2 and CO are produced. With coal represented as CH, 0.5 extra mol of H_2 is produced per mol C reacted, so the H_2/CO becomes roughly 1. If the feed enters below reaction temperature, more oxygen will have to be added to increase reaction (7), and thus compensate for the enthalpy changes associated with raising the temperature of reactants. The H_2/CO ratio will then become even lower.

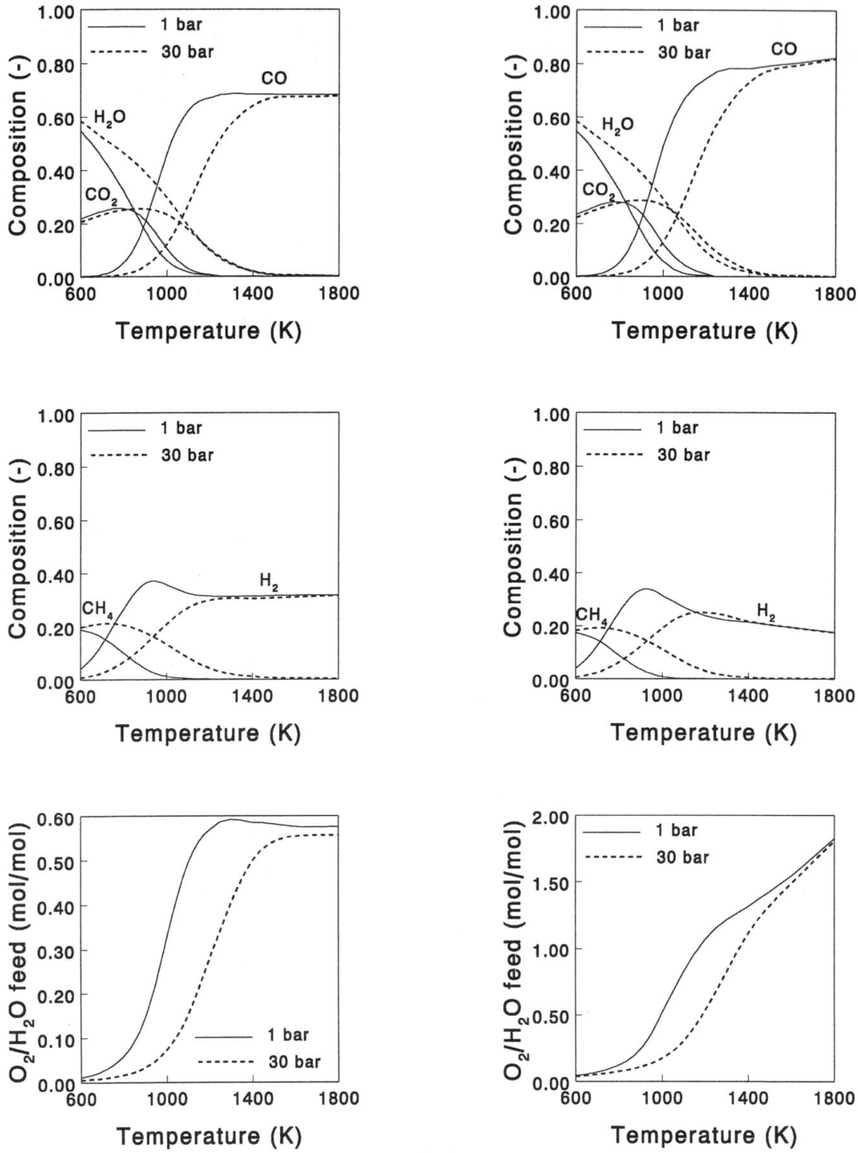

Fig. 8. Equilibrium composition (top: CO, CO_2, H_2O; middle: H_2, CH_4) and O_2/H_2O feed ratio (mol/mol) in the autothermic C-H-O system as a function of temperature and pressure; left: minimum; right: maximum (see text for explanation); coal = C

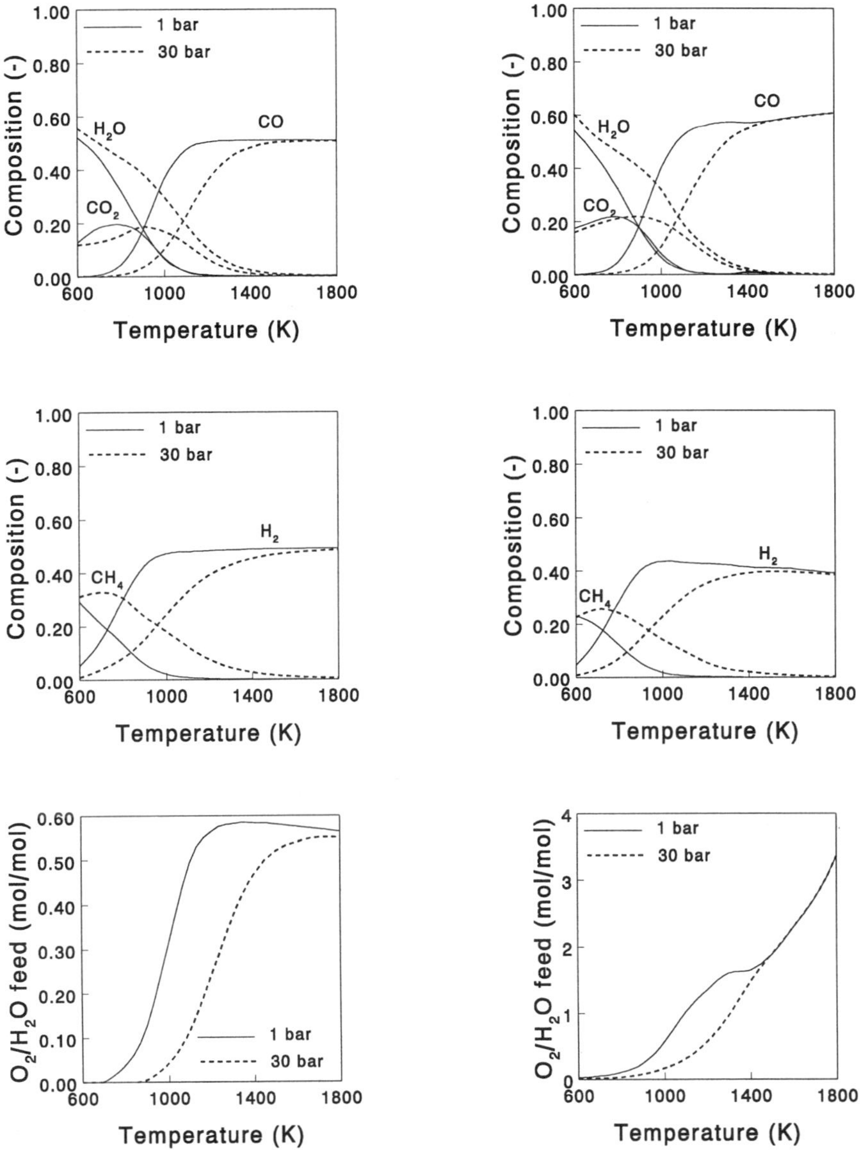

Fig. 9. Equilibrium composition (top: CO, CO_2, H_2O; middle: H_2, CH_4) and O_2/H_2O feed ratio (mol/mol) in the autothermic C-H-O system as a function of temperature and pressure; left: minimum; right: maximum (see text for explanation); coal=CH

Just as for the allothermic system the enthalpy changes associated with the gasification reactions and heating of the reactants together determine the total energy input required, these enthalpy changes determine the O_2/H_2O feed ratio required for the system to become autothermic.

Figs. 8 and 9 show the equilibrium compositions and the O_2/H_2O molar feed ratio required for the C-H-O reaction system to become autothermic, with coal represented as C and CH, respectively. The production of CO and CO_2 (top graphs) is higher compared to the allothermic system as a result of oxygen addition. Consequently, the production of H_2 and CH_4 (middle graphs) is lower. This effect becomes more appreciable if the reactants are at lower temperature initially, as can be seen when comparing the left (minimum O_2/H_2O ratio) and right ('maximum' O_2/H_2O ratio) sides of Figs. 8 and 9.

At higher pressure less oxygen is needed for the system to become autothermic. This can be explained by the increased formation of methane through the exothermic reactions (2), (3), and (5) at high pressure, which produce the heat otherwise furnished by reactions with oxygen. The 'maximum' O_2/H_2O feed ratio shows a similar trend to the 'maximum' enthalpy change (Fig. 7). The relative contribution of the oxygen required to raise the temperature to reaction temperature is even higher because the extra amount of oxygen required also has to be heated, thus counterbalancing part of the effect of the exothermic reactions.

4 Industrial Gasifiers

A large number of processes have been developed for coal gasification. Different applications ask for different gas compositions, and thus different operating conditions. Furthermore, the properties of coals vary, and not all coals can be processed in every type of reactor. In the discussion of reactor types, the aspects concerning coal properties will not be addressed in detail. We will rather focus on the composition of the products. A comparison is made with the calculated equilibrium compositions. Only autothermic gasification is considered.

4.1 Reactors

Three principal reactor types are applied for autothermic coal gasification: the moving bed, the fluidized bed, and the entrained flow reactor (see Fig. 10). The reactor type strongly influences the temperature distribution, and thus, the composition of the products. The reaction temperature varies from about 1000 to 2000 K, with each type of gasifier covering a specific range [12].

The moving bed reactor is operated countercurrently. The coal enters the gasifier at the top and is slowly heated and dried on its way down while cooling the product gas as it exits the reactor. The coal is further heated and devolatized as it descends. In the gasification zone, part of the coal is gasified in steam and carbon dioxide, which is formed in the combustion zone. Here, the highest temperatures are reached.

All that remains of the original coal is char and ash. Endothermic reaction of the char with steam keeps the temperature at the bottom of the reactor below the slagging temperature of the ash, so the produced ash is dry. The coal is supported by a grate where the ash is cooled by releasing heat to the entering steam and oxygen.

The fluidized bed gasifier is a back-mixed reactor in which feed coal particles are well mixed with coal particles already undergoing gasification. This gasifier operates at moderate uniform temperature. Char particles which leave the reactor with the product gas are recovered in cyclones and recycled to the reactor. Dry ash leaves the reactor at the bottom.

The entrained flow gasifier is a plug-flow system in which pulverized coal reacts cocurrently with steam and oxygen. The residence time is a few seconds. The temperature has to be high in order to maximize coal conversion. Ash is removed as molten slag.

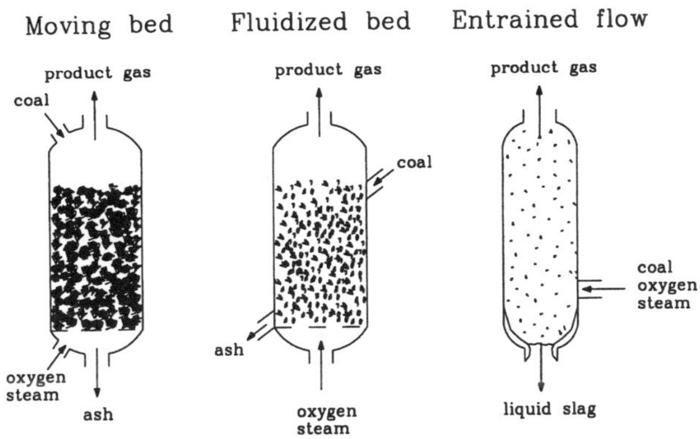

Fig. 10. Basic types of coal gasification reactors (after van Heek [1])

The choice of gasifier is for a large part determined by the type of coal to be processed. Not all types of coal can be processed in a moving bed reactor. Some coals get sticky upon heating and form large agglomerates leading to increased pressure drop, and even plugging of the reactor. Coals showing this behaviour are called 'caking coals'. The fluidized bed gasifier is suitable only for reactive coals, because due to its back-mix character and the moderate temperatures, conversion of less reactive coals is too low. The entrained flow reactor can handle all types of coals. General characteristics of the three reactor types and the coals they can handle are given in Tables 4 and 5.

Once a particular gasifier type has been specified, the gasifier temperature is dictated by factors such as ash melting temperature, and residence time in the reactor. In autothermic gasification, the temperature is controlled by the amount of oxygen and steam fed to the gasifier. For instance, the maintenance of a relatively low combustion temperature is essential for moving bed gasifiers, since if the ash melts it cannot be handled on a supporting grate. Therefore, excess steam is added, and very little oxygen. A novel development is the so-called slagging Lurgi reactor in which the temperature in the combustion zone is above 2000 K, resulting in slagging of the ash. Entrained flow reactors have to operate at high temperature in order to ensure complete conversion of the coal in the short time it is in the reactor. A high amount of oxygen is required, and only a small amount of steam is added.

Table 4. Coal properties for moving bed, fluidized bed and entrained flow gasifiers

	examples	throughput ton/day	d_p mm	coal type
moving bed	Lurgi dry-ash	1500	20	non-caking
fluidized bed	Winkler	3000	2	reactive
entrained flow	Shell, Texaco	5000	< 0.1	all

Table 5. Characteristics of moving bed, fluidized bed and entrained flow gasifiers

	$T_{reactor}$ K	$T_{exit, gas}$ K	P bar	O_2-consumption kg/kg coal
moving bed	1250-1350	700	20-30	0.2 - 0.5
fluidized bed	1250-1400	1150	1-80[1]	0.5 - 0.8
entrained flow	1600-2200	1300	1-80[1]	0.8 - 1.3

1 Early gasifiers operate at about atmospheric pressure, often with air as the oxygen supplier, more recent gasifiers have an operating range of 20 to 80 bar.

4.2 Gas Composition in Relation to Gasifier Type

From the foregoing it is clear that the product gas from the different gasifiers will differ markedly. Fig. 11 shows the equilibrium composition obtained by autothermic gasification of carbon (with the reactants at gasification temperature) at 30 bar as a function of temperature. The operating ranges of the industrial gasifiers are indicated.

The moving bed reactor which operates at the lowest temperature produces relatively large quantities of methane. The excess steam, which is added to moderate the temperature, does not participate in the reaction and is recovered from the product gases as rather polluted water. Of course, in general this water is recycled to the gasifier. The entrained flow reactor operates at high temperature, so the product gas mainly contains carbon monoxide and hydrogen. Fluidized bed reactors yield intermediate gas compositions.

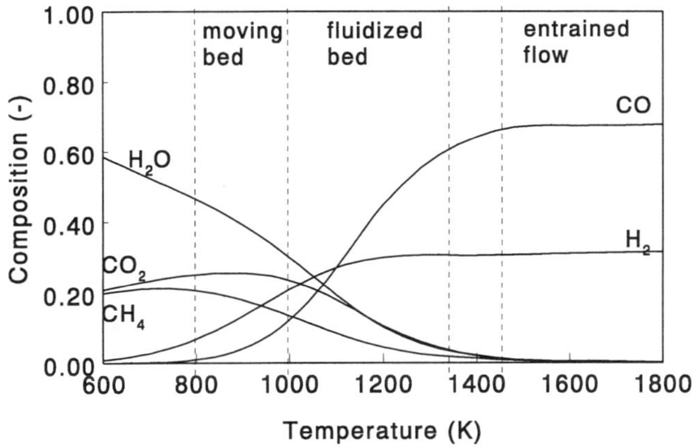

Fig. 11. Equilibrium composition of autothermic carbon gasification as a function of temperature and operating ranges of industrial gasifiers

The compositions shown in Fig. 11 only give an indication of actual industrial compositions. Factors which have not been accounted for are the coal composition (C, H, O, and others such as S), the degree of preheat, whether the coal feed is dry, etc.

5 Concluding Remarks

Thermodynamic equilibrium data help determine optimal conditions for coal gasification processes. The ideal conditions derived can serve as a relative guide but not as an absolute measure of performance. As is always the case in chemical reactions, kinetic limitations, and heat and mass transfer limitations (see for instance reference [2]), prevent the reaction system from reaching chemical equilibrium. Furthermore, practical constraints such as the reactivity and caking characteristics of the coal, product throughput, the melting temperature of the ash produced, and material and mechanical limitations, influence the choice of process and its performance. In the end, economic aspects dominate. These include the local price of coal, investment costs, etc.

References

1 Van Heek, K.H., 1986, General Aspects and Engineering Principles for Technical Application of Coal Gasification, in: Figueiredo, J. L., Moulijn, J.A. (eds.), *Carbon and Coal Gasification*, pp. 383-401, Martinus Nijhoff Publishers

2 Kapteijn, F. and Moulijn, J.A., 1986, Kinetics of Catalysed and Uncatalysed Coal Gasification, in: Figueiredo, J. L., Moulijn, J.A. (eds.), *Carbon and Coal Gasification*, pp. 291-360, Martinus Nijhoff Publishers

3 Tromp, J.J. and Moulijn, J.A., 1988, Slow and Rapid Pyrolysis of Coal, in: Yürüm, Y. (ed.), *New Trends in Coal Science*, pp. 305-338, Kluwer Academic Publishers

4 D.R. Stull and H. Prophet (eds.), 1971, *JANAF Thermochemical Tables*, NSDRS, NBS-37, US Government Printing Office, Washington D.C.

5 Smith, J.M., Van Ness, H.C., 1987, *Introduction to Chemical Engineering Thermodynamics*, 4th edition, McGraw-Hill

6 Kirov, N.Y., 1965, BCURA, Month. Bull., **29 (2)**, pp. 33-57

7 Reid, R.C., Prausnitz, J.M., Poling, B.E., 1987, *The Properties of Gases and Liquids*, 4th edition, McGraw-Hill

8 Johnson, J.L., 1981, Fundamentals of Coal Gasification, in: Elliot, M.A., *Chemistry of Coal Utilization*, chapter 23, John Wiley and Sons

9 Johnson, J.L., 1979, *Kinetics of Coal Gasification*, John Wiley and Sons

10 Tromp, P.J.J., Kapteijn, F., Moulijn, J.A., 1989, Characterization of Coal Pyrolysis, *Fuel Processing Technology*, **23**, pp. 63-74

11 Lopez-Peinado, A.J., Tromp, P.J.J., Moulijn, J.A., 1989, Quantitative heat effects associated with pyrolysis of coals, ranging from anthracite to lignite, *FUEL*, **68**, pp 999-1004

12 Probstein, R.F., Hicks, R.E., 1982, *Synthetic Fuels*, chapter 4, McGraw Hill

Application of IGCC Technology to Power Generation

R.E. Ayala

GE Corporate Research and Development, One Research Circle, Schenectady, New York, 12309 USA

Abstract. Improved efficiency and low cost are two of the objectives in the development and commercialization of power generation cycles. With the advent of today's commercial advanced gas turbines and high-temperature gas cleanup systems, the Integrated Gasification Combined-Cycle (IGCC) technology is a power generation cycle being commercialized that shows economic gains against conventional coal steam plants. IGCC has demonstrated improved performance to meet current and future environmental regulations. This paper reviews the status of the IGCC technology, highlighting its technical, economic, and environmental performance, and overviews the status of current large-scale projects being developed towards application to power generation.

1 Introduction

Global power generation markets have shown a steady penetration of gas turbine/combined cycle technology into oil and gas fired applications as the technology has matured. The low cost, improved reliability and efficiency advantages of combined cycles can now be used to reduce the cost of electricity and meet environmental acceptance of fuels other than natural gas such as coal, heavy oil, petroleum coke and waste products.

Several technologies have been proposed to convert coal to electricity, including combustion (atmospheric and pressurized), liquefaction, and gasification. An overview of gasification technology has been given in the accompanying paper by the same author in these proceedings [1]. Within gasification, the Integrated Gasification Combined-Cycle (IGCC) is now moving towards being a commercial technology. In the global marketplace, this shift is being demonstrated using various gasification technologies to produce a clean fuel for the combined cycle. Early plants in the 1980's demonstrated the technical/environmental features and suitability for power generation. Economics, however, were disappointing until advances in gas turbine technology were first used in the early 1990's. Since 1993, a number of IGCC projects have been ordered with GE turbine technology for commercial application in the mid- to late-1990's. These applications utilize seven different gasification technologies to meet specific application needs.

Building on the excellent environmental performance of commercial low-temperature cleanup systems, hot gas cleanup is being developed to further improve the overall IGCC cycle efficiency by removing coal gas contaminants at temperatures more compatible with those required by the gasification and gas turbine cycles. The selection and optimization of each of the IGCC components (gasification, gas cleanup, and turbines) for higher efficiencies affects the process conditions and the overall power

NATO ASI Series, Vol. G 42
Desulfurization of Hot Coal Gas
Edited by A.T. Atimtay and D. P. Harrison

generating system efficiency. As an example, regenerable sorbents for hot gas cleanup (HGCU), as part of the fuel cleanup, are affected by the operation of the gasification process. The choice of one sorbent or another is greatly determined, for instance, by the coal gas temperature, pressure, and composition, just to name a few.

It is the objective of this paper to review the current application of IGCC technology to power generation, highlight its needs and advantages, and summarize the status of recent demonstration projects worldwide. In the process, an attempt will be made to put in perspective the overall process needs for the development of hot gas cleanup regenerable metal-oxide sorbents, which is the focal point of this NATO symposium.

2 Power Cycles based on Gasification Technology

Many technologies have been proposed over the years to convert coal to electricity. Among them, coal gasification is considered an advanced technology capable of generating power in a more efficient way while meeting current and future environmental regulations. Two major areas within gasification have received a great deal of interest in the last decade (Figure 1): fuel cell cycles (i.e., IGFC) and gas turbine cycles, including the integration of the gasification with the gas turbine/steam turbine combined cycle (IGCC) and cycles using steam addition to the gas turbine (STIG, ISTIG, and HAT cycles). Each of them is important in that they fulfill the needs for a given sector of the coal utilization market. This paper will focus on the IGCC technology, and only a brief overview will be given of the other technologies.

In the IGFC cycle, the gasification is integrated with fuel cells, where electricity is generated via electrolytic oxidation of the fuel gas (primarily H_2 and CO) after gas cleaning (Figure 2). The fuel cell cycle gives a very high thermodynamic efficieny, and the exothermic nature of the reactions allows to recuperate some of the wast heat to drive a bottoming gas turbine or a steam turbine. The degree of development is less than that of IGCC systems, and fuel cell-based systems have been built and demonstrated only below 100 MW. Two types of fuel cells have received a higher degree of development. One is the molten carbonate fuel cell (MCFC), the other is the solid oxide fuel cell (SOFC). Fuel cells have more stringent requirements for gas cleaning than gas

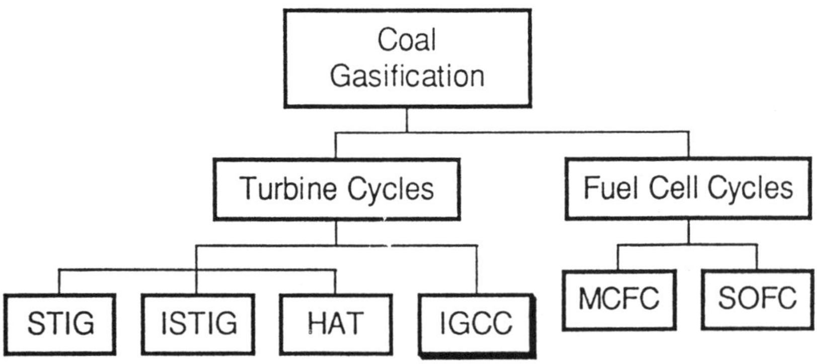

Figure 1. Gasification technologies for conversion of coal to electricity

turbines, thus a sulfur guard is generally required in addition to hot gas cleanup for MCFC operation. Reactivities of catalytic electrodes in MCFC's are quite sensitive

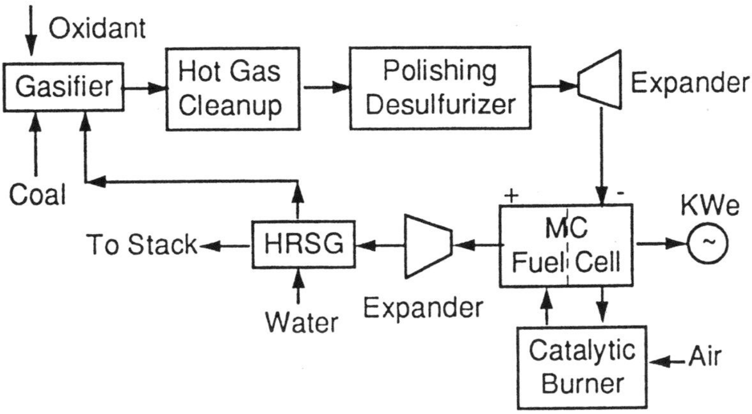

Figure 2. Schematic diagram of a molten carbonate fuel cell (MCFC) power cycle

functions of several impurities in coal gasifier effluent, the most serious problem species being H_2S, COS and HCl. Total sulfur species must be less than 1 ppmv and HCl must be below 0.2-10 ppmv [2].

In gas turbine and combined cycle power generation, coal is usually gasified under high pressure and high temperature. After gas cleaning, the high pressure and temperature gaseous fuel are used to drive a gas turbine. The residual heat from the combustion gases can be used to generate steam. If steam is injected to the gas turbine, in addition to the fuel, the cycle is called steam injected gas turbine (STIG) cycle, shown in Figure 3, where details of the integration of gas turbine, gas combustor, gas compressor and HRSG are given. The injected steam, with its higher specific heat capacity and mass flow, contributes to a large increase in efficiency and power output in the STIG cycle. If an intercooling step between two stages of the gas turbine compressor is used, the modification of the STIG cycle is called intercooled steam-injected gas turbine (ISTIG). A more recent cycle proposed for development by EPRI is called the humid air turbine (HAT). It is a variation of the ISTIG cycle where the air is humidified to a water content of 20-40% and heated to around 538 °C prior to being introduced in the gas turbine. ISTIG and HAT cycles offer the potential for improvements in efficiency and lower capital costs, but the higher risk, because of being in the early stages of development, may hinder the rapid evolution towards a commercial system.

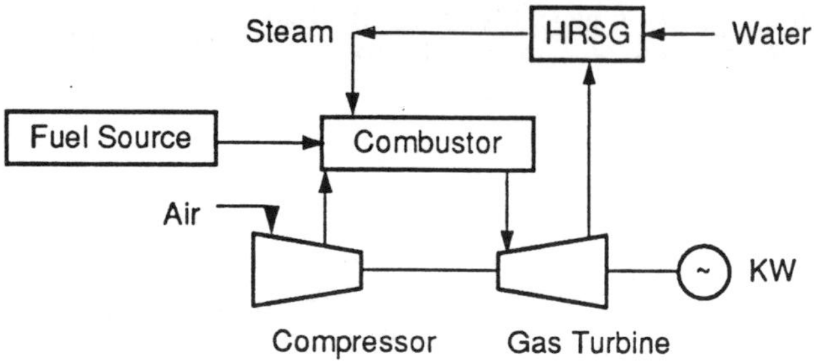

Figure 3. Schematic diagram of a STIG cycle

When the steam generated from heat exchanging with the gas turbine exhaust gases is used for driving a steam turbine and generate additional power, the process becomes the integrated gasification with the combined cycle (IGCC). This cycle will be described in greater detail next in Section 3 – The IGCC Process.

Gasification, gas cleanup, PFBC, gas turbines, and fuel cells are all technologies which have significant improvement and growth potential. Furthermore, gas turbines can be designed to handle both natural gas and coal-derived gaseous fuels.

3 The IGCC Process

Several studies have been made that support the expectation of great reductions in environmental emissions at lower costs by the use of IGCC technologies instead of conventional coal-fired boilers. Two of such reports have been published by the Electric Power Research Institute [3,4]. The IGCC takes advantage of thermodynamic efficiencies by integrating the gasification cycle with the combined cycle for power generation. The end result is a more efficient system [5]. Several fuels can be gasified, including coal, heavy oils, petroleum coke and waste products. The emphasis of the description of the IGCC will be on coal as a fuel, even though examples of applications using the other fuels will be given from time to time to give a broader view of the commercialization of the IGCC technology.

In power generation technology, the combined cycle consists of the gas turbine cycle (Brayton cycle) and the steam turbine cycle (Rankine cycle). The combination of both results in a more efficient configuration than each of the two cycles taken separately.

The major components of the coal-based IGCC cycle, shown in schematic form in Figure 4, are:

- Coal handling and gasification
- Gas cleanup
- Gas turbine cycle
- Steam turbine cycle

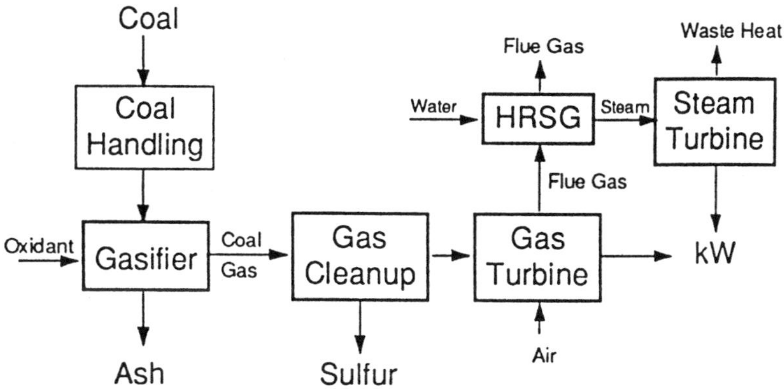

Figure 4. Schematic diagram of the IGCC system

In practice there are many other sub-elements of the IGCC configuration that support the operation of the major components enumerated above.

Coal handling includes unit operations such as grinding, coal cleaning, size screening, slurry preparation, transportation, etc. Once the coal is available in a suitable form for gasification, it is introduced into the gasifier. Here, the coal fuel is contacted with the oxidant, air or oxygen, in addition to steam or other medium for temperature control, to produce a gaseous fuel composed primarily of hydrogen, carbon monoxide, carbon dioxide, and steam. If air has been used for gasification, the product coal gas will also contain nitrogen as a diluent.

Other components present in coal in minor concentrations – such as sulfur, chlorine, organic nitrogen, and heavy metals – will end up in the coal gas as contaminants in the form of hydrogen sulfide, H_2S, hydrogen chloride, HCl, ammonia, NH_3, and trace element contaminants. Most of these contaminants will have to be removed prior to combustion in the gas turbine or prior to venting of the flue gases to the atmosphere. Cleaning of the coal gas can be accomplished either by existing commercial low-temperature cleanup (LTCU) technologies or at high temperature via hot gas cleanup (HGCU) systems.

The clean coal gases are then fully combusted to carbon dioxide and water in a gas turbine, producing electricity in the process. Some of the thermal energy of the exiting flue gas from the turbine can be recovered by heat exchange in a heat recovery steam generator (HRSG), thus producing steam, which finally goes through a steam turbine for further electricity production. The combination of a gas turbine, HRSG, and steam turbine, is commonly called the combined cycle, because it combines the Brayton cycle of the gas turbine with the Rankine cycle of the steam turbine, with the HRSG acting as a link between the two cycles.

Some of the advantages of the IGCC configuration with respect to coal-fired combustion systems are:

- High cycle efficiency, which is even higher with hot gas cleanup (vs. low temperature cleanup)
- Excellent environmental performance
- Smaller amounts of solid wastes
- Lower water consumption
- Capability of staggered/phase construction (i.e., modular construction)

In principle, the IGCC system operates as a two-stage combustion system, where the gasifier acts as a partial combustor producing a fuel of hydrogen and carbon monoxide, and the gas turbine which completes the combustion of the coal gases to carbon dioxide and water. Depending on the plant size, degree of complexity or efficiency needed, and the environmental regulations that need to be met, the IGCC can be operated in a minimal configuration called the simplified IGCC system, to be described below.

3.1 The Simplified IGCC System

The U.S. Department of Energy (USDOE) has proposed an air-blown gasification system with hot gas cleanup called the simplified IGCC. This configuration is very efficient and cost effective for small size plants. Several studies have been published [6, 7, 8] that show promise of achieving both significant reductions in air emissions and capital costs of under $1000/kWe through the use of a simplified IGCC approach. Figure 5 shows the schematic representation of a simplified IGCC system showing the gasification, gas cleanup, and power generation block using a gas turbine with a heat recovery steam generator.

As summarized by Notestein [9], the goals of lower costs in the simplified IGCC

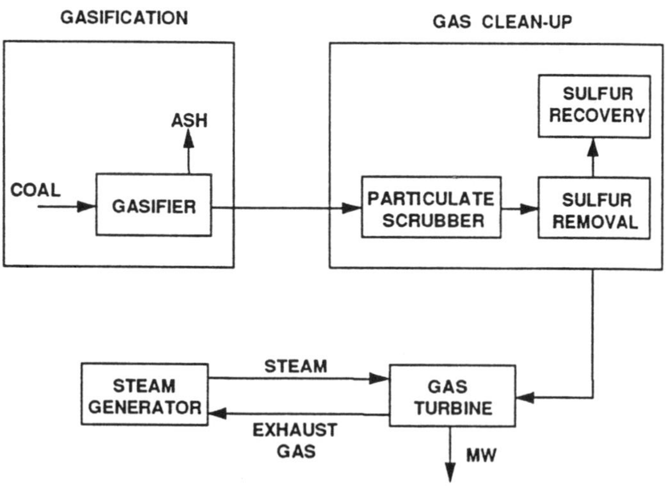

Figure 5. Schematic diagram of a simplified IGCC configuration

configuration are met by 1) maintaining nearly constant temperature and pressure through the fuel gas preparation portion of the system as a means of minimizing the number of operational units within the overall system, and 2) sizing the components such that shop fabrication (as opposed to field) and contemporary classes of gas turbine can be utilized. Figure 6 shows the details of of the simplified IGCC concept using hot gas desulfurization and the details of the power generation block using a gas turbine, an HRSG, and a steam turbine. In addition to sending the steam to a steam turbine, the option of injecting steam in the gas turbine for additional power augmentation is also included. The important features of the simplificed IGCC are:

- It uses air from the gas turbine compressor for the air-blown gasifier.
- The coal gas from the gasifier is cleaned and then combusted in the gas turbine at temperatures and pressures similar to those at the exit of the gasifier.
- No waste-water streams are produced.
- In general, the level of emissions from a simplified IGCC system are not as low as those obtained from a full-blown, low-temperature, liquid-based, fuel gas scrubbing system, but the resulting emissions reductions are cost effective.

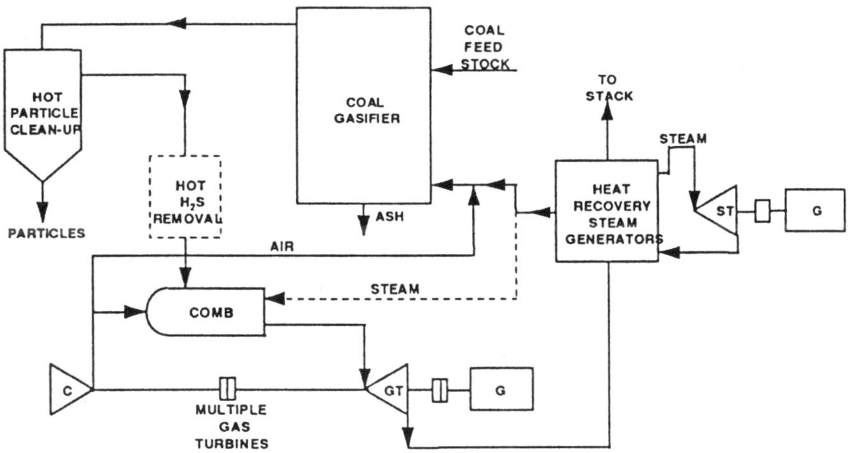

Figure 6. Detailed diagram of the simplified IGCC concept

Because of the above features, the simplified IGCC usually contains fewer operations than the full-blown IGCC system. Table 3.1 shows the comparison between two somewhat extreme cases: a conventional IGCC system using an oxygen-blown entrained flow gasifier with a CGCU, and a simplified IGCC configuration using an air-blown, fixed-bed gasifier using HGCU. The number of operations and complexity of the conventional IGCC case are obviously higher, especially if full integration is desired, but the operability, environmental compliance, and efficiency may vary if intermediate combinations of gasifier, coal gas cleanup system, and level of plant integration are chosen. Although the comparison of the two systems shown is for one of the bare-bones configuration of simplified IGCC system, the number of unit operations in the simplified IGCC system may increase and the large differences shown in Table 3.1 may be much less for other hybrid systems. For instance, the use of a fluidized-bed gasifier in a simplified IGCC may require also coal grinding and convective cooling operations, just as in a conventional IGCC system. For this reason, the magnitude of the advantages (in process complexity and costs) of one system over the other is highly dependent on the

Table 3.1. Comparison of unit operations in conventional and simplified IGCC configurations

Unit Operation	Conventional	Simplified IGCC
Coal Preparation	Coal Grinding Coal Slurry Preparation Oxygen Plant	
Gasification	O_2-blown, Entrained-Flow Gasifier	Air-blown, fixed-bed gasifier Booster compressor
Heat Recovery	Radiant Cooling Convective Cooling	
Gas Cleanup	Gas Scrubbing Low-temperature gas cooling Sulfur Removal Fuel Gas Saturation Waste Water Treatment Tail Gas Scrubbing	Hot Gas Particulate Removal Hot Gas Sulfur Removal Sorbent Regeneration Sulfur Recovery
Power Generation	Gas Turbine Heat Recovery Steam Generator Steam Turbine	Gas Turbine Heat Recovery Steam Generator Steam Turbine

configuration chosen and the benefits may be significant or modest. In addition, hot gas cleanup is not commercially available yet as the low-temperature cleanup system is, so the comparisons do not include risk assessments.

3.2 Integration of Components in the IGCC System

Because there are several modular components in the IGCC cycle, integration of the components (Figure 7) is of primary importance to synergistically increase the overall efficiency of the system above that of the isolated components without integration. For instance, oxygen-blown gasifiers use the oxygen from the air separation unit. The nitrogen byproduct from the air separation unit can be injected to the gas turbine for additional mass flow throughput and added power generation. The flue gases from the gas turbine are sent to the heat recovery steam generator (HRSG) to produce steam for the steam turbine. Some of the steam can be injected as additional steam to the gasifier. As one can see, the overall process optimization becomes very important and intricate when all the alternatives are considered.

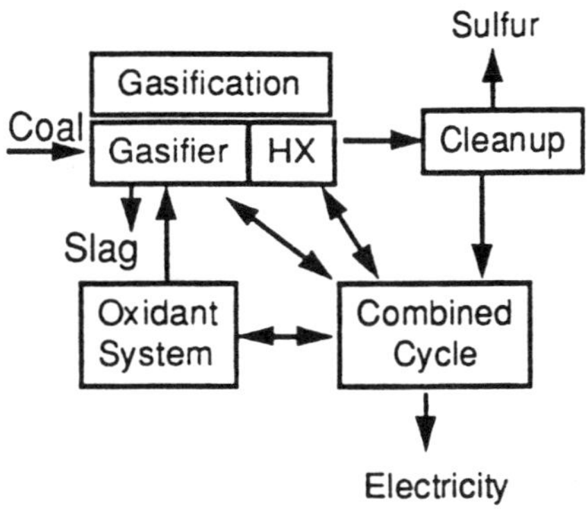

Figure 7. Integration of components in the IGCC configuration

Following the concept of integration of components, as shown in Figure 7, a more elaborate diagram showing an example of an IGCC system with a higher degree of integration is shown in Figure 8. The same subsystems of the schematic diagram of Figure 7 (oxidant system, gasification, gas cleanup, and combined cycle) are present, but more detail is given to some of the potential integration stream lines available. Integration of components is achieved in various ways. The oxidant system is shown as an air separation unit (ASU) which provides the oxygen needed for the operation of the gasifier. The nitrogen by-product from the air separation unit is utilized into the gas turbine for higher mass flow rate and, hence, higher power output. The gasification island is shown for an oxygen-blown, entrained-flow gasifier using coal slurry as a fuel, coupled to a train of cold gas cleanup (CGCU), such as a conventional acid gas removal system. Cold gas cleanup technologies are commercially available (described in Section 5), while those for hot gas cleanup are currently being tested for commercial feasibility (see Section 6). The dual CGCU/HGCU configuration will be used at the Polk Station in Florida to obtain process data for both types of cleanup systems. See Section 6.3 for a description.

High pressure steam produced in the gasification island is also utilized in the power generation block (gas turbine, HRSG, and steam turbine). Condensate from the grey water processing stream is heated in the process of low-temperature gas cooling and reintroduced into the gasifier. In these ways, the above mentioned streams that would otherwise be discarded are used to increase the heat rate of the integrated system. In addition, the low-temperature acid gas removal system utilizes a low-temperature gas cooling system (composed of convective heat exchangers), a sulfur recovery system and tail gas treating units, which can benefit of the integration of waste liquid streams. Instead of CGCU one could use a HGCU system. The HGCU system, comprising hot particulate and desulfurization units, also utilizes sulfur recovery system but without generation of liquid streams. Sulfur by-products of the cleanup system can be elemental sulfur or sulfuric acide, depending on the market requirements for saleable byproducts.

Despite the obvious advantages for developing a more efficient IGCC system by means of higher levels of integration, a recent publication [10] has warned that even though each integration step can enhance energy efficiency and environmental performace, it also brings sacrifices in complexity and operating flexibility, for instance.

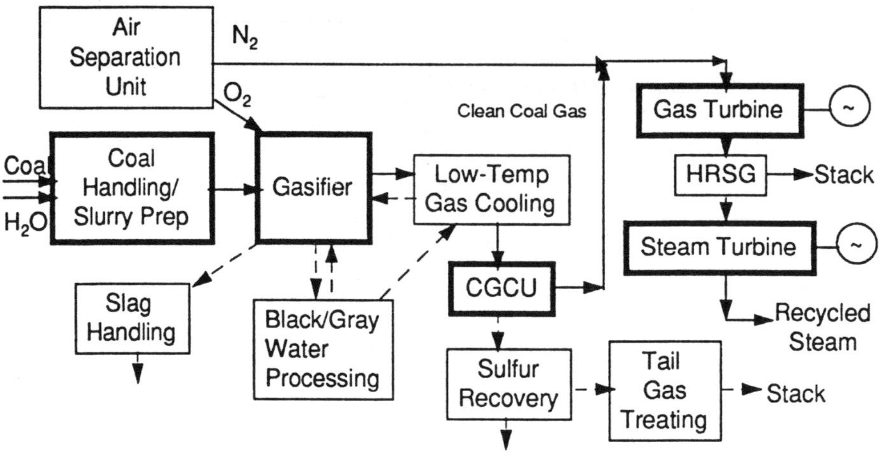

Figure 8. Generalized flow diagram of an IGCC system with CGCU

4 IGCC Efficiency and Economic Goals

The ultimate objective of the IGCC system is to maximize the overall cycle efficiency, which is calculated from the individual contributions of at least four general components:

- fuel conversion (gasifier) efficiency
- gas turbine cycle efficiency
- steam cycle efficiency
- generation efficiency

At the same time, the parasitic energy consumption and the cost of ancillary operations (e.g., gas cleaning processes) must be minimized. In mathematical terms, the overall system efficiency, E_{sys}, is given by [11]

$$E_{sys} = E_{con}\left[E_{gt}(1-E_{sc})(1-H_{bp})+E_{sc}\right]E_{gen} \quad (4.1)$$

where:
E_{con} = fuel conversion efficiency
E_{gt} = gas turbine cycle efficiency
E_{sc} = steam cycle efficiency
H_{bp} = heat by-pass ratio
E_{gen} = generation efficiency

The overall system efficiency, E_{sys}, represents the difference in energy between the fuel and the net electricity production, after subtracting parasitic energy losses to operate the system. E_{con} includes both the change in energy from gasification as well as the difference in energy from gas cleaning. The gas turbine efficiency, E_{gt}, represents the efficiency by which the energy from the fuel is converted into shaft power at the gas turbine. Equivalently, the steam cycle efficiency, E_{st}, represents the corresponding efficiency of the complete steam cycle system as energy is converted to shaft power, including the net efficiency of the heat recovery steam generator, and other heat exchanger subsystems. The term heat by-pass ratio, H_{bp}, defines the amount of potential energy in the fuel that does not contribute to the power generation in the gas turbine (hence "by-passing" the gas turbine directly to the steam turbine). As an example, a boiler furnace would have a heat by-pass ratio of 1.0 since there is no gas turbine and all the energy from the combustion gases go directly to steam generation.

The first IGCC plant in the U.S.(the Cool Water demonstration facility) was built in the mid 1980's at a capital cost of around US$3000/kW. The efficiency at that time was comparable to coal-fired plants (typically around 33%). In the U.S., the U.S. Department of Energy/Morgantown Energy Technology Center (USDOE/METC) is sponsoring several programs to demonstrate reduction in capital costs and increases in thermal efficiency of IGCC systems [12].

Rath et al.[13], have summarized several case studies of expected efficiencies of IGCC systems using GE's "F-Technology" (i.e., 1260-1288 °C, or 2300-2350 °F, turbine inlet temperatures) and producing a nominal 250 MW of electricity (Table 4.1). Variations within each configuration were allowed to take advantage of each gasifier's unique characteristics in order to optimize performance. For instance, limestone injection was considered in the cases of the fluidized-bed and 2-stage entrained flow gasifiers, but not used in the fixed-bed slagging and slurry feed entrained-flow gasifiers. Variations in gas cleanup included CGCU and HGCU using zinc titanate sorbents. Overall efficiencies were in the vicinity of 43-44% (HHV), with net power productions at 260 MW or above except for the case of the fixed-bed slagging gasifier configuration. The reasons for the lower performance of the latter case were: 1) the inability to utilize raw-gas heat recovery (due to the presence of tars in the raw gas), and 2) higher miscellaneous power usage of the partial air separation unit (ASU) supplying oxygen and cold gas cleanup. The overall environmental performance of the IGCC systems utilizing hot gas cleanup was excellent relative to a pulverized coal plant with low NO_x burners and scrubbers. The overall environmental performance of the IGCC system utilizing low temperature cleanup was predicted to be close or equal to a natural gas combined cycle (without carbon dioxide externalities).

Table 4.1 IGCC Efficiency using F-Type Technology

Gasifier type	Fluidized	2-Stage Entrained	Fixed-Bed Slagging	Slurry feed Entrained
Oxidant	Air	Air	O_2 (partially integrated ASU)	O_2 (partially integrated ASU
Gasifier C Conversion, %	96.5	98.7	99.2	98.9
Gasifier Pressure	27.6 bar (400 psia)	22.4 bar (325 psia)	27.2 bar (395 psia)	42.4 bar (615 psia)
Gas cleanup	654 °C, ZnTi, HGCU, fixed bed	666 °C, ZnTi, HGCU, fluid bed	38 °C Selexol, Claus/Scot CGCU	677 °C, ZnTi, HGCU, Moving Bed
Sulfur Recovery	Atm sulfation (Disposal)	Sulfuric acid	Elem. sulfur	Sulfuric acid
Net Power	275 MW	260 MW	248 MW	268 MW
Overall Eff., % of HHV	43.3	43.9	41.8	42.8
COE, mill/kW, const US$	55.6	53.9	64.4	58.5

During the mid to late 1990's, the reduction in capital costs for IGCC are expected to be down to US$1500-2000/kW and the increases in efficiency are expected to be up to 40 to 45% (HHV). Over the next twenty years, and up to the year 2010, the advances in current research and development technology will bring further reduction in capital costs to near US$1050/kW and further increase efficiencies as high as 52% (HHV). The 27% (relative) gain in efficiency over those IGCC systems undergoing demonstration today is expected to translate into 28% COE reductions. These advances in cost and efficiency require, of course, that the environmental performance of the plants are also being improved to meet or exceed regulations at that time.

In order to accomplish these goals, USDOE/METC is sponsoring R&D aimed at accomplishing the needed product improvements shown in Figure 9. Some of the strategic and environmental benefits to be gained from this sustained effort are: continued used of eastern U.S. coals, reduced sulfur and nitrous oxide emission rates, reduced CO_2 emissions of 33 million tons per year, energy savings equivalent to 75 million barrels of oil, 20 million person-year construction jobs, and 350 thousand permanent jobs [12].

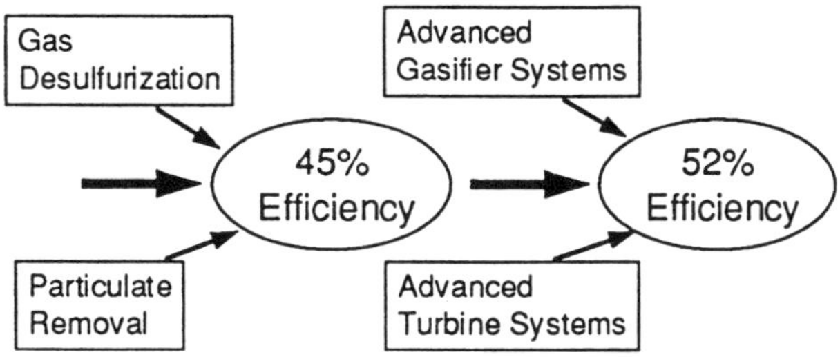

Figure 9. Subsystem Product Improvements needed in IGCC systems [10]

Advances in particulate removal and desulfurization will increase the IGCC efficiency towards 45% (HHV) and could be achieved before the end of the century. IGCC operation will need cyclones for particulate removal under reducing coal gas environments at temperatures above 800 °C and be integrated with hot gas desulfurization systems, rather than low-temperature cleanup systems.

The longer-term objective of achieving 52% efficiency (HHV) in an IGCC system is expected to be accomplished by the use of advanced gasifiers and advanced gas turbine systems (ATS). The concept of an advanced gasifier, such as the PYGAS or transport gasifiers would exhibit higher carbon conversions and, in the case of a transport gasifier, higher limestone utilization during desulfurization. Table 4.2, taken after Rath et al.[13], shows the achievable efficiencies of three of the IGCC systems previously shown in Table 4.1 but now using advanced concepts in gasification and ATS technology. These advances will require operation of the gas turbine at more demanding conditions for both the gas turbine and the steam turbine. For instance, the gas turbine will operate with improved cooling of the first stator airfoils, higher turbine inlet gas temperatures and higher pressure ratios; the steam turbine will also require operation at higher pressures and temperatures using superheated steam. All these improvements will require metallurgy developments over the next decade that are not commercially available at the present time.

Table 4.2 IGCC Efficiency using ATS Technology [13]

Gasifier type	Fluidized Bed	2-Stage Entrained Flow	Fixed-Bed Slagging
Oxidant	Air	Air	Oxygen
Gasifier C-conversion	96.5	99.0	99.2
Gas Cleanup/ Sulfur Recovery	HGCU, Sulfation	HGCU, Sulfuric Acid	CGCU, Elemental Sulfur
Net Power, MW	441	423	378
Overall Efficiency, % of HHV	51.2	52.2	48.1

Based on economic comparisons drawn from previous and planned IGCC plants, gas turbine "E" technology, such as the one used in the Cool Water project, can only compete economically where fuel costs are low and multiple products are produced. IGCC plants based on gas turbine "FA" technology can compete on an equal basis where environmental standards are important or where repowering of existing plants is desired. Finally, IGCC systems planned using an advanced turbine system (ATS) technology will be the clear choice for many future applications [14].

5 Environmental requirements for IGCC

Environmental issues of greatest concern to coal-fired power generation have included emissions of acid rain precursors and particulate matter. The disposal of solid and liquid residues, and the cost of complying with environmental regulations in coal-fired plants is not insignificant. In Germany, compliance with environmental regulations is estimated to increase the capital costs of conventional coal-fired power generation plants by approximately one third [15].

There are two goals for emissions controls in IGCC systems: one is for process-related contaminant control, the other is for control of emissions dictated by environmental regulations. Examples of process-related undesired contaminants are, for instance, contaminants that affect the performance of i) combustion turbines, such as alkali metals, vanadium, and sulfur (in the form of alkali sulfates) that induce high temperature corrosion in gas turbine components, and ii) the heat recovery steam generator (HRSG), which is affected by chlorides in the fuel gas that induce metal corrosion. Calcium, magnesium, and particulates must be kept low to reduce ash deposition, erosion and corrosion (DEC). Examples of contaminants that are regulated as environmental emissions include dust and particulate matter, sulfur species (acid rain precursors), ammonia or NO_x, alkali and halogens, organic and trace elements, and greenhouse gases (CO_2, N_2O). An excellent review of technologies for gas cleaning in advanced coal-based power generation systems is given by Thambimuthu [16]. In general, coal gas cleaning technologies fall under the following categories:

- Dust and particulate matter
- Sulfur species
- NO_x-precursors, including fuel-bound nitrogen and ammonia

- Alkali and halogens
- Organic and trace elements (heavy metals)
- Green house gases (CO_2, N_2O)

Table 5.1 Typical filtration efficiencies for several particulate removal devices

Filtration velocity	Filtration device	Typical efficiency
4-10 m/sec.	High-efficiency cyclones (1 or 2 stages)	92-99.5%
0.6-1.0 m/sec	Electrostatic precipitators Fluidized beds	94-99%
0.001-0.04 m/sec	Fabric filters Rigid media filters	>94% >98%

Particulate removal systems have been under development for many years, starting with coal-fired combustion systems. They fall under the following categories:

- inertial separation and agglomeration (cyclones, impact separators, dust agglomerators)
- electrostatic precipitators (high temperature/high pressure)
- fabric and fabric filters (ceramic fabric filters, metallic fabrics and nested filters)
- rigid-barrier filters (candle filters, ceramic cross-flow filters, metallic gap filters)
- granular-bed filters (fixed/moving-bed filters and fluidized-bed filters)

The efficiency of filtration devices varies from about 92% to 99+% depending on the filtration velocity. Typical examples of filtration efficiencies are given in Table 5.1. Efficiencies, of course, depend on the particle size distribution. Development of secondary and tertiary fine particulate filters is the most important issue affecting availability of high-temperature/high-pressure particulate removal devices.

Sulfur removal processes have been classified into low-temperature (cold) gas cleanup (LTCU or CGCU) systems, and high-temperature (hot) gas cleanup (HTCU or HGCU) systems. CGCU systems operate below 300 °C, more typically around 20-120 °C, and some even require cryogenic stages (Table 5.2). They are based on wet scrubbing, physical or chemical absorption processes, and are proven commercially available technologies. The limitation is that CGCU technologies are energy intensive and the application to a particular process depends on the partial pressures of H_2S and CO_2. A good review of the applicability of CGCU processes to coal gas cleaning is given by Thambimuthu [16].

Table 5.2 Examples of CGCU processes

Type	Solvent	Temp.	Examples
Physical solvent	methanol NMP DMPEG	-10/-70 °C -20/40 °C -40 °C	Rectisol Purisol Selexol
Chemical	Amines	20-120 C	MEA, ADIP Amine Guard Econamine
Inorganic chemical	Na, K, carbonates	20-120 °C	Benfield and versions
Physical/Chemical solvent	mixture of organic solvents and amines	5/40 °C	Sulfinol Amisol
Liquid phase oxidation	carbonates disulfonic acids		Stretford Unisulf Sulferox

HGCU desulfurization systems operate above 400 °C, and are based mostly on fast chemical reactions. The technologies are under development or in the early stages of commercialization. They have a cost advantage over CGCU systems, but the degree of long-term reliability is still to be proven. HGCU systems are based on using mixed-metal oxide sorbents, particularly zinc-based sorbents and have been developed under sponsorship of the U.S. DOE and other governmental organizations around the world. With respect to environmental compliance, IGCC systems have proven that they are superior than most other conventional and advanced coal-based technologies.

6 Demonstration Scale and Commercial Applications of IGCC

The performance of the gasification system is a key factor affecting the overall power generating efficiency of any IGCC system. It is important, therefore, to ensure that the gasification and gas cleanup islands convert the coal into clean fuel with the greatest possible efficiency.

6.1 Early IGCC Demonstration Plants

Three major demonstration plants with complete commercial-scale IGCC systems have been constructed until 1990 [11]: Lünen (Germany), Cool Water (USA), and Plaquemine (USA).These plants are considered the pioneers in the demonstration of feasibility of the IGCC concept. A short comparison of the three plants is given in Table 6.1.

Lünen was the first commercial-scale IGCC demonstration power plant. It was built by Steinkohlen Elektrizitäts AG (STEAG) at its Kellermann station, Lünen, Germany in 1972. The plant operated for a period of approximately five years with the longest continuous on-line run time being about 2400 hours [17]. An air-blown Lurgi gasifier was used with the exiting raw gas treated in scrubbers to remove particulate matter, condensables, and tars. A standard KWU gas turbine generating appproximately 74 MW of electricity was used while the steam turbine generated the remaining 96 MW. Details of the plant operation are provided by Takematsu and Maude [11]. The net system efficiency was estimated to be around 34.4% (HHV). Given the efficiency of available commercial gas turbines at that time, this overall power generating efficiency was regarded as an excellent performance.

The Cool Water project operated in California, USA for five years. Coal gas was first made on May 7, 1984, following a two and one half year construction period. Cool Water was a US $274 millon project that utilized a 1200-ton per day, oxygen-blown Texaco gasifier producting medium heating value gas. The gasifier utilized a radiant and convection system and a separate 100% quench system for coal gas cooling. The cleanup system was a low-temperature process using a Selexol plant for sulfur removal, a Claus process for sulfur recovery, and a Scot process for tail gas processing. The gasification and cleanup systems were then integrated with a GE STAG 107E combined-cycle power plant [14].

Table 6.1 Comparison of Early Demonstration IGCC Plants

	Lünen	Cool Water	Plaquemine
Gasifier	Lurgi	Texaco	Dow
Gasifier Operation	Air blown, Dry coal feed Moving bed	Oxygen blown, Slurry feed Entrained flow	Oxygen blown, Slurry feed Entrained flow
Carbon conversion, %	91.4	98.9	99
Coal gas efficiency, %	76.3	74.3	77.0
Power output, MW	170	93	160
Overall system efficiency, %	34.4	31	36

Southern California Edison was the host utility and, along with the partners, successfully demonstrated that a standard 80 MW gas turbine could operate in IGCC configuration on a utility system. The plant operated for a a period of five years with various coal types and operating conditions. During that period, it demonstrated production of 118 MW of electricity, above the initial target 100 MW. The overall system efficiency was 31%, somewhat less than that of the Lünen plant, partly because the plant was primarily designed to demonstrate the IGCC concept on a commercial scale at a reasonable cost and, purposely, did not pursue optimizing efficiency as the main objective.

At the Cool Water project, emissions were near one-tenth of standards and NO_x levels on coal of 20 ppmvd (Table 6.2). N_2O, a greenhouse and ozone depletion pollutant sometimes produced in coal boilers, is not an issue with IGCC because of its very low levels in the stack gas. Most importantly, the ash was non-hazardous and saleable.

The third major demonstration plant of coal-based IGCC technology was built at Plaquemine, Louisiana, USA. Full two-stage operation of the gasifier started in August 1987 [11]. The plant is operated by the Dow Chemical Company and uses an oxygen-blown, slurry-fed, Dow gasifier. Cyclones provide hot gas particulate removal from the coal gas and an amine scrubber provides for the H_2S and some CO_2 removal. The steam produced in the combined cycle is used for power and process steam. Equivalent power generation from the unit is approximately 160 MWe. The higher efficiency of the

Table 6.2 Cool Water plant emissions

Air Emissions	Actual
SO_2	97% Removal
NO_x	20 ppm
Particulates	1/10 of EPA Standards
N_2O	0.2 ppm
Waste Ash	Non-leachable, non-hazardous Meets California Standards Saleable
Water usage:	70% of conventional plant

Plaquemine plant over Lünen and Cool Water plants is in part due to the two-stage gasifier operation. However, it has similar disadvantages due to slurry feeding and oxygen-blown arrangement.

6.2 HGCU R&D Programs in the US

The ultimate goal for research and development of new IGCC system configurations is to increase the overall system efficiency at a reasonable cost. One approach to increase overall efficiency is to utilize a HGCU process for gas cleaning.The current effort in the US is aimed at supporting HGCU in the current generation of demonstration-scale IGCC projects (described in Section 6.3 – IGCC Projects in the US). In the past, it was considered that the higher the desulfurization temperature above 500 °C, the greater the efficiency and cost advantages for the overall IGCC plant. Recent system studies of IGCC technology [18, 19] have suggested the temperature range of 350 °C to 550 °C, instead of temperatures above 650 °C, is a more desirable range to remove coal gas contaminants, particularly sulfur species. In general, the benefits to be gained at these temperatures (e.g., lower plant costs, hardware reliability, and engineering feasibility) outweigh the slight loss in desulfurization efficiency and still result in a lower overall cost of electricity. At temperatures above 650 °C risks due to higher capital costs and reliability of HGCU and process gas handling become a concern.

The U.S. Department of Energy/Morgantown Energy Technology Center (DOE/METC) has sponsored development of various configurations of high-temperature desulfurization systems using mixed metal oxide sorbents over the last decade [20]. Fixed beds, moving beds, fluidized beds, and transport reactors are four of the configurations that have been considered for HGCU desulfurization. Although the fixed-bed approach using metal oxides has been shown to reduce the H_2S content of the cleaned gas to very low levels, the operation was not continuous and it suffered from operational problems such as unsteady production of sulfur dioxide or elemental sulfur byproduct. To overcome these problems, US DOE has continued sponsoring fluidized-bed/transport reactor as well as moving- bed hot gas desulfurization research more heavily in recent years. The implementation of hot coal gas desulfurization heavily relies on the development of regenerable sorbent materials exhibiting high sulfur capacity and efficient sulfur removal (from several thousand ppmv down to a few ppmv) over many cycles of sulfidation/regeneration. Structural stability and good mechanical strength are also desirable features in a sorbent. Not surprisingly, only a few metal oxides can meet all these stringent requirements simultaneously. Iron oxide [21], zinc oxide [22], copper oxide, as well as mixed-metal oxide compounds such as zinc titanates [23, 24] and other copper-based [25] compounds have been investigated for use as regenerable sorbents in HGCU applications.

GE is developing a moving-bed, high-temperature desulfurization system for IGCC power systems in which zinc-based regenerable sorbents are currently being used as desulfurization sorbents [26, 27]. Zinc titanate and other zinc-based oxides are being considered as sorbents for use in the Clean Coal Technology Demonstration Program at Tampa Electric Co.'s (TECO) Polk Power Station I. A key to success in the development of high-temperature desulfurization systems is the matching of sorbent properties, namely sustainable desulfurization kinetics, sulfur capacity, and mechanical durability over multiple cycles, with existing plant operating conditions.

Several other reasons related to sorbent operability favor desulfurization below 650 °C. Although the reaction rates for hot gas desulfurization using metal oxides increase with increasing temperatures making it desirable to operate above 500 °C, the lower thermal stresses and propensity to sintering experienced by the sorbent at lower temperatures can lead to better sorbent reactivity and mechanical durability over longer periods of operation with minimal losses in performance. The net effect is a reduction in the sorbent replacement cost. In turn, the benefit to be gained by lower temperature operation results in lower overall cost of electricity. No data on long-term operation of a commercial-scale HGCU is available to date.

6.3 IGCC Projects in the US

There are currently three active IGCC demonstration projects in progress under the sponsorship of the U.S. Department of Energy's Clean Coal Technology (CCT) programs in the U.S. (Table 6.3). All three US CCT programs use GE "F" gas turbines. Two of them will have HGCU (Tampa Electric Co's Polk Station and Sierra Pacific Power Co.'s Piñon Pine), and are expected to achieve a 42% (LHV) overall efficiency.

Table 6.3. U.S. Clean Coal Technology IGCC projects in progress

Project	Technology	Efficiency % LHV	MWe	Application	Start-up
PSI/Destec -Wabash River	Destec O_2 blown CGCU GE 7 FA	40	270	Repower	Mid 1995
Tampa Electric Co. -Polk Station	Texaco O_2 blown CGCU+ MB HGCU GE 7 FA	42	260	Greenfield	Mid 1996
Sierra Pacific -Piñon Pine Station	KRW Fluid bed air blown TR HGCU GE 6 FA	42	100	Existing site	Early 1997

The PSI Energy/Destec project at the Wabash River power station (Terra Haute, Indiana) is a joint venture between PSI Energy Inc., Plainfield, Indiana, and Destec Energy Inc., Houston Texas. The project repowers a 100 MW coal-fired steam turbine/generator with a gas turbine firing coal gas (Figure 10). The plant will produce a net 270 MW of electricity. This is the first plant using a Destec (formerly Dow) gasifier specifically designed for the local high-sulfur, bituminous coal and producing a medium-Btu syngas. The technology being used is an extension of the experience gained from the operation of the Plaquemine demonstration plant operated by Louisiana Gasification Technology, Inc. (LGTI). The system uses a cold gas cleanup desulfurization system. The synthetic fuel gas (syngas) is sent to a GE MS 7001F high-temperature combustion turbine. A heat recovery steam generator recovers the gas turbine exhaust heat to produce high-pressure steam; this steam and that from the gasifier feed the steam turbine-generator [28]. The plant is rated for an overall 40% efficiency (LHV). The plant has been designed to substantially outperform the standards for the Clean Air Act Amendments for the year 2000. Some of the important features of the design are:

- A hot/dry particulate removal and recycle system will be demonstrated at full commercial scale (prior Destec's operation at Plaquemine had utilized a wet scrubber system to remove particulates).
- The slag fines recycle system recovers most of the carbon present in the slag byproduct stream, and recycles it back to the gasifier for improved carbon conversion.
- An advanced design oxygen plant will produce 95% pure oxygen so that overall system efficiency will be increased due to lower power requirements of the air separation unit.
- Carbonyl sulfide removal is accomplished via conversion to H_2S so that total sulfur will be removed as 99.7% pure.

The power generation facilities will also optimize the advancements in combined cycle technology. The combustion turbine is a GE MS7001F gas turbine, operating at higher firing temperatures and higher pressure ratios, generating approximately 192 MW. The steam produced by the HRSG will drive the steam turbine producing an additional 109 MW. Repowering of the existing steam turbine involves upgrading the

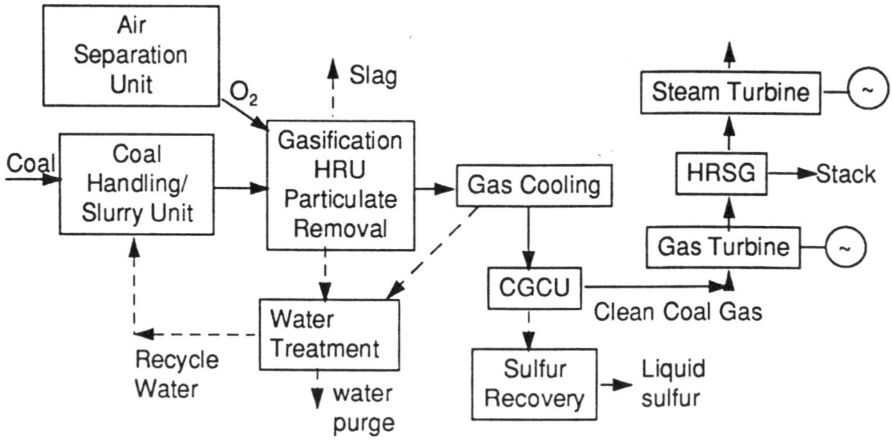

Figure 10. Diagram of the Wabash River Repowering IGCC Project

unit's capacity to accept increased steam flows from the HRSG. Parasitic plant auxiliaries (e.g., oxygen plant, gasification area) will consume approximately 33 MW for a nominal net power generation of 270 MW.

In contrast to the repowering project at Wabash River, the Tampa Electric Project at the Polk Station is a greenfield project and includes both a commercial cold gas cleanup system and a demonstration moving-bed HGCU system [29]. The plant is located in Polk County, Florida, southeast of the city of Tampa. Some of the salient characteristics of the plant are (Figure 11):

- A commercial Texaco, oxygen-blown, entrained-flow gasifier is employed to utilize about 2300 tons per day of coal (dry basis). The coal is slurried in water to 60-70% solids.
- The oxygen for the gasifier is produced by the air separation unit (ASU), giving the gasifier a greater than 95% carbon conversion in a single pass. The coal syngas will be fed into one radiant syngas cooler where the temperature is reduced from about 1370 °C to about 700 °C.
- After the radiant cooler, the gas is split into two parallel convective coolers, where the temperature will be further reduced to approximately 482 °C.
- Ninety percent of the flow will be directed to the cold gas cleanup (CGCU) system, while ten percent will go the hot gas cleanup (HCGU) system. The HCGU system is a moving-bed configuration developed by GE Environmental Services, Inc. (GEESI), of Lebanon Pennsylvania. Instead of cooling the gas prior to sulfur removal, the HGCU system will accept gas at 482-500 °C, thus providing higher efficiency to the system.
- The HGCU system will use a zinc-based sorbent, which will produce a concentrated SO_2 stream during sorbent regeneration. The SO_2 will be fed to a sulfuric acid plant for production of saleable acid by-product.

- For particulate removal, high-temperature cyclones and a high-temperature barrier filter will be tested as part of the HGCU demonstration.
- The use of sodium bicarbonate in a circulating fluidized-bed system will also be tested for the removal of halogen species (i.e., hydrogen chloride and fluoride) from the coal gas.
- Coal gas will be combusted in a GE 7F gas turbine generating 192 MW of electricity, with the balance of the total power generation being produced by the steam turbine in a combined cycle configuration.

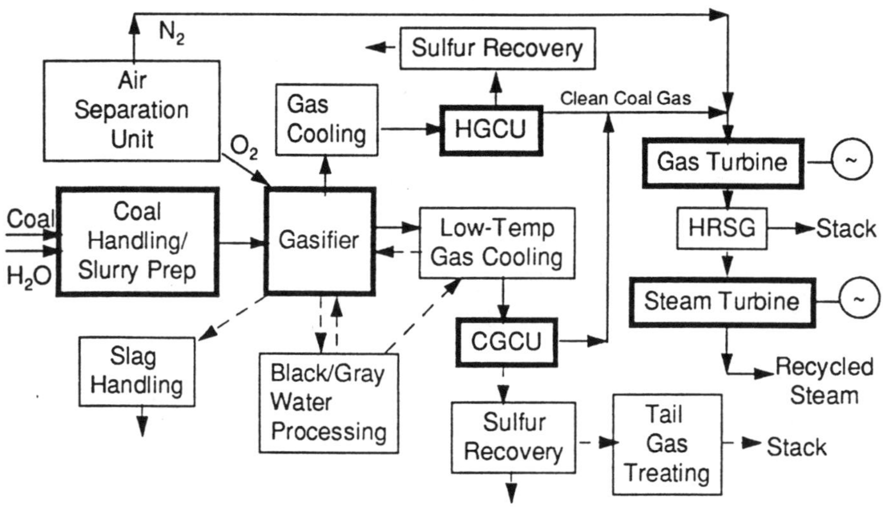

Figure 11. Diagram of Tampa Electric Co.'s Polk Station IGCC plant

- Plant startup is proceeding as scheduled and first generation of syngas is targetted for July 1996.

The third active CCT project in the US, the Sierra Pacific Power Co's Piñon Pine project, is located at the Tracy Station, east of Reno, Nevada. It is being built on an existing site to generate approximately 100 MW of electricity. The key features of the project are (Figure 12):

- An air-blown KRW gasifier with in-bed desulfurization utilizing calcium sorbents.
- 100% hot gas particulate removal system via cyclones and a ceramic carrier filter.
- A polishing external HGCU desulfurization system using zinc-based sorbents and KRW's fluidized/transport reactor technology operating at near 550 °C.
- No sulfur byproduct recovery system; rather, gypsum-quality ash will be recovered.
- Regeneration SO_2 gases from the polishing HGCU are sent to the sulfator where they react with lime and air to form calcium sulfate and are recycled to exit with the gasifier ash.

- A GE 6FA gas turbine will combust the coal gas while the steam turbine generator will produce the balance of electricity for a total of 100 MW (net), 107 MW (gross).
- A portion of the steam produced by the HRSG is used in the gasifier island, gas quenching, and sorbent regeneration.

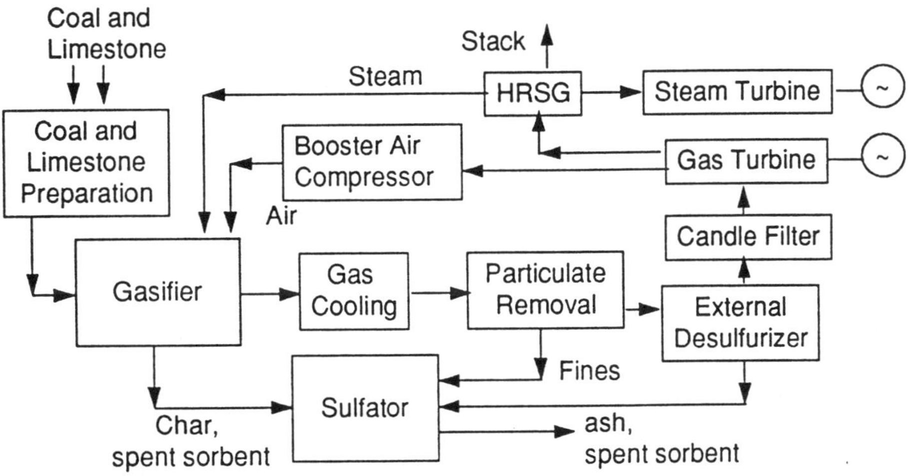

Figure 12. Diagram of Sierra Pacific Co's Piñon Pine IGCC plant

Construction and development of the Piñon Pine project is proceeding as scheduled, with a projected startup by early 1997.

Other projects originally proposed under early rounds of the Clean Coal Technology Program, such as the ABB/Combustion Engineering IGCC Repowering Project and the Tampella/U-Gas Toms Creek IGCC Demonstration Project have been discontinued under the CCT Program due to lack of suitable sites.

6.4 IGCC Projects in Europe

In Europe, several major gasification combined cycle projects are also underway (Table 6.4). All European projects are greenfield applications and use commercial cold gas cleanup technologies and Siemens gas turbines. Reported expected efficiencies range from 43 to 45% (LHV).

The major IGCC project outside of North America, the 235-MW Buggenum station in Haelen, The Netherlands, completed the construction stage in October 1993, and several runs were made at 50% to 75% load. The system utilizes a shell gasifier with CGCU and a Siemens V.94.2 gas turbine. According to a recent publication [30], operation at design conditions and full validation of design parameters is still in progress due to encountered vibration problems in the gas turbine and damaged ceramic filters in the gasification system.

Table 6.4 European IGCC projects in progress

Project	Technology	Efficiency % LHV	MWe gross/net	Application	Start-up
SEP Buggenum, Netherlands	Shell O_2 blown CGCU V.94.2	43	284/253	Greenfield	Early 1994
ELCOGAS Puertollano, Spain	Prenflo O_2 blown CGCU V.94.3	45	335/300	Greenfield	Mid 1996
RWE KOBRA Hurth, Germany (on hold)	HT Winkler Fluid bed air blown CGCU V.94.3	45	357/312	Greenfield	Post 2000

The ELCOGAS project, located in Puertollano, Spain, is a net 300 MWe IGCC plant utilizing an oxygen-blown, Prenflo gasifier followed by a cold gas cleanup system. The power generation block includes a Siemens V.94.3 gas turbine. The overall efficiency of the system is rated at 45% and startup is scheduled for mid 1996.

The third major IGCC project in Europe, the RWE KOBRA, was originally selected for location in Germany, but currently the project is on hold. This project was planned to use a high-temperature, air-blown, fluid-bed HT Winkler gasifier coupled to a low-temperature gas cleanup system and a Siemens V.94.3 turbine. Other European countries with IGCC demonstration projects under consideration are, for instance, Italy, Sweden, and the UK [11].

6.5 IGCC Projects in Asia

Even though the major IGCC demonstration projects are underway in the U.S. and Europe, activity in power generation is also reported in Asia, particularly Japan and China.

China is the largest producer of coal in the world, and consequently, coal is one of its major domestic assets. Early plans by KRW to build a gasifier in Fulargi, China, have been place on hold due to reorganization of the Chinese government agency coordinating the effort [11]. Texaco has announced plans for gasification projects in Lunan (for an ammonia plant) and Shougang (fuel gas production) locations in China [31]. However, there are no short-term plans for construction of a major IGCC demonstration project in China.

In parallel to the efforts in China, Japan is vigorously pursing development of IGCC. The curent state of development for IGCC with hot gas cleanup is at the pilot-plant scale [32]. Since 1991, tests at the Nakoso power plant have been sponsored by the Japanese government (MITI/NEDO) and the Japanese utility power companies (IGC Research Association). The IGCC demonstration-scale plant configuration, code named "Hercules", is scheduled for operation in the year 2000, and be commercial by the year 2010. "Hercules" will consist of an air-blown, two-stage entrained gasifier being developed by Mitsubishi Heavy Industries (MHI), and a dry gas cleanup system.

Currently, the 200 ton/day pilot plant is testing a parallel train of dry gas cleanup: one train with a granular bed filter and a fluidized-bed desulfurization system, and the other train with a candle ceramic filter and a fixed-bed honeycomb desulfurization system. The honeycomb structure uses supported iron oxide as a sorbent and is capable of removing H_2S down to 20-50 ppm for this gasifier coal gas; however, other zinc and iron-based sorbents are being studied in the laboratory since it is expected that a different sorbent will be required for oxygen-blown coal gases. Development of the hot gas cleanup system is being pursued by MHI and Kawasaki Heavy Industries (KHI). Expected gross efficiencies are 46% and 50% (LHV) for the demonstration plant and the commercial plant, respectively.

7 Summary

The IGCC process has proven to be the cleanest of many advanced power generation systems using solid fuels such as coal or heavy oils. Compared to conventional coal-fired systems, IGCC promises significant reductions in environmental emissions at a lower cost of electricity. Sulfur removal levels of 98% or greater, as well as low NO_x emissions levels and solid waste that is non-hazardous, have been achieved. With the advent of today's commercial advanced gas turbines, the IGCC technology can also show economic gains against conventional coal steam plants.

Current demonstration plants using HGCU are targetted to deliver up to 45% (LHV) efficiencies, and future systems to achieve 52% efficiency by the year 2010. An alternative to the full-blown IGCC system has been proposed as the simplified IGCC configuration. In general, the level of emissions from a simplified IGCC system are not as low as those obtained from a full-blown, low-temperature, liquid-based, fuel gas scrubbing system, but the resulting emissions reductions are cost effective. For any case chosen, proper integration of HGCU sorbent operation in the power generation cycle increases the overall efficiency (above that of CGCU systems), by up to 2 percentage points, and consequently, optimization of HGCU must be included in the integration analysis of IGCC and other high-efficiency power generation cycles.

The current three CCT demonstration projects in the US are proceeding on schedule and promise to deliver the expected improvements in efficiency and cost. Projects in Europe and Asia are also moving at a fast pace with results of full-scale operation expected in the next few years. Overall, IGCC technology has been moving very rapidly and it is expected that it will be part of the rise in worldwide power generation in the next century.

8 Acknowledgments

The author acknowledges A. Furman, D. Najewicz, and S.G. Kimura at GE Corporate Research and Development for reviewing the manuscript and encouraging participation in this NATO/ASI symposium. Special thanks to D. Todd, GE Power Generation, for providing sources of information and references on the combined-cycle efficiency, status of commercialization of IGCC technology in power generation, and operation of IGCC components. Also, the help of R. Gupta, Research Triangle Institute, in providing some reference information is also acknowledged. Finally, my appreciation to the staff of the U.S. Dept. of Energy, Morgantown Energy Technology Center/Gasification and Gas Cleanup Division for sponsoring the development and demonstration of GE hot gas cleanup and IGCC technology in the U.S.

References

1 Ayala, R.E. 1996. Assessment of Coal Gasification Processes – Relevance to Sorbent Development. In *Proceedings of the NATO/Advanced Study Institute Conference - Desulfurization of Hot Coal Gas with Regenerable Metal Oxide Sorbents: New Developments*, Kusadasi, Turkey, July 7-19 1996.

2 Ham, D. A. Gelb, G. Lord, and G. Simons. 1984. Hot-Gas Cleanup for Molten Carbonate Fuel Cells-Dechlorination and Soot Formation. Final Report to the U.S. DOE. DOE/MC/16242-1554. NTIS/DE84003072. Springfield, Va.: National Technical Information Service.

3 Electric Power Research Institute. 1988. *Proceedings: Seventh Annual EPRI Contractors' Conference on Coal Gasification*. EPRI AP-6007-SR. Palo Alto, CA: Electric Power Research Institute.

4 Electric Power Research Institute. 1990. *Proceedings: Proceedings: 1989 Conference on Technologies for Producing Electricity in the Twenty-First Century*. Co-sponsored by Electric Power Research Institute (SA), Kerforschungsanlage Juelich (FRG), and New Energy and Industrial Development Organization (Japan). EPRI GS-6691. Palo Alto, CA: Electric Power Research Institute.

5 Corman, J.C. 1986. *System Analysis of Simplified IGCC Plants*. General Electric Company. DOE/ET/14928-2233, NTIS/DE87002508. Springfield, VA: National Technical Information Service.

6 Notestein, J.E. 1989. Update on Department of Energy Hot Gas Cleanup Programs. In *Eight Annual EPRI Conference on Coal Gasification*, Paper No. 14. EPRI GS-6485. Palo Alto CA: Electric Power Research Institute.

7 Pitrolo, A.A., and T.F. Bechtel. 1988. Simplified IGCC: Coal's Adam Smith Response to a Changing World. In *Proceedings: Seventh Annual EPRI Conference on Coal Gasification*. EPRI AP-6007-SR. Palo Alto CA: Electric Power Research Institute.

8 Bajura, R.A. K.R. Craig, K.R. Stone and L.H. Berkshire. 1986. Gasification Island Concept Systems. In *Proceedings: Third U.S.A.-Korea Joint Workshop on Coal Utilization Technology*. Pittsburgh, PA: Korea Institute of Energy and Resources and (U.S. DOE) Pittsburgh Energy Technology Center.

9 Notestein, J.E. 1990. Commercial Gasifier for IGCC Applications. Study Report.

10 Swanekamp, R. 1996. Startup of large-scale projects casts spotlight on IGCC. Power, pp. 59-68. June 1996.

11 Takematsu, T. and Maude, C. 1991. Coal Gasification for IGCC Power Generation. IEACR/37. IEA Coal Research, London, 1991.

12 Schmidt, D. 1994. Integrated Gasification Combined Cycle - A View to the Future. In Proceedings of the *Coal-Fired Power Systems 94 – Advances in IGCC and PFBC Review Meeting*, p. 61. DOE/METC-94/1008. NTIS/DE94012252. Springfield, Va.: National Technical Information Service.

13 Rath, K.L., T.G. Lee, and P.H. Le. 1994. IGCC System Studies. In Proceedings of the *Coal-Fired Power Systems 94 – Advances in IGCC and PFBC Review Meeting*, p. 65. DOE/METC-94/1008. NTIS/DE94012252. Springfield, Va.: National Technical Information Service.

14 Todd, D.M. 1994. Clean Coal and Heavy Oil Technologies for Gas Turbines. GE Power Systems technical report. GER-3650D. General Electric Co., Schenectady, New York.

15 Schilling, H.D. 1990. Environmentally orientated structural changes in the power plant technology of the Federal Republic of Germany. *International Journal of Global Energy Issues*; 2 (3); pp. 138-161.

16 Thambimuthu, K.1993. Gas Cleaning for Advanced Coal-Based Power Generation. IEACR/53. IEA Coal Research, London, 1993.

17 Hozumi, S. 1986. Integrated coal gasification and combined cycle power generation. *Enerugi Shigen:* 7(5): 42-48 (1986).

18 Droog, H.A., A. Delwel, and R. Assink. 1993. Comparison of High-Temperature Gas Treatment Options with Low-Temperature Gas Treatment in IGCC's following Entrained Flow Gasification. In *Proceedings 12th EPRI Conference on Gasification Power Plants.* San Francisco, CA, Oct 27-29, 1993

19 Schreurs, H. 1994. Hot Clean Up of coal Gas. An Overview of Applicability. Netherlands Agency for Energy and the Environment Novem. August 31, 1994. Report No. 94/E0705/229100.

20 McDaniel, M.H., D.J. Mollot, and V.K. Venkataraman. 1995. Editors. In *Proceedings of the Advanced Coal-Fired Power Systems '95 Review Meeting,* Vol I and II. DOE/METC-95/1018. NTIS/DE95009732. Springfield, Va.: National Technical Information Service.

21 Tamhankar, S. S., Hasatani, M. and Wen, C. Y., *Chem. Eng. Sci.* 36, 1181-1191 (1981).

22 Lew, S., Jothimurugesan, K. and Flytzani-Stephanopoulos, M., *Ind. Eng. Chem. Res.* 28, 535 (1989).

23 Gangwal, S. K., Harkins, W. M., Stronger, J. M., Bossart, S. J., "Testing of Novel Sorbents for H_2 Removal From Coal Gas," *Environ. Prog.*, 8, (1) 26 (1989).

24 Ayala, R.E. T.L. Chuck, and R.P. Gupta. 1995. Moving-Bed Sorbents. In *Proceedings of the Advanced Coal-Fired Power Systems '95 Review Meeting,* Vol II, pp 591-600. DOE/METC-95/1018 Vol 2. NTIS/DE95009733. Springfield, Va.: National Technical Information Service.

25 Abbasian, J., Hill, A. H., and Stephanopoulous, M.F. 1993. Development of Novel Copper-Based Sorbent for Hot Gas Desulfurization." Paper presented at the AIChE 1993 Annual Meeting, St. Louis, Mo.

26 Bevan, D.J. Najewicz, R.E. Ayala, A. Feitelberg, and A. Furman. 1995. Integrated Operation of a Pressurized Gasifier, Hot Gas Desulfurization System and Turbine Simulator. In *Proceedings of the Advanced Coal-Fired Power Systems '95 Review Meeting,* Vol I, pp. 187-201. DOE/METC-95/1018. NTIS/DE95009732. Springfield, Va.: National Technical Information Service

27 Bevan, S., D.J. Najewicz, E. Gal, A.H. Furman, R. Ayala, and A. Feitelberg. 1994. Integrated Operation of a Pressurized Gasifier, Hot Gas Desulfurization System and Turbine Simulator. In Proceedings of the Coal-Fired Power Systems 94 – Advances in IGCC and PFBC Review Meeting, p. 222. DOE/METC-94/1008. NTIS/DE94012252. Springfield, Va.: National Technical Information Service.

28 Mann, G.J., P. Amick, J.J. Cook, R. Fisackerly, and R.C. Spears. 1992. Wabash River Coal Gasification Repowering Project. In *Proceedings of the Twelfth Annual Gasification and Gas Stream Cleanup Systems Contractors Review Meeting,* p.34. DOE/METC-92/6128. NTIS/DE93000228. Springfield, Va.: National Technical Information Service.

29 Black, C. 1993. A Utility's Perspective of the Market for IGCC. In Proceedings of the *Coal-Fired Power Systems 93 – Advances in IGCC and PFBC Review Meeting,* p. 8-18. DOE/METC-93/6131. NTIS/DE93000289. Springfield, Va.: National Technical Information Service.

30 Swanekamp, R. 1996. Startup of large-scale projects casts spotlight on IGCC. Power, 59-68, June 1996.

31 Epstein, M. 1990. Commercial Size Gasification Projects. Domestic and Foreign. US DOE Contractors Meeting Proceedings. 1990

32 Kaneko, S., T. Hashimoto, T. Furuya, S. Uchida, A Hashimoto. 1996. Japan tests air-blown IGCC. *Modern Power Systems Review. 1996*

Hot Gas Cleaning - An Overview

Ian R Fantom, Phillip Cahill and Peter W Sage

Coal Technology Development Division, British Coal Corporation, PO Box 199, Stoke Orchard, Cheltenham, Glos, GL52 4ZG, England

Abstract. A number of coal gasification based combined cycle power generation systems are being developed throughout the world. Performance and environmental criteria demand that gas cleaning is employed to reduce contaminant levels in the fuel gas. Conventional techniques are based on wet gas cleaning systems which can have high capital costs, an adverse effect on cycle efficiency and produce a liquid effluent that is expensive to treat. Dry gas cleaning at high temperatures using sorbents and catalysts offers a lower cost alternative. This paper reviews briefly some of the techniques and issues relating to the development of HGCU systems and provides examples of recent results primarily from test work at the CTDD of British Coal Corporation UK .

1 Introduction

The Coal Technology Development Division of British Coal (CTDD) is developing components of a coal gasification based combined cycle power generation system known as the Air Blown Gasification Cycle (ABGC) [1,2]. The ABGC is being developed jointly by an industry led group comprising GEC Alsthom, PowerGen, Mitsui Babcock Energy and British Coal Corporation, towards a demonstration of the technology. The cycle offers significant increase in power generation efficiency compared to conventional coal fired systems with the added advantage that this can be achieved at reduced capital cost. A number of work programmes are being undertaken at CTDD to develop the component parts of the cycle ie, high pressure gasifier, char combustor, gas cleaning and gas combustion systems. The information from the various work programmes will be used to design a Prototype Integrated Plant (PIP).

For such advanced power generation systems there is the need to remove dust, sulphides, ammonia, cyanides, halides and trace metals from the fuel gas. Existing technologies to remove these pollutants are based on wet gas cleaning systems which

NATO ASI Series, Vol. G 42
Desulfurization of Hot Coal Gas
Edited by A. T. Atimtay and D. P. Harrison

have an adverse effect on cycle efficiency, can have high capital cost and can produce an aqueous effluent requiring expensive treatment before disposal. The development and introduction of hot gas cleaning systems offers the potential of a lower cost approach to pollutant control with associated cycle efficiency advantages. The hot gas systems which will replace the conventional wet gas technologies will need to operate at 300-600°C and at pressures of 10-25 bar_a. In order to realise fully the cost and environmental advantages, it is essential that the systems developed give not only efficient removal of the respective pollutant but also have the reliability and availability to match the respective wet gas system. This paper gives an overview of techniques for the removal of fuel gas contaminants and describes briefly results from CTDD's Hot Gas Cleaning research programme.

2 Hot Gas Clean-up

Hot gas cleaning (HGCU) including particulate removal, sulphur, chlorine and ammonia reduction can improve plant efficiency and simplify cycle configuration when compared to conventional wet gas cleaning techniques. Hot gas cleaning is normally defined as being carried out above the water condensation temperature, although higher temperatures are usually used so that other components such as ammonium chloride and tars do not condense and cause fouling problems. The upper temperature is limited to around 600°C because of limitations on materials for construction of components in the hot gas cleaning system. Below 600°C corrosive alkali metals will condense and be collected on the filter fines. (See section 7)

Techno-economic studies to evaluate the benefits of hot gas cleaning have been carried out for many Integrated Combined Cycle (IGCC) processes and hybrids such as the ABGC. For the ABGC a number of cases namely 400°C and 600°C gas filtration, and cold gas cleaning using wet scrubbing have been evaluated. In this study the HGCU system comprised of a cooler, to raise high pressure steam, and filter system. The cold gas cleaning system comprised a cooler, wet scrubber and reheat system. The study showed that HGCU results in a simpler cycle configuration; gives an improvement in overall plant efficiency of 1 % and produced a capital cost saving of 6 %. There was little difference in cost or efficiency between the 400°C and 600°C cases. It should be noted that wet gas cleaning systems will remove some chloride and heavy metals which are not removed by filtration alone. The overall cost of electricity generation is very sensitive to plant availability and therefore it is essential that the HGCU components are proven and reliable. The advantages from HGCU should apply to all gasification combined processes.

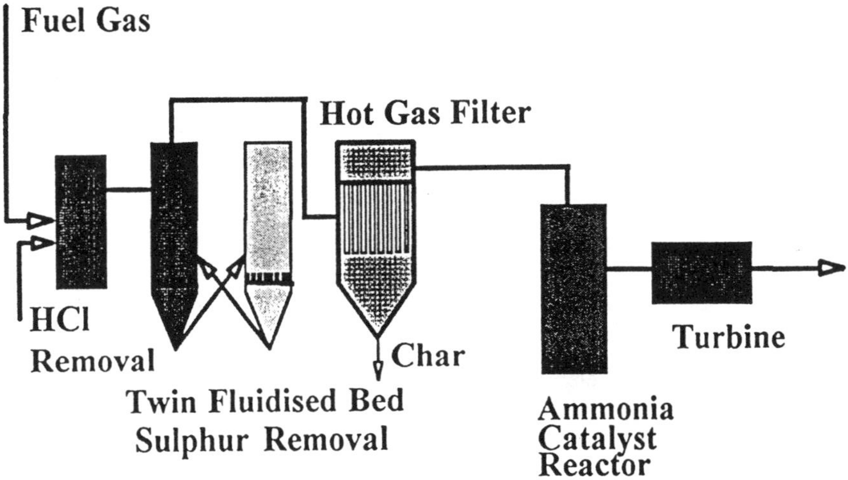

Fig.1 Hot gas cleaning system for the ABGC

At CTDD development of the individual components of a conceptual hot gas cleaning system is being undertaken. Fig.1 shows a schematic of that HGCU process. In the longer term there will be economic and efficiency benefits from combining the individual processes eg having a bed to remove sulphur and ammonia simultaneously.

3 Particulate Removal

Coarse removal of particles elutriated from gasifiers is undertaken using cyclones. However, for high performance gas turbines almost complete removal of dust is required using barrier filters. For HGCU the main focus has been on ceramic candle filters. Considerable interest has been shown for application to Pressurised Fluidised Bed Combustion (PFBC) where there is a desire to use less expensive gas turbines rather than the ruggedised turbines currently being used. The filtration temperature for PFBC application is around 800-900°C where the ash may become sticky. A number of demonstration filters on PFBC plants have been built and operated including Grimethorpe [3], Karhula [4], Tidd [5] and Wakamatsu [6] and some element breakages and candle cleaning problems have been observed.

For IGCC type systems, however, the operating temperature is lower, 250°C - 700°C and the dust is fly slag or carbonaceous (for partial gasifiers). There has been significant success with filters operating under these conditions in work performed by Shell[7] and Rheinbraun where long periods of operation have been achieved with stable pressure drops and few problems.

A programme of work was formulated and is underway to develop a hot gas filter system for the ABGC over the temperature range 400°C to 600°C [8]. The specific objectives are given below but apply to any filter system being used for IGCC or related technologies. The work programme was divided into two areas:-

a. To investigate critical components of the filter including:-

- Life of the filter elements with respect to thermal shock and permeance
- Integrity of hold down devices used to secure and seal the elements in the tube sheet.
- Performance of materials used in the critical components of the filter
- Performance of the fast opening valves and the use of hot pulse gas to clean the elements
- Design of the pulse system using hot gas.

b. To investigate the operating and control conditions for the filter including:-

- Start up philosophy.
- Cleaning strategy to giver good dust removal, minimum damage to elements and minimum use of pulse gas
- Effect of filtration temperature and pressure
- Effect of gasifier coal and sorbent feedstock
- Effect of changing filtration velocity
- Effect of plant load changes on operation.

To date over 4000 hours of operation has been achieved on a high pressure test rig and an atmospheric pressure Filter Development Rig(FDR). A schematic of the FDR is shown as Fig.2. The tests, mainly at 600°C, include continuous operation over periods of 250 - 300 hours with no incidents of dust leakage or candle breakage. Operational pressure drops of 100-150 mbar were achieved on the pressurised filter system operating at 12 bar. Dust loadings as high as 100,000 mgm_n^{-3} have been experienced and stable operation achieved despite this. Typically the dust contains over 60 % carbon and is in the size range 5-8 microns. The durability of the clay bound ceramic elements has been tested in a thermal cycling rig. Using hot gases at pressure accelerated tests to simulate extended operation by quick succession pulsing were undertaken. The tests show that even for long duration pulses strength losses due to thermal shock are acceptable.

The results are currently being used to design, in association with an industrial engineering company, a commercial scale filtration system.

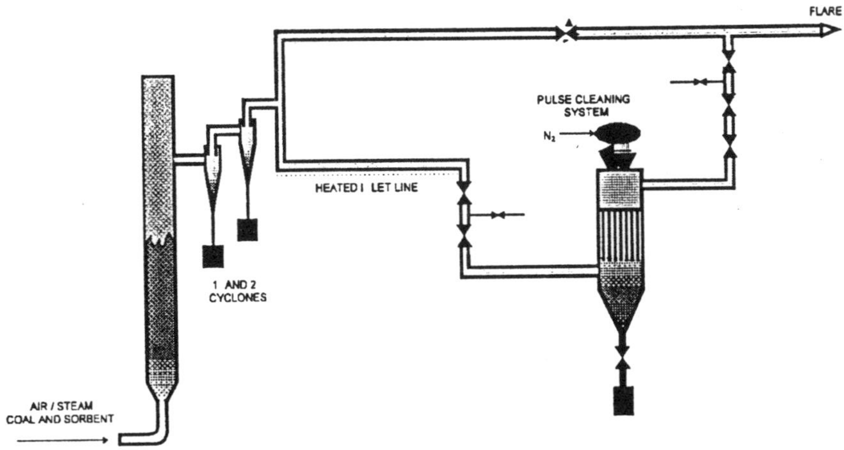

Fig.2 CTDD filter development rig

4 Desulphurisation

Current wet gas cleaning techniques comprise a string of processes including carbonyl sulphide hydrolysis, amine scrubbing, hydrogen sulphide stripping, followed by Clause and Scott type processes. These items may typically be as large as the rest of the IGCC plant and represent a significant part of the capital and operating cost.

An alternative is the use of regenerable metal oxides. This subject forms the purpose of this conference. The issues relating to sorbent development and testing are discussed in detail in many paper in these proceedings. Within this paper the key issues of sorbent choice, sorbent durability, contacting method, regeneration gas and sulphur recovery system are outlined from an industrial point of view.

4.1 Sorbent and Contacting Method

The choice of primary metal oxide for the sorbent depends on the temperature of interest and the degree of sulphur removal required. To date formulations based on calcium, zinc, iron, copper, molybdenum, nickel, manganese and tin have been reported in the literature for use on coal gases in the temperature range 250°C to 900°C. Additives are often incorporated to stabilise the primary metal oxide and to suppress side-reactions. The current developmental status of many of these sorbents is discussed elsewhere in these proceedings and also at recent Pittsburgh Coal Conferences [9,10]. To date only zinc and iron based materials have been tested in real coal gas systems. [11,12]. In real coal gas systems not only must the sulphidation reaction be monitored but also side reactions. Methanation and water gas shift reactions can be catalysed by additives to the sorbents resulting in excess bed temperatures and sorbent sintering, carbon deposition may plug the pores and reaction with hydrogen chloride could produce volatile metal chlorides.

Moving bed, fluidised bed and transport reactor systems are being investigated and suitable sorbent sizes and formulations manufactured for each. Each process has advantages and disadvantages over the others and as yet there is no clear indication as to which is likely to have the greatest long term success. Large scale demonstration projects are currently on going at Pinon Pine, USA (transport reactor) and Polk County, USA (moving bed).

For economic viability of the process the sorbents must retain their activity as they pass round and round the system from sulphider to regenerator. The sorbents must also retain mechanical stability with minimal attrition losses. Initial calculations on a polishing system for the ABGC based on a twin fluidised bed reactor suggest that if the sorbent performs for 1000 cycles the additional cost of the process on electricity production will be minimal. The calculations assumed that the sorbent picked up around 4 % wt sulphur from the sulphider in each cycle.

The performance of sorbents is often compared on the basis of reduction in capacity over a number of cycles. However, caution should be exercised. For example in a moving bed a cycle may last several hours with the sorbent picking up 10 % wt S in each cycle, but in transport reactor the sorbent may collect only a fraction of a percent wt sulphur and a cycle will last only 5 -10 seconds.

4.2 Regeneration Gas, Sulphur Recovery and Disposal

The sorption process can not be treated in isolation as the regeneration is the most critical step of the whole process. The temperature has to be maintained within limits to ensure good regeneration, to prevent sulphate formation and sintering all of which limit the sorption ability of the sorbent.

The regeneration gas may be air, air\steam (or other inert), flue gas or steam depending on the process chosen and sorbents used. The gasification process must be able to produce the regeneration gas eg using air in nitrogen would not be an option for an air blown system where there is no air separation plant. However, high pressure air is available from the gas turbine compressor and steam from the steam cycle. If air alone is used then heat extraction may be needed from the regeneration bed to moderate the temperature.

The "light off" temperature may also be significant. The sorbents require a minimum temperature to sustain regeneration. Typically, this is around 550°C. Developments are underway to lower the "light off" temperature. This has two advantages, firstly little additional heating of the regeneration gas is required saving on efficiency and secondly higher air concentrations can be used as a higher temperature rise is required in the regeneration process.

To date most of the sorption processes produce concentrated SO_2 streams in regeneration. Simple low cost systems are required to collect the sulphur. Options include direct sulphur recovery by reduction with fuel gas over a catalyst to produce elemental sulphur, production of sulphuric acid, flue gas desulphurisation or in the case of the ABGC collection in the CFBC by addition of extra calcium based sorbent. The optimal process in terms of cost and efficiency is cycle specific and may be affected by local issues such as the market for sulphur or sulphuric acid.

4.3 Other Approaches

As well as the use of regenerable metal oxides other hot gas cleaning techniques for the removal of hydrogen sulphide are being studies. The use of high temperature ceramic membranes is being considered to selectively remove hydrogen sulphide. At Rheinbraun [13] a process whereby air is added to the fuel gas immediately prior to the filter is being developed. The air reduces the hydrogen sulphide to sulphur in a catalytic process involving the lignite char. The sulphur is then taken out in the filter on the char.

5 NOx Reduction

NOx emitted from gasification based combined cycles is produced in the gas turbine by three mechanisms. Firstly from oxidation of molecular nitrogen (thermal-NOx) in the oxygen rich zone of the flame. This is normally only a problem where high CV gases and premium fuels are used. For low CV fuels with peak burning temperatures of $< 1600°C$ thermal NOx is not an issue. Secondly, is prompt-NOx formed from atmospheric nitrogen through intermediates as results of the hydrocarbons in the fuel. Finally fuel-NOx formed from bound nitrogen components such as ammonia and hydrogen cyanide.

NOx formation in the gas turbine can be minimised by combustor design. Where thermal NOx is the dominant mechanism low NOx burners are used to reduce the flame temperature. For systems where fuel-NOx is the primary mechanism of NOx formation low NOx burners are used to create a rich-lean air staging in order to promote conversion of fuel nitrogen to molecular nitrogen [14]. However, use of these systems is unlikely to be sufficient to meet stringent NOx limits. In order to reduce emissions levels further, removal of fuel nitrogen compounds prior to combustion is required. Where wet gas cleaning techniques are used the ammonia is largely removed by dissolution in the aqueous scrubber (the ammonia then has to be stripped and treated separately). For hot gas cleaning two potentially much simpler processes have been identified.

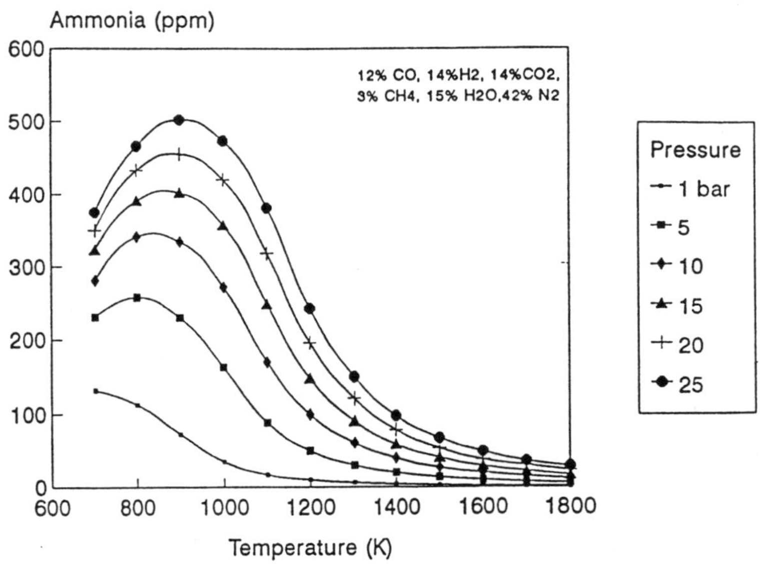

Fig.3 Equilibrium ammonia levels for the ABGC

5.1 Decomposition Catalysis

Thermodynamic equilibrium calculations show that for coal gasification processes the ammonia levels in the fuel gas are often above the equilibrium value. Fig.3 illustrates this for the ABGC. In principle a catalyst should be able to decompose the ammonia by the following reaction:-

$$2\,NH_3 \rightarrow N_2 + 3\,H_2$$

A number of catalysts formulated from nickel and iron were tested by Gavin [15] in a laboratory scale reactor. Good activity was observed over the temperature range 600-800°C in clean gas. However, when the nickel based catalysts were tested in real gas at around 600°C no activity was measured [16]. The activity of these materials appears to be reduced at these temperatures due to sulphur and chlorine poisoning and carbon deposition. Similar observations were made by Mojtahedi who tested nickel and ruthenium catalysts on real fuel gases [17]. At higher temperatures say > 900°C these problems could be reduced as nickel sulphide formation and carbon deposition are less favourable [18].

5.2 Selective Oxidation

An alternative approach is to oxidise the ammonia to harmless products using a variety of oxidants by a reaction such as:-

$$2\,NH_3 + 1.5\,O_2 \rightarrow N_2 + 3\,H_2O$$

The oxidants could be air or NO derived from the flue gases or compressor air. To be of any benefit the reaction has to be highly selective. At KEMA a molybdenum oxide on silica sorbent was tested and gave 90% reduction when 2 % oxygen was added at 450°C[19].

In Finland ammonia gasification of biomass leads to ammonia levels in excess of 4000 vpm [20]. At VTT nitrogen oxide and air were added to fuel gas over an aluminum oxide bed in laboratory scale texts. At temperatures below 450°C over 80 % reduction of ammonia was reported.

At CTDD equimolar amounts of oxygen and nitrogen oxide were added to a fuel gas containing 2000 vpm ammonia. NO reduction was observed between 500°C and 800 °C . However, at 450 °C and at 900°C up to 90% reduction was measured.

The costs of these approaches must be balanced carefully against the cost of installing Selective Catalytic Reduction (SCR) technology. SCR is now commonly used in power stations to remove NOx from flue gases. In APG systems catalyst beds could easily be installed in the Heat Recovery Steam Generator (HRSG) system at the outlet of the gas turbine. The price of this technology has decreased markedly in recent years.

6 Hydrogen Chloride Removal

In almost all coal gasification processes the chlorine present in the coal is released into the vapour phase as hydrogen chloride gas. Coal chlorine contents vary widely (100 - 200 ppm wt) although UK coals tend to have a higher chlorine content (up to 8700 ppm wt) than typical US, Australian and European coals. There is little legislation governing the emissions of chlorine from fossil fuel plants. However, tighter legislation is expected and already low emissions levels have been set for plant fired on waste fuels. In addition to potential environmental problems hydrogen chloride removal could be required to protect sulphur and ammonia sorbents, to reduce corrosion of hot gas path components and to suppress the volatility of toxic and corrosive trace elements. For hot gas cleaning applications in the range 300 - 600°C calcium and sodium based sorbents have been identified as thermodynamically favourable . Typical reactions are shown below.

$$CaCO_3 + 2\ HCl \rightarrow CaCl_2 + H_2O + CO_2$$
$$CaO + 2\ HCl \rightarrow CaCl_2 + H_2O$$
$$Ca(OH)_2 + 2\ HCl \rightarrow CaCl_2 + 2\ H_2O$$
$$Na_2CO_3 + 2\ HCl \rightarrow 2\ NaCl + CO_2 + H_20$$
$$NaHCO_3 + HCl \rightarrow NaCl + CO_2 + H_20$$

As with other contaminants the sorbents could be contacted with the fuel gas using fixed, fluidised, or transport reactors. At CTDD an approach of injecting the sorbent prior to the filter has been adopted [21]. Calcium and sodium based materials have been entrained in the fuel gases on an atmospheric gasifier prior to a hot gas filter. Hydrogen chloride reduction of up to 80 % was measured. Krishnan [22] used a laboratory scale fixed bed reactor to test various sodium based sorbents including Nahcolite, Dawsonite, Shortite and Katalco. The sorbents were able to reduce HCl levels from 300 vpm down to less than 1 vpm.

This technology remains to be tested at a significant scale on real coal gasification plant. Of particular concern is the possible formation of sticky calcium or sodium chlorides. If formed these could blind the filter or deposit in the duct work causing a blockage.

7 Trace Metals

Trace metals present in the coal can be volatilised during the gasification process and be transported into the hot fuel gas path and subsequently into the gas turbine, waste heat boiler and to atmosphere. Elements passing the filter such as sodium, potassium, and zinc may present a corrosion hazard to the gas turbine. Other toxic metals such as selenium, arsenic, and mercury could be considered a problem if emitted to atmosphere.

Thermodynamic calculations have been performed to assess the potential problems of various trace metals [23]. The calculations were undertaken on a thermodynamic equilibrium package, MTDATA, supplied by the National Physical Laboratory, UK. Whilst the fuel gas composition used was based on the ABGC gas the conclusions are relevant to other IGCC systems. In general 4 categories of elements were identified:-

- Trace metal compounds remaining condensed at gasifier operating conditions, 1000°C, 25 bara. These include chromium, vanadium, nickel and calcium.
- Elements that initially vaporise but condense on cooling to 600°C or have a residual vapour pressure of around 1 ppb wt or less. These include sodium and zinc. Fig.4 illustrates the fate of zinc vapours on cooling from 1000°C to 400°C in an ABGC coal gas environment.
- Elements that have appreciable vapour pressures at 600°C but are wholly condensed or have negligible vapour pressure at 400°C. Lead and tin are amongst elements in the group.
- Highly volatile trace metals that have large vapour pressures even at 400°C. The highly toxic heavy metals arsenic, cadmium, mercury and selenium are in this group.

Piguard and Helble [24] undertook a similar study and concluded that As, Se, Sb, Zn and Pb were the trace elements most likely to be emitted from coal gasification processes.

In recent years processes have been developed to collect volatile elements from the hot fuel gases to protect gas turbines. Trace metals such as sodium, potassium, lead and zinc can deposit liquid sulphates on turbine blades causing accelerated corrosion. Bachovchin [25] used emathlite, Laughlin [26] used Fullers Earth and Lee [27] used activated bauxite to remove sodium and potassium from both fuel gases and combustion gases at temperatures of up to 900°C. In each case the sorbent was able to reduce total alkali levels to below those required by the gas turbine manufacturers (24 ppb wt in the turbine inlet gas). Work is currently being performed to apply these systems for the removal of air toxins such as lead [28].

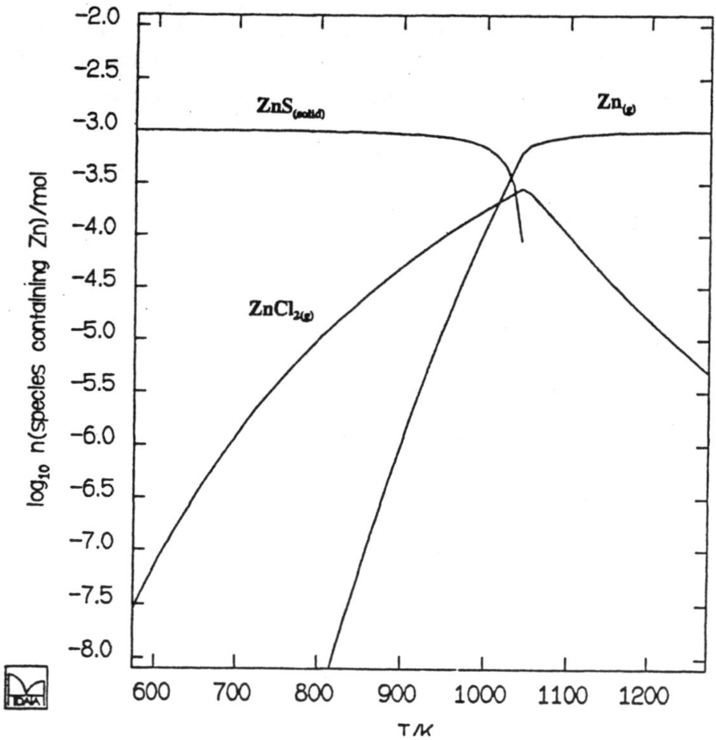

Fig.4 Speciation of zinc compounds

8 Conclusions

- Hot Gas clean-up for IGCC systems offers significant cost and efficiency benefits compared to conventional wet cleaning techniques.

- Hot gas filters have already been operated on demonstration size IGCC power plants. Within the range 250°C to 600°C few operational problems have been experienced and the technology can be considered commercially available.

- Techniques for the removal of hydrogen sulphide have been developed and are now being demonstrated on commercial scale gasification plants. Removal systems for ammonia and hydrogen chloride still require substantial laboratory scale development before they can be applied on a demonstration plant.

References

1 Sage P W , Development of the Air Blown Gasifier, Paper to ETSU Coal Contractors Meeting, Nottingham, UK, November 1994.

2 Sage P, Mills S, Overview of Clean Coal Technologies and Current Status of the Air Blown Gasification Cycle, Desulphurisation of Hot Coal Gas with Regenerable Metal Oxide Sorbents : New Developments, NATO ASI, Kusadasi, July 1996.

3 Burnard et al., Operation and performance of the EPRI Hot Gas Filter at Grimethorpe PFBC Establishment 1987-1992, 2nd Int Symp Hot Gas Cleaning, Surrey 1993.

4 Isaksson J, ACTF Filter Experiences , 2nd EPRI Workshop on Filtration of Dust from Coal Derived Reducing and Combustion Gases at High Temperatures, San Francisco,USA, 1992.

5 Lippert T, Field test results obtained on the integrated operation of ceramic barrier filters in PFBC testing facilities, International Joint Power Generation Conference, Kansas City, USA, 1993.

6 Fujita M, Start-up of the Wakamatsu PFBC Plant, EC/JAPAN Workshop, Tokyo, 1994.

7 Phillips J N, Dries H W A, Filtration of Fly Slag from the Shell Gasification Process using Porous Ceramic Candles, 2nd Int. Symp. Hot Gas Cleaning, Surrey 1993.

8 Cahill P, Dutton M, Tustin M, Rasmussen G, Development of Hot Gas Filtration for Air Blown Gasification Plant, 12th Annual International Pittsburgh Coal Conference, Sept 1995.

9 Pittsburgh Coal Conference, various papers, 1995.

10 Pittsburgh Coal Conference, various papers, 1996.

11 Fantom I, Radford S, Performance of Regenerable Sulphur Sorbents on Coal Derived Fuel Gas, Paper to be presented at Gas Cleaning at High Temperatures Conference, Karlsruhe, Sept 1996.

12 Kobyashi S, Fundamental Studies and Present Status of IHI Hot Gas Desulphurisation Process, International Conference on Coal Science, Elsevier, 1987.

13 Meyer B, Walter H, Schiffer H-P, Adlhoch W, Sulphur Removal from coal Based Fuel Gas, Paper to be presented at Gas Cleaning at High Temperatures Conference, Karlsruhe, Sept 1996.

14 Kelsall G J, Cannon M F, Combustion of Low Heating Value Gas in a Turbine, Power Production from biomass II, Espoo, Finland, March 1995.

15 Gavin D G et al., Minimisation of Nitrogenous Pollutants during Gasification and Combustion of Coal, ECSC Project No 7220-EC\010, Final Report, April 1995.

16 Fantom I, Removal of Ammonia in Coal Gasification Systems, UKDTI Contract No.C\00089, August 1996.

17 Mojtahedi W et al., Catalytic Decomposition of Ammonia in Fuel Gas Produced in Pilot Scale Pressurised Fluidised Bed Gasifier, Fuel Processing Technology 45, 1995.

18 Krishnan G N et al., Study of Ammonia in Fuel Gas in Coal Gasification Processes, Final Report to USDOE/METC, Contract No.DE-AC21-86MC23087, 1988.

19 Bolt N, Meijer R, et al, Haalbaarheidsbeoordeling Hoge Temperatur Gasreiniging KV-STEG, KEMA, 1993.

20 Leppalahti J, Koljonen T, Hupa, H, Kilpinen P, Application of Selective Catalytic Oxidation (SCO) to NH_3 Removal in Gasification, Paper to be presented at Gas Cleaning at High Temperatures Conference, Karlsruhe, Sept 1996.

21 Duo W, et al., Kinetics of HCl Reactions with Calcium and Sodium Based Sorbents for IGCC Fuel Gas Cleaning, Chem.Eng.Sci., 51(11), p2541, 1995.

22 Krishan G N et al., Removal of Hydrogen Chloride from High Temperature Coal Gases, Journal of American Chemical Society, 33, 1, 1988.

23 Reed G, Fantom I, Cahill P, Brain S, Control of Trace Elements in Gasification - A Measurement Methodology to Validate Thermodynamic Predictions, To be published in Fuel 1996.

24 Pigeard A, Helble J J, Trace Metal Emissions for IGFC, Coal Fired Power Systems 94 - Advances in IGCC and PFBC Meeting, DOE/METC-94/1008, 1994.

25 Bachovchin D, A Study of High Temperature Removal of Alkali from Pressurised Gasification Systems, Final Report, Westinghouse R&D Centre, Chemical Sciences Division, Pittsburgh, 1986.

26 Laughlin J, The Removal of Volatile Alkali Salt Vapours from Hot Coal Derived Gases, PhD Thesis, University of Surrey, 1990.

27 Lee S, Measurement of Sodium and Potassium in PFBC of Beulah Lignite, J.Inst.Energy, 53(457), 1990.

28 Shadman F et al., Environmental Science and Technology, Vol.29 No.6, 1995.

Survey of Desulfurization Processes for Coal Gas

A.G.J. van der Ham, R.H. Venderbosch, W. Prins and W.P.M. van Swaaij

Department of Chemical Technology, Twente University of Technology, P.O. Box 217, 7500 AE Enschede, The Netherlands

Abstract. An overview is given of recent developments in the field of regenerative, high temperature, coal gas desulfurization. The results of a wide variety of sorbents and reactor types either tested on lab-, bench- as well as on pilot-plant scale are presented. The sorbents discussed are mainly oxides of transition metals either pure, mixed or deposited on an inert carrier. Regeneration performance of the sorbent and the regeneration off-gas composition are also taken into account.

1 Introduction

In the last decades the importance of coal as an energy carrier is, compared to gas or oil, increasing due to the fact that the proven coal stocks significantly exceed the proven gas or oil stocks. Coal is usually burned in (circulating) fluidized bed combustors using the heat produced to generate steam for power production in a steam turbine. The large amount of flue gas which contains SO_2, NO_x etc., is cleaned at the end of the process just before entering the stack. Due to the low concentration levels and the high removal degree demanded, the clean-up of this gas is rather expensive.

During last decade IGCC technology is receiving more and more attention because of its promise of higher thermal efficiency and reduced environmental impact compared to coal combustion plants. Current IGCC plants operate with an efficiency of about 43%, compared to 35% for a conventional coal plant [1]. With improved gas turbines and high temperature gas clean-up technology, efficiencies exceeding 50% may be achieved. A system study conducted by NOVEM [2] has shown that the maximum benefit, i.e. a rise of 2 to 3% in power production efficiency, will be gained when the low-temperature wet-scrubbing of coal gas is replaced by a dry gas cleaning process operated at 350 °C minimum. Higher temperatures will result in a slight rise in overall efficiency but will also increase the mechanical problems and cost of the material of construction for the gas cleaning units. Corman [3] already reported an optimal temperature range of 340-540°C based on a reasonable compromise between cycle efficiency, capital cost and operating cost.

Compared to flue gas, the type of contaminants in coal gas are different (e.g. SO_2 vs. H_2S etc.) while their concentrations are much higher. For IGCC the ability for sulfur removal is higher than 99%, compared to 90% for conventional coal plants using scrubbers [1]. Unfortunately, these high temperature coal gas desulfurization processes are still investigated on lab-scale, pilot-scale and demonstration scale but not yet for full scale applications.

NATO ASI Series, Vol. G 42
Desulfurization of Hot Coal Gas
Edited by A. T. Atimtay and D. P. Harrison

2 Sorbent Screening

Sulfur removal is one of the main process steps of the dry gas-cleaning system and has already been investigated extensively. Literature reviews are given by Furimsky and Yumura [4], Gäng and Löffler [5], Heesink [6] and Jansen [7]. At present a wide variety of sorbents and reactor types are being developed and tested on a lab-, bench- as well as on pilot-plant/demonstration scale. Sorbents investigated are mainly regenerable oxides of transition metals either pure, mixed or deposited on an inert carrier material to increase reactivity and/or mechanical stability.

The regenerative process combines an absorption step, in which the metal oxide reacts with H_2S (and possibly COS) producing the metal sulfide. In a succeeding regeneration step the oxide is regained next to a sulfur component (either H_2S, SO_2 or elemental sulfur). The concentration level at which these components become available is of crucial importance, because it should be high enough to ensure further economical processing. The important prevailing sulfidation and regeneration reactions can schematically be represented as:

Sulfidation: $Me_xO_y + x\,H_2S + (y\text{-}x)\,H_2 \text{ ---> } x\,MeS + y\,H_2O$

$Me_xO_y + x\,COS + (y\text{-}x)\,CO \text{ ---> } x\,MeS + y\,CO_2$

Regeneration: $x\,MeS + (x+y/2)\,O_2 \text{ ---> } Me_xO_y + x\,SO_2$ or

$MeS + H_2O \text{ ---> } MeO + H_2S$ or

$x\,MeS + (y/2)\,SO_2 \text{ ---> } Me_xO_y + (y/2)\,S$

Success or failure of the proposed regenerative process will strongly depend on the following sorbent parameters:

- chemical: stable, sufficient fast reaction rate for sulfidation and regeneration
  sulfur removal efficiency (thermodynamic and kinetic restriction)
  no unwanted side reactions like reduction to pure metal, metal evaporation, sulfate formation etc.
  sulfur capacity (wt% S),
  regenerator offgas composition
- thermal: sintering,
  loss of porosity and contact area
  deterioration
- mechanical: attrition
  cracking

Thermodynamic calculations are often used as a first indication of the feasibility of a suggested regenerative process. However, these results should be interpreted carefully since they do not include the reaction time needed to approach equilibrium. In 1976 Westmoreland and Harrison [8] published the results obtained from a study concerned with thermodynamic screening of the high temperature desulfurization potential of 28 elements, primarily present as metal oxides.

A typical low-BTU coal gas composition (see Table 1) was considered in a temperature range from 360-1400 °C. Finally, eleven candidate solids based on the elements Fe, Zn, Mo, Mn, V, Ca, Sr, Ba, Co, Cu and W were selected for their thermodynamic feasibility, defined as a H_2S removal efficiency of more than 95% and the existence of thermal stable components (i.e. no melting). Fig. 1 shows the temperature range required for the existence of a stable solid phase. In the temperature range 350-450 °C, the metal oxides of the following eight elements can, in principle, be used: Co, Cu, Fe, Mn, Mo, V, W and Zn. From this group of sorbents most attention has been paid to the materials containing Fe, Mn and Zn.

A recent thermodynamic study similar to the one of Westmoreland and Harrison, on metal oxides selection for desulfurization of high-BTU coal gas (the composition is shown in the third column of Table 1) in the temperature range 350-450°C and a sulfur removal efficiency of >99% using the flowsheeting program ASPEN Plus with solids data derived by Barin & Knacke [9], selected the same 8 elements. Furthermore, the study also showed that tin oxide is a potentially suitable sorbent.

Different forms of metal oxides have been tested, e.g. single (pure) oxides, combination of oxides, oxides combined with solid supports and natural sorbents with a high content of these oxides. Each of them will be discussed separately. Additionally, basic information on the sorbent is presented in Table 2 together with the operation conditions applied and major remarks.

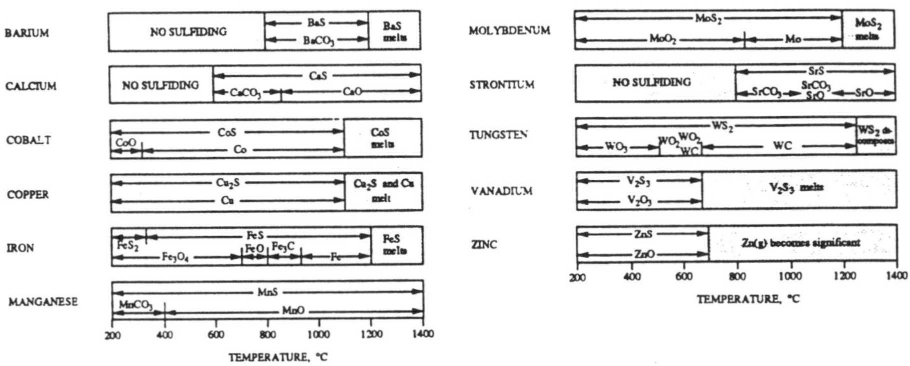

Fig. 1. Sorbents selected by Westmoreland and Harrison [8] for high temperature desulfurization as a function of their operational temperature window. Shaded area shows non applicable temperature range.

Table 1. Composition of the low BTU coal gas considered by Westmoreland and Harrison [8] and a high BTU coal gas obtained from the Shell gasification process [2].

Gasification process	Low BTU	Shell	Shell 1)
Pressure (bar)	20	30	30
Temperature (°C)	360-1560	1700	400
composition (mol%):			
CO	8 - 33	64	63
H_2	5 - 37	30	33
CO_2	1 - 20	2	4
H_2O	4 - 7	2	0.2
$N_2 + Ar$	35 - 44	1	1
H_2S	0.7-1.0	0.1-0.4	-
COS	0.02-0.08	0.01-0.02	-
NH_3	0-0.04	0.03	-
HCl/HF	-	0.02-0.08	-

1) composition of gas according to shift equilibrium at 400 °C.

3 Iron Based Sorbents

In this case one of the stable forms of iron oxide i.e. Fe_2O_3, Fe_3O_4, FeO, or elemental Fe reacts with H_2S to produce FeS according to the following reactions:

$$Fe_xO_y + x\,H_2S + y\text{-}x\,H_2 \dashrightarrow x\,FeS + y\,H_2O$$

or $$Fe_xO_y + x\,H_2S + y\,CO \dashrightarrow x\,FeS + y\,CO_2$$

or $$Fe + H_2S \dashrightarrow FeS + H_2$$

The oxidic form present depends on thermodynamic factors (e.g. temperature, pressure and coal gas composition) but also on the kinetics of the successive reduction reactions, going from Fe_2O_3 to FeO and finally Fe. The equilibrium concentration of H_2S strongly depends on the H_2, H_2O and CO_2 contents of the coal gas.

3.1 Iron Oxide Without Support

Iron oxide is probably one of the oldest sorbents used for H_2S removal. In the beginning of this century it was already applied as 'Dry Boxes' in the gas factories to desulfurize the 'town-gas'. In the seventies it was extensively investigated in the USA. Either pure iron oxide, or mixtures of iron oxide with fly ash [10] or silica [11, 12, 13, 14, 15], or gasifier ash with large amounts of iron [16] were investigated. These materials, however, were shown to be thermally instable, resulting in attrition losses, decrepitation of the sorbent and a decreased lifetime. In the eighties it became clear that a more efficient, reactive, thermally stable sorbent was required which could reduce the sulfur emission levels even further.

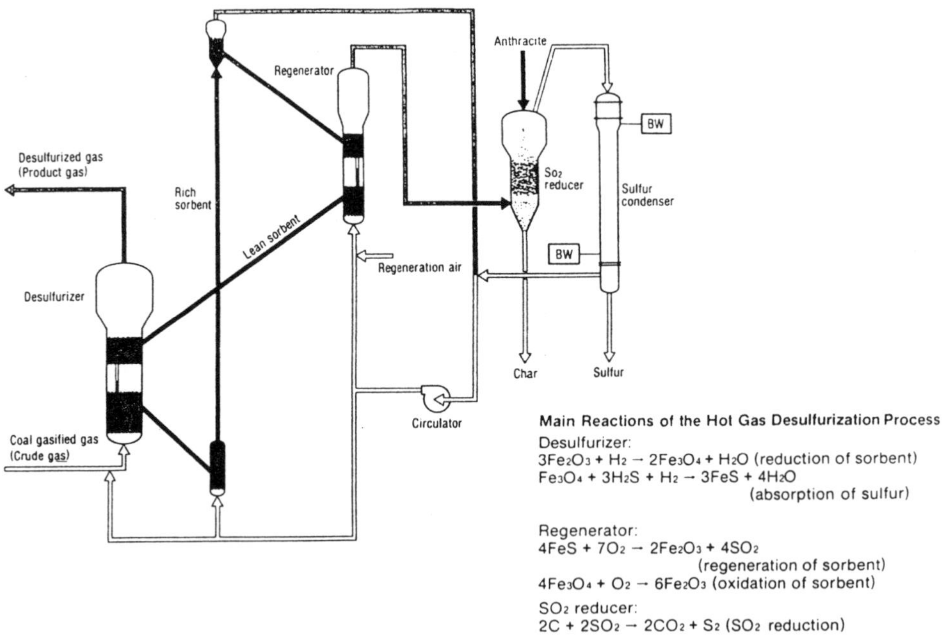

Fig. 2. Schematic representation of the desulfurization process developed by IHI and NEDO (copy from a brochure distributed by IHI).

Nevertheless, the unsupported iron oxide is the sorbent applied in the recent developments on a pilot and demonstration scale in Japan. Ishikawayama-Harima Heavy Industries (IHI) together with NEDO developed a process on basis of crushed iron ore (see Fig. 2).

The crude coal gas enters a two stage bubbling fluidized bed desulfurizer at the bottom where H_2S reacts with the Fe_3O_4 in the sorbent to produce FeS. In the upper fluid bed the coal gas is further desulfurized by contacting with lean sorbent, transferred from the regenerator but also reduces Fe_2O_3 to Fe_3O_4. The partially converted sorbent is drawn from the lower bed and lifted vertically by transport gas to the regenerator, also a bubbling fluidized bed. Here, FeS is oxidized to Fe_2O_3 by air diluted with circulation gas which serves as temperature moderator. The sulfur in the sorbent is converted to SO_2 and led to the reducer. In the SO_2 reducer (a moving bed with anthracite as reducing agent) elemental sulfur is produced. Experimental results obtained at a 40 t/d coal through-put scale were reported by Kobayashi et al. [17] and showed sulfur removal efficiencies of 95% at 460 °C as long as the concentration of water was below 10%. The preliminary results obtained at the 200 t/d coal through-put scale also showed a satisfactory desulfurization performance, stability of operation and reliability of equipment [18].

3.2 Iron Oxide With Support

In order to overcome problems like thermal instability, several sorbents have been developed consisting of iron oxide deposited on a carrier (SiO_2, Al_2O_3, TiO_2 etc). Van der Wal [19] prepared sorbents according to the deposition-precipitation method, consisting of a silicium oxide carrier loaded with iron oxide up to 50 wt%. The effects brought about by precipitating iron oxide on silica were: i) the inability to reduce iron oxide to elemental iron and hence preventing soot formation (Boudouard reaction), ii) a high resistance against deactivation and iii) a high surface area (±200 m^2/g) which remains practically the same over many cycles of testing. The desulfurization process studied in the temperature range of 300 to 500 °C consists of a reduction step to FeO (stabilized by the carrier) and a H_2S absorption step to FeS. Regeneration with highly diluted air streams (<0.5% O_2) yielded mixtures of SO_2 and elemental sulfur in the temperature range 250-400 °C. The maximum sulfur capacity of the sorbent amounts to 9 wt% S for a 50 wt% Fe_2O_3 sorbent, which corresponds with half of the theoretical capacity.

The present author prepared an iron oxide based sorbent analogous to the method applied by Van der Wal. To increase the reactivity and mechanical strength further the iron oxide was deposited on a carrier material (FCC) according to the deposition-precipitation method. A mechanical strong sorbent is essential for the continuous process proposed by the authors and schematically shown in Fig. 3 [20]. It consists of a circulating fluidized bed unit as absorber for H_2S, and a fluidized bed unit as regenerator. Both are interconnected by a continuous flow of loaded and regenerated sorbent. The product stream from the regenerator contains SO_2 which has to be treated further. Available options are the conversion to sulfuric acid or elemental sulfur, the latter for instance with the Direct Sulfur Recovery Process (DSRP) described by Gangwal et al. [21].

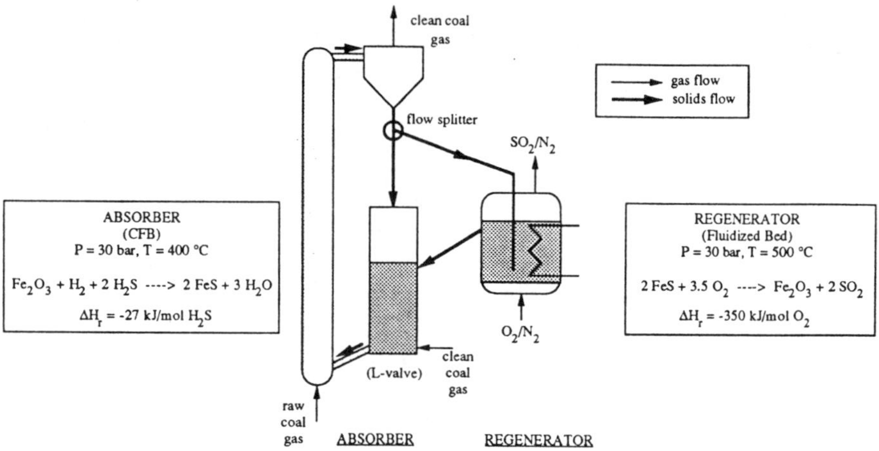

Fig. 3. Schematic representation of the proposed regenerative desulfurization process [20].

The absorption capacity and reactivity of the sorbent towards H_2S was investigated in an atmospheric packed bed reactor equipped with a continuous monitor for the H_2S outlet concentration. Breakthrough curve analysis was used to interpret the experimental data. The regeneration of the loaded sorbent with an air or a diluted air stream was investigated in a similar experimental set-up, but now equipped with a fluidized bed instead of a packed bed. Such a fluidized bed reactor was required to control the very exothermal oxidation reaction. Results showed that the optimal regeneration temperature was 500 °C, as a compromise between sulfate production (yielding SO_2 during the successive sulfidation step) and reactivity and activity of the sorbent. Unfortunately, high oxygen concentrations during regeneration (>15%) resulted in (reversible) deactivation of the sorbent [22]

In order to obtain more favorable regeneration conditions such as higher O_2 concentrations and no iron sulfate production, Van Yperen [23] continued the work of Van der Wal and modified the iron oxide sorbent through addition of a second metal oxide (MoO_3) and selection of a different carrier (precipitated alumina phosphate). Molybdenum oxide and iron oxide showed synergism: Mo prevented the production of iron sulfate even in a SO_2 atmosphere while iron oxide prevented MoO_3 reduction to Mo and subsequent evaporation. Moreover, MoO_3 reacted with H_2S to produce MoS_2 and removes H_2S to very low concentration levels (< 10 ppm) even for gases with a high water content (8 wt%). The main advantage of this process will be the regeneration of the sulfided sorbent with a diluted O_2 in SO_2 stream, yielding (almost) pure SO_2 as the final product, without producing excessive amounts of sulfates.

Venderbosch [24], in co-operation with the Energy Research Foundation (ECN), applied a mixed iron oxide/molybdenum oxide on alumina sorbent, which was prepared according to the method of Van Yperen [23], in a batch-wise operated turbulent fluidized bed. More details of the turbulent fluid bed regime are listed in our second contribution to this book. The regeneration was initially focused on the effect of the oxygen concentration (2-20 vol%) on the succeeding absorption reaction. Fig. 4 shows a typical H_2S break-through curve and its regeneration curve. The maximum bed temperature increase during regeneration is shown to be a linear function of the inlet oxygen concentration and equals 8 °C/vol% O_2.

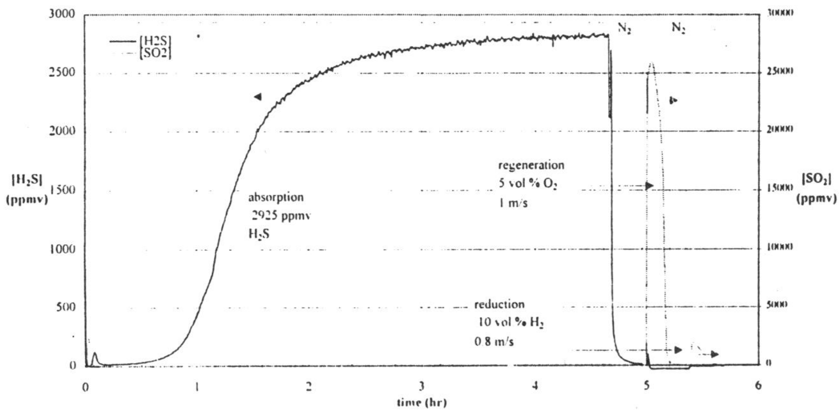

Fig. 4. A typical H_2S breakthrough curve and regeneration curve [24]
absorption: 0.8 m/s; 2925 ppm H_2S, 10% H_2; 400 °C
reduction: 0.8 m/s; 10% H_2
regeneration: 1 m/s; 5% O_2; 400 °C

The following preliminary conclusions of the turbulent fluid bed experiments can be drawn:

1. reproducible, very high initial H_2S conversion over more than 50 cycles, even at high gas velocities (outlet H_2S concentrations < 20 ppm)
2. regeneration under severe conditions (20 vol% O_2) showed no effect on absorbent capacity and reactivity
3. limited sulfate formation, about 5% of FeS is oxidized to sulfate.

KEMA [49] developed and patented a supported desulfurization sorbent which was applied in a two bubbling fluidized bed reactor system, one of which operates as absorber and the other as regenerator. The primary regeneration product is elemental sulfur. Laboratory and bench scale tests were conducted to validate the performance of the sorbent. Experiments also showed that the sorbent is capable of removing HCN and NH_3 simultaneously with H_2S.

In Japan Kawasaki Heavy Industries (KHI) developed a demonstration-scale regenerative moving bed process for the hot coal gas cleanup [25]. The main feature of the absorber, a cross flow moving bed reactor, is the simultaneous removal of sulfur and dust particles. An appropriate granular sulfur sorbent with iron oxide as the sulfur absorbing component and with enough strength for such a moving bed operation, was developed. It consists of an alumina-silica based porous carrier with sufficient pore volume to hold enough iron oxide to obtain a sufficient sulfur absorbing capacity. In the regenerator, a counter-current moving bed reactor, the FeS is oxidized by oxygen to Fe_2O_3 and SO_2. The SO_2 in the diluted regenerator off-gas stream is directly reduced with coal gas to elemental sulfur by means of a Co-Mo catalyst. A test facility (4 t/d) for the cleaning of 1000 Nm^3/h of coal gas at 420 °C and 2.5 MPa was designed, constructed and tested during 1700 hours [50]. Sulfur and dust particle removal efficiencies up to 98% and 99.8% respectively were obtained. Major problem was the iron sulfate formation during regeneration and the consequent reduction and sulfur release in the absorber. This resulted in a deterioration of the desulfurization performance. Installation of a sorbent reducer between the regenerator and the absorber largely improved the performance.

Mitsubishi Heavy Industries (MHI) developed a fixed bed type hot gas clean-up process on a 2 t/d and on a 20 t/d pilot plant scale. An iron oxide-titanium oxide honeycomb type absorbent is applied [26]. The desulfurization unit consists of three reactors which alternately are in the sulfidation, the regeneration and the reduction mode. SO_2 generated by the regeneration reaction is recovered as elemental sulfur by the gas reduction method. The results of the 2 t/d pilot unit showed a desulfurization efficiency of 95% corresponding with less than 100 ppm of outlet sulfur compounds. Satisfactory operation characteristics of the honeycomb type absorbent was observed during 500 cycles.

4 Zinc Based Sorbents

Zinc oxide shows superior thermodynamic properties and is capable of reducing the H_2S concentration in coal gas to very low levels. However, in strongly reducing atmospheres at high temperatures, zinc oxide is slowly reduced to volatile metallic zinc. To prevent reduction ZnO is combined with other metal oxides.

The two best known zinc based mixed sorbents are zinc ferrite ($ZnFe_2O_4$) and zinc titanate ($ZnTiO_2$). Zinc ferrite is an equimolar mixture of ZnO and Fe_2O_3 and was developed at the Morgantown Energy Technology Centre (METC) in the beginning of the eighties [28]. The general idea behind the zinc ferrite sorbent is to combine the favorable properties of both individual oxides ZnO (excellent thermodynamics) and Fe_2O_3 (high absorption rates and easily regenerable) to obtain a sorbent with superior thermodynamic properties (a H_2S removal efficiency down to the ppm level), a high theoretical sulfur capacity and a high reactivity. Although the theoretical sorbent capacity (based on stoichiometry, see reactions below) is 39 wt%, practically measured values never exceeded 40 to 50% of this theoretical capacity [29].

Overall absorption reaction:

$$ZnFe_2O_4 + 3\,H_2S + H_2 \Longleftrightarrow ZnS + 2\,FeS + 4\,H_2O$$

Overall regeneration reaction:

$$ZnS + 2\,FeS + 5\,O_2 \Longleftrightarrow ZnFe_2O_4 + 3\,SO_2$$

Favorable conditions for zinc ferrite are temperatures ranging from 500 °C to 650 °C and low-BTU gas with 15% steam [5]. The temperature range is limited because the desulfurization rate and sulfur capacity at temperatures below 500 °C is insufficient whereas at temperatures exceeding 650 °C Fe_2O_3 tends to reduce to FeO or even worse to Fe (resulting in carbon formation and sintering). At these temperatures the ZnO also reduces to Zn which starts to evaporate. The sulfidated zinc ferrite is regenerated with dilute O_2/N_2 mixtures yielding SO_2 as the product (1-3 vol%). Promising lab-scale results have led to slip stream tests in a process development unit of the KRW (Kellog-Rust-Westinghouse) fluidized bed gasifier. Outlet concentrations of $H_2S+COS < 20$ ppm were obtained which corresponds with a sulfur removal efficiency of >99.9%.

Lab-scale tests for a simulated KRW fuel gas mixture were less successful; severe decrepitation of the sorbents occurred. This phenomenon was attributed to the reduction of iron oxide to FeO and metallic iron with successive iron carbide formation. To maintain favorable sulfur absorption conditions, steam should be added to the coal gas (minimum water content ±20%). Gupta et al. [30] reported that the use of zinc ferrite is limited to a maximum temperature of ±550 °C and to moderately reducing coal-gas (e.g. a low BTU gas containing at least 15% steam). The search for an alternative that would be applicable to highly reducing gases and/or higher temperatures (up to 760 °C) culminated into the development of zinc titanate sorbents. On the other hand, Akyurtlu and Akyurtlu [31, 32] tried to improve the iron ferrite sorbent through addition of various amounts of V_2O_5. Experiments showed that the sorbent containing about 5 wt% vanadium has the highest conversion and sulfur capacity of all the tested vanadium promoted sorbents, and exhibited higher conversion than zinc ferrite and higher capacity than zinc titanate after five sulfidation (at 600°C) - regeneration (at 700°C) cycles using low moisture (7%) simulated fuel gas. When sulfided at 650°C, its performance deteriorated to the level of other vanadium-promoted sorbents and zinc ferrite. Addition of more vanadium should increase the stability of the sorbent but it was not possible to prepare it successfully yet. Significant amounts of elemental sulfur (8 to 14 % of total sulfur) were observed during regeneration of vanadium containing sorbents and only small amounts of sulfate were observed. Sulfate formation became significant for regeneration temperatures below 700 °C. Akyurtly and Akyurtlu [32] indicate that sorbents with high surface areas and high porosities containing 15wt% vanadium will have stabilities comparable to those of zinc titanate but with a much higher sulfur capacities.

Zinc titanate is a mixture of ZnO and TiO_2 in different molar ratios. Unlike zinc ferrite, only the zinc oxide present reacts with H_2S. The TiO_2 provides the stability to ZnO to withstand reduction by CO and/or H_2. The overall reactions during desulfurization and regeneration are given by:

Sulfidation:

$$Zn_2TiO_4 + 2\,H_2S \Longleftrightarrow 2\,ZnS + TiO_2 + 2\,H_2O$$

$$ZnTiO_3 + H_2S \Longleftrightarrow ZnS + TiO_2 + H_2O$$

Regeneration:

$$2\,ZnS + TiO_2 + 3\,O_2 \Longleftrightarrow Zn_2TiO_4 + 2\,SO_2$$

The theoretical sulfur capacity of this sorbent varies from 17 to 26 wt% depending on the Zn/Ti ratio which is, compared to the zinc ferrite sulfur capacity, much lower since the titanate present is inert. Experimentally determined sulfur capacities for $(ZnO)_{1.5}TiO_2$ vary from an initially value of 13.4 wt% (60% capacity utilization) down to 9 wt% at the 50th absorption/regeneration cycle. Zinc titanates are regenerated with diluted air streams yielding a gas mixture containing 1-3 vol% SO_2 [30]. Pilot plant tests with several zinc titanate sorbents having Zn/Ti ratios of 0.8 and 1.5, were conducted downstream of the Texaco gasifier in Montebello and showed favorable desulfurization performance (>99% H_2S removal) for multiple cycles. However, severe physical degradation was observed in the fixed bed pilot plant tests and also in some of the bench scale tests [33]. Sulfate formation might be responsible for this degradation.

Mojtahedi and Abbasian [34] tested two zinc titanate sorbents (UCI-2 and UCI-4) both with a Zn/Ti ratio of 1.5 in a high pressure, high temperature fluidized bed reactor system. Both sorbents were subjected to cyclic sulfidation and regeneration tests. The H_2S breakthrough curves and the sulfur load at H_2S breakthrough concentrations of 50 and 100 ppm for the sorbent UCI-2 indicate that the reactivity of the sorbent gradually decreases during successive cyclic tests. UCI-4 was tested at 650°C and showed analogous behavior. It was concluded that the sorbents do not appear to stabilize after a finite number of cycles. However, their excellent attrition resistance makes them physically suitable for fluidized bed applications.

Swisher at al. [35] managed to improve mechanical properties and durability of zinc titanate over many sulfidation-regeneration cycles. The approach was to contain the zinc titanate in a structural matrix of excess titania. Experimental results showed that the optimal composition was Zn_2TiO_4 plus 50% excess TiO_2, which corresponds to 33.5% ZnO and 66.5 % TiO_2. The performance of the sorbent was evaluated during 10 cycles in a fluidized bed as well as in a packed bed reactor. Results showed an increase in breakthrough time from 41 min to 92 min for the first and the tenth cycle respectively. The latter corresponds with a sulfur load of 15wt%, which is near the theoretical limit. Furthermore, the regeneration procedure was improved. If air is used the amount of heat generated form the exothermic reaction will sinter and degrade the sorbent material. Swisher et al. [35] suggests regeneration with O_2-plant off-gas and gas turbine exhaust gas. Tests with diluted exhaust gas gave good results.

Since zinc titanate became to brittle after several absorption and regeneration steps, Ayala and Jain [36] added MoO_3 (2.7 wt%) to zinc titanate (Zn/Ti = 2). This was shown to give a sufficiently strong and still chemically active, sorbent for long-term operation in a moving-bed system. Its strength was believed to be caused by the molybdenum, which modified the pellet pore structure by removing the micropores during calcination. This would yield wider and fewer pores and consequently thicker and stronger pore walls. The chemical reactivity is decreased by Mo-addition, but only to a limited extent if the porosity was controlled by organic burn-out additives. Regeneration at temperatures below 650°C and with high O_2 and SO_2 concentrations may lead to zinc sulfate formation. This is undesirable since the sulfate occupies more pore volume than the sulfides (producing more internal pellet stress) and decomposes during the subsequent absorption step while releasing SO_2. Hence, a regeneration scheme in which the temperature is above 680°C at the end of the regeneration is necessary.

In several very recent publications the Phillips Z-Sorb sorbents were evaluated by Khare et al. [37], Ayala et al. [38], Campbell and Henningsen [39] and several others researchers. Z-Sorb is based on zinc oxide supported in a porous matrix and contains a nickel oxide promoter. It was among others tested at Research Triangle Institute (RTI) in their bench-scale high temperature, high pressure, fluidized bed reactor system and at General Electric in their moving-bed, high temperature desulfurization system.

The RTI test program consisted of 50 absorption/regeneration cycles with an absorption temperature of 540 °C and a regeneration temperature of 650 °C [37]. Diluted air (2.5% oxygen in nitrogen) was applied.The results show that the initial high sulfur capacity of 20 wt% S decreases after 50 cycles of operation to a value of about 10 wt% S and is still decreasing. The experimental results obtained at General Electric [38] initially show an entirely satisfactory reactivity and sulfur capacity. However, within 200 hours of pilot plant testing Z-Sorb III lost more than 50% of its reactivity and sulfur capacity. While the mechanism of sorbent deactivation has not been positively identified, interaction between the sorbent and steam in the coal gas (and/or regeneration gas) appear to be a likely cause for the loss in reactivity according to Ayala et al. [38].

5 Calcium Based Sorbents

The thermodynamic calculations have shown that also CaO and $CaCO_3$ have the potential to absorb sulfur. These components are present in cheap natural sorbents like the mineral dolomite (i.e. $CaCO_3.MgCO_3$) and limestone ($CaCO_3$).

A regenerative desulfurization process on the basis of half or fully calcined dolomite has first been proposed by Squires [40]. In this process the following steps are distinguished:

- sulfidation of the half or fully calcined dolomite:

  $CaCO_3.MgO + H_2S/COS \text{ ---> } CaS.MgO + CO_2 + H2O/CO_2$

  $CaO.MgO + H_2S/CO \text{ ---> } CaS.MgO + H_2O/CO_2$

- Regeneration of the sulfided dolomite with a mixture of carbon dioxide and steam:

  $CaS.MgO + CO_2 + H_2O \text{ ---> } CaCO_3.MgO + H_2S$

- Calcination of the regenerated dolomite, only if fully calcined dolomite is applied:

  $CaCO_3.MgO \text{ ---> } CaO.MgO + CO_2$

The American company Conoco Development Co. developed a regenerative desulfurization process in the 70's based on half calcined dolomite. This process was tested on pilot plant scale. Both desulfurization and regeneration were carried out at a pressure of 15 bar. As the coupled Conoco gasification process delivered a coal gas stream containing much CO_2 and H_2O (8 and 13 vol% resp.), sulfidation had to be carried out at the relatively high temperature of 900 °C to reach the desired removal efficiency of 95%. Regeneration was carried out at 700 °C with a mixture of 35 vol% H_2O and 65 vol% CO_2. Though the Conoco process was capable of removing 95% of the sulfur, the process was never scaled up, because [52,53]: i) the H_2S content of the regenerator off-gas was too low (1-3 vol%) for further processing, ii) the fear for alkali metal evaporation from dolomite during absorption forming liquid deposits on downstream equipment and thereby inducing hot corrosion, and iii) the attrition rate of the dolomite was expected to be rather high in a full scale installation.

Heesink [41,51] investigated the Conoco process but now operated at a lower temperature level. According to his thermodynamic analysis regeneration carried out at 600 °C and 30 bar with a mixture of 50 vol% CO_2 and 50 vol% H_2O yielded a H_2S concentration in the regenerator off-gas up to 25 vol% (or 37 vol% after condensation of the not reacted steam). Advantages associated with a reduction of the absorption temperature are the expected drop in the emission of alkali metals and the higher absorption capacity obtained upon cycling. However, from the thermodynamic point of view, a decrease of the absorption temperature has a negative effect on the H_2S removal efficiency (92 to 98% of the sulfur present in Shell Coal gas). On the other hand, a significant part of the HCl present in Shell coal gas might be removed simultaneously. Though dolomite is relatively sensitive to attrition its use may be reconsidered in the light of new developed dust removal equipment [42] because the economic incentive to use dolomite instead of a synthetic sorbent is considerable. The price of dolomite amounts to about 30 ECU per ton (including transport) whereas supported synthetic sorbents typically cost 20 thousand ECU per ton.

Van der Ham et al. [43] investigated an alternative regenerative, high-temperature desulfurization process for coal gas using calcined limestone (CaO) as a sorbent. The process includes three steps (see Fig. 5). First H_2S reacts with CaO towards CaS (sulfidation), which is partly oxidized with SO_2 in a second step to obtain $CaSO_4$ and elemental sulfur. Finally, in a third step (regeneration) the solid mixture of CaS and $CaSO_4$ remaining after the oxidation step is decomposed to produce CaO and SO_2 which are both re-used in the first and second step respectively. The proposed process thus yields elemental sulfur as the final product. Several sulfidation-oxidation-regeneration cycles (up to nine) were carried out in a thermogravimetric analyzer (TGA) using various calcined limestones. The results are promising, showing a constant absorption capacity after two cycles of 30-50% already. On the basis of kinetic experiments and thermodynamic considerations, the optimum temperatures were determined at 815-845 °C for the desulfurization of Shell coal gas, 600-700 °C for oxidation and 1100-1200 °C for regeneration. Unfortunately, the regeneration reaction could only be examined up to the maximum attainable temperature in the TGA set-up, viz., 980°C.

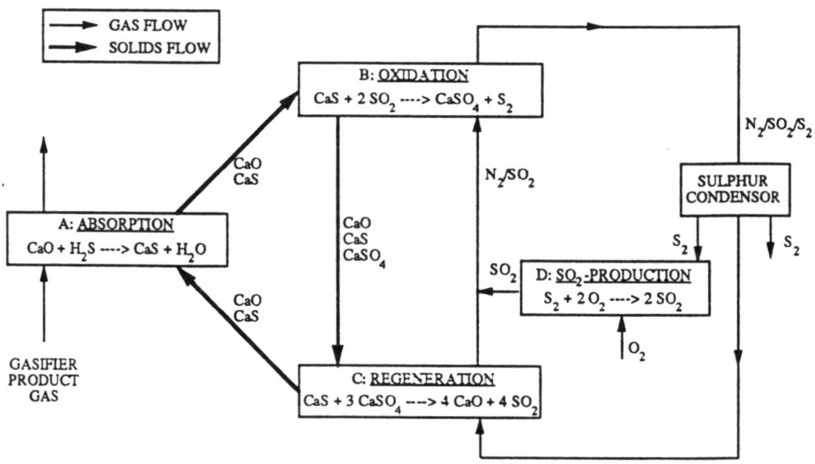

Fig. 5. Schematic representation of regenerative, high-temperature desulfurization process, based on the sorbent CaO, proposed for high-BTU, high-pressure coal gas.

6 Miscellaneous Sorbents

6.1 MnO or FeO on γ-Al_2O_3

The main advantage of both sorbents [44] is the regeneration route; the sorbents are namely regenerated with steam by which H_2S is produced at a high concentration level (15 vol% dry).

$$MnO/\gamma\text{-}Al_2O_3 + H_2S \Longleftrightarrow MnS/\gamma\text{-}Al_2O_3 + H_2O$$

The product gas differs from the gas obtained with the previously discussed sorbents (H_2S versus SO_2) and can easily be converted into elemental sulfur (e.g. in a Claus plant) without using reducing gases. In principle, the sorbent is a spinel formed by the metal oxide MnO or FeO and the carrier γ-Al_2O_3. Consequently the metal loading cannot be too high (5-10 wt%): the theoretical maximum sulfur capacity remains limited to 1.5 wt%. Typical break-through capacities measured, were 0.35 wt% S for the Mn-based and 0.20 wt% for the Fe-based sorbents. Furthermore, COS was also effectively removed and trace components like HCl had only a small negative influence on the break-through capacity.

Bakker et al. [45] developed a rotating monolith reactor for application of this reaction system in coal gas desulfurization. The monolith is coated with a γ-Al_2O_3 washcoat layer on which 1.4 - 32 wt% manganese was impregnated. Optimal operation temperature is between 850-950°C. Experiments showed only minor deactivation during 65 subsequent sulfidation and regeneration cycles.

6.2 SnO_2

Højlund Nielsen and Sigurdardóttir [46] investigated sorbents based on tin oxide. Characteristic features of these sorbents are i) the utilization of the reversible gas solid reaction

$$H_2 + H_2S + SnO_2 \Longleftrightarrow 2\,H_2O + SnS$$

in the temperature range 350-500°C, ii) its catalytic activity for carbonyl sulfide conversion, iii) the vanishing catalytic activity for the water gas shift and the hydrocarbon synthesis reaction and iv) it's mechanical strength. However the application of tin oxide has one serious drawback, i.e. SnO_2 may be reduced to Sn which is a liquid above 232°C. Coal gases containing water are easily handled whereas gases without water may require the addition of steam. CO does not react with SnO_2 at temperatures below 500°C. The presence of water also suppresses the formation of tin chlorides. Regeneration with steam yields a 2-3% H_2S containing product stream which is concentrated by steam condensation. The authors state that regeneration with steam has a number of advantages compared to e.g. regeneration with air/oxygen. The regeneration is almost thermo-neutral compared to the strongly exothermic regeneration with O_2/N_2 mixtures, the product of the regeneration is (almost) pure H_2S instead of diluted SO_2 and finally steam and coal gas do not form an explosive mixture opposed to oxygen and coal gas.

Recently Højlund Nielsen and Sigurdardóttir [47] reported on the ‘sandwich’ sorbent developed by Topsoe for deep sulfur removal in coal gases. By a combination of a SnO_2 based sorbent with a ZnO based sorbent a stagewise desulfurization process is obtained with a sulfur removal efficiency down to a few ppm at 400°C. SnO_2 takes care of the bulk desulfurization whereas ZnO performs the final polishing step. Regeneration is conducted at 450-550°C with steam containing a limited amount of oxygen. The system has been tested both in laboratory and on pilot plant scale using simulated coal gases. During regeneration, all sulfur was recovered and zinc sulfate formation was avoided below a few hundred ppm. The sulfur capacity and mechanical strength was pertained during a six-cycle test.

6.3 Copper Oxide

Zinc based sorbents suffer from zinc volatilization at elevated temperatures leading to sorbent deterioration. Copper-based sorbents do not suffer form this problem due to the high melting point of copper. However, bulk copper oxide is generally reduced to metallic copper in reducing fuel gas leading to thermodynamic limitations, resulting in insufficient level of desulfurization. According to Abbasian et al. [48] the reduction stability and therefore the desulfurization performance of copper oxide sorbents can be significantly improved by combining copper oxide with other oxides in a supported form or as bulk mixed metal oxides. The sorbents studied included copper oxide with oxides of chromium, cerium, aluminium, magnesium, manganese, titanium, iron and silicon. Initial results indicated that mixed binary oxides of copper with chromium are the most promising sorbents. The sorbent shows an excellent desulfurization efficiency (<10 ppm) at 750 °C. However, the capacity decreases gradually in the cyclic process. Therefore further testing in the temperature range 650-750° is necessary.

6.4 Various mixed Sorbents

Recently, the opinion on the optimum temperature range for desulfurization has been changed in the USA. Four research groups joint forces to develop advanced sorbents for desulfurization of coal gas in the temperature range 350 -550 °C [54]. This range is a reasonable compromise between cycle efficiency, capital costs and operation costs. The sorbents developed consist of three functional parts: i) a main component for the desulfurization, ii) a secondary component for the chemical stability and porosity, and iii) a inert structural component to maintain durability during recycling. Out of over 70 sorbents prepared, four sorbents (composition not specified) were finally selected based on the criteria of sufficient reactivity and mechanical strength for fluidized bed reactor testing in the temperature range 350-550 °C.

7 Regenerator Off-Gas Treatment

Regeneration of the sulfided sorbent should aim at producing elemental sulfur or H_2S and otherwise concentrated SO_2. Unfortunately, in most of the processes discusses in this paper, the regenerator off-gas produced is in general a diluted SO_2 stream, and this has to be processed in an economically and environmentally acceptable manner. Methods available are: i) recycling to the gasifier; which is only practical for fluidized bed gasifiers with in-situ sulfur capture, ii) sulfuric acid production and iii) reduction to elemental sulfur either direct with anthracite as in the iron-ore process (IHI), or catalytically like the reduction of SO_2 with coal gas over Mo-Co catalyst (KHI), or like in the Direct Sulfur Recovery Process [21]. The DSRP process operates with gases containing up to 12.4% SO_2, at pressures above 20 bar and temperatures in the 400-700 °C range. In this process an overall sulfur removal efficiency of more than 99% can be reached. However, it consumes some additional coal gas for the reduction of the SO_2 present.

Table 2. Overview of desulfurization processes and their experimental conditions

Sorbent type	Reference	particle diameter & preparation method	absorption conditions	gas composition $H_2/CO/CO_2/H_2O$ H_2S/N_2 (vol%)	purification degree (ppm H_2S) sorbent capacity (wt% S)	Regeneration condition	Regeneration gas/products	Status/Reactor	Remarks
pure iron ore	Kobayashi [17] IHI/NEDO	diam.: 0.15 mm grinded iron ore	420-480°C 16-20 bar	13/12/11/15 0.15/balance	< 80 ppm > 95% cap.: 5-8 wt%	600-800 °C 16-20 bar	4-8% O_2/N_2 ±2-4% SO_2/N_2	Pilot plant scale, two stage fluid bed	- capacity pilot plant: 1.2 and 11 Nm^3/s - strong influence of water on purification degree - continuous process with two stage fluidized bed absorber and regenerator - SO_2 is reduced with anthracite to elemental sulfur (purity > 99%) in moving bed (550-800°C) - catalyses water-gas shift reaction - removes also COS
Fe_2O_3 on FCC	Van der Ham et al. [22]	diam.: 0.1 mm, deposition-precipitation of iron oxide on FCC	300-600 °C 1 bar	10/0/0/0 0.1-0.7/balance	< 10 ppm (detection limit) cap.: 4 wt%	400-500 °C 1 bar	≤20% O_2/N_2 <12% SO_2/N_2	lab-scale; packed, and (circulating) fluidized bed reactors	- high reactivity, $\tau_{packed\ bed}$< 0.05 and τ_{cfb} < 1 s. - stable during more than 25 desulf./reg. cycles, - optimal desulf. temp. is 400-500 °C, above 600 °C sorbent deactivation, - optimal regeneration with 15% O_2/N_2 at 500 °C - sulfate formation ± 10% of absorbed H_2S.
Fe_2O_3 / MoO_3 on Alumina phosphate 12 and 6 wt% respectively	Van Yperen [23]	diam.: 0.15-0.5 mm carrier: precipitated, MeOx: deposited, particles: pressed & calcined	400 °C, 1 bar	various: H_2 : 1 - 60 % CO: 1 - 60 % H_2O: 0 - 8% H_2S: 1%	< 10 ppm cap: ±4 wt%	± 530°C 1 bar	2% O_2 in 50% SO_2/N_2 >50% SO_2/N_2 + sulfur	lab-scale fixed bed (>200 cycles)	- no sulfate produced during oxidation at T>530 °C for 12wt% Fe_2O_3 6wt% MoO_3 - iron oxide stabilizes Mo-oxide - high removal efficiency even with 8% water - catalyses COS shift reaction - Mo more expensive than Fe - regeneration gas with very high SO_2 content, even O_2 diluted with pure SO_2 was applied.
Fe_2O_3 / MoO_3 on Alumina 9 and 9 wt% respectively	Venderbosch ECN [24]	diam.: 0.17 mm	350-450 °C 1 bar	10/0/0/0 0.3-0.8/balance	< 10 ppm (detection limit) cap.: 7 wt%	400 °C 1 bar	≥20% O_2/N_2 <12% SO_2/N_2	lab-scale; turbulent fluidized bed reactor (ϕ=0.07 m)	- highly reactive sorbent, - high gas velocities, contact time < 0.5 s. - regeneration with air, ΔT of bed 170 °C - stable up to 50 cycles, even without pre-reduction before desulfurization, - limited sulfate formation, ≤5% of H_2S absorbed.
Fe_2O_3 on alumina-silica carrier	Ishikawa KHI [25]	diam.: 1mm, sperical shape	400-427°C 1 bar	13-24/0/15/2-3 0.3-0.4/balance	< 20 ppm	450 °C 1 bar	10-12% O_2/N_2 10%SO_2/3%O_2 in nitrogen	bench-scale; absorber: cross flow moving bed regenerator: counter flow moving bed	- simultaneous removal of particles and sulfur - COS is rapidly converted to H2S in presence of iron oxide, - sorbent stable during repeated sulfidation and regeneration (up to 2000 times)
Fe_2O_3 - TiO_2 honeycomb	MHI/CRIEPI MHI/IGC Suehiro et al. [26]	honeycomb structure	300-500°C 16 bar	1000 ppm H2S	<100 ppm cap: 10-15 wt%	±670°C 16 bar	1.5% O_2/N_2 1% SO_2/N_2	bench-scale fixed bed (500 cycles)	- sorbent capacity decreases to 50% of initial capacity - not much data available

Table 2. Overview of desulfurization processes and their experimental conditions (continued)

Sorbent type	Reference	particle diameter & preparation method	absorption conditions	gas composition $H_2/CO/CO_2/H_2O$ H_2S/N_2 (vol%)	purification degree (ppm H_2S) sorbent capacity (wt% S)	Regeneration condition	Regeneration gas/products	Status/Reactor	Remarks
zinc ferrites	Jain et. al [28]	0.15-0.2 mm granulated zinc oxide/ iron oxide (1:1)	530-800°C 15 bar	9/6/5/19 0.45/39.5	< 20 ppm 1th cycle: 16 wt% 100th cycle: 3 wt% average: 5 wt%	700-760 °C	2-3% O_2/N_2 1-2% SO_2/N_2	Bench scale, high pressure, fluidized bed (100 cycles)	- additional steam needed to prevent iron oxide reduction, carbide formation and Zn vaporization - attrition losses (42 wt%) mainly due to chemical transformations - sorbent may not be suitable for T>550 °C - sorbent not suitable for highly reducing syngas
zinc titanates	Gupta et al. [30]	0.1-0.3 mm granulates Zn/Ti: 1.5	750 °C 15 bar	Texaco syngas with 1.1% H_2S	< 20 ppm 1th cycle: 13 50th cycle: 9	760-785 °C 15 bar	2.5% O_2/N_2 1.6% SO_2/N_2	Bench scale high pressure, fluidized bed (100 cycles)	- negligible Zn vaporization at T ≤ 750 °C - negligible physical loss of sorbent - no deleterious effect of HCl (500 ppm) and NH_3 in Texaco syngas.
zinc ferrites + vanadium oxides	Akyurtlu and Akyurtlu [31,32]	diam.: 20/40 mesh zinc ferrite particles impregnated with V_2O_5 followed by calcination at 600 or 750°C	600-700°C 1 bar	20/-/-/7-20 1/72	< 1 ppm 1th cycle: 26 wt% 5th cycle: 24 wt% (for 5 wt% V)	700°C 1 bar	10% O_2/N_2 mixture of $S/SO_2/N_2$	lab scale packed bed reactor	- optimum amount of Vanadium is ±5 wt%, more V increases stability but decreases capacity - break through concentration initially <1 ppm - sulfidation performance at 650°C is less due to zinc loss, increase of amount of Vanadium helps - during regeneration 8-14% of sulfur is obtained as elemental sulfur - regeneration at 650°C yields sulfate production
zinc titanates	Robin et al. [33]	extrudates Zn/Ti: 1.5	540-600 °C 24 bar	Oxygen blown syngas	>99% removal 1-5 cycles: 15	± 700 °C 15 bar	1% O_2/N_2 <0.7% SO_2/N_2	Pilot scale, high pressure, fixed bed (1ft^3)	- significant degradation in terms of loss of zinc, pore volume, surface area and crush strength - high pressure causes severe cracking of sorbent due to enhanced zinc sulfate formation during regeneration - sulfate formation can be minimized and prevented by correct reactor choice.
zink titanate Zn/Ti = 1.5	Mojtahedi and Abbasian [34]	diam.: 0.38 mm (UCI-2) 0.27 mm (UCI-4)	550-650°C 20 bar	13/18/8/11 0.15/balance	< 50 ppm 1th cycle: 4-5 wt% 30th cycle: 1-2wt%	750 °C 20 bar		high pressure, high temperatue, fluidized bed reactor (ϕ=0.075m)	- superf. gas velocities of 0.2-0.3 m/s at 0.1-0.3 m fluidized bed heights. - no stabilization of sorbent performance, probably due to zinc migration and deterioration of sorbent - excellent attrition resistance
zinc titanate + 50% excess TiO2 Zn/Ti = 0.5	Swisher et al. [35]	diam.: 0.1-0.3 mm agglomorated, calcinated, clushed and sieved	650 °C 15 bar	28/39/13/19 1/0	< 100 ppm cap. ≤ 15 wt% S	720 °C 15 bar	2.5% O_2/N_2 1.8 % SO_2/N_2	bench-scale fluidized bed reactor	- sorbent shows high strength and durability which is obtained for 2 μm oxide powder and sintering at 1000 °C. - little sulfate formation - no degradation due to sulfate formation or sintering

Table 2: Overview of desulfurization processes and their experimental conditions (continue)

Sorbent type	Reference	particle diameter & preparation method	absorption conditions	gas composition $H_2/CO/CO_2/H_2O$ H_2S/N_2 (vol%)	conversion or purification degree (ppm H_2S) sorbent capacity (wt% S)	Regeneration condition	Regeneration gas/products	Status/Reactor	Remarks
zinc titanates + molybdenum oxide	Ayala and Jain [36]	diam.: 5-6 mm rounded, calcined pellets of ZnO, TiO_2, MoO_3 Zn/Ti = 2 Mo = 1.7 wt%	540 °C	simulated coal gas varying from low BTU to oxygen blown syngas with 1% H_2S	< 10 ppm cap.: 3 wt%	initial: 540 or 680°C final: 760°C (to decompose the sulfate)	1-21% O_2/N_2 dilute SO_2/N_2	bench scale packed bed reactor	- sorbent meant for moving bed operation - better performance in strength result of Mo, which yields a higher mean pore diameter - negligible levels of C after sulfidation and sulfate after regeneration, detected - no Mo losses when concentration ≤ 1.7 wt% - 10wt% organic burnout (for porosity control) and 2-3wt% bentonite binder added
Z-Sorb (ZnO in a porous matrix with nickel oxide promotor)	Khare et al. [37]	diam.: 0.1-0.3 mm	540 °C 20 atm	14/23/6/7 0.5/balance	< 50 ppm cap.: 20 (cycle 1) 10 (cycle 50)	650 °C 20 bar	2.5% O_2/N_2 SO_2/N_2	bench-scale high pressure high temperature fluidized bed reactor	- very low sulfate levels at regenerator exit - sorbent shows excellent activity and regenerability - after 50 cycles half of the sulfur capacity is lost and still decreasing.
Z-Sorb III	Khare et al. [37] Ayala et al. [38]	diam.: 4 mm	540 °C 19 bar	16/10/12/25 0.3/balance	<50 ppm (ζ = 0.98-0.99)	730 °C 19 bar	staged, air/recycle gas SO_2/N_2	moving-bed hot gas cleanup system	- very low attrition losses - very low sulfate levels at regenerator exit - sulfur capacity dropped within 200 hours to a quarter of the initial capacity
$CaCO_3.MgO$ half-calcined Lhoist dolomite	Heesink [41]	diam.: 0.1-0.2 mm half-calcined at 800 °C in 20% CO2 in N2 at 8 bar.	750-850°C 8 bar	14/0/24/24 1.5/balance	ζ= 0.92-0.98 (at 750 °C, 30 bar for Shell coal gas according to thermodyn.) cycle 1: 23 wt% cycle 15: 2 wt% (stable)	600-650 °C 15 bar	50% CO_2/ 50% H_2O $H_2S/H_2O/CO_2$ (thermo: 25% H2S)	HPTGA	- cheap, natural based sorbent, - good desulfurization performance in lab-scale fluidized bed operated at a high gas velocity of 0.7 m/s at 700 °C, static bed height 0.05 m, - neg. influence of CO_2 on desulf. performance (probably sintering) - no influence of water.
natural based Calcium oxide	Van der Ham et al. [43]	diam.: 0.15-0.21mm grinded, sieved, and calcined at 850 °C during 20 min.	300-850°C 1 bar	5-6/0/0/0 1-2.4/balance	TGA was used cap.: 30-57 wt%	600-860 °C and >1000 °C 1 bar	10% SO_2/N_2 S_2/S O_2/N_2	lab-scale; thermogravimetric analyzer setup	- regenerative cycle consists of a sulfidation, oxidation and regeneration step. - testedupto 8 absorption/regeneration cycles - sorbentcapacity stabilizes on 15-30 wt% - optimal operation conditions for Shell coal gas: sulfidation at 815-845 °C oxidation at 700 °C, regeneration at 1100 - 1200 °C
tin oxide	Højlund Nielsen and Sigurdardóttir [46]	SnO_2 pellets and extrudates	350-500°C 20-30 bar	70/10-20/10-20/- 0.05-0.7/balance	> 16 ppm (Thermo, at 400°C, 3% H_2O, 25% H_2)	350-550°C 10-30 bar	steam 2-3% H_2S in steam	up-flow and down-flow fixed bed reactor	- catalytic activity for carbonyl sulfide conversion - vanishing catalytic activity for water gas shift and hydrocarbon synthesis reaction - no deterioration of strength and mechanical breakdown - few % H_2O needed for SnO_2 stabilization against reduction and chlorination.

8 Conclusions

A limited overview of the recent developments in the field of regenerative, high temperature coal gas desulfurization is given. The basic information of nearly every sorbent/process discussed is summarized in Table 2.

Desulfurization process developments are more and more directed towards testing on a pilot-plant/demonstration scale. This inevitably leads to a major interest in the process economy (i.e. capital costs and operating costs) resulting in a consensus about the operating temperature range (350-550 °C), operating mode (continuous) and the sorbent characteristics. The sorbent should consist of a mixture of metal oxides supported by a inert carrier yielding a mechanical strong particle suitable f.i. for fluidized bed operation. Furthermore, the regeneration reaction and, even more important, the production of a regenerator off-gas mixture suitable for economic processing is also more and more under investigation. On the other hand, the influence of other contaminants present in the coal gas (HCl, HF, HCN etc.) on the regenerative desulfurization performance of the sorbent/process is often not taken into account.

References

1. Harrison, D.P. (1995). Proceedings of the twelfth Annual International Pittsburgh Coal Conference, 1047-1052.
2. NOVEM (1990). Systeemstudie hoge temperatuur gasreiniging bij KV-STEG-systemen. Ref.nr. 90-310, Dossiernr. 8725-21421/500.
3. Corman, J.C. (1986). System Analysis of simplified IGCC plants, U.S. Department of Energy Topical Report DOE/ET/14928-2233.
4. Furimsky, E. and M. Yumura (1986). Erdöl und Kohle, 39(4), 163-172.
5. Gäng, P. and F. Löffler (1993). Gas Cleaning at High Temperatures, Blackie Academic & Professional, London, p 522-540.
6. Heesink, A.B.M. (1991). Development of systems for regenerative desulfurization of CG-gases with a separate FB or CFB absorber. Progress report no 1 of contract no. JOUF-0051-C(TT).
7. Jansen, D. (1990). Systeemstudie hoge-temperatuur-gasreiniging, deelstudie 2.1: H_2S/COS-verwijdering, Netherlands Energy Research Centre (ECN), the Netherlands. ECN-C-90-050.
8. Westmoreland, P.R. and D.P. Harrison (1976). Env. Sci. Techn., 10(7), 659-661.
9. Barin, I. and O. Knacke (1973). Thermochemical properties of inorganic substances. Springer-Verlag, Berlin.
10. Oldaker, E.C. and A.M. Poston Jr. and W. L. Farrior Jr. (1975). Am. Chem. Soc., Div. of Fuel Chemistry, 20(4), 227-237.
11. Oldaker, E.C. and D.W. Gillmore (1976). Am. Chem. Soc., Div. of Fuel Chemistry, 21(4), 79-90.
12. Tamhankar, S.S., M. Hasatani and C.Y. Wen (1981). Chem. Eng. Sci, 36, 1181-1191.
13. Tamhankar, S.S. and C.Y. Wen (1983). Industrial Gas Separation, ACS Symp. Series, 223, 255-279.

14. Tamhankar, S.S., S. Garimella and C.Y. Wen (1985). Chem. Eng. Sci, 40(6), 1019-1025.
15. Tseng, S.C., S.S. Tamhankar and C.Y. Wen (1981). Chem. Eng. Sci, 36, 1287-12914.
16. Schrodt, J.T. and C.P. Mohan (1981). AIChE, Symp. Series (no. 211), 77, 1-8.
17. Kobayashi, A. (1990). Test results of Yubari hot gas desulfurization pilot plant. Coal Conversion Technologies Department, IHI Co., Ltd, Japan.
18. Kimura, H. (1994). Test results of hot gas clean-up facility for 200 t/d entrained flow coal gasification pilot plant. Publication of Ishikawajima-Harima Heavy Industries Co, Ltd.
19. Van der Wal W.J.J. (1987). Desulfurization of process gas by means of iron oxide-on-silica sorbents. Ph.D. thesis, University of Utrecht, the Netherlands.
20. Van der Ham, A.G.J., W. Prins and W.P.M. van Swaaij (1994). Circulating Fluidized Bed Technology IV, Pergamon Press, Canada, 657-664.
21. Gangwal, S.K., W.J. McMichael and T.P. Dorchak (1991). The direct sulfur recovery process Environmental Progress, 10(3), 186-191.
22. Van der Ham, A.G.J., W. Prins and W.P.M. van Swaaij (1994). Regenerative, high temperature desulfurization of coal gas in a circulating fluidized bed. Ph.D. thesis, University of Twente, the Netherlands.
23. Van Yperen, R. (1994). On the high-temperature desulphurizaiton of coal gas; the development of a regenerable absorbent. Ph.D. thesis, University of Utrecht, the Netherlands.
24. Venderbosch, R.H. (1997). Ph.D. thesis, University of Twente, the Netherlands (to be published).
25. Ishikawa, K., N. Kawamata and K. Kamei (1993). Development of a simultaneous sulfur and dust removal process for IGCC power generation system. Proc. of 3rd Int. Symp. on Gas Cleaning at High Temperatures, 419-435.
26. Suehiro, M., and H. Fujishima (1993). Fixed bed type hot gas cleanup technology for IGCC power generation. Fifth Japan / USA joint technical meeting on surface coal gasification, October 13, Tokyo, Japan.
27. Ishikawa, K., C. Krueger, M. Flytzani-Stephanopoulos, W. Ji, R.J. Higgins, B.A. Bishop and R.L. Goldsmith (1995). Proceedings of the twelfth Annual International Pittsburgh Coal Conference, 1195-1200.
28. Jain, S.C., R. Gupta and S.K. Gangwal (1991). Development of zinc ferrite sorbents for desulfurization of hot coal gas in a fluid-bed reactor. Paper presented at AIChE Summer National Meeting, Pittsburgh, Pennsylvania.
29. Gangwal, S.K. et al. (1991). Reaction kinetics and simulation models for novel high-temperature desulfurization sorbents - final report. RTI-report.
30. Gupta, R., W.J. McMichael, S.K. Gangwal, S.C. Jain and T.P. Dorchak (1992). Development of a hot-gas desulfurization system for IGCC applications. Paper presented at the AIChE 1992 Annual Meeting.
31. Akyurtlu, J.F. and A. Akyurtlu (1992). Hot gas desulfurization with sorbents containing mixed metal oxides. Paper presented at the AIChE 1992 Ann. Meeting.
32. Akyurtlu, J.F. and A. Akyurtlu (1995). Gas Sep. Purif., 9(1), 17-25.
33. Robin, A.M., J.S. Kassman, T.F. Leininger, J.K. Wolfenbarger and P.P. Yang (1993). Gas Cleaning at High Temperatures, Blackie Academic & Professional, London, p 438-453.
34. Mojtahedi, W. and J. Abbasian (1995). Energy & Fuels, 9, 429-434.

35. Swisher, J.H., J. Yang and R.P. Gupta (1995). Ind. Eng. Chem. Res., 34, 4463-4471.
36. Ayala, R.E. and S.C. Jain (1992). Development of durable mixed-metal oxide sorbents for high-temperature desulfurization of coal gases in moving-bed reactors. Paper presented at the AIChE 1992 Annual Meeting.
37. Khare, G.P., G.A. Delzer, D.H. Kubicek and G.J. Greenwood (1995). Environmental Progress, 14(3), 146-150.
38. Ayala, R.E., A.S. Feitelberg and A.H. Furman (1995). Proceedings of the twelfth Annual International Pittsburgh Coal Conference, 1053-1058.
39. Campbell, W.M. and G.B. Henningsen (1995). Proceedings of the twelfth Annual International Pittsburgh Coal Conference, 1059-1064.
40. Squires, A.M. (1967). Adr. Chem. Ser., 69, 205-229.
41. Heesink, A.B.M. (1994). High temperature coal gas desulphurization. Ph.D. thesis, University of Twente, the Netherlands.
42. Clift, R. and J.P.K. Seville (1993). Gas Cleaning at High Temperatures, Blackie Academic & Professional, London.
43. Van der Ham, A.G.J., A.B.M. Heesink, W. Prins and W.P.M. van Swaaij (1996). Ind.Eng.Chem.Res., 35, 1487-1495.
44. Wakker J.P. (1992). Development of a high temperature steam regenerative H_2S removal process based on alumina supported MnO and FeO. Ph.D. thesis, Technical University of Delft, the Netherlands.
45. Bakker, W.J.W., M. Vriezendorp, J.M. Mugge, A. Cybulski, F. Kapteijn, J.A. Moulijn (1996). Sorbent development of continuous regenerative H_2S removal in a rotating monolith reactor. Poster presented at SON meeting.
46. Højlund Nielsen, P.E. and I.D. Sigurdardóttir (1993). Gas Cleaning at High Temperatures, Blackie Academic & Professional, London, p 454-469.
47. Højlund Nielsen, P.E. and I.D. Sigurdardóttir (1995). Proceedings of the twelfth Annual International Pittsburgh Coal Conference, 1074-1079.
48. Abbasian, J., A.H. Hill, F.S. Lau, M. Flytzani-Stephanopoulos and F.I. Honea (1995). Proceedings of the twelfth Annual International Pittsburgh Coal Conference, 1189-1194.
49. Meijer, R., F.J.J.G. Jansen, F.L. Farina and J.W.H. Hellendoorn (1996). Third Int. Symp. and Exh. on Gas Cleaning at High Temperatures, Karlsruhe, 18-20 sept., 638-647.
50. Ishikawa, K., K. Kamei, T. Shoji, K. Shindo and N. Kawamata (1996). Third Int. Symp. and Exh. on Gas Cleaning at High Temperatures, Karlsruhe, 18-20 sept., 316-338.
51. Heesink, A.B.M. and W.P.M. van Swaaij (1996). Third Int. Symp. and Exh. on Gas Cleaning at High Temperatures, Karlsruhe, 18-20 sept., 528-542.
52. O'Neill, E.P. and D.L. Keairns (1977). A.I,Ch.E. Symp. Ser., 73, 100-107.
53. Morrison, G.F. (1979). Hot gas cleanup, Report no. ICTIS/TR 03, IEA Coal Research, London.
54. Abbasian, J., R.B. Slimane, J.R. Wangerow, R.E. Ayala, S. Venkataramani, M. Flytzani-Stephanopoulos, D.C. Cicero (1996). Third Int. Symp. and Exh. on Gas Cleaning at High Temperatures, Karlsruhe, 18-20 sept., 607-617.

Removal of H_2S from coal gas: effect of various parameters on the desulfurization efficiency

R.H. Venderbosch, W. Prins, A.G.J. Van der Ham

Department of Chemical Engineering, Twente University of Technology,
P.O.Box 217, 7500 AE Enschede, The Netherlands

Abstract. After a decade of extensive research, a number of solid sorbents have been identified as being suitable for the high-temperature removal of H_2S from coal gas. Current activities are aimed at the development of a continuous desulfurization process through reactor studies and pilot-plant testing. In this paper, the influence of various parameters related to the type of sorbent, the type of reactor, and the required process conditions, are discussed, and elucidated for the promising iron-oxide based sorbents as a typical example.

1 Introduction

Integrated gasification combined cycle (IGCC) power plants have been the subject of considerable research over the past decades throughout the whole world [a.o. 1, 2, 3, 4, 5]. In the second half of the eighties, it was decided in the Netherlands to investigate the option of electricity production by an IGCC system more intensely. In 1989, the SEP started to build a IGCC demonstration plant of 250 MW(el) at Buggenum, the Netherlands, which was completed in November 1993. Several runs on mixtures of coal gas and natural gas were realized in the beginning of 1994. Integration tests are still ongoing. A system study, initiated by the Netherlands Agency for Energy and the Environment (NOVEM) concluded that the application of a dry gas cleaning process at a high temperature level (> 623 K) improves the overall efficiency of the IGCC plant significantly, if compared to an installation with conventional wet gas cleaning. One of the major process steps involved in dry gas cleaning is the high temperature desulfurization, i.c. the removal of H_2S and COS from the hot coal gas.

In high-temperature desulfurization, a good match between a (regenerable) sorbent and the applied reactor is required. Extensive reviews of possible sorbents, based on the original paper by Westmoreland and Harrison [6], are presented by Heesink [7], Van der Ham [8], and (in this book) Van der Ham *et al.* [9]. Generally, such sorbents can be distinguished into the group of expensive, synthetic sorbents having a long life-time, and the cheap, natural sorbents. The choice of reactor configuration strongly depends on the sorbent

NATO ASI Series, Vol. G 42
Desulfurization of Hot Coal Gas
Edited by A. T. Atimtay and D. P. Harrison

type aimed at. In this paper the effect of parameters related to the type of sorbent and reactor, on the desulfurization efficiency, will be discussed. As an example, the iron-oxide sorbent system will be reviewed and used as a reference.

2 Reactor Selection for H_2S Removal

An interesting strategy for multiphase reactor selection is proposed by Krishna and Sie [10]. According to these authors the 'ideal' reactor configuration should meet as close as possible the requirements of three strategy levels: one on catalyst design, the second on injection and dispersion, and a third one on the choice of the hydrodynamic flow regime. Many of the parameters in their selection model are inter-related and consequently, an iterative approach should be undertaken to select the 'ideal' reactor. If this strategy is applied in H_2S removal from coal gas, the following relevant process wants can be listed:

- The degree of H_2S-removal should be high. A maximum concentration of approximately 20 ppmv H_2S is allowed in the turbine gas, which corresponds to a removal degree of at least 99%. For the past decades, H_2S removal has been focused on the use of metal oxides sorbents. They must react selectively and reversibly with H_2S in a reducing coal-gas environment at severe conditions. The required *removal efficiency* is determined by thermodynamic considerations, which reduces the possible high-temperature desulfurization sorbents to potentially 9 elements (Co, Cu, Fe, Mn, Mo, Sn, V, W and Zn), primarily in the form of metal oxides [6, 7, 8, 9].
- The H_2S absorption should take place at the operating pressure of the gasification process (up to 30 bar), and at temperatures above 600 K, in order to increase the efficiency of an IGCC plant substantially. In recent economic studies however, it is indicated that too high temperatures also carry significant penalties in terms of the costs of construction materials. With respect to the *optimum temperature range*, a reasonable compromise between cycle efficiency, capital costs and operating costs, now leads to values between 600 and 800 K [5].
- Because the desulfurization process is carried out at elevated pressure, the reactor diameter must be kept as small as possible; therefore *high gas velocities are preferred.*
- The sorbent should preferentially yield elemental sulfur. If that is not feasible, then H_2S and eventually (concentrated) SO_2 is desired as the *product of regeneration.* To overcome the difficulties of pressure (or temperature) staging in a continuous process, the regeneration should preferably be carried out at the same pressure (temperature) level as the absorption pressure (temperature).

In the beginning of the desulfurization process and sorbent development, relatively high operation temperatures (>800 K) were preferred to maximize the increase in overall thermal efficiency of the IGCC process. For iron-oxide sorbentia however, such a high temperature level is unfavourable because the equilibrium is then shifted too much to the reactant side [11, 12]. The H_2S equilibrium concentration over iron oxide rises rapidly at temperatures above 800 K, by which the possible removal degree becomes too low. Therefore, other sorbents, especially those based on zinc, were considered in the early 1980s. Pioneering work led to the development of zinc ferrite, $ZnFe_2O_4$ [13], zinc titanate, $ZnTiO_2$ [14], a combination of molybdenum oxide (MoO_3) and zinc titanate [15], and Z-Sorb, zinc oxide in a porous matrix containing a nickel oxide promotor [f.i. 5, 16]. All these sorbents were examined for application at temperatures above 800 K. However, the opinion on the optimum temperature range has been changed lately and desulfurization temperatures in the order of 600-800 K are currently preferred. At that temperature level, iron oxide is again an attractive sorbent; apart from its high reactivity, it has the following obvious advantages:

- cheap,
- non-toxic,
- high H_2S removal efficiency,
- easy regenerable,

A detailed description of the various desulfurization sorbents is beyond the scope of this contribution (included in other parts of this book).

In early papers concerning desulfurization of coal gas, iron oxide is often mentioned as a sorbent material; it was used in the so-called 'Dry Boxes' of the former gas factories to desulfurize town gas. The absorption of H_2S yields ferrous sulfide as the solid reaction product, according to:

$$Fe_2O_3 + 2\,H_2S + H_2 \leftrightarrow 2\,FeS + 3\,H_2O$$

Upon selection of a proper desulfurization reactor in combination with a suitable sorbent, one should be aware of all chemical, physical and hydrodynamic factors possibly affecting the rate of H_2S absorption. In general, the following rate limiting steps can be identified:

1. diffusion of H_2S from the bulk of the gas to the external particle surface,
2. diffusion of H_2S through a product layer formed around the unreacted core of the sorbent particle,
3. the intrinsic rate of the reaction between H_2S and the solid reactant,
4. back-diffusion of the product gas out of the solid particle
5. back-diffusion of products from the particle to the bulk of the gas.

It should be noticed that the chemical reaction rate and the internal diffusion processes depend on the composition and structure of the sorbent material (and consequently on the applied preparation method), while the gas-to-particle, external mass transfer resistances are only determined by the gas and particle properties, the reactor geometry and the applied operation conditions.

Table 1. Comparison of reactors for large scale H_2S removal from hot coal gas.

	Packed bed	moving bed	bubbling fluid bed	turbulent fluid bed	riser of a CFB
construction/ technology	cheap and simple	fairly cheap and simple	complex	complex	complex
operation mode	batch / swing, changing product composition	continuous	continuous	continuous, recirculation of entrained solids	continuous, recirculation of all solid material
heat -exchange	troublesome	troublesome, controllable by proper gas flow or solids circulation	good, in- or external or by addition of solids	excellent, see BFB	good, see BFB
temperature distribution	large temp. gradients may occur	large temp. gradients may occur	almost uniform	uniform	uniform
gas-to-particle mass transfer	poor	poor	poor, additional bubble phase resistance	good, despite the formation of stagnant clusters	good, especially in case of high slip velocities
solids hold-up	0.6	0.6	0.3 - 0.55	0.1 - 0.3	0.01 - 0.1
gas velocity (m/s)	0 - 0.7	0 - 0.7	0.01 - 0.4	0.4 - 1.3	2 - 20
relative gas residence time (m^3_s . s / m^3_g)	1 - 5	1 - 5	1 - 5	0.3 - 1	0.01 - 1
solids residence time	days	hours - days	hours	minutes-hours	seconds
particles	2-10 mm	2-10 mm	50-500 μm	50 - 500 μm	50 - 300 μm
particle utilization efficiency	small	small	relatively high	high	maximal
pressure drop	relatively low	see PB	acceptable	see BFB	low
flow behaviour	plug flow of gas, stationary solids	plug flow of gas and solids: co-, counter- or cross-current	ideal solids mixing, back-mixing of gas	See BFB	co-current plug flow of solids and gas
disadvantages with respect to H_2S removal and sorbent regeneration	discontinuous, low specific surface area, low heat transfer coefficient, dust not acceptable	low specific surface area, low heat transfer coefficient, particle attrition, dust not acceptable	particle attrition, erosion gas by-passing	particle attrition, erosion, entrainment losses	see TFB

Pioneering work on high-temperature desulfurization, was mainly concerned with sorbent development and characterization (reactivity, capacity, regenerability, deactivation) in small and simple, cyclic operated packed bed reactors. However in practice, *continuous operation* is required, and nowadays the interest has shifted towards process development work for which moving beds, fluidized beds, and risers are considered as possible large-scale reactors.

Generally, the objective of reactor design is to maximize the product yield and selectivity while keeping the reactor easy operable and as small as possible. In case of H_2S removal from hot coal gas, a high removal efficiency is required without the formation of any undesirable solid or gaseous byproducts, and also without significant loss of activity upon repeated regeneration of the sorbent. If i) the thermodynamic conditions for a high removal efficiency are favourable, and ii) the intrinsic reaction rate is fast, it is then important to increase the particle conversion rate by a proper sorbent design, until the regime of external mass transfer controlled conversion is reached. This is a situation in which intra-particle diffusion limitations are minimized by using a sorbent consisting of small, porous particles. The problem of reactor selection and optimization is then reduced to creation of good mass transfer conditions, i.c. high gas-to-particle transfer coefficients in combination with a large specific surface area.

In this contribution, iron-oxide based sorbents will be taken as a typical example for elucidating the effects of various parameters like sorbent particle properties, reactor choice and regeneration route, on the overall H_2S removal efficiency. Table I shows a number of characteristic features for various reactors, which are important in large scale operation. Generally, packed beds are easy and cheap to construct. But they have many drawbacks when applied for high-temperature coal gas cleaning. Fluid bed technology is more complicated but proven for many other applications. A fluid bed seems to have several advantages if considered as a reactor for H_2S removal.

Hereafter we will discuss the suitability and applicability of the various reactors listed in Table I in more detail.

3 Packed Bed (PB)

Packed beds are the most simple reactors for gas-solids reactions. Normally, the gas flow passing through a bed of solid particles is approximately in plug flow. Back-mixing of gas does not occur, and conversions close to the thermodynamic maximum are feasible. Particle attrition rates are negligible because the sorbent particles are not transported in or out of the reactor but fixed in a stationary bed. A packed bed can be used for slow reactions of a gaseous component with a non- or slowly deactivating solid material. Space velocities are chosen such that H_2S removal efficiencies are largely dictated by the thermodynamic conditions. For iron-based sorbents, the operating temperatures should be kept below 800 K, to guarantee a sufficiently high removal efficiency. The operating pressure has no significant effect upon the equilibrium conversions [11]. Swing operation must be applied with at least two reactors involved, when the solids need to be regenerated after a while. The active material can either be used directly, whether or not mixed with an inert material (grains, pellets), or be precipitated in the interior of porous support particles for enhanced reactivity and mechanical strength. The carrier can also consist of packing materials like random packings, spheres and cylinders, or the uniformly structured ones, as for instance parallel or cross corrugated plates, and honeycombs

3.1 Miscellaneous Iron-Oxide Based Sorbents

Lab-scale experiments for high temperature desulfurization in a fixed bed have been carried out already in the early seventies at Morgantown Energy Research Center of the Bureau of Mines (MERC). Extruded pellets, prepared from a mixture of iron oxide and fly ash obtained from a real coal gasifier (the fly ash contains magnetite Fe_3O_4 and hematite Fe_2O_3), were successfully applied in lab-scale units at high temperatures (> 850 K) and pressures up to 20 bars [17]. H_2S was removed up to near-equilibrium concentrations (1000 ppm, corresponding with 90% H_2S removal). The sulfur uptake at 10% breakthrough was in the order of 40 to 80 wt.%. Lower temperatures, which are favorable with respect to the (thermodynamically) achievable H_2S removal efficiency, reduced the sulfur capacity to an unacceptable level due to lower sorbent reactivities. Unfortunately, the sorbent strength appeared to be quite low. Besides, the high exothermicity of the applied regeneration reaction with oxygen imposes operational restrictions such the necessary dilution.
Based on criteria like a sufficient absorption capacity, mechanical strength and regenerability, laboratory studies at MERC eventually led to the selection of various mixtures of iron oxide and silica as a supporting material. As far as we know, the MERC fixed-bed concept has not yet been commercially applied.

Schrodt *et al.* [11] and later Schrodt and Best [18] showed in a lab-scale fixed bed facility that also coal ash (d_p= 4 mm) can be used to desulfurize coal gas at temperatures of 700-1090 K, pressures of 6.5-11.4 bar, and space velocities of 800-2000 hr^{-1}. At 1000 K and a H_2S inlet concentration of 1.25 vol.%, the coal ash removed more than 99% of the H_2S, a result which was in accordance with the equilibrium concentration prevailing in the gas outlet. Sorbent capacities higher than 60 wt.% were observed, increasing with temperature. Spent beds were regenerated up to seven times (by diluted air, at 920 K) without any significant effect on the sorbent quality. Decrepitation of the sorbent was not mentioned by these authors. They continued their research with experiments in a lab-scale bubbling fluidized bed (see section 5.2).

Reaction rate measurements by Tamhankar *et al.* [19], performed with mixtures of iron oxide and porous silica in a laboratory thermogravimetric analyzer (875 - 1175K), indicated that after an initial fast reduction (by H_2 and CO) of Fe_2O_3 to a spongy iron, a slow reaction occurred in which the hydrogen sulfide was bound to the iron to form iron sulfide. This result has been confirmed by Van der Wal [12] who showed in lab-scale fixed bed measurements that for unsupported iron oxide particles of 0.5 to 1 mm diameter, the reaction rate is indeed low during the first sulfidation. However, immediately after the first regeneration the absorption capacity and the reaction rate is increased significantly due to the creation of additional surface area. This is caused by i) breakage of the particle structure and a corresponding increase in porosity, due to differences in the molar volume of the solid reactants and products (Fe_2O_3, Fe_3O_4, FeO, FeS, $FeSO_4$) involved in the various possible reactions, and ii) the redistribution of FeO in a finely dispersed form. Unfortunately, for unsupported iron oxide the physical strength is lost completely and the structure of the particle collapses during subsequent absorption/regeneration cycles.

3.2 Iron-Oxide Impregnated Sorbents

A solution for the problems of low reactivity and risk of particle decrepitation has been proposed by Van der Wal [12]. He prepared an iron oxide-on-silica sorbent by deposition-precipitation of iron (III) ions in a suspension of the silica carrier. This yields a strong and stable sorbent with the iron oxide finely dispersed in the porous structure of the silica carrier. Van der Ham *et al.* [9] provide some information on the fixed bed laboratory work by Van de Wal in another chapter of this book.

In subsequent work at the University of Twente, the reactivity and the mechanical strength of the sorbent were further increased by depositing the iron-oxide on a mechanically very strong carrier material, viz. fluid cracking catalyst [8]. The average diameter of the carrier material was chosen such that it could be directly applied in future, continuously operated (fluid bed related) reactors (d_p=100 μm). Initial work to determine its reactivity and capacity was carried out again in a lab-scale packed bed reactor at atmospheric pressures with a simplified, artificial coal gas. At moderate operating temperatures of 673 K, the sorbent exhibited a high removal efficiency (> 99%, for H_2S inlet concentrations ranging from 1000 to 7500 ppm), and very high initial reactivities (contact times of 0.04 s) during more than 25 absorption/ regeneration cycles. It was also shown that, despite the small average particle diameter applied, for the fresh material gas-to-particle diffusion was still the rate-limiting step, which means that the intrinsic reactivity must be extremely high. Regeneration was conducted under mild conditions, using a 0.5-0.7 vol.% O_2 in N_2 stream. At a temperature level higher than approximately 873 K, a considerable loss in reactivity due to sintering was observed.

Obviously, fixed bed reactors have been applied mainly on a small scale, viz. to develop and characterize various sorbent materials. A shift in interest can now be noticed from unsupported sorbents (iron ore, fly or bottom ash from coal gasifiers, and mixtures) towards sorbents on basis of a strong porous carrier material impregnated with iron oxide whether or not in combination with other metal oxides. As the iron-oxide impregnated sorbents have a considerably higher reactivity and stability, they can indeed be used in the desired temperature window of 600 to 800 K.

Large-scale packed bed desulfurization units have been realized only in Japan. After successful operation of a hot gas cleanup unit for 2 tons coal per day, in 1993 a pilot-plant packed bed installation was realized in Nakoso, for cleaning of hot gas slip stream corresponding to 20 tons coal per day [20, 21], see figure B. This unit consisted of three reactors, viz. for sulfidation, regeneration, and reduction, and was directly connected to a 200 t/d entrained flow gasification power generation plant. To cope with the risk of sorbent decripitation, a honeycomb type fixed bed was applied, containing an iron-oxide/titanium-oxide sorbent. In the separate reduction step, the generated SO_2 was recovered as elemental sulfur.

Fixed beds are characterized by a low effective heat conductivity and small heat transfer rates to the wall (see table I). In case of reactions with a large heat effect, this can lead to undesired temperature gradients, culminating in hot spots or a moving heat wave which may ruin the sorbent. Besides, a continuous fixed bed system must be operated in swing mode, and switching of the gas flow between two or more beds demands manifolds and valves to be applied at high temperatures and pressures. Recently, the attention in process development has been focused more on reactors which can be operated continuously by solids transportation through the absorber and the regenerator. The increased degree of solids mixing in such a system also allows a better control of the reactor temperature. In figure A to G a number of possible configurations (including fluid beds) is presented.

4 Moving Bed (MB)

A simple form of continuous operation is the moving bed mode, see figures A [4] and G [21]. Basically, it is a fixed bed of coarse solid particles which are slowly added and removed from the bed. Transport of the solids takes place by gravity, and gas is fed either counter- or co-currently, or in cross flow with respect to the solids main-flow direction. In general, gas and solids are moving essentially in plug flow, while relatively large particles have to be applied. The maximum size is limited by the requirement of an acceptable conversion rate per unit of sorbent volume, which decreases for larger particles due to the increasing effects of external and internal mass transfer limitations. The minimum particle size is determined by factors like pressure drop or, in case of counter-current operation, the risk of particle fluidization in the reactor. Major disadvantages are the possible blockage of the particle flow (bridging), and the need for a higher sorbent strength than possible in a fixed bed. On the other hand, scale-up of moving beds is still relatively simple.

The only large-scale moving bed reactor application with iron oxide sorbents was, as far as we know, undertaken in Japan [21, 22, 23]. Iron oxide on an alumina-silica based porous carrier was selected as a sorbent material after extensive testing in a fixed bed lab-scale facility (I.D. 0.221 m). On laboratory scale, no drop in breakthrough time was observed for repeated sulfidation and regeneration during up to 2000 cycles at atmospheric pressures and temperatures of 773 K. A pilot plant moving bed reactor, corresponding to a 4 ton coal per day gasification system, was installed at the same Nakoso plant as mentioned already in section 3. Tests were performed in three (parallel) cross-flow moving beds at 26 bar and 673 K. The exit H_2S concentrations were in the range of 35-130 ppm, corresponding to a 78-95 % sulfur removal efficiency. Regeneration temperatures above 673 K were necessary for complete regeneration. Unfortunately, the applied high oxygen partial pressures decreased the regeneration efficiency by promoting the formation of sulfates, which appeared to be stable up to 840 K. In a subsequent absorption step these sulfates can be reduced to FeS by the coal gas thus limiting the absorption capacity. To eliminate this problem, a sorbent reductor decomposing the sulfates in the regenerated sorbent was additionally installed.

5 Fluidized Beds

A fluid passing upwards through a bed of particles, initially percolates through the void spaces between the (stationary) particles. At increased flow rates, particles begin to move apart, and if all the solid particles are suspended by the gas (i.e. the frictional forces between the particle and fluid counterbalance the weight of the particles) the bed is considered to be at minimum fluidization conditions; the corresponding gas velocity is called the minimum fluidization velocity. For type A powders, the bed first remains homogeneous in nature. It becomes heterogeneous slightly beyond U_{mf}, at the minimum bubbling velocity U_{mb}. For B/D powders, the bed is immediately fluidized heterogeneously at superficial gas velocities above U_{mf}. Another important difference between A and B powder fluidization deals with the bubble growth. In beds of type A particles, bubbles grow quickly to a few centimeters in size, and remain then constant in size as a result of the equilibrium between coalescence and splitting. However, rising bubbles in B-type fluid beds keep growing continuously, and in small-diameter beds these bubbles can easily reach a size equal to the bed diameter, which results in slugging behaviour.

Main disadvantages of fluidized beds are:

- complex flow behaviour of the gas which is divided over a bubble and an emulsion (or dense) phase,
- possible gas by-passing through rising bubbles due to a limited gas exchange between the bubble and dense phase (especially for A powders),
- difficult scale-up,
- high rates of solids attrition losses,
- erosion of internals and vessel walls.

However, some major advantages are also generally recognized:

- transport of particles in or out the fluid bed reactor is possible, and can be controlled easily,
- rapid mixing of solids through the action of rising bubbles, resulting in nearby isothermal conditions over the entire reactor volume,
- high (gas-to-particle and bed-to-wall) heat and mass transfer rates,
- small particle sizes, which allows an effective use of the catalyst or sorbent.

In industry, a fluid bed is commonly used when a continuous gas-solid reactor is required. Different types of fluidization are nowadays recognized: homogeneous, bubbling, turbulent or fast fluidization, and pneumatic transport. Besides the gas properties and the applied superficial gas velocity, the particle properties are decisive for the type of fluidization. Geldart [24] came up with four clearly recognizable powder types in a classification which is still often referred to. A brief summary may be useful here.

- Group C: cohesive, or very fine powders (typical $d_p<20$ μm, $\rho_s<\approx1000$ kg/m^3). Normal fluidization of these particles is extremely difficult, because of inter-particle forces. Examples are starch and flour.
- Group A: aeratable, and easily fluidizable particles, with homogeneous fluidization at low gas velocities, and relatively small bubbles at higher velocities ($30<d_p<100$ μm, $\rho_s<\approx2000$ kg/m^3). Typical examples are FCC catalyst, and small γ-alumina particles.
- Group B: sand-like particles ($100<d_p<800$ μm, $1400<\rho_s<4000$ kg/m^3), showing vigorous bubbling with large bubbles.
- Group D: spoutable, and large (dense) particles (typical $d_p<1$mm). Large, fast growing bubbles, are observed. Examples are coffee beans, and some roasting ores.

For a detailed description of the characteristics of these materials, the reader is referred to the original paper of Geldart, or the more recent work carried out by Grace [25]. For details on fluidization engineering, the extensive review by Kunii and Levenspiel [26] is recommended.

In the following sections, the different fluidization regimes will be considered in relation to H_2S removal from hot coal gas, while going from small to high superficial gas velocities.

5.1 Bubbling Fluidized Bed (BFB): laboratory work

Schrodt and Mohan [27] applied their coal-ash sorbent in a small fluidized bed (I.D. 2 cm and a height of 10 cm) to remove hydrogen sulfide from a simulated (low caloric value) fuel gas at atmospheric pressures. Three average particle diameters were applied, varying from 90 to 200 μm (type B powders). The operating temperature was varied from 750 to 1000 K. At the applied superficial gas velocities (up to twice the minimum fluidization velocity: 5-10 cm/s), the bubbles in the bed are still relatively small in number and size, which implies poor mixing of the bed and (at industrial scale) an unrealistic good mass exchange between bubble and dense phase. The authors reported that desulfurization started at the entrance of the sorbent bed, and progressed under a concentration wave of H_2S that moved very slowly within a reaction zone through the bed. This indeed indicates poor mixing, due to the low gas velocities applied and the corresponding lack of rising bubbles (which are known to induce the solids mixing). The consequence of this bad solids mixing however, was the reduction of the Fe_2O_3 in front of the reaction zone, where a strong reducing gas is present ($H_2/H_2S \gg 100$), to Fe_3O_4 and sometimes even FeO (wüstite). It is interesting to note that a similar phenomenon, but with more severe consequences, was observed for ZnO-absorbens in packed beds: due to the reduction of ZnO in strongly reducing atmospheres at high temperatures, metallic zinc volatilizes, which results in a loss of absorption capacity [28].

Referring to the rate limiting steps 1 to 5 mentioned in section 2, an additional mass transfer step must be introduced if a fluidized bed is considered, viz. the resistance for transfer of the reactant from the bubbles to the emulsion phase. However, in the fluidized bed set-up used by Schrodt and Mohan [27], very low gas velocities were applied, just beyond the minimum bubbling point. The bubble fraction is then likely to be small and the gas passes through the reactor mainly through the emulsion phase. At the same time, the bubbles are, on the average, small in size and have a large specific surface area available for mass exchange. Therefore, the bubble-to-emulsion mass transfer resistance has presumably been negligible in their experiments. Altogether, it is not surprising that sulfur removal efficiencies of more than 95% have been recorded (for a H_2S inlet concentration of 0.5-1 vol.%). The observed sulfur capacity ranged from 55 to almost 90% at 10 % breakthrough, and increased with increasing temperature. Regeneration of the sorbent was carried out by fluidizing with a 3 vol.% O_2 in nitrogen gas mixture. In contrast to the observations of investigations in fixed beds [f.i. 19], no significant decrepitation of any of the coal ashes used was measured after 10 days of fluidization at room temperatures, nor after 60 days at the high temperature level of 750 to 1000 K.

Most of the work concerning H_2S removal in fluidized beds has been carried out on laboratory scale. A typical set-up is presented in figure E [29]. Unfortunately, on this small scale, major practical problems related to gas by-passing, solids segregation effects, heat transfer, etc. are minimized. It remains often unclear how results obtained in small-scale reactors can be applied for the design of commercial installations.

5.2 Bubbling Fluidized Bed (BFB): pilot plant / commercial scale work

Despite difficulties related to fluid bed disadvantages as listed above, a process using pulverized iron oxide at high temperatures for removal of hydrogen sulfide from a coke oven was already applied on pilot plant scale (3,000 m^3/hr) in the 1950s in England by the South Western Gas Board [30]. In this fluid bed process, desulfurization was accomplished at operating temperatures between 600 and 675 K, and space velocities of 3000 hr^{-1}. It resulted in H_2S removal efficiencies over 95%. As this pilot-plant project proved the technical feasibility of the process, a full-scale commercial plant (40,000 m^3/hr) was installed afterwards, to treat coal-derived gases at the Exeter Works of the South Western Gas Board. Pulverized iron oxide particles, varying in diameter from 0,15 to 1 mm, was applied as the sorbent material. To avoid excessive gas by-passing through rising bubbles and a corresponding decrease in sulfur removal efficiency, staging was applied: the absorber was built up of four fluidized bed sections. Regeneration of the spent sorbent was carried out with 8-10 vol.% O_2 (presumably diluted air). Operation of the commercial plant revealed some technical problems, but none of them was believed to be insurmountable. Unfortunately, decrepitation and attrition of the sorbent occurred after frequent cycling. Overall oxide losses were appreciably higher than had been anticipated. Besides, low revenues from the produced sulfuric acid were obtained. Eventually, economic concerns largely dictated the shutdown of the commercial scale plant in 1965. Operation of the pilot plant desulfurization plant has also been terminated.

In Yubari, Japan, iron oxide was applied as the sorbent material in a pilot-scale multi-stage fluidized bed to desulfurize coal gas derived from a fluidized bed gasifier (40 ton coal per day). Subsequently, Ishikawayama-Harima Heavy Industries (IHI) and the New Energy and Industrial Technology Development Organization (NEDO) developed a commercial scale desulfurization process (46,300 Nm^3/hr, corresponding with 200 ton coal per day) on basis of a two-stage bubbling fluidized bed desulfurizer. Relatively mild operating temperatures of 700 K and pressures of 26 bars were applied. A schematic flow-sheet is presented in figure C [21]. The process includes an absorber, a regenerator, a SO_2-reducer and a sulfur condenser. Unsupported iron ore (Geldart type B, d_p=150 μm) was taken as a sorbent because it is inexpensive and readily available. As this type of fluidized bed material yields large, fast rising bubbles, the H_2S removal capacity may be reduced due to the formation of an extra (bubble-to-dense phase) mass transfer resistance. Such effects were minimized by reducing the absorber bed height, and carrying out the absorption in a two stage process. At 100-460 ppm inlet conditions, the exit H_2S concentration ranged from 25 to 50 ppm (the design specification being 100 ppm), which corresponds to a removal efficiency range from 87 to 92%. These figures are limited by the large water content in the coal gas, which increases the attainable H_2S (equilibrium) concentration in the coal gas. The overall removal efficiency could not be improved by using a mixture of Fe_2O_3 and TiO_2 as a new sorbent material. Besides, the attrition losses appeared to be more than five times higher than for the pure Fe_2O_3. The sorbent is regenerated by air in a separate bubbling bed. The produced SO_2 is then reduced very efficiently to elemental sulfur (sulfur recovery > 99%) in a fluid bed of anthracite.

Table 2. Operating conditions of the Japanese fluid bed plants [2, 20, 21, 22, 23].

	laboratory		40 t/d pilot plant		200 t/d pilot plant	
	abs.	reg.	abs.	reg.	abs.	reg.
P (bar)	1-11	1-11	14-20	14-20	≈ 24	
T (K)	730-925	975	700-760	875-1025	615-690	910
u_g (m/s)	0.065	0.0945	0.3	0.3	0.4[a]	
I.D. bed (m)	0.04	0.04	0.84	0.32	1.87	
bed height (m)	0.04-0.12	0.0318	0.7-1.0	0.55-1.2		
d_p (μm)	127-150		150		150[a]	
gas comp. (vol.%)						
H_2	12-13	-	12-13	-	9.4	-
CO	10-16	-	10-13	-	22.5	-
CO_2	10	-	10-13	-	5.4	-
H_2O	10-30	-	10-30	-	-	-
H_2S	0.1-0.3	-	0.1-0.3	-	0.03	-
N_2	bal.	bal.	bal.	bal.	62.2	bal.
O_2	-	6	-	4-8	-	2.4-5.3

[a]estimated.

Table II shows the experimental conditions of the Japanese installations. An important requirement for the success of the process development is to reduce the sorbent attrition and entrainment rates. Current research is devoted mainly to the development of supported sorbentia [21].

Despite the success of fluidized bed processes, and of supported iron-oxide particles as a sorbent material, some major problems still remain. They may range from attrition losses for unsupported sorbents, to the formation of sulfates (both occurring in the IHI/NEDO plant), and a loss in sorbent activity due to sintering (T> 900 K; 8, 31). The risk of sorbent sintering is obviously related to the presence of hot spots, which may even occur in a fluidized bed reactor.

On pilot-plant and commercial scale, sulfur removal efficiencies of fluidized beds are usually significantly lower than expected on basis of small-scale experiments. This effect is well known from the (petro)chemical industry; scale-up of fluidized bed reactors has been a research topic for many decennia and various design models have been proposed in the literature [see f.i. 32, 33]. The origin of any difficulties in scaling-up fluidized bed reactors is in their complex hydrodynamics. For instance, when A- or B-type particles are applied for a bubbling bed, the mass transfer between the bubbles and the dense phase of the bed may be a limiting factor for the conversion of a gaseous reactant. If that is

true, the total mass exchanging surface (specific bubble surface) becomes a crucial factor. It is not easy to predict this surface, as it depends on the properties of the gas and bed particles, the type of gas distributor plate, and the bed geometry. The progress in computational fluid dynamics may enhance the understanding of bubble formation and growth in two phase systems, and eventually contribute to improvement of scaling rules for fluidized beds [34].
Some problems which are typical for a bubbling bed of A or B powders, can be overcome by applying a *turbulent* bed [26], which is operated at relatively high superficial gas velocities (up to 10 times the terminal falling velocity of the individual particles). A turbulent bed has excellent heat transfer properties and excludes the problem of possible gas by-passing through bubbles.

5.3 Turbulent Fluidized Beds (TFB)

It is a widespread misunderstanding that the use of fine particles (with an average diameter smaller than 100 μm) in fluidization would lead to extreme low superficial gas velocities and avoiding of excessive entrainment of bed material. Contrary, such particles can in fact be fluidized at velocities far beyond the terminal falling velocity of the individual particles. However, a totally different behaviour then emerges. The surface of the bed becomes diffuse, and a turbulent motion of solids clusters and voids of gas of various sizes and shapes is observed. This high-velocity regime (up to 1 m/s superficial gas velocity for 100 μm diam. particles) for fine powders is referred to as the turbulent fluidized bed regime. In this regime, relatively small amounts of solids are entrained over the top of the column, while there is still a dense bed retained in the bottom part. The (local) solids hold-up is only a function of the applied superficial gas velocity. For steady state operation, the entrained particles have to be collected by cyclones, and returned to the bed. Despite its widespread use in commercial operations (f.i. the regenerators in modern cat-cracking units), the advantages over conventional fluidized bed reactors are still not always recognized. The most significant ones are:

- a higher gas throughput,
- a shorter and more uniform gas phase residence time,
- a shorter and more uniform solids residence time,
- higher (bulk)gas-to-particle mass and heat transfer rates,
- higher heat transfer rates to submerged surfaces, and vessel walls.

Nowadays, it is more and more recognized that the onset of turbulent fluidization is marked by u_c, the gas velocity for which a maximum is observed in the pressure fluctuation [35, 36, 37, 38, 39]. The transition point towards turbulent fluidization depends, amongst others, on particle type, reactor configuration and gas properties. As a rule of thumb, for increasing particle sizes the ratio of u_c/u_t ranges from 6 to 3 for type-A particles. This ratio decreases when the fluidized particles become heavier and/or larger; typical values for type B particles are 3 to 1. Although the turbulent fluid bed operation is nowadays widely applied in FCC regenerators, only a single laboratory study has considered this regime for batch-wise absorption and regeneration of a desulfurization sorbent [40]. As was shown also in the Japanese work discussed before [23], regeneration of the sorbent can have serious consequences for the sulfidation efficiency, in case of simultaneous formation of sulfates. This happens especially at high oxygen concentrations and results in a loss of sorbent activity. Van Yperen [31] overcame this problem by applying an improved sorbent material, viz. a mixture of molybdenum oxide and iron oxide on amorphous aluminium phosphate as carrier material, supported by the Netherlands Energy Research Foundation ECN. The combination of these three material in various weight ratios (5.8-25 wt.% Fe_2O_3 and 0-11 wt.% MoO_3) proved to have the following advantages:

- the addition of molybdenum oxide suppresses the formation of iron sulfate, (formation of molybdenum sulfate was never observed),
- the activity of the sorbent is less sensitive towards the presence of water in the coal gas,
- no iron-silicates are formed,
- the carrier material is thermally much more stable than silica.

In the laboratory study of Venderbosch *et al.* [40], also supported by the ECN, sulfidation/regeneration experiments were carried out for this sorbent material. A Geldart B-type material (170 μm, u_t= 0.65 m/s) was fluidized in a lab-scale set up (I.D. 0.05 m) at superficial gas velocities of 0.2 to 1.3 m/s (see figure D). Unfortunately complete turbulence could not reached due to the specifications of the available sorbent and the limitations of the experimental set up. Yet, the applied velocity range is already far beyond the terminal falling velocity of the individual particles. Excellent fluidization was accomplished without excessive particle entrainment. The sorbent appeared to have a sulfur capacity of 5 wt.%. At a temperature of 675 K, well reproducible and high values for the initial H_2S removal degree (> 99.5%) were observed, even at the highest gas velocities, and over more than 50 cycles. Until break-through, the H_2S outlet concentration remained always less than 10 ppm. Complete regeneration (also at 675 K) of the sorbent was obtained with oxygen in concentrations up to 20 vol.%, without any effect on sorbent capacity and reactivity for over 50 cycles. In some separate experiments, reduction of the sorbent with hydrogen has been carried out to estimate the degree of iron sulfate production. The remaining sulfate formation for this improved sorbent was limited to about 5 wt% of the total absorbed amount of sulfur.

Future research for still smaller sorbent particles (Geldart type A powder) and higher velocities must demonstrate if the sulfur removal degree can be further increased in the regime of full turbulent fluidization where mass transfer resistances are minimized.

5.4 Risers of Circulating Fluidized Beds (CFB)

At still higher superficial gas velocities, that is beyond the so-called transport velocity u_{tr} of the specified particles, no dense bed can be maintained anymore: all the particles are blown over the top of the reactor column. A correlation for the estimation of this transport velocity is presented by Bi and Fan [42]. In literature, various terms have been used for this fluidization regime; in this paper it will be indicated as the riser regime. Because the (local) solids hold-up (ranging from 1 to 10 vol.%) is a function of the gas velocity and the solids flux, smooth and steady circulation to the riser bed through a standpipe is essential for good operation.

The same advantages as listed above for a turbulent fluidized bed (high throughput, short residence times, excellent gas-to-particle mass and heat transfer) are even more valid for the riser regime. An exception is the bed-to wall heat transfer, which is a little worse due to the decreased particle concentration in the column. An additional advantage in comparison with the turbulent regime is the decreased solids back-mixing; plug flow conditions are closely approached for both gas and solids. Besides, a riser reactor is well suited to deal with very cohesive powders which are otherwise difficult to fluidize [43]. It is nowadays generally accepted that in the developed region of the riser (viz. the region above the so-called acceleration zone) differences between the gas velocity and the solids velocity (the so-called slip velocity) can be much higher than those calculated on basis of the terminal falling velocity of the individual particles. Besides, it seems that the relatively high (bulk-to-particle) mass and heat transfer rates are still lower than expected on basis of the individual particle approach. Both observations can be explained by segregation of solids over the reactor diameter, and/or the formation of agglomerates or clusters of particles.

From some preliminary experiments in the small packed bed by Van der Ham [8], it was already concluded that reaction rates for the desulfurization of (simulated) coal gas over his supported iron-oxide are so high, that gas residence times in the order of seconds are sufficient. On basis of this unique feature, a circulating fluid bed was considered as a potential absorber, allowing high gas throughputs (avoiding also unfavourable large reactor diameters of high-pressure vessels) and good gas-solids contacting [44]. Sulfidation and regeneration experiments with iron-oxide on FCC carrier as a sorbent material were successfully performed in the set-up presented in figure F. Reaction rates for the fresh material appeared to be so fast, that at nearly all conditions gas-to-particle mass transfer limitation occurred. The CFB is a suitable reactor for the desulfurization process, although the gas-solids contacting is still not ideal [40].

In high-velocity fluidization, careful consideration should be made regarding the entrainment of particles, especially if expensive sorbent materials, like the supported iron-oxide based sorbents, are applied. It should be noticed that the experimental work concerning desulfurization in high velocity fluid beds is, until now, limited to laboratory scale. Scale-up of these reactors for desulfurization is still a problem to be considered in future research.

6 Possible Regeneration Routes

It seems now justified to conclude that three major parameters dictate the achievable sulfur removal efficiency: i) the choice of sorbent, and its thermodynamic properties, ii) the reactor choice, which should be directly linked to sorbent design, and iii) the sorbent regeneration method. In literature, most attention is directed towards the sorbent design, especially in relation to the desulfurization reaction. However, during the last decade, it is well recognized that the regeneration method can have serious complications for the desulfurization processes. Generally, there are 3 possible regeneration routes:

1) The first one is to recover elemental sulfur from the (by-)products of the desulfurization and regeneration reactions. Sulfur dioxide is obtained by regenerating part of the sulfided sorbent with air or oxygen, and then passed over the remaining FeS at temperatures between 875 and 1275 K [18, 45].

$$3\,FeS + 2\,SO_2 \leftrightarrow Fe_3O_4 + 2.5\,S_2$$

The sulfur production rate is controlled by equilibrium conditions and the maximum possible conversion of sulfur dioxide to sulfur at 1000 K is limited to approximately 5 %. Besides, the sulfur production rate drops sharply with increasing conversion. Consequently, for industrial purposes, high solids residence times (large reactors) are required, in combination with a large recycle of SO_2 containing gases (the recycle flow must be cooled and reheated to separate the elemental sulfur!). These factors indicate that recovery of elemental sulfur in this way is both technically and economically unattractive. For iron oxide, this regeneration route is certainly not yet a good option.

2) A second possibility explored by Tamhankar *et al.* [46] is the reaction of the sulfided sorbent with steam and steam-air mixtures, the latter yielding H_2S and SO_2. Then the most likely overall (parallel) reactions are:

$$3\,FeS + 4\,H_2O \leftrightarrow Fe_3O_4 + H_2 + 3\,H_2S$$
$$FeS + 3\,H_2O \leftrightarrow FeO + SO_2 + 3\,H_2$$

Results in a TGA showed that the conversion rates are much too slow, the rate being controlled predominantly by the intrinsic chemical reaction. Consequently, regeneration with steam is neither a good option.

3) Regeneration of the sorbent with (diluted) air is perhaps the most feasible option. This strongly exothermic reaction yields SO_2 and Fe_2O_3:

$$4\,FeS + 7\,O_2 \leftrightarrow 2\,Fe_2O_3 + 4\,SO_2$$

In this reaction no elemental sulfur is obtained, although in laboratory experiments the formation of elemental sulfur has been observed occasionally [1, 12] under quite special conditions. As the regeneration reaction of the sorbent with oxygen is fast and complete, this regeneration route offers, up till now, the best potential for the iron based sorbents. High SO_2 concentrations are preferred and undiluted air must be considered as a reactant gas for the regeneration. Because the heat of the regeneration reaction is proportional to the oxygen feed concentration, and therefore also directly related to the required SO_2 outlet concentration, it is of utmost importance to select a reactor in which the released reaction heat is smoothed out rapidly over the reactor volume. A good solids mixing is required to suppress the occurrence of hot spots and ensure a uniform reactor temperature. Moreover, the rate of heat transfer to cooling surface (wall and/or internals) must be as high as possible. A fluid bed reactor is known for its good mixing and heat transfer characteristics, and therefore emerging as a suitable reactor for regeneration. The rate of heat transfer to the wall or immersed surfaces is especially high in fluid beds of relatively fine particles [47].

At the University of Twente, some regeneration work has been carried out in a small fluid bed set-up. Van der Ham *et al.* [8] performed successive regeneration and absorption cycles for his FCC-supported iron oxide sorbent. A major drawback of the iron sulfide was identified: during the regeneration of the absorbent, iron sulfate is formed according to the overall reaction:

$$2\,FeS + 4\,O_2 \leftrightarrow 2\,FeSO_4$$

To overcome the difficulties involved with this side reaction (such as the corresponding decrease in sorbent capacity and the possible production of SO_2 during a subsequent absorption cycle due to the presence of reducing gases), the regenerated sorbent has to be reduced with H_2 in a separate reduction step, according to:

$$3\,FeSO_4 + 2\,H_2 \leftrightarrow Fe_3O_4 + 3\,SO_2 + 2\,H_2O$$

The Netherlands Energy Research Foundation ECN supported research at the Universities of Utrecht [31] and Twente [40], aimed at the development of a sorbent improved with respect to the undesired sulfate formation. Van Yperen showed that the addition of molydenum-oxide to the iron-oxide on an amorphous aluminium phosphate carrier suppresses the formation of the iron sulfates. Regeneration of this sorbent material with undiluted air in a turbulent fluidized bed set-up, was carried out batch-wise by Venderbosch. No loss in sulfur capacity was observed up to 40 cycles, without intermediate reduction.

7 Conclusions

In the eighties, lab-scale fixed bed reactors have been applied to develop and characterize suitable sorbents for high temperature desulfurization of coal gas. Important sorbent-type related parameters are the reactivity, equilibrium conditions, capacity, regenerability and the occurrence of undesired side reactions. In the few larger-scale tests, another important parameter appeared to be the mechanical strength of the sorbent. This was the reason for the development of supported or impregnated sorbents. In case of iron oxide a shift in interest could be noticed from the inexpensive unsupported materials like ore, and mixtures of iron oxide with coal ash, to artificial sorbents prepared by impregnation of iron ions on a porous carrier material (silica, alumina or FCC particles).

The application in continuous processes requires solids transportation from the absorber to a regenerator, if swing operation is to be avoided. This leads to the selection of fluid beds as attractive reactors. Fluid bed technology is well known and widespread through industry; it is applied especially when large heat effects are involved in the process. A major drawback of a bubbling fluid bed is the possible by-passing of reactant gas through rising bubbles. This can be avoided by the application of high-velocity fluid beds (a turbulent bed or a riser), which is possible for attrition-resistant and very reactive sorbent particles. An additional advantage of such a high-capacity fluid bed reactor is that its diameter is minimized, which is especially attractive for high-pressure application.

The overall H_2S removal efficiency of a continuous desulfurization process is not only determined by the sorbent particle properties and the reactor selection, but also by the method of regeneration. In case of iron-oxide based sorbents, regeneration with air (which yield a gas stream containing up to 20 vol.% SO_2) seems the best option. The regeneration reaction is however highly exothermal and can only be carried out properly in a fluid bed with its good heat transfer characteristics. Another problem is the formation of iron sulfate during regeneration, leading to a reduction in sorbent capacity and possible SO_2 formation during absorption.

The development of a strong and stable sorbent type (a mixture of molybdenum oxide and iron oxide on amorphous aluminium phosphate) in The Netherlands, by ECN in cooperation with the Universities of Utrecht and Twente, seems to be a good starting point for the development of a fluid bed desulfurization process (similar to the well known catalytic cracking process) in a circulating system, with a riser for the absorption step and a turbulent fluid bed as the regenerator.

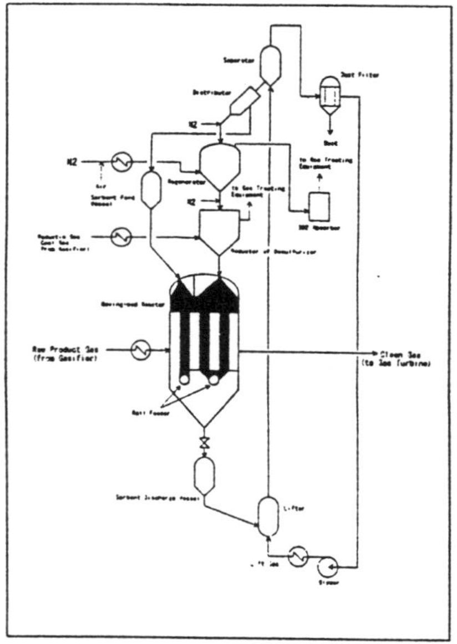

A. Multi-stage Moving Bed (Nakayama *et al.*, 1996). Pilot plant scale

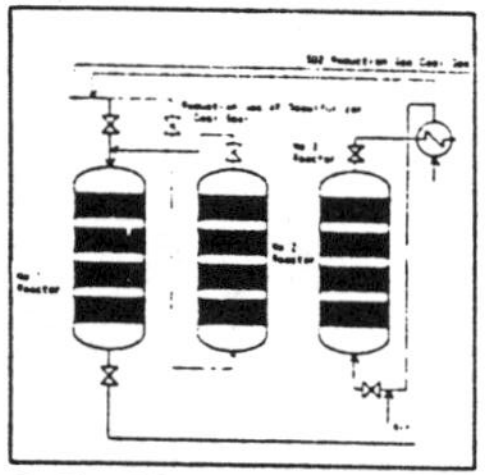

B. Fixed Bed (Nakayama *et al.*, 1996). Pilot plant scale

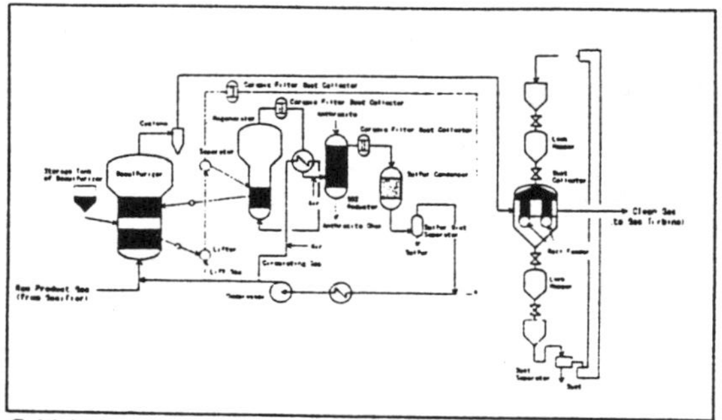

C. Multi-stage Bubbling Fluid Bed (Nakayama *et al.*, 1996). Commercial scale

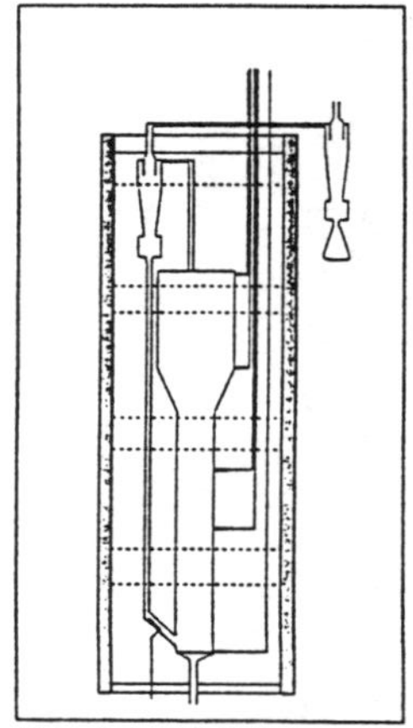

D. Turbulent Fluid Bed (Venderbosch *et al.*, 1997). Lab-scale

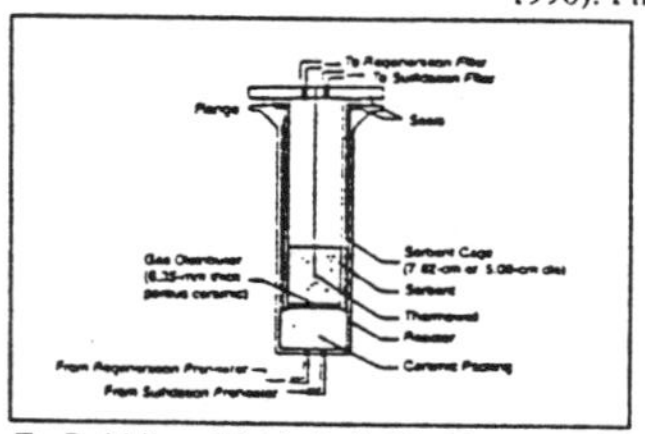

E. Bubbling Fluid Bed (Swisher *et al.*, 1995).

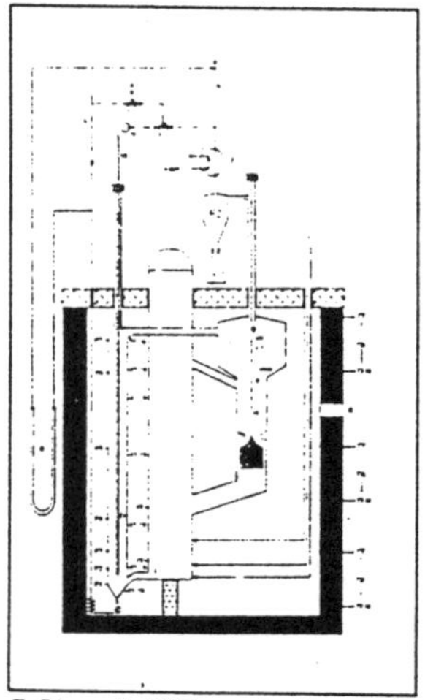

F. Riser of a CFB (van der Ham *et al.*, 1994). Lab-scale

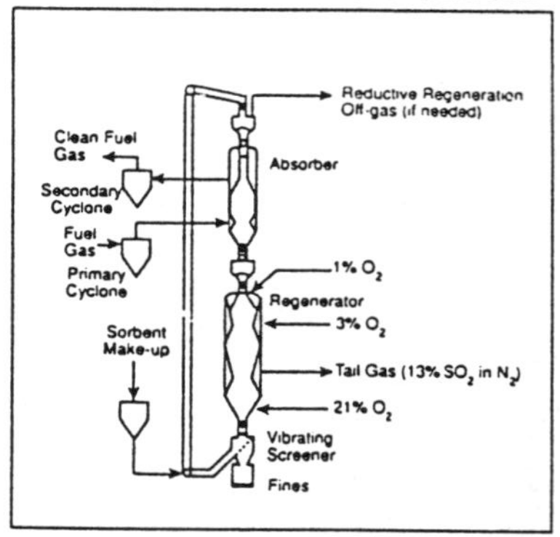

G. Moving Bed (Gangwal, 1991). Pilot plant scale

References

1. Schrodt J.T., 1980, Hot desulfurization I: use of gasifier ash in fixed bed process, US DOE Report DOE/ET/10463-Ti, Washington DC
2. Sugitani T., Kakai J., Yamamoto H., Kawai I., Kobayashi S., Ishii T., Hozumi S., 1987, Fundamental studies and present status of IHI hot gas desulfurization process, International Conference on Coal Science (edited by J.A. Moulijn et al.), 915
3. Kobayashi S., 1990, Test results of Yubari Hot gas desulfurization pilot plant, Coal Conversion Technology Department Ishikawajima-Harima Heavy Industries Co. Ltd
4. Gangwal, S.K., 1991, Hot-gas desulfurisation sorbent development for IGCC systems, in: ICHEME Symp.Series, 123, 159
5. Ayala R.E., Feitelberg A.S., Furman A.H., 1995, Development of a high-temperature moving bed coal gas desulfurization system, in: Proceeding of the 12th annual international Pittsburgh coal conference (edited by Chiang S.H.), University of Pittsburgh, 1053
6. Westmoreland, P.R., Harrison D.P., 1976, Evaluation of candidate solids for high-temperature desulfurization of low BTU gases, Env. Sci.,Techn., 10(7), 659
7. Heesink, 1994, High temperature coal gas desulphurization: applying calcareous sorbents and fluidized bed contactors, PhD thesis University of Twente
8. Van der Ham, 1994, Regenerative high temperature desulfurization of coal gas in a circulating fluidized bed, PhD thesis University of Twente
9. Van der Ham A.G.J., Venderbosch R.H., Prins W., van Swaaij W.P.M., 1997, Survey of Desulfurization Processes for Coal Gas, in these proceedings
10. Krishna R., Sie S.T., 1995, Strategies for multiphase reactor selection, Chem. Engng. Sci., 24A, 4029
11. Schrodt J.T. Hilton G.B., Rogge C.A., 1975, High-temperature desulfurization of low-CV fuel gas, Fuel, 54, 272
12. Van der Wal W.J.J., 1987, Desulfurization of process gas by means of iron-oxide-on-silica sorbents, PhD thesis University of Utrecht, the Netherlands
13. Jain S.C., Gupta R., Gangwal S.K., 1991, Development of zinc ferrite sorbents for desulphurization of hot coal gas in a fluid-bed reactor, A.I.Ch.E. Summer Meeting, Pittsburgh, Pennsylvania, USA
14. Mojtahedi W., Abbasian J., 1995, H_2S removal from coal gas at elevated temperature and pressure in fluidized beds with zinc titanate sorbents: 1. cyclic tests, Energy and Fuels, 9, 429
15. Ayala R.E., Jain S.C., 1992, Development of durable mixed-metal oxide sorbents for high-temperature desulphurization of coal gases in moving-bed reactors, A.I.Ch.E. Annual Meeting, no. 239d
16. Khara G.P., Delzer G.A., Kubicek D.H., Greenwood G.J., 1995, Hot gas desulfurization with Philips Z-Sorb sorbent in moving and fluidized bed reactors, Environmental Progress, 14, 146
17. Edwards M.S., 1979, H_2S-removal processes for low-BTU coal gas, Department of Energy/Fossil contract no. W-7405-eng-26
18. Schrodt J.T., Best H.E., 1978, Sulfur recovery from fuel gas desulfurization sorbents, AIChE. Symp.Ser., 74(175), 184
19. Tamhankar S.S., Hasatani M., Wen C.Y., 1981, Kinetic studies on the reactions involved in the hot gas desulfurization using a regenerable iron oxide sorbent-I, Chem. Engng. Sci., 36, 1181
20. Suehiro M., Fujishima H., 1993, Fixed bed type hot gas cleanup technology for integrated coal gasification combined cycle power generation, 5th Japan/USA technical meeting on surface coal gasification, October 13, Tokyo, Japan
21. Nakayama T., Araki S., Takahata E., Takahashi A., 1996, Development of hot gas cleanup technology for IGCC- technical trends of 3 types hot gas cleanup process., J. Japanese Inst. of Energy, 75, 351
22. Ichikawa K., 1996, Test results of the IGCC system by the 200 t/d Nakosa pilot plant, paper presented at the 6th Japan Australia Joint Technical Meeting on Coal, Agency of Industrial Science & Technology, 4-6 june 1996, Hoikado

23. Ichikawa K., Kuwamata N., Kamei K., 1993, Development of a simultaneous sulfur and dust removal process for IGCC power generation system, Proc. of 3rd Int. Symp. on Gas Cleaning at High Temperatures, 419
24. Geldart D., 1973, Types of gas fluidization, Powder Technol., 7, 285
25. Grace J.R., 1986, Contacting modes and behaviour classification of gas-solid and other two-phase suspensions, Can.J.Chem.Engng, 64, 353
26. Kunii D., Levenspiel O., 1991, Fluidization Engineering, Butterworth-Heinemann, Boston
27. Schrodt J.T., Mohan C.P., 1981, Fuel gas desulfurization in fluidized-beds of gasifier waste ashes, AIChE. Symp.Ser., 77(211), 1
28. Harrison D.P., 1995, Control of gaseous contaminants in IGCC processes, an overview, in: Proceedings of the 12th annual international Pittsburgh Coal Conference, (edited by Chiang S.H.), Pittsburgh Coal Conference, Pittsburgh, 1047
29. Swischer J.H., Yang J., Gupta R.P., 1995, Attrition-resistant zinc titanate sorbent for sulfur, Ind. Engng. Chem. Res., 34, 4463
30. Bureau A.C., Olden M.J.F., 1967, The operation of the Frodingham desulfurising plant at Exeter, Chem. Engng., March, 55
31. Van Yperen R., 1994, On the high temperature desulphurization of coal gas: the development of a regenerable absorbent, PhD thesis University of Utrecht, The Netherlands
32. Van Swaaij, 1978, The design of gas-solids fluid bed and related reactors, ACS Symp.Ser., 72, 193
33. Werther, 1978, Mathematische Modellierung von Wirbelschichtreaktoren, Chem. Ing. Techn., 50, 850
34. Kuipers J.A.M., Prins W., Van Swaaij W.P.M., 1991, Theoretical and Experimental Bubble Formation at a Single Orifice in a Two-dimensional Gas Fluidized Bed, Chem. Engng. Sci., 46, 11, p.2881
35. Chehbouni A., Chaouki J., Guy C., Klvana D., 1994, Characterization of the flow transition between bubbling and turbulent fluidization, Ind. Engng. Chem. Res., 33, 1889
36. Chehbouni A., Chaouki J., Guy C., Klvana D., 1995, Effets de differents parametres sur le vitesses de transition de la fluidisation en regime turbulent, Can. J. Chem. Engng., 73, 41
37. Bi H.T., Grace J.R., 1995, Effect of measurement method on the velocities used to demarcate the onset of turbulent fluidization, Chem. Engng. J., 57, 261
38. Bi H.T., Grace J.R., Lim K.S., 1995, Transition from bubbling to turbulent fluidization, Ind. Engng. Chem. Res., 34, 4003
39. Arnaldos J., Casal J., 1996, Prediction of transition velocities and hydrodynamical regimes in fluidized beds, Powder Technol., 86, 285
40. Venderbosch, 1997, PhD thesis University of Twente, The Netherlands, to be published
41. Kohl A.L., Riesenfeld F.C., 1974, Gas Purification, 2nd ed., Gulf Publishing Co., Houston, 343
42. Bi H., Fan L.S., 1992, Existence of turbulent regime in gas-solid fluidization, A.I.Ch.E.J., 38, 297
43. Contractor R.M., Chaouki J., 1991, Circulating fluidized bed as a catalytic reactor, in: Circulating Fluidized Bed Technology III, Pergamom Press, Oxford, 39
44. Van der Ham, A.G.J., Prins W., van Swaaij W.P.M., 1994, Regenerative, high temperature desulfurization of coal gas in a lab-scale circulating fluidized bed set-up, in: Circulating Fluidized Bed Technology IV, Pergamon Press, Canada, 657
45. Tseng S.C., Tamhankar S.S., Wen C.Y., 1981, Kinetic studies on the reactions involved in the hot gas desulfurization using a regenerable iron oxide sorbent-II, Chem. Engng. Sci., 36, 1287
46. Tamhankar S.S., Garimella S., Wen C.Y., 1985, Kinetic studies on the reactions involved in the hot gas desulfurization using a regenerable iron oxide sorbent-III. Reactions of the sulfided sorbent with steam and steam-air mixtures, Chem. Engng. Sci., 40, 1019
47. Prins, W., Harmsen G.J., de Jong P., van Swaaij W.P.M., 1987, Heat transfer from an immersed fixed silver sphere to a gas fluidized bed of very small particles, in Fluidization VI, Engineering Foundation, Banff, 677

Hot Gas Cleaning, Sulfiding Mechanisms in Absorption of H_2S by Solids

W.J.W. Bakker, J.C.P. van Rossen, J.P. Janssens and J.A. Moulijn

Department of Chemical Engineering, Delft University of Technology, Julianalaan 136, 2628 BL Delft, The Netherlands.

Abstract. A short introduction is given on coal gasification processes and gas cleaning. Special attention is paid to sorbents for high temperature desulfurization and the selection of these sorbents. Sulfiding mechanisms are explained using temperature programmed sulfiding profiles of MoO_3/Al_2O_3, WO_3/Al_2O_3, reduced and oxidic Ni/Al_2O_3, several supported and unsupported manganese oxides, and $MnAl_2O_4$. Also the regeneration of $MnAl_2O_4$ and MnO with steam is discussed in relation to the sulfidation behavior.

1 Coal Gasification in Electric Power Generation

About one-third of the world's energy fuels consumption is in generating electricity of which almost 60 % is produced from fossil fuels. The role of coal in electricity production (and the production of chemicals) is expected to become more and more important as the world supply of oil and gas is diminishing. It is foreseen that the world demand for electricity will increase by a factor 2.5 in the first half of the next century to $20\text{-}35.10^{12}$ kWh. Should 40% be regenerated by coal, then 1700-2500 GW_e coal based capacity must be installed, which is the equivalent of 3000-4000 units of 600 MW_e each. So, there is a huge potential market for clean and more efficient coal technologies to penetrate.

In conventional coal fired plants (direct combustion) about two-third of the fuels heating value is lost during the generation of electricity, hence there is an important opportunity to improve energy efficiency. The overall efficiency of conventional coal combustion systems is limited by the relatively low inlet temperature of the combustion gases. Promising technologies to improve the usage of chemical coal energy more effectively (up to over 50 %) are amongst others, Integrated Coal Combined Cycle power generation (IGCC), and the so called "Topping cycle" or hybrid cycles, which have been developed in particular by British Coal.

NATO ASI Series, Vol. G 42
Desulfurization of Hot Coal Gas
Edited by A.T. Atimtay and D. P. Harrison

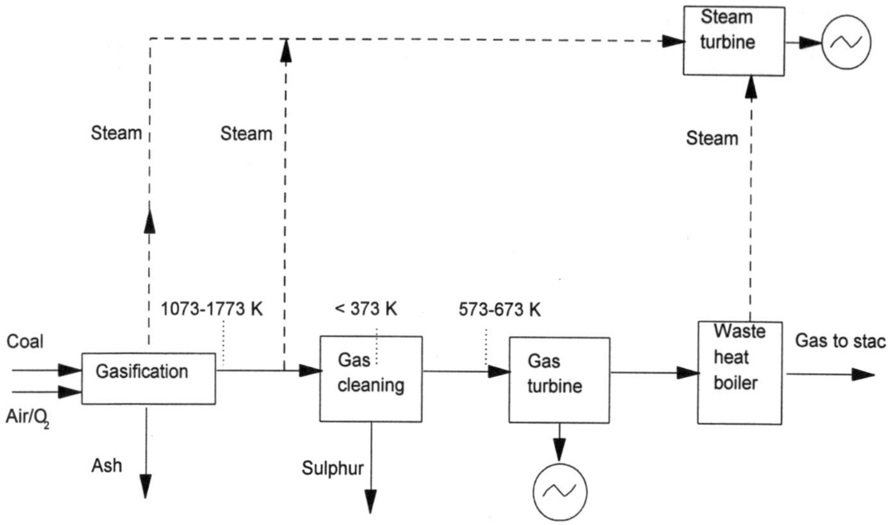

Fig. 1.1. Schematic block diagram for Integrated Gasification Combined Cycle (IGCC) Power generation

In an IGCC unit, as schematically depicted in Fig. 1.1, both a gas and steam turbine are used. Fuel gas is generated in the gasifier and purified in the gas clean-up system. Gas is sent to the gas turbine where it is burned with compressed air to provide a stream of hot gas which drives the turbine to generate electricity. The exhaust gases from the turbine pass to a waste heat boiler where steam is raised at high temperature and pressure. The steam passes to the steam turbine which generates electricity. Integration with chemical process technology, adapted from other applications, opens a number of possible routes for coal utilization, which are capable to meet very stringent environmental standards. A 250 MW_e IGCC demonstration plant has been started up in Buggenum (The Netherlands) recently, and some other demonstration processes will be started up in the near future.

In one of the possible topping cycle configurations, which is schematically shown in Fig. 1.2, pressurized fluidized bed combustion (PFBC) and coal gasification are coupled. In this system of gasifier and combustor, 60-80 % of the coal is gasified by air. The resulting char is added to the rest of the coal and passed to the combustor. The fuel gas (1275 K) is cleaned and then burned in the combustor to raise the inlet temperature of the turbine to 1535 K. In early proposals, fuel gas from the gasifier was used to "top up" the gas temperature at the inlet to the gas turbine, thus enhancing the temperature of gases leaving a PFBC combustor. More recent proposals use a topping combustor to convert non-gasified char. In this case, high carbon conversion in the gasifier becomes less critical [1].

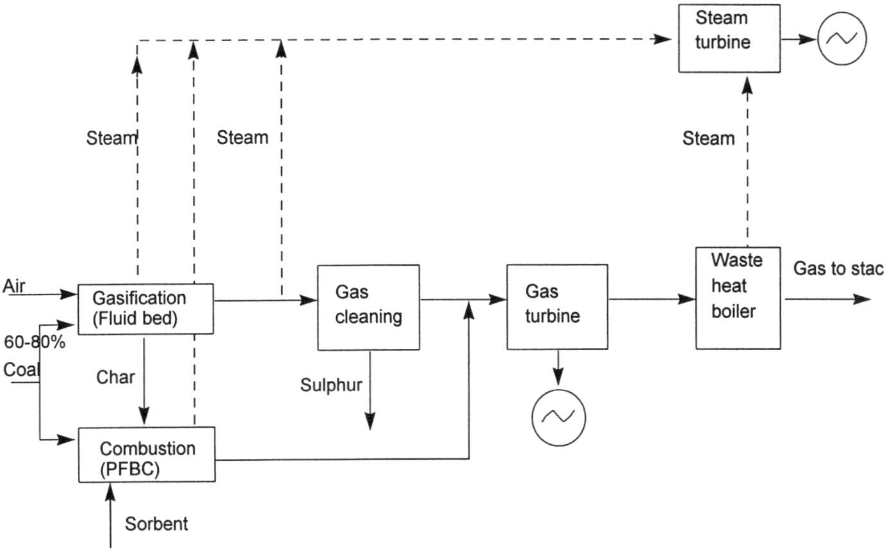

Fig. 1.2. Schematic block diagram for the British Coal Topping Cycle

Both options described above provide a significantly higher energy efficiency than conventional, pulverized coal-fired, combustion plants. In Table 1.1 and Fig. 1.3 some processes are compared (Topper et al. [2]). Many studies on the performance of several design options have been conducted [*e.g.* 2, 3, 4, 5]. Apart from the influence of the system configuration, other parts of coal utilization technology also have a large influence on the efficiency and capital cost.

Table 1.1. Plant size, efficiency, capital cost and environmental impact data based on UK installation, 1993 [2]

Technology	Conventional plants	PFBC	IGCC	Topping Cycles
	200	200	250	330
Nominal size (MWe)				
Net efficiency (%)	38.8	41.4	43	46.9
Specific capital cost ($/kW$_e$)	1780	1460	1520	1355
CO_2 (kg/kWh)	87	82	78	72
Sulfur retention (%)	90	90	99	90
NO_x @6% O_2 (mgNO_x/m^3)	500-650	150-300	120-300	200-300
Particulates @ 6% O_2 (mg/m^3)	40	10	-	30

Most important are:

- Gasification technology (Fluid bed, fixed/moving bed, entrained flow)

- Gasification oxidant (air/oxygen) [5]
- The extent of integration of different parts of the process (e.g. air separation plant)
- Gas turbine technology and steam cycle configuration/operation
- Used coal type and type of feed system
- Scale of the process

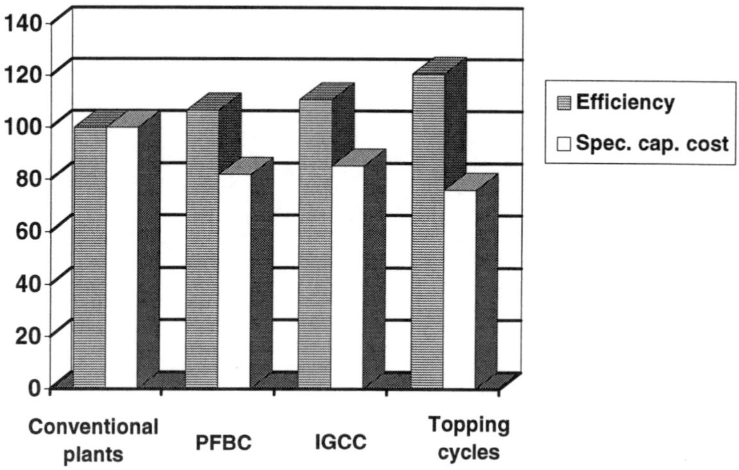

Fig 1.3. Relative efficiency and specific capital cost ($/kWh) for three different coal conversion processes compared to conventional plants (adapted from [6])

1.1 Coal Gasification

Gasification uses 20-30% of the oxygen theoretically required for complete combustion to carbon dioxide and water. Carbon monoxide and hydrogen are the principal products, in particular at high temperature. In this book, a detailed description of the heterogeneous and homogeneous reactions and the thermodynamics of coal gasification is given by Van Diepen and Moulijn.

The composition of the coal gas is variable and dependent of the type of coal, feed system, the type of gasifier, the oxidant and the gasification temperature. The hydrogen to carbon monoxide proportion can range from 0.4 to 1, see Table 1.2.

Generally two feed systems can be distinguished: slurry feed systems (*e.g.* Texaco and Dow) and dry feed systems (*e.g.* Shell). The main difference between these two types is the amount of water present in the feed mixture.

Either air or oxygen is used as gasification oxidant. Air blown gasifiers have, in their *specific* application, a somewhat higher net efficiency and are mostly used in lower temperature processes such as the moving bed and fluidized bed gasifier. High temperature processes, such as the entrained flow gasifier, generally use oxygen. In these processes the equipment is somewhat smaller and the carbon conversion somewhat higher. The typical gas composition of the gas leaving the gasifier of 4 different processes is summarized in Table 1.2.

Table 1.2. Composition and gas cleaning goals for four gasification processes [3, 4, 6, 7]

Process	Air blown, various processes	Britisch Gas/Lurgi slag.	Shell	Texaco
Conditions				
Oxidant	Air	Oxygen	Oxygen	Oxygen
Feed type	Mainly Dry	Dry	Dry	Slurry
Bed type	Mainly fluidized	Moving	Entrained flow	Entrained flow
Typical outlet temperature (K)	1073-1273	723	1173	973
Typical pressure (MPa)	1-3	2.5	3	4-4.5
Component (mol%)				
CO	15-30	54-57	62-65	45
H_2	18-25	29	28-30	30
CO_2	2-15	4.5	1.6-2	5-8
H_2O	3-11	[1])	0.4-2	18
N_2	45-55	3	0.7-3.1	0.8
Typical Impurities (ppm)				
H_2S *(max. H_2S + COS = 15000 ppm)*	750-7000	6000-7000	1260-4000	1100-3500
COS	25-500	[1])	75-240	30-100
NH_3	50-500	[1])	300	800
HCN	0.2-50	[1])	150	40
HCl/HF	<50	[1])	200-830	170-690
Dust (mg/m^3)		[1])	17000	8300
Specifications				
Components	Max. allowable conc. in fuel gas	Determined by:		
H_2S + COS (ppm)	20-100	Boiler kettle corrosion		
NH_3/HCN	< 90 g NO_x/GJ_{th}	Emission standards		
HCl/HF (ppm)	60	Emission standards		
Dust (mg/m^3)	4.5	Protection gas turbine		
Alkali (mg/m^3)	0.5	Protection gas turbine		

[1]not reported

1.2 Gas Cleaning

As shown in Table 1.2, coal gas contains several impurities, of which dust, the sulfur compounds hydrogen sulfide and carbonyl sulfide, ammonia, hydrogen chloride halides, hydrogen cyanide, and hydrogen fluoride are the most important. Downstream requirements (*e.g.* protection of the gas turbine) and environmental regulations dictate that the raw fuel gas produced by a coal gasifier must be cleaned before further use. A wide variety of existing commercial and well developed gas processing systems is available gas for the removal of sulfur, ammonia and other undesirables from the fuel gas.

The removal of the various components is mainly achieved by wet scrubbing at temperatures generally below 350 K, whereas sulfur is often removed using liquid absorbents based on chemical or physical absorption of hydrogen sulfide at low temperatures. Most liquid absorbents are based on solved amines or mixtures of amines. During regeneration of the H_2S loaded sorbent the temperature of the solvent is increased while the gas and pressure is kept low. By stripping with steam the complexes are decomposed and the acid gas components set free.

Table 1.3. Commercial low temperature processes for the removal of H_2S [8, 9, 10]

Name	Class[1)]	Sorbent	T (K)	Prod.
Alkanol amines	CS	methyldiethylamine (MDEA)	300	H_2S
Girbotol	CS	monoethanolamine (MEA)	295-315	H_2S
SNPA-DEA	CS	diethanolamine (DEA)	295-315	H_2S
Econamine	CS	diglycolamine (DGA)	315-355	H_2S
ADIP	CS	diisopropanolamine (DIPA)	305-315	H_2S
Benfield,Catacarb	ICS	hot potassium carbonate (K_2CO_3+water)	355-395	H_2S
Vacuum carbonate	ICS	sodium carbonate & catalyst	293-313	H_2S
Sulfinol	PS/CS	mixture of 2N-diisopropanolamine (DIPA) or MDEA, sulfolane and water	295-335	H_2S
Amisol	PS/CS	mixture of methanol & MEA, DEA, diisopropylamine or diethylamine	278/313	H_2S
Selexol	PS	dimethylether of polyethylene glycol	295	H_2S
Rectisol	PS	methanol	245-355	H_2S
Purisol	PS	n-methyl-2-pyrolidone	310-355	H_2S
CNG	PS	carbon dioxide	220	H_2S
Sepasolv	PS	n-oligoethylene glycol & methyl isopropyl ethers	310-355	H_2S
Fluor Solvent	PS	propylene carbonate	<293	H_2S
Stretford <Localt, Sulferox, Sulfint>	O/DC	sol. of Na-salts of H_2CO_3 anthraquinone-2-sulfonic acid, vanadic acid (iron based chelate replacing sodium metavanadate)	295-315	S
Takahax	O/DC	solution of Na-salts of H_2CO_3, 1,4-naphtoquinone-2-sulfonic acid	295-315	S
Townsend	O/DC	aqueous solution of triethyleneglycol	340-395	S

[1)] CS = chemical solvent, PS = physical solvent, ICS = inorganic chemical solvent, PS/CS = both chemical/physical solvent, O/DC = oxidation/direct conversion

After the acid gas mixture is removed, the concentrated acid gas mixture is usually sent to a Claus unit, where H_2S is converted towards elemental sulfur. Other processes use direct conversion of H_2S to elemental sulfur, thus eliminating the use of a Claus unit. Several commercial low temperature H_2S removal processes are given in Table 1.3.

A major drawback of gas cleaning at low temperatures is an efficiency loss because of the cooling of the coal gas, from 1000-1600 K to below 350 K, and the heating up of the gas after the gas cleaning to 673 K. System studies show that the efficiency of an IGCC-plant can be raised (2-4%, absolute) and the capital cost can be reduced (6%) in case of high temperature gas cleaning. Therefore, attention has been given to the development of hot cleaning techniques that can be applied in the "gasification bed" or directly at the gasifier outlet.

The optimal temperature for hot gas cleaning is subject to debate. It appears that the benefit of hot gas cleaning compared to low temperature scrubbing is most clearly present up to say 575-675 K. The reason is that phase transitions are omitted. At high temperatures materials are increasingly costly and, as a consequence, it is often stated that 675 K is the optimal temperature. On the other hand a dedicated hot gas cleaning process might lead to large savings because heat exchange equipment is omitted. A good example is the radiation gas cooler present in high temperature gasifiers - gas turbine set-ups. In that case a temperature equal to the gas inlet temperature might well be optimal. With the ongoing development of high temperature resistant turbine blades the trend is towards increasing gas turbine inlet temperatures.

2 High Temperature Desulfurization

For the removal of sulfurous gases at high temperatures practically all studies use gas-solid absorption processes. Sulfurous compounds can either be removed by sorbents mixed with the coal, so called in-bed removal, or after the gasification in an external sorbent bed. A combination of both is of course also a possibility. In most cases the desulfurization efficiency should be above 98.5 %, in order to remove sulfur containing compounds to an acceptable level (see Table 1.2). Below, we will discuss some points which are important for sorbent selection.

2.1 Sorbent Selection

Equilibrium Constant

The main requirements for (external) sorbents are that they are stable and selectively react with H_2S and COS in a reducing coal-gas environment at 1 to 4 MPa and 675-1175 K. In general, the following reaction should be favorable:

$$MeO(s) + H_2S(g) \quad \Leftrightarrow \quad MeS\,(s) + H_2O(g) \qquad (1)$$

where Me refers to any metal.

When the solid phase is in excess one can assume that the activity of the solid is constant and, as a consequence, the equilibrium constant can be expressed as the ratio of the gas phase concentrations.

$$K_{eq} = \frac{[H_2O]}{[H_2S]} \tag{2}$$

One can, however, argue that the surface sites on the acceptor should also be included in this expression, and an equilibrium constant K^i_{eq} can be defined as:

$$K^i_{eq} = \frac{[H_2O]}{[H_2S]} \cdot \frac{[\text{sulfided surface sites}]}{[\text{free surface sites}]} \tag{2a}$$

To obtain good desulfurization K_{eq} should be sufficiently high. Evaluations to determine the metal oxides to be used in high temperature removal of H_2S has been carried out by several investigators. Westmoreland *et al.* [11, 12] showed that the oxides of Ba, Ca, Co, Cu, Fe, Mn, Mo, Sr, W, V, and Zn are thermodynamically feasible candidates. Also feasible in principle are Ce and Sn.

Mixed Metal Oxides

Mixed metal oxides offer a large number of possibilities and may be very interesting for improvement of existing sorbents; *e.g.* a higher absorption capacity, ability to decrease H_2S concentrations to a lower level, better (thermo-) stability, less deactivation *etc.*). The thermodynamics of most mixed oxides are unknown and (strongly) dependent on the concentration of the metal in the sorbent. Moreover, besides bulk properties also the characteristics of the surface phase are important. Up to now the thermodynamical data are lacking for most surfaces. This makes a selection based on thermodynamics very time consuming, if not impossible. In this book, see chapter on performance analysis of Mn based sorbents, we will discuss the thermodynamics of manganese based sorbents in more detail.

Supported vs. Unsupported Sorbent

In general, supported metal oxide acceptors have advantages over unsupported metal oxides for high temperature desulfurization because of higher reaction rates. The adsorption capacity is, however, in most cases somewhat less than for bulk sorbents. The support has a large surface area and the metal oxide is spread out on this surface. If desired, stabilization of the support is possible in many cases. When the reaction takes place exclusively at the surface, stabilization is essential for stable operation. At higher temperatures sintering of the support occurs which slows down the reaction considerably. However, the following should not be overlooked. Solid state diffusion is an activated process and, as a consequence, with increasing temperature diffusion in the solid phase (partly) compensates the decrease in reactivity by the loss of surface area.

Reactivity
Elements which were rejected because their oxides (or carbonates) are stable relative to the element and not reactive throughout the temperature range from 600-1200 K include Al, Cr, K, Li, Mg, Na, Ti, Zr. Some elements were rejected because of their inadequate desulfurization potential (< 95%) even though their equilibrium is in the sulfided state: Ag, La, Ni, and Sb.

Reduction to a Low Melting Metal
The melting point of a metal is relatively low compared to its metal oxide. Melting of the sorbent results in most cases in a deactivation of the sorbent. Therefore, the thermodynamics and/or kinetics of the following reduction reactions should be unfavorable:

$$MeO\ (s) + H_2\ (g) \Leftrightarrow Me\ (s) + H_2O\ (g) \quad (3)$$

and

$$MeO\ (s) + CO\ (g) \Leftrightarrow Me\ (s) + CO_2\ (g) \quad (4)$$

Elements which can be rejected because their metal oxide can be reduced to a low melting metal are: Bi, Cd, Pb and Sn. The latter element however deserve a special treatment when the gas phase contains sufficient H_2 and/or H_2O, in that case Sn is not present as the metal and the problem of melting does not occur. Moreover stabilizers have been developed.

Coke and Sulfate Formation
Coke and sulfate formation should be minimized. Some metals catalyze the formation of coke; *e.g.* Fe. Especially at temperatures below 1050 K coke formation can occur. Small amounts of coke may however be removed during regeneration. Sulfate formation generally leads to volume expansion and swelling of the sorbent causing them to spell and crack. Moreover SO_2 formation will occur during the sulfiding step which is undesirable.

Volatile Metal Chlorides,Fluorides and Carbonyls
One should be aware of the possibility of extensive volatilization of the sorbent at too high operating temperatures. The presence of HCl and HF causes the formation of metal chlorides or fluorides which are much more volatile than the corresponding metal oxides. Loss of metal by volatilization of metal chlorides or fluorides causes deactivation of the sorbent. Large differences exist in melting points between different metal chlorides. $ZnCl_2$ has a boiling point of 1005 K whereas $MnCl_2$ has a boiling point of 1504 K The carbonyl formation which leads to efficient evaporation and therefore the possibility of this process should be checked.

Stability and Mechanical Strength
Regenerable sorbents should perform for at least 100 cycles to make them economically competitive to disposable sorbents. For fixed bed operation even a higher durability is needed. Depending on the application of the sorbent, it should be

resistant towards attrition. Especially when the sorbent is applied in a fluidized bed reactor this is an important factor.

Regeneration

Regeneration of the sorbent requires an oxidation step in which a sulfur ion is exchanged with an oxygen ion. Most effective would be the production of elemental sulfur during regeneration. Such processes are still in the developing stage. When no elemental sulfur can be made, H_2S is preferred over SO_2. At this moment most sulfided metal sorbents are regenerated with oxygen. The reaction with oxygen is however very exothermic (see chapter on performance analysis of Mn based sorbents). To avoid local overheating the oxygen or air is diluted, resulting in diluted SO_2 as regeneration product. Reducing gases are needed to convert SO_2 in H_2S and the H_2S stream should be concentrated to say > 15 mol %, before entering a Claus unit. It should be noted that catalytic reduction processes are feasible for more diluted streams.

For some sorbents regeneration is possible with H_2O under the formation of H_2S. A good regenerability with H_2O means also a high sensitivity for H_2O during the sulfidation. These sorbents can only be applied for the desulfurization of dry coal gas or in combination with a sorbent which is not sensitive for H_2O. For the direct production of elemental sulfur regeneration with SO_2 is an option. Sulfate formation is, however, in many cases a problem.

2.2 Examples of Sorbents

In-bed sorbents often form an essential part of the gasification process. Cheap and abundantly available materials like limestone (calcium carbonate) and dolomite (a mixture of calcium and magnesium carbonates) are often used. Sorbents for external use are usually based on regenerable metal oxides.

Zinc ferrite and zinc titanate have received much attention as external sorbents over the past years because, from a theoretical point of view they have desirable properties: high sulfur capacity, ability to remove H_2S from coal derived fuel gas to levels of a few ppmv, and the feasibility of regeneration with air.

Various other mixed oxide sorbents are currently under development. Some of these sorbents are based on iron and zinc oxides, but are stabilized on other inert carriers such as alumina or zeolites. In other applications, oxides of cerium, copper, cobalt, manganese and tin are used in the sorbent formulations. Most of these sorbents are currently being evaluated in laboratory or bench-scale tests, to establish the optimum chemical formulation, or to develop improved physical properties and stability for repeated use in multiple absorption/regeneration cycles. The greater use of these sorbents will depend on their improved performance relative to zinc ferrite's or titanates and/or ability to use simple and cost effective schemes for treating regenerator off gas. In the paper by Van Dam in this book, a summary is given of several regenerable metal oxide acceptors which are currently under investigation or being applied.

3 Sulfiding Mechanisms

3.1 Sulfiding Reactions

Depending on the temperature the following types of reactions might be expected:

Sulfiding of metal oxides by O-S exchange:

$$MO_x + H_2S \Leftrightarrow MO_{x-1}S + H_2O \quad (4a)$$

Sulfiding of metal oxysulfides by O-S exchange:

$$MO_xS_y + H_2S \Leftrightarrow MO_{x-1}S_{y+1} + H_2O \quad (4b)$$

Reduction of metal oxides:

$$MO_x + H_2 \Leftrightarrow MO_{x-1} + H_2O \quad (5a)$$

Reduction of metal oxysulfides:

$$MO_xS_y + H_2 \Leftrightarrow MO_xS_{y-1} + H_2S \quad (5b)$$

Reduction of metal sulfides:

$$MS_x + H_2 \Leftrightarrow MS_{x-1} + H_2S \quad (6)$$

Sulfiding of metals:

$$M + H_2S \Leftrightarrow MS + H_2 \quad (7)$$

In the next sections typical examples of TPS (Temperature Programmed Sulfiding) patterns are given for metal oxide samples. The sulfiding reactions and mechanisms are determined from TPS profiles of the important building blocks of commercial hydrotreating catalysts (CoMo-, NiMo-, NiW/Al_2O_3) and some manganese oxides (MnO_x) and manganese aluminate ($MnAl_2O_4$). The latter are used for hot gas desulfurization.

3.2 Experimental

In order to study the sulfiding reactions and mechanisms of bulk or supported metal oxides Temperature Programmed Sulfiding (TPS) proved to be an excellent technique for this purpose [14].

A simplified scheme of the TPS apparatus is shown in Fig. 3.1. The samples are held in a quartz tube between two quartz-wool plugs. The composition of the sulfiding gas mixture is 2-4 mol% H_2S, 25-50 mol% H_2 and 46-72 mol% Ar. An UV (ultra violet) spectrophotometer is used to determine H_2S consumption or production, a TCD (thermal conductivity detector) to determine the H_2 consumption or production. In some cases a mass spectrometer is used to monitor the H_2O concentration in the effluent gas. The sulfiding is divided in an isothermal sulfiding at room temperature, followed by the actual temperature programmed sulfiding stage and an isothermal stage at the end of the temperature program.

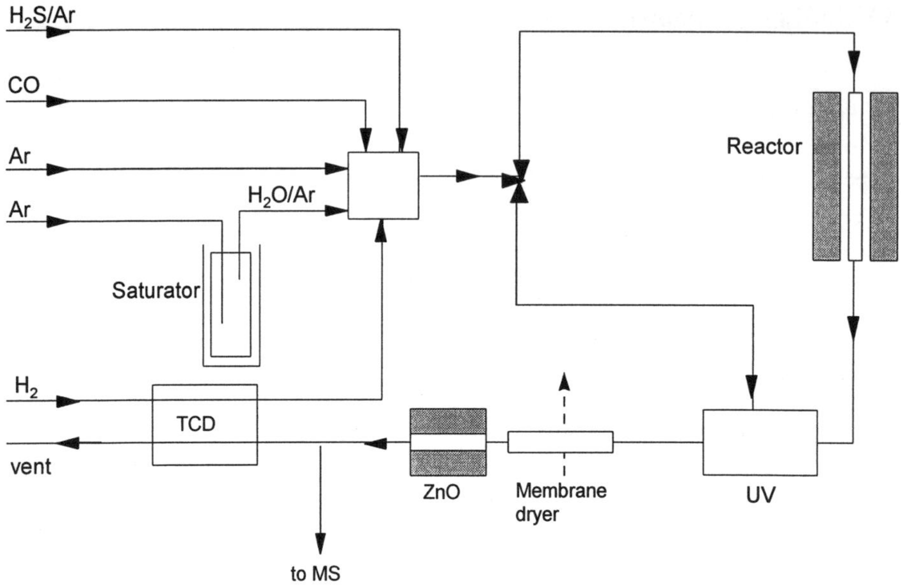

Fig. 3.1. Simplified scheme of the TPS apparatus used

The metal oxides discussed here were home made. To obtain supported manganese oxides, a γ-Al_2O_3 support was impregnated with a manganese acetate solution and calcined. TPS profiles were recorded from samples calcined at 575 (MnO_2) and 875 K (Mn_3O_4). From the sample calcined at 875 K another TPS profile was recorded after several cycles in which this sample was subsequently sulfided and regenerated with steam at 1123 K ($MnAl_2O_4$). The same procedure was followed for a pure manganese oxide sample which was obtained by calcination of pure manganese acetate at 875 K.

XRD pointed out that the sample calcined at 673 K contained MnO_2 and the other samples Mn_3O_4 and Mn_2O_3, where the pure manganese oxide contained only Mn_2O_3. After several sulfidation-regeneration cycles the XRD revealed that the supported sample was converted to manganese aluminate ($MnAl_2O_4$) and the unsupported to manganese oxide (MnO).

For two $MnAl_2O_4$ samples, containing 16 and 35 wt% Mn respectively, and an unsupported MnO sample also temperature programmed regeneration with steam (after several sulfidation-regeneration cycles) was measured. Details of the preparation of other compounds are given elsewhere (14-17).

4 TPS Profiles of Different Metal Oxides

4.1 Sulfiding of MoO_3/Al_2O_3 and WO_3/Al_2O_3 Catalysts

Fig. 4.1 shows a typical TPS pattern of a MoO_3/Al_2O_3 catalyst [14]. Already at room temperature H_2S consumption takes place due to adsorption on, or sulfiding of specific sites, probably electron deficient Mo sites (this uptake is not depicted in the figure). In the low temperature region (below 500 K) a major H_2S consumption is observed which correlates to a H_2O production. No H_2 consumption is observed in this region. This shows that sulfiding takes place by O-S exchange on Mo(VI) (reaction *(4a, b)*). At approximately 500 K, H_2S is produced in a sharp peak, coupled with H_2 consumption.

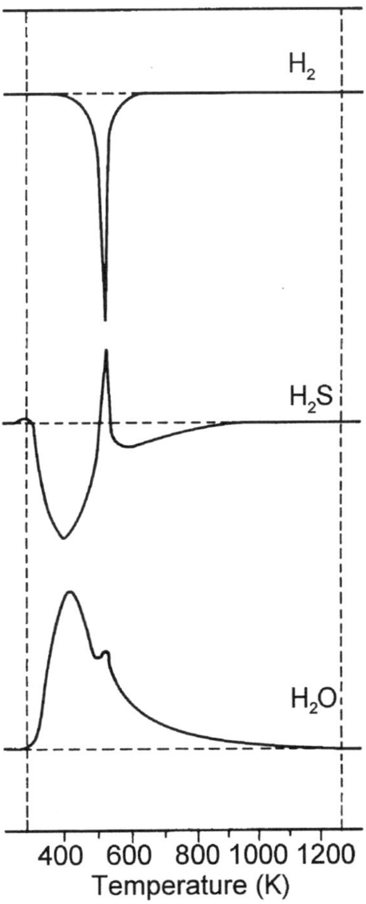

Fig. 4.1. TPS patterns (H_2S, H_2O, H_2) of MoO_3/Al_2O_3 [14]

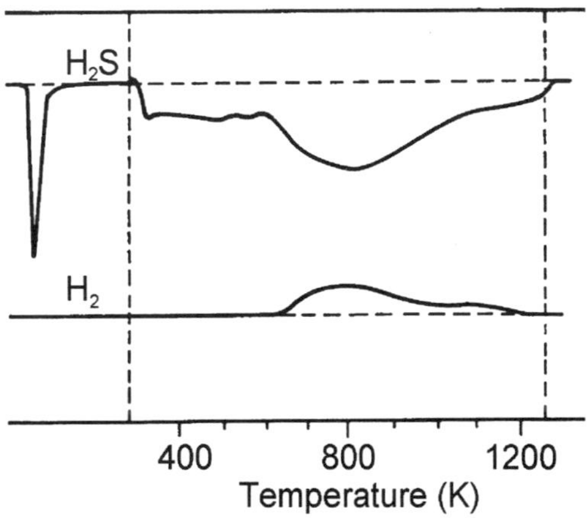

Fig. 4.2. TPS patterns (H_2S, H_2) of WO_3 [15]

This can be explained by the breaking of Mo(VI)-S bonds, resulting in the reduction of Mo(VI) to Mo(IV) and the formation of H_2S (reaction *(5b)*). This should not be taken too literally. H_2 might well assist the Mo-S bond breaking. At high temperatures (above 500 K) sulfiding is completed to form MoS_2, again via O-S exchange (reaction *(4b)*).

A typical TPS pattern of a WO_3/Al_2O_3 catalyst is shown in Fig. 4.2 [15]. Again a H_2S consumption at room temperature is observed indicating that sulfiding already takes place. In the temperature programmed stage two regions can be discerned. Sulfiding takes place at a steady rate below 600 K with a broad maximum in the rate of H_2S. H_2 uptake is seen in the high temperature region. Sulfiding below 600 K is not coupled to reduction (H_2 consumption) and therefore proceeds through the formation of W(VI) oxysulfides via O-S exchange (reaction *(4a, b)*). Above 600 K the sulfiding is accelerated (onset of H_2 consumption). This suggests that H_2 removes oxygen from the oxysulfide, and thereby creates sites where sulfiding can continue (reaction *(4b, 5a)*).

From the TPS patterns, the following sulfiding mechanism for MoO_3/Al_2O_3 catalysts can be deduced:

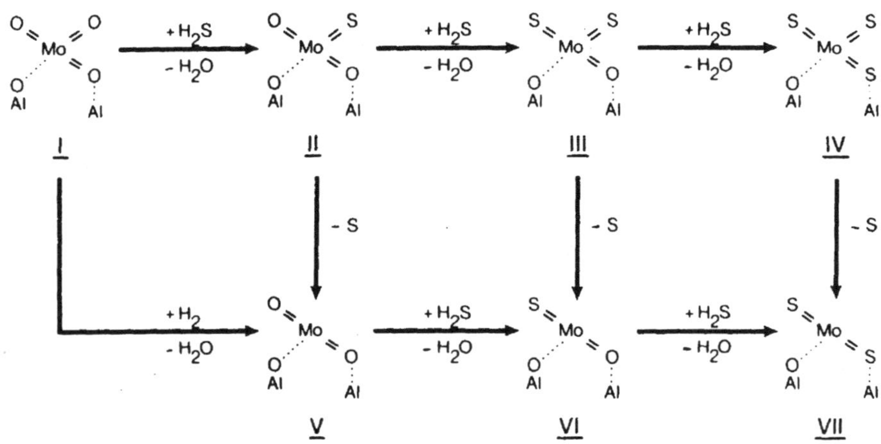

Fig. 4.1a. Schematic representation of the sulfiding mechanism for MoO_3/Al_2O_3 catalysts

In comparison with MoO_3/Al_2O_3 catalysts, WO_3/Al_2O_3 catalysts are more difficult to sulfide. This lower sulfidability is ascribed to the difference in the binding energy of the metal-oxygen bond. W and O are more strongly bonded than Mo and O.

4.2 Sulfiding of Oxidic and Reduced Ni/Al_2O_3 Catalysts

The TPS patterns of an oxidic and reduced Ni catalyst are shown in Fig. 4.3 and 4.4 [16]. The oxidic Ni catalyst shows an isothermal H_2S uptake at room temperature followed by a small desorption of H_2S upon temperature increase. The major H_2S consumption peak is found between 350 and 500 K, which demonstrates sulfiding by O-S exchange (reaction *(4 a,b)*. Between 500 and 600 K a H_2S production coupled with a H_2 consumption is found indicating the reduction of the nickel sulfide (reaction *(6)*). From the patterns it can be calculated that formally sulfiding takes place from NiO in NiS to Ni_3S_2. Therefore, sulfiding of the oxidic Ni takes place according to:

$$NiO \underset{-H_2O}{\overset{+H_2S}{\longrightarrow}} NiS \underset{-H_2S}{\overset{+H_2}{\longrightarrow}} Ni_3S_2 \tag{8}$$

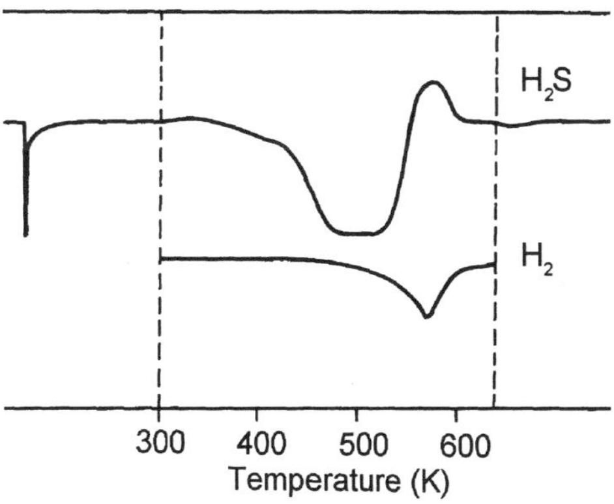

Fig 4.3. TPS patterns (H_2S, H_2) for oxidic Ni/Al_2O_3 catalyst [16]

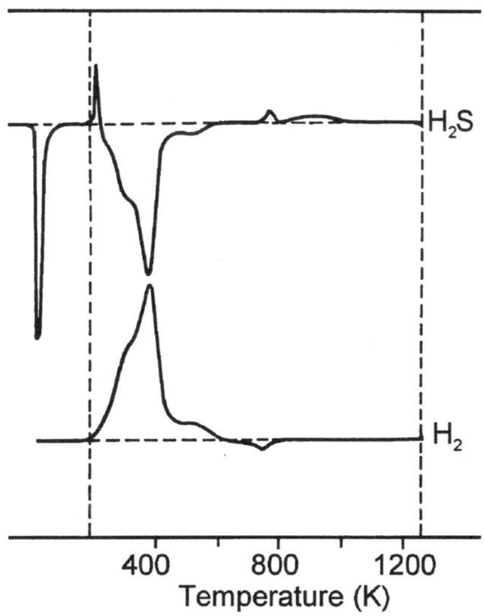

Fig 4.4. TPS patterns (H_2S, H_2) for reduced Ni/Al_2O_3 catalyst [17]

Reduced Ni/Al_2O_3 catalysts, which consists of alumina supported Ni^o-crystallites, show a different sulfiding behaviour than oxidic Ni/Al_2O_3 catalysts. The reduced catalysts show besides the isothermal sulfiding at 210 K and a small H_2S desorption slightly above 210 K, large H_2S consumption peaks at 270, 300 and 380 K coupled with a H_2 production (reaction *(7)*). This indicates that the sulfiding reaction takes place according to:

$$Ni^0 \underset{-H_2}{\overset{+H_2S}{\longrightarrow}} NiS \underset{-H_2S}{\overset{+H_2}{\longrightarrow}} Ni_3S_2 \tag{9}$$

An interesting difference between the sulfiding of oxidic and reduced Ni catalysts is that the reduced catalysts sulfide at a significantly lower temperature. This is explained by a higher dissociation rate of H_2S on the reduced nickel surface and a higher solid state diffusion rate in the reduced nickel core.

4.3 Sulfiding and Regeneration of (Supported) Manganese Oxides and Manganese Aluminate

4.3.1 Temperature Programmed Sulfiding

A typical TPS profile of supported MnO_2 is given in Fig. 4.5. Already at room temperature H_2S consumption takes place. Physically adsorbed H_2S is released at the beginning of the temperature program. A broad absorption band is observed from 375 to 575 K. At 675 K some H_2S is produced, corresponding with H_2 consumption.

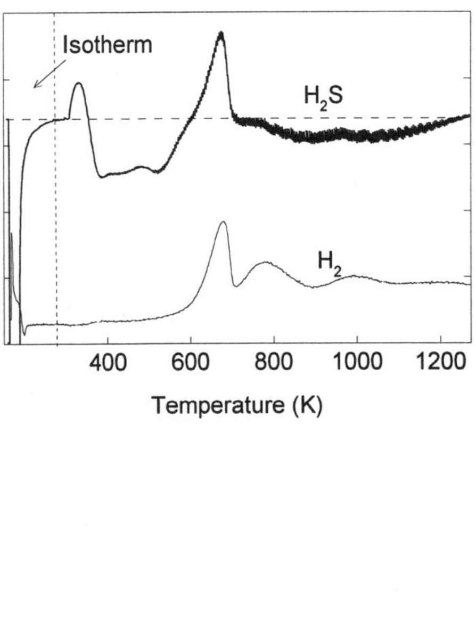

Fig. 4.5. TPS pattern and hydrogen consumption of γ-alumina supported MnO_2 (calcined at 573 K)

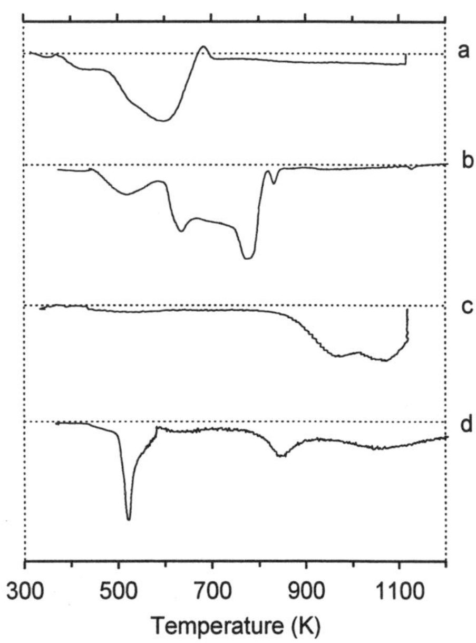

Fig. 4.6. TPS patterns of different supported and unsupported manganese oxides:
a) Mn_3O_4 on γ-Al_2O_3; b) unsupported Mn_2O_3; c) $MnAl_2O_4$; d) unsupported MnO

Fig. 4.6 shows TPS patterns of different supported and unsupported manganese oxides, during the temperature programmed stage of the experiment. In this Fig. the H_2 signals are not depicted but, if measured, they will be discussed in the text.

Fig. 4.6a shows the TPS curves of alumina supported Mn_3O_4. Absorption occurs between 375 and 650 K. A minor absorption peak is observed around 425 K and a major absorption peak around 600 K. At 675 K a H_2S production peak is observed. From thermodynamic calculations [18] it follows that below 625 K both MnS_2 and MnS are formed. At higher temperatures Mn^{4+} is reduced to Mn^{2+} under the formation of MnS and S or H_2S if hydrogen is present. This explains the H_2S production and the H_2 consumption (only depicted in Fig. 4.5) around 675 K.

Roughly, the same sulfiding pattern as described above is observed for the unsupported manganese oxide (Mn_2O_3) sample although the moment at which absorption occurs shifts to higher temperatures (see Fig. 4.6b). This can be explained by the relatively slow diffusion in a bulk manganese oxide which causes part of the H_2S absorption to occur at higher temperatures. No H_2S production peak is observed but from the TPS pattern it follows that the H_2S production at 675 and 800 K likely coincides with simultaneous H_2S absorption.

$MnAl_2O_4$, depicted in Fig. 4.6c, shows two major absorption peaks around 1000 K. XRD revealed the formation of MnS and γ-Al_2O_3 during sulfiding. No H_2 consumption or production was observed.

Unsupported manganese oxide, given in Fig. 4.6d, shows its major absorption peak at 500 K, besides a broad band from 800-1200 K. The broadness of latter adsorption is due to severe diffusion limitation in the bulk of this material which causes part of the sulfiding to occur at a much higher temperature than 500 K. No H_2 consumption or production was observed.

A metal aluminate generally absorbs at higher temperatures than the corresponding metal oxide because of the stronger metal-oxygen bonding in the aluminate spinel structure. This explains the large difference in absorption temperature between $MnAl_2O_4$ and MnO. As during the TPS no reduction of Mn^{2+} can occur it is self-evident that no H_2S production or H_2 consumption was observed during the sulfidation of MnO and $MnAl_2O_4$.

4.3.2 Temperature Programmed Regeneration (TPReg) of MnS

Temperature programmed regeneration of MnS with steam was carried out for supported samples with respectively 16 and 35 wt% Mn, and an unsupported sample. The results of the temperature programmed regeneration are given in Fig. 4.7. At low loading (16 wt% Mn) three main H_2S production peaks are to be seen at 475, 730 and 1050 K. A sample with 35 wt% Mn has its main H_2S production between 1020 and 1220 K. This broad peak (shouldered at 975 K) has its maximum at 1175 K. The difficult regeneration of unsupported MnS is clear. The sample was kept at 1273 K for 1 hour and still was not completely regenerated.

The observed trend is that the supported samples are easier to regenerate than the unsupported. This can be explained from thermodynamics. During regeneration of the alumina supported samples $MnAl_2O_4$ is formed while the unsupported sample forms the thermodynamically less stable compound MnO. This results in a more difficult regeneration of the latter as the driving force for regeneration is smaller. The shift to a higher regeneration temperature of the 35 wt% sample compared to the 16 wt% sample can be explained by the occurrence of less diffusion limitation in the 16 wt% sample. The regeneration peak observed for the low loading sample at 450 K are surface sites which regenerate relatively easy. In the 35 wt% sample the contribution of surface sites is relatively low as its surface area compared to the 16 wt% sample is lower. Moreover the surface sites of the 35 wt% sample partly consists of MnO-like sites which are more difficult to regenerate than $MnAl_2O_4$ surface sites [18].

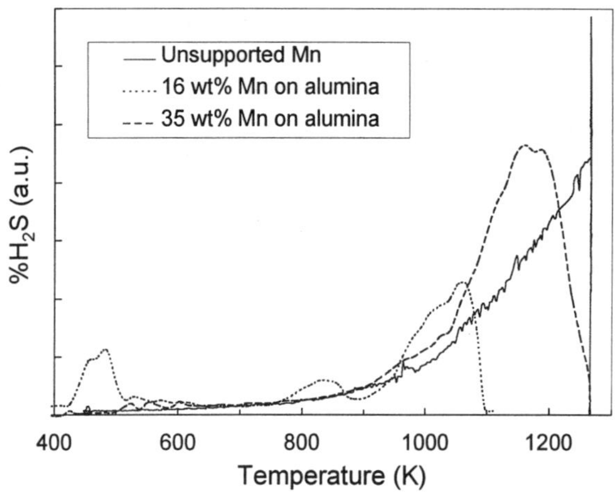

Fig. 4.7. Temperature programmed regeneration of MnS for of two supported and one unsupported sample

Comparing the temperature programmed regeneration profiles of supported MnS and unsupported MnS with the corresponding TPS profiles, being respectively $MnAl_2O_4$ (Fig. 4.6c) and MnO (Fig. 4.6d) it can be concluded that MnO is sulfided easily and difficult to regenerate and that $MnAl_2O_4$ is relatively difficult to sulfide but easy to regenerate. Besides thermodynamics, diffusion limitation at low temperatures plays also a role as thermodynamically both components should be sulfided at room temperature. The driving force for sulfidation of $MnAl_2O_4$ is less than for MnO which results sulfidation of $MnAl_2O_4$ to occur at higher temperatures.

4.3.3 Reactions During Sulfiding of Manganese Oxides

From TPS and TPR experiments on the manganese oxides and manganese aluminate, as described in the previous section an XRD analysis, a number of reactions can be derived.

During acceptation and regeneration with steam O-S exchange takes place (Eq. 4a, 10-13). Furthermore reduction of Mn^{4+} to Mn^{2+} occurs during sulfiding around 600 K under the formation of H_2S (Eq. 6 and 11). The specific reactions for different components are summarized below.

Manganese Oxides
Up to 600 K:

$$MnO_2 + 2\ H_2S \Leftrightarrow MnS_2 + 2\ H_2O \quad (10)$$
$$Mn_2O_3 + 3\ H_2S \Leftrightarrow MnS + MnS_2 + 3\ H_2O \quad (11)$$
$$Mn_3O_4 + 4\ H_2S \Leftrightarrow 2\ MnS + MnS_2 + 4\ H_2O \quad (12)$$
$$MnO + H_2S \Leftrightarrow MnS + H_2O \quad (13)$$

Above 600 K:

$$MnS_2 + H_2 \Leftrightarrow MnS + H_2S \quad (14)$$

Manganese Aluminate

$$MnAl_2O_4 + H_2S \Leftrightarrow MnS + \gamma\text{-}Al_2O_3 + H_2O \quad (15)$$

5 Concluding Remarks

It has been shown that temperature programmed techniques are a very useful tool to clarify sulfiding mechanisms. Metal oxides sulfide mainly via O-S exchange and metals via oxidation. In reduction of oxides H_2S is usually more important than H_2. During sulfidation the dispersion (crystal size) plays a major role. Solid state diffusion is slow although it is strongly activated, therefore at very high temperatures, typically 1200 K, the thermodynamic equilibrium composition can be applied as a reference point.

6 References

1 C. Maude, Advanced power generation - a comparative study of design options for coal, IEA Coal Research, London (March 1993).
2 J.M. Topper, P.J.I. Cross and S.H. Goldthorpe, Clean coal technology for power and cogeneration, *Fuel*, 1994, **73**(7), 1056-63.
3 P. Gäng and F. Loeffler, Cleaning of hot gases from coal gasifiers in a dry combined process, *Gas Clean. High Temp., [Pap. Int. Symp.]*, 2nd (1993), 522-40
4 T. Takematsy and C. Maude, Coal gasification for IGCC power generation, IEA Coal Research, London (March 1991).
5 J. Klosek, J.C. Sorensen and M. Wong, Luft oder Sauerstoff für Vergasungs-Kombi-Kraftwerke (GCC), *VGB Kraftwerkstech.* (1994), **74**(12), 1066-71

6 P.T. Anderliesten and others, Systeemstudie hoge temperatuur gasreiniging bij KV-STEG installaties, NOVEM, November 1990.

7 R. van Yperen, On the High-temperature desulphurization of coal gas, The development of a regenerable absorbent, Phd. Thesis, Utrecht University, The Netherlands, 1994.

8 J.P. Wakker, Development of a high temperature steam regenerative H_2S removal process based on alumina supported MnO and FeO, Phd. Thesis, TU Delft, The Netherlands, 1992.

9 K. Thambimuthu, Gas cleaning for advanced coal-based power generation, IEA Coal Research, London (1993).

10 T.H. Soerawidjaja, Steam regenerative removal of H_2S at high temperatures using metal oxide on alumina acceptors, Phd. Thesis, TU Delft, The Netherlands, 1985.

11 P.R. Westmoreland, J.H. Gibson and D.P. Harrison, Evaluation of candidate solids for high temperature desulphurization of low-Btu gases. *Environ. Sci. Technol.* (1976), **10**, 659-661.

12 P.R. Westmoreland, J.H. Gibson and D.P. Harrison, Comparative kinetics of high temperature reaction between H_2S and selected metal oxides, *Environ. Sci. Technol.* (1997), **11**, 488-491.

13 A.B.M. Heesink, High temperature coal gas desulphurization, Phd. Thesis, Twente University, The Netherlands, 1994.

14 P. Arnoldy, J.A.M. van den Heijkant, G.D. de Bok and J.A. Moulijn., Temperature-programmed sulfiding of molybdenum(IV) oxide/alumina catalysts, *J. Catal.* (1985), **92**(1), 35-55.

15 B. Scheffer, P.J. Mangnus and J.A. Moulijn, Temperature-programmed sulfiding study of nickel oxide-tungsten trioxidide/alumina catalysts, *J. Catal.* (1990), **121**(1), 18-30.

16 J.P. Janssens, A.D. van Langeveld, R.L.C. Bonne, C.M. Lok and J.A. Moulijn, Role of sulfiding procedure on activity and dispersion of nickel hydrotreating catalysts, Prepr. - Am. Chem. Soc., Div. Pet. Chem. (1994), **39**(4), 571-4.

17 P.J. Mangnus, Characterisation of hydrotreating catalysts with temperature programmed techniques, A search for improved catalysts, Phd. Thesis, University of Amsterdam, 1991.

18 W.J.W. Bakker, J.C.P. van Rossen, J.M. Mugge, F. Kapteijn, J.A. Moulijn, Performance of Manganese-based Sorbents in High-Temperature Coal Gas Desulfurization, this book.

Kinetics of Sulfidation Reactions Between H_2S and Bulk Oxide Sorbents

M. Flytzani-Stephanopoulos and Z. Li

Department of Chemical Engineering, Tufts University, Medford, MA 02155, U.S.A.

Abstract. This paper considers the sulfidation kinetics of bulk single oxides as well as mixed metal oxides that have promising overall properties for application to regenerative hot gas desulfurization. The hot fuel gas typically contains a large excess of reductant gases (H_2, CO), hence the kinetics of the potential parallel reduction reaction of the oxide sorbent are also considered. Effects of water vapor, typically present in various amounts in coal-gas streams, on the sulfidation/reduction kinetics are addressed as well. In addition to gas composition, the impact of sorbent physical properties and composition on the sulfidation reaction kinetics is examined in this paper. Two major classes of bulk oxide sorbents, namely; ZnO- and CuO-based materials, are used to illustrate the important issues in sulfidation of bulk oxides.

1 Introduction

Over the past two decades significant effort has been devoted to the development of regenerable hot-gas desulfurization sorbents. Regenerable sorbents would allow the desulfurization of coal-derived fuel gas streams at high temperatures, in the range of 550-750 °C, in emerging advanced coal-fired power generation such as the integrated gasification combined-cycle (IGCC) power plants. Hot gas desulfurization offers energy efficiency gains over low-temperature liquid scrubbers of H_2S and avoids costly wastewater treatment. The projected cost of hot gas cleanup is approximately half that of commercially-available cold gas desulfurization; hence, the high interest in developing the new process. However, the development of a low-cost, highly efficient, rapidly reacting, regenerable sorbent, which also provides high sulfur capacity and durability still eludes the art. All sorbents studied to date lack one or more of the above features. As a consequence, they have limited range of application, e.g. to narrow gas composition-temperature domains, and may require frequent cycling and/or make-up rates that incur high operating costs.

Bulk metal oxides have been studied more extensively than supported oxides as potential hot gas desulfurization sorbents. This is due to lower manufacturing costs and higher sulfur capacity (defined as g sulfur per 100 g of total sorbent) of bulk oxides than for the corresponding supported oxides. The theoretical sorbent capacity, however, can only be realized if the sulfidation rate is not limited by poor intrinsic kinetics and/or slow gas diffusion through the sorbent pores or slow diffusion through the solid sulfide product layers. Thus, the sorbent utilization depends critically on the sulfidation kinetics and

NATO ASI Series, Vol. G 42
Desulfurization of Hot Coal Gas
Edited by A. T. Atimtay and D. P. Harrison

physical properties (pore structure, grain size) of the sorbent and the change of physical properties in cyclic operation.

Justifiably, initial sorbent screening studies focused on the thermodynamic equilibria of sulfidation, which determine the H_2S removal efficiency. From a large number of elements, Westmoreland and Harrison in 1976 arrived at eleven metals and their oxides as feasible H_2S sorbents on the basis of favorable thermodynamics of sulfidation (1). Another criterion applied to regenerable sorbents is the thermal stability of the corresponding sulfate that may be formed during regeneration. Sorbents forming thermally stable sulfates are not practical for regenerative desulfurization.

Kinetic investigations of reduction/sulfidation coupled with studies of associated solid structural changes have been undertaken sporadically over the years. This was due primarily to lack of understanding of their importance in the overall sorbent performance, but also due to premature process development efforts using promising, but insufficiently studied sorbent compositions.

Early kinetic studies of sulfidation of metal oxides include the half-calcined dolomite system (2), the oxides: MnO, ZnO, CaO and V_2O_3 (3), and iron oxides (4-8). The intrinsic kinetics of sulfidation of Fe_2O_3 are rather slow (in the absence of H_2 or CO) while those of metallic iron are faster (8). The sulfidation of iron has slower kinetics than ZnO, CaO or MnO, even though the opposite has often been quoted in the literature on the basis of an early report (9), which used inconsistent data to calculate the kinetic constant of Fe_2O_3 sulfidation. However, as reported later by Tamhankar et al. (8) who studied iron oxide dispersed in a silica matrix in a simulated coal gas mixture, Fe_2O_3 is quickly reduced to spongy iron, which then reacts with H_2S to form iron sulfide.

Reduction during sulfidation is an important issue for all reducible oxide sorbents. At the high temperature of coal gas desulfurization, the reduced metal oxide or the metal itself may be the stable phase predicted by thermodynamics, and in addition the kinetics of reduction may be faster than the sulfidation kinetics. For example, depending on the molar ratio of $(H_2 + CO)/(H_2O + CO_2)$ and the temperature of the coal gas, iron may exist in Fe_2O_3, Fe_3O_4, FeO or Fe form, while copper oxide will be reduced to Cu metal in typical fuel gas compositions. Even non-reducible oxides, such as ZnO, can be prone to reduction effects, in this case magnified by the immediate vaporization of any elemental zinc that might be formed. Thus, in general, the reduction reaction should be considered along with the sulfidation of the sorbent:

$$MO_x + H_2S + H_2\,(CO) \rightarrow MS_y + H_2O + (CO_2); \quad y \leq x$$

More importantly, reduction kinetics should be studied separately from the sulfidation kinetics to obtain reliable kinetic parameters for each reaction.

In this paper, we present an overview of kinetic studies of reduction/sulfidation of single and mixed oxide sorbents that have appeared in the recent literature, and then consider two main classes of regenerable sorbents, ZnO-based and CuO-based materials, from a kinetics viewpoint as examined by our group to illustrate the importance of sulfidation kinetics and structural changes on the overall sorbent performance.

2 Overview of Bulk Sorbent Sulfidation Kinetics

Because of its low cost and its earlier use for desulfurization of town gas, iron oxide was the first material to be examined as a regenerable sorbent for coal gas desulfurization. The sulfidation equilibrium constants of the various iron oxides and of the metal are rather low, with corresponding H_2S concentrations in the hundred parts per million range for a composition representative of low Btu coal-gas and temperature in the range of 500-800 °C. The kinetics of reduction and sulfidation of iron oxide were reported by Tamhankar, *et al.* (8) in a thorough investigation of granular Fe_2O_3 dispersed in a porous matrix of SiO_2 (45wt% Fe_2O_3 - 55wt% SiO_2). The grain model was used to correlate the data. Experiments with powders at low conversions were used to obtain apparent rate constants, while pellets were used to estimate the effective diffusivities. The sulfidation rate constant of metallic iron was expressed as: $k' = 158 \exp(-3300 \text{ cal/RT}) \text{ sec}^{-1}$, and the reaction rate was first order in H_2S. The overall rate for any particle size over the temperature range of 600-900 °C was adequately represented by the summation of two terms, chemical reaction, and pore diffusion.

In a comparative study of sulfidation kinetics of CaO, MnO, ZnO and V_2O_3 powders in a thermogravimetric analyzer (TGA), Westmoreland, *et al.* (3), found that the intrinsic sulfidation rate of MnO was approximately one order of magnitude greater than the rate of either ZnO or CaO and two orders of magnitude greater than the reaction rate of V_2O_3. All reactions were first order in H_2S.

MnO is much inferior to ZnO in terms of its thermodynamic H_2S removal efficiency. It also forms a rather stable sulfate (9). Thus, manganese oxide has been considered only for application to very high temperature (>750 °C) coal-gas desulfurization where the use of ZnO sorbents is not practical due to Zn vaporization. More about the MnO-H_2S interaction may be found elsewhere in this volume in the papers by Van der Ham, *et al.* and Moulijn, *et al.*

CaO has much better sulfidation equilibria than MnO; actually it was ranked second only to ZnO in a study of twelve selected oxides by Furimsky and Yumura (10) over a wide temperature range. CaO is also very resistant to reduction. However, as mentioned before, due to the stability of its sulfate, it has been examined as a non-regenerable sorbent for in-bed desulfurization in fluidized-bed coal gasifiers, effectively removing a large fraction of H_2S and lowering the load to the external (regenerable) sorbent. In such an application, the bed material should be stabilized to $CaSO_4$ form, which is more suitable for disposal in landfills. However, problems exist with both the overall reaction rate of sulfidation which is limited by pore and product layer diffusion and with stabilization of CaS to $CaSO_4$. More on the reactivity of limestones and calcines can be found in this volume in the papers by Sotirchos, *et al.* and related publications (11, 12), while the papers by Abbasian, also in this volume, cover the in-situ desulfurization of coal-gas by CaO and stabilization of spent sorbents.

In regards to vanadium oxides, the thermodynamic data initially considered and reported in (9) were erroneous (13). V_2O_3 does not have attractive sulfidation equilibria. In fact, it was reported later that only chemisorption of H_2S, but no bulk sulfidation occurs on V_2O_5 (14, 15). Another property of V_2O_5 is that it forms eutectics with the oxides of zinc or copper. This was explored as a means of enhancing the kinetics of sulfidation of the main oxide, ZnO or CuO, by gas diffusion in a liquid (eutectic) rather than a solid (15, 16). The potential loss of vanadium by vaporization of V_2O_5 which would be corrosive to the turbine blades was another barrier to further development of vanadium-containing sorbents.

Zinc oxide combines a high sulfidation equilibrium constant with good reactivity. While commonly used as a throwaway sorbent in "guard-beds" (protecting downstream catalysts from sulfur impurities) ZnO was not considered as a regenerable sorbent until the early 80s

(17, 18). ZnO is of higher cost than iron oxides and forms a more stable sulfate. The regenerability of ZnO is restricted by loss of surface area (sintering) at high regeneration temperatures and formation of zinc sulfate at low regeneration temperatures. Another limitation of ZnO, not recognized in the initial sorbent development work, is its appreciable rate of reduction to elemental zinc, a liquid above 419°C, which has considerable vapor pressure at the hot gas cleanup temperatures (>550°C). Thus, zinc can be lost to the system, or react in the gas phase with H_2S and deposit and grow as a dense ZnS layer on the outside surface of the reacting particles (19-21).

Kinetic studies of sulfidation using single pellets of zinc oxide were conducted in the early 1980s (22, 23). The importance of sorbent structural properties and changes occurring during sulfidation were recognized in that work. Zinc sulfide product has higher molar volume (1.64:1) than zinc oxide, thus, grain growth occurs as reaction proceeds. Depending on the solid pore structure and gas composition and temperature, the global rate of reaction may contain more than one terms, i.e. chemical reaction (intrinsic kinetics), product-layer diffusion, pore diffusion and gas-film diffusion. With ZnO pellets of low porosity, utilization below 20% was measured due to pore closure and dense sulfide layer formation (17, 22). The grain model (24) and, more recently, the overlapping grain model (25) and generalized random pore model (26) have been used to analyze the sulfidation of zinc oxide powders and pellets (22, 23, 27, 28).

It is instructive to consider at this point a comparison of the intrinsic kinetics of sulfidation of single metal oxides or the corresponding reduced metals as they exist in the literature. Table 1 shows the kinetic constants calculated from initial rate measurements in TGA assemblies using sorbent powders. In all the cases shown, the reaction was found to be first order in H_2S. For easier comparison, literature data have been converted to express the k constants in the same units (cm/s). The Arrhenius constants $k_{s,o}$ and E_a listed in Table 1 come from the following expression for k_s:

$$k_s = k_{s,o} \exp [-E_a/RT]$$

where $k_{s,o}$ is the pre-exponential factor (cm/s) and E_a is the activation energy of sulfidation. An Arrhenius plot of the sulfidation rate constants of the sorbent oxides and metals considered here is shown in Figure 1.

Table 1. Arrhenius Constants for the Sulfidation Reaction of Various Oxides

	ZnO[a]	MnO[a]	CaO[a]	Fe[b]	$Cu-CeO_{2-x}$
E_a (cal/mol)	7240	5690	5160	3300	3967
$k_{s,o}$ (cm/s)	0.11	0.47	0.039	0.0023	0.075

(a: ref. (3); b: ref. (8); c: ref. (30))

Included in Table 1 and Figure 1 are k_s constants for the system $Cu-CeO_{2-x}$, the sulfidation kinetics of which were recently measured in this lab (29,30) after reduction of bulk CuO-CeO_2. In a coal gas atmosphere, CuO is rapidly reduced to copper metal, which is then sulfided to Cu_2S. Metallic copper has inferior sulfidation equilibria to ZnO, but under most conditions it is superior to iron and manganese oxides. However, CuO cannot be used alone (uncombined) as a sorbent, because of copper metal formation and sintering of Cu_2S product layers. Large expansion of CuO pellets with concomitant formation of onion-like shells of Cu_2S was reported in a recent sulfidation study (31). In combined form, however, with other active or inert oxides, CuO has many desirable features as discussed below. CeO_2 is known to disperse copper oxide, without forming any compounds or solid solutions with it, so it was used as a matrix to study the kinetics of Cu sulfidation after reduction of the bulk CuO-CeO_2 material. Additional information on this system is provided below.

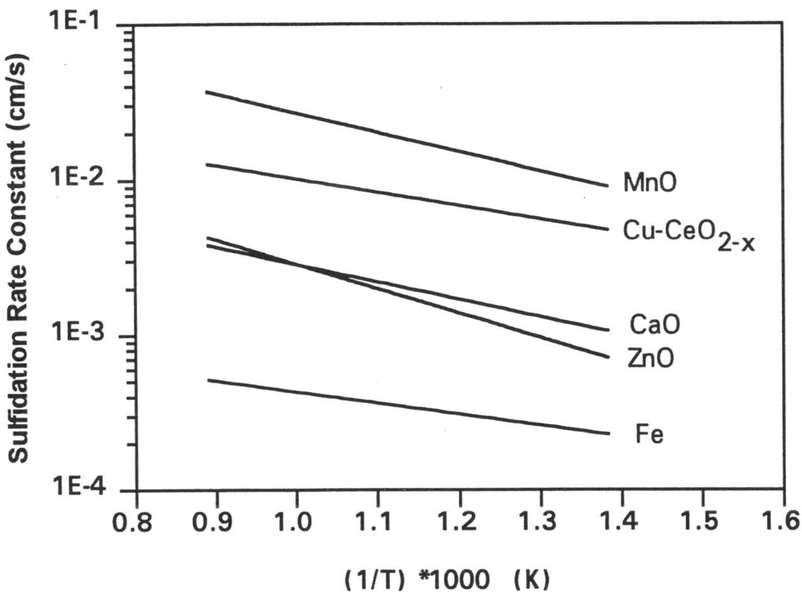

Figure 1. Arrhenius plots of the sulfidation rate constant, k_s, of various oxides; ref. (3, 8, 30).

The low values of the activation energies for the sulfidation reactions, typically 3-8 kcal/mole, need some explanation. The intrinsic reaction mechanism is not known. It may be hypothesized, however, that an adsorption-desorption mechanism is involved in the sulfidation reaction, according to which H_2S is dissociatively adsorbed on the metal or metal oxide surface and hydrogen is desorbed as a product gas. Alternatively, one could assume that the E_a values shown in Table 1 are apparent activation energies, where $E_a = E_{true} - Q$, with Q denoting the heat of adsorption of H_2S. For high values of Q, the apparent activation energy can have a small value (32).

On the basis of the intrinsic reaction kinetics shown in Figure 1 and Table 1, one could rank the sorbents as Fe < ZnO ≅ CaO < Cu-CeO_{2-x} < MnO in order of increasing activity. However, the intrinsic sulfidation kinetics is only one of several guidelines we need to employ in ranking the overall sorbent activity, which is often limited by product layer diffusion and/or pore diffusion. Structural changes during reduction/sulfidation and subsequent regeneration may affect the steady-state reactivity of the sorbent. These issues are discussed in the following for ZnO- and CuO-based sorbents.

3 Sulfidation Kinetics of Mixed Metal Oxide Sorbents

In the following we discuss in more detail the kinetics of reductive sulfidation including structural effects using mixed metal oxide sorbents. It has become apparent over the years that mixed metal oxides comprising at least one additional oxide to the main active phase are better suited for various hot gas desulfurization applications and reactor designs. The second phase (active or inert) can provide dispersion to the active sorbent phase or form compounds with it that have lower reducibility than the uncombined active oxide and/or improved sulfidation efficiency through mixed sulfide product formation (33). Other major reasons for developing mixed oxide sorbents are (i) to improve sorbent regenerability and structural stability in cyclic operation, and (ii) to increase the strength and attrition resistance of the sorbent; the latter is very important for certain sorber designs, such as fluidized-bed and entrained-flow designs. The evolution in development of bulk mixed metal oxide sorbents is described elsewhere in this volume by Akyurtlu, *et al.* Also, the sulfidation/regeneration performance of supported metal oxides is presented in a paper by Atimtay in this volume.

In this paper, we consider the sulfidation of zinc titanate sorbents, studied first by our group as alternatives to zinc oxide sorbents (19, 34, 35) for high-temperature coal gas desulfurization. Also, we illustrate the positive effect of mixing CuO with other metal oxides on the reactivity of the former.

3.1 Comparative Kinetics of Reductive Sulfidation of Zinc Oxide and Zinc Titanates

As stated above, development of mixed oxide sorbents is aimed at improving one or several properties of the active sorbent phase. Iron oxide was the first additive considered for mixing with zinc oxide. The compound formed, zinc ferrite, has undergone extensive studies under sponsorship of the U.S. Department of Energy/Morgantown Energy Technology Center (METC), where work on zinc ferrite was initiated (18). In the mid-eighties, however, in work carried out at the Jet Propulsion Lab (15, 36) and then at the Massachusetts Institute of Technology (19, 34, 37) it was unambiguously shown that iron oxide provides no stabilization to ZnO against reduction. As a result, in high-Btu gases or temperatures above 550°C, irreversible structural damage occurs to zinc ferrite pellets (37, 33). Thus, the sulfided pellets would develop a subsurface enriched in zinc, due to zinc migration, while the surface (~10 μm depth) would consist of dense iron sulfide due to vaporization of surface zinc. This is detrimental to sorbent utilization and also causes structural damage (cracking, spalling) in cyclic operation (37). There were some other reports in the eighties of mixing effects of various oxides on ZnO. For example, zinc chromite and zinc aluminate, both poor H_2S sorbents, were studied with the objective of producing elemental sulfur during regeneration (38). Mixed ZnO-CuO was studied for its better resistance to surface area loss (39), while addition of small amount of CuO into zinc ferrite was found to improve the H_2S removal efficiency of the latter, apparently due to the existence of cuprous oxide, an excellent H_2S sorbent (40, 41). However, none of these combinations could meet the two main criteria of a good sorbent, namely; to combine good overall reactivity with structural stability in cyclic operation.

The breakthrough came with the development of zinc titanates as hot coal gas desulfurization materials (19, 21, 34, 35, 37). TiO_2, an inert oxide in desulfurization, is mixed with ZnO to form compounds (zinc titanates), which combine excellent sulfidation equilibria with lower reducibility than ZnO neat. TiO_2 is clearly superior to Al_2O_3, SiO_2, and Cr_2O_3 oxides, all of which form compounds with ZnO that are very poor H_2S sorbents. TiO_2 is also superior to Fe_2O_3, in that zinc titanates are more stable in the reducing coal gas atmosphere than zinc ferrite, which is reduced to ZnO and Fe_3O_4 (or a lower iron oxide). Another attractive feature of zinc titanates is their superior mechanical strength to zinc oxide. Also, in regeneration of ZnS (TiO_2), formation of zinc titanate rather than $ZnSO_4$ is favored (42). For further information on the development of zinc titanate sorbents by contractors of the U.S. DOE in recent years, the reader is referred to the paper by Harrison on the performance analysis of ZnO-based sorbents in this volume.

3.1.1 Experimental Methods

Preparation and Characterization of Solids. In the work reported by Lew, *et al.* (35, 20, 21), bulk mixed oxide powders of zinc and titanium were prepared by a known method for synthesizing highly dispersed mixed oxides from amorphous citrate precursors (43, 44). The preparation of Zn-Ti oxides consists of mixing a 2:1 volume ratio solution of glacial acetic acid and titanium (IV) isopropoxide with an aqueous solution of zinc acetate and citric acid monohydrate. Typically, an equal mole ratio of citric acid to metal ions (zinc and titanium) was used in preparation of the solution. The final solution was first dehydrated rapidly (15-30 min) in a rotary evaporator at 65-75 °C under vacuum to form a viscous liquid and then dehydrated slowly (4-6 h) in a vacuum oven at 70-80 °C to form a porous solid foam. The solid foam was calcined in air in a muffle furnace at 720 °C for 12 h producing a porous, homogeneous mixed metal oxide. This was crushed and sieved to 90-125 μm size particles.

The solids were characterized by several bulk and surface analysis techniques. The elemental composition (zinc and titanium) of the solids was verified by atomic absorption spectroscopy (Perkin Elmer 360 spectrophotometer) of the solids dissolved in a hot HF-HCl-H_2O solution (~90 °C). X-ray diffraction (XRD) for identification of crystalline phases in the mixed oxides was performed with a Rigaku RU300 instrument using CuK_α radiation. Scanning electron microscopy (SEM) with energy-dispersive X-ray analysis (EDS) using a Cambridge Stereoscan 250 MK3 instrument were used to observe the surface morphology, crystallite size, and compositional variation of the solids. Surface areas were measured by a Micromeritics Flow Sorb III 2300 BET apparatus using N_2 gas, while pore volumes and pore size distribution were measured by a Micromeritics Autopore 9200 mercury porosimeter.

Physicochemical Properties of Bulk Zn-Ti-O Sorbents. Three distinct zinc titanate phases, namely Zn_2TiO_4, $Zn_2Ti_3O_8$, and $ZnTiO_3$, can be formed through solid preparation by the citric acid complexation method from zinc acetate and titanium (IV) isopropoxide precursors followed by pyrolysis in air at different time-temperature conditions. The type of phases present depends on the Zn/Ti atomic ratio and the calcination temperature. The observed phase transformation with increasing temperature is $Zn_2Ti_3O_8 \rightarrow ZnTiO_3 \rightarrow Zn_2TiO_4$. At high calcination temperature (≥ 1000 °C), Zn_2TiO_4 is the only stable mixed oxide phase for all Zn-Ti-O solids, coexisting with either ZnO or TiO_2 phases depending on the solid stoichiometry (Zn/Ti ratio). Also, Zn_2TiO_4 is the stable phase for solids with Zn/Ti ≥2 calcined at temperatures ≥700°C for long periods of time (≥12 h). At calcination temperatures below 800°C, all three zinc titanate phases may be present (35, 21).

For the sorbents calcined at 720°C for 12 h, the type of phases present depended on the Zn/Ti ratio. Decreasing the Zn/Ti ratio of the solids produced phases in the order ZnO → Zn_2TiO_4 → $ZnTiO_3$ and $Zn_2Ti_3O_8$ → TiO_2 (35, 21). Table 2 lists the properties of the solids discussed in this paper.

Table 2. Chemical Properties of Sorbents Used in Sulfidation Experiments[a]

sorbent	(Zn/Ti) (atomic ratio)	crystalline phases[b] (wt %) ZnO	Zn_2TiO_4	$ZnTiO_3$	$Zn_2Ti_3O_8$	TiO_2 (rutile)
Z3T7[c]	3/7	0	0	69	0	31
Z2T3-a	2/3	0	0	65	16	19
Z2T3-b	2/3	0	0	83	0	17
ZT	1/1	0	20	45	35	0
Z3T2	3/2	0	68	18	14	0
Z2T-a	2/1	0	100	0	0	0
Z2T-b[d]	2/1	0	100	0	0	0
Z3T	3/1	28	72	0	0	0
ZnO		100	0	0	0	0

[a]Prepared from zinc acetate and titanium (IV) isopropoxide with 1:1 mole ratio of metal ions to citric acid (unless otherwise noted). All solids calcined at 720°C for 12 h except for ZnO, which was calcined for 4 h. [b]Identified by X-ray diffraction. [c]Prepared with titanium tetrachloride. [d]Prepared with 1:2 mole ratio of metal ions to citric acid.

Effects of varying the Zn/Ti atomic ratio on the physical properties of the Zn-Ti-O solids, i.e., surface area and pore volume, are shown in Figure 2. Solids with up to 50 mol% TiO_2 are characterized by higher surface area and pore volume than ZnO neat. Addition of small amounts of TiO_2 into ZnO has the largest effect, with a maximum in surface area and pore volume shown for $(Zn/Ti)_{atomic} = 9/1$. These data indicate that TiO_2 disperses ZnO, effectively preventing ZnO particle growth (sintering). High levels of TiO_2, however, and compound formation (e.g., $ZnTiO_3$, $Zn_2Ti_3O_8$) reduce the overall surface area.

As a result of the preparation method, the sorbents were produced in a highly macroporous form. Macropores (>1 μm diameter) typically comprised more than 70% of the pores of the 90-125 μm-size particles. Such pore structures allow for kinetic studies in the absence of pore diffusion limitations even with relatively large particles (~100 μm).

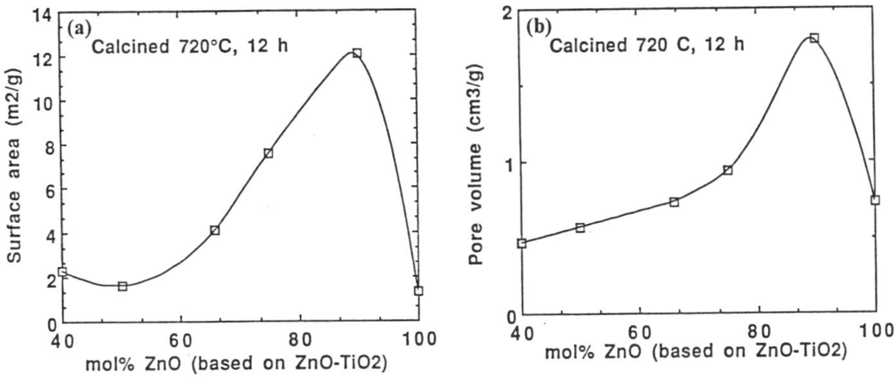

Figure 2. Effect of Zn-Ti-O composition on the physical properties of the solid particles (90-125μm): (a) surface area; (b) pore volume.

Apparatus and Procedure. Kinetic experiments with solids containing various Zn/Ti atomic ratios were performed in a Cahn System 113-X thermogravimetric analyzer (TGA) equipped with a Cahn 2000 electrobalance. The solid was pretreated in a vacuum oven at 90°C for 1 h to remove any absorbed H_2O before it was reacted in the TGA.

Gas-flow rates were set by passing the gases through mass-flow controllers. A gas flow rate of 485 or 350 cm^3 min^{-1} (STP), respectively, was used in the reduction and sulfidation experiments. In sulfidation, approximately half the gas flow (containing H_2S, H_2, and N_2) entered the reactor (TGA) through a side arm. The other portion of the gas containing only N_2 entered the balance section of the TGA serving both to protect the balance from the corrosive H_2S and as a diluent to the reactant gas. Water vapor was added to the gas by bubbling nitrogen and hydrogen through a water saturator. The saturated gas stream entered the apparatus side arm through heated lines. A thin layer of solid particles (1-3 mg and 90-125 μm size) was placed on a quartz pan suspended by a quartz hang-down wire. Isothermal reduction experiments were performed at temperatures between 550 and 1050°C, while the sulfidation experiments were run at 400-800°C.

It was experimentally verified by varying the gas flow rate, quantity of sample, and particle size that these experiments were performed in the absence of both mass transfer and pore diffusional limitations. Thus, the measured rate was due only to the intrinsic reduction or sulfidation kinetics and product layer diffusion. Typically, each sulfidation experiment was repeated at least once to verify reproducibility.

3.1.2 Initial Reduction Rates

The initial reduction rate of various Zn-Ti-O materials was measured in H_2-N_2-H_2O mixtures. Details are given by Lew, *et al.* (20, 35). The initial surface reaction rate was calculated by:

$$R_o = \frac{(dW/dt)_{t=0}}{(81.3794)A_o}$$

where R_o is the initial rate (mmol cm^{-2}s^{-1}), $(dW/dt)_{t=0}$ (mg s^{-1}) is the slope of the weight vs time profile at $t = 0$, 81.3794 is the molecular weight of ZnO, and A_o (cm^2) is the initial surface area of the reacting solid. This expression for R_o is used for both ZnO and Zn-Ti-O solids, since the products of reduction are vaporized zinc and H_2O (and unreacted TiO_2 for the titanates). TiO_2 was not reducible under the conditions of these experiments.

In dry reduction, Zn-Ti-O solids had lower reduction rate than ZnO at temperatures in the range of 550-1050°C. The activation energy of reduction was 37 and 24 kcal mol^{-1}, respectively, for Zn-Ti-O and ZnO. All zinc titanate phases (i.e. Zn_2TiO_4, $ZnTiO_3$, and $Zn_2Ti_3O_8$) had the same initial rate of reduction.

Water vapor inhibited the initial reduction rate of ZnO and Zn-Ti-O solids. However, increasing the water vapor content from 1 to 8 mol% did not cause any further decrease in the reduction rate. Addition of water vapor was accompanied by changes in the activation energies to 44 kcal mol^{-1} for both ZnO and Zn-Ti-O solids. Increasing the amount of TiO_2 from zero mol% to 60 mol% was accompanied by a corresponding decrease in the reduction rate.

Comparative Arrhenius plots of ZnO, Z2T, and Z2T3-b in the presence of H_2O are shown in Fig. 3. The difference in reactivity lies in the pre-exponential constant, $k_{r,o}$, calculated from the equation:

$$k_r = k_{r,o} \exp [-E_a/RT],$$

where k_r is the reduction rate constant, calculated from the surface reduction rate of the solid,

$$R_o = k_r C^n_{H2}$$

and where C_{H2} (mmol cm^{-3}) is the gaseous concentration of H_2 in the gas. The wet reduction of both zinc oxide and zinc titanate was first order in H_2 (20). The values of $k_{r,o}$ (cm s^{-1}) are 9.27 x 10^5, 3.36 x 10^5 and 1.31 x 10^5, respectively, for ZnO, Z2T and Z2T3-b.

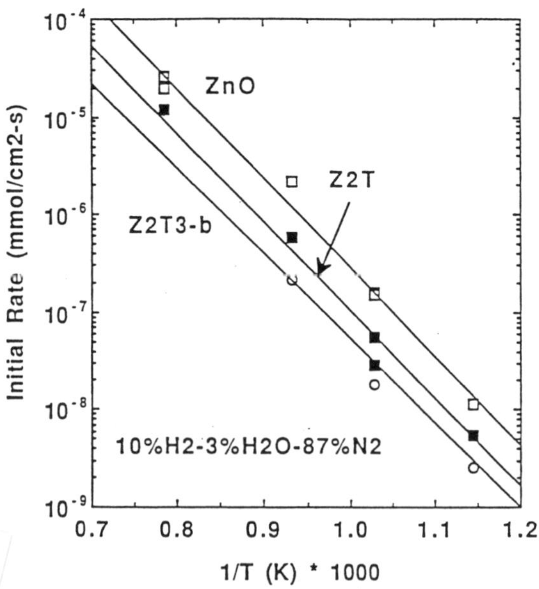

e plots of the initial reduction rates of ZnO, Z2T and Z2T3-b in 10% H_2-3% H_2O-87% N_2.

3.1.3 Reduction Conversion Profiles

Typical conversion-time plots for ZnO and Zn_2TiO_4 reduction are shown in Fig. 4. The fractional conversion is defined as

$X = (W_i - W) / (W_i - W_f)$

where W is the instantaneous weight, W_i is the initial weight, and W_f is the final weight at complete conversion. For ZnO, W_f is equal to zero. W_f is the weight of the remaining TiO_2 for Zn-Ti-O solids. For both ZnO and Z2T, the reaction rates remained relatively constant up to almost complete conversion. For Z2T reduction, a product layer of TiO_2 is formed. Since the ratio of the solid product (TiO_2) molar volume to the solid reactant (Zn_2TiO_4) molar volume is only 0.41, a porous product layer is formed. Based on the conversion profiles, this product layer poses no hindrance to reaction.

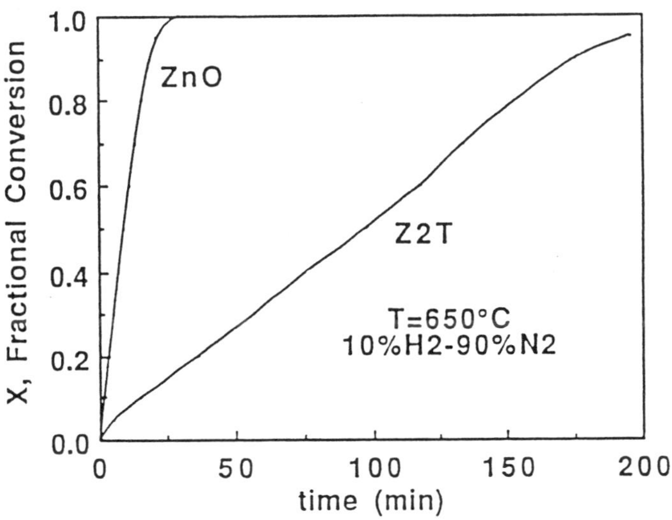

Figure 4. Comparison of the experimental reduction profiles of ZnO and Z2T at 650°C in 10% H_2-90% N_2.

The chemical phase transformations of the Zn-Ti-O solids during reduction were followed by XRD analysis of partially reacted solids (20, 35). No free ZnO was detected. $ZnTiO_3$ and $Zn_2Ti_3O_8$ were reduced directly to Zn (g) + TiO_2 (s). Zn_2TiO_4 reduction followed the direct route as well as the indirect route through transformation into $Zn_2Ti_3O_8$. SEM analysis of partially reduced solids showed void formation (caused by Zn evaporation) throughout the solid particle. Thus, reduction of zinc titanates is not limited by product layer diffusion. Figure 4 exemplifies this; the fractional reduction conversion of both ZnO and Z2T sorbents is linear with time up to completion of the reduction.

The overall conclusion from the reduction of zinc titanates is that titanium suppresses the reduction rate of ZnO by eliminating sites for hydrogen adsorption (geometric interaction). Because the addition of TiO_2 does not change the activation energy of wet reduction, the effect is not very pronounced. In other words, zinc titanates would also be limited to desulfurization temperatures not exceeding 650-700°C, the exact temperature dictated by both the gas phase and sorbent composition.

3.1.4 Sulfidation Kinetics

Initial Reaction Rates. Initial sulfidation experiments in the TGA were performed under isothermal conditions to determine the sulfidation reactivity of the solids listed in Table 2. The gas mixture consisted of 2% H_2S-1% H_2-97% N_2. The small amount of H_2 was added to the gas mixture to prevent the decomposition of hydrogen sulfide. The effects of H_2 and H_2O on the initial sulfidation rate were examined separately.

The initial rate of sulfidation was measured at 600 and 700°C and was approximately 1.5-2.0 times higher for ZnO than for titanates containing ≥25 mol% TiO_2. No bulk titanium sulfide was formed (21, 35). The initial sulfidation rate was similar for the various titanates despite the fact that sorbents with different Zn-Ti-O compositions formed different zinc titanate phases as shown in Table 2.

In the absence of both diffusional and mass-transfer resistances, the irreversible surface chemical reaction can be described as

$$R_o = k_s C^n_{H2S}$$

where R_o is the initial molar rate of ZnS formation per unit surface area of the solid reactant [$mmol cm^{-2} s^{-1}$], k_s is the sulfidation rate constant, C_{H2S} is the molar concentration of hydrogen sulfide [$mmol\ cm^3$], and n is the reaction order. The reaction orders (n) for ZnO and Z2T-a sulfidations were both determined to be 1. This agrees with the kinetic data reported by Westmoreland *et al.* (3) for zinc oxide. An Arrhenius relationship can be used to express the intrinsic sulfidation rate constant as

$$k_s = k_{s,o} \exp[-E_a/RT]$$

where $k_{s,o}$ is the Arrhenius frequency factor [$cm\ s^{-1}$], E_a is the activation energy [$kcal\ mol^{-1}$], R is the gas constant [1.987×10^3 $kcal\ mol^{-1}K^{-1}$)], and T is the temperature [K].

The Arrhenius dependence for ZnO, Z2T-a, and Z2T3-a sulfidation reactions was determined by measuring the initial sulfidation rate as a function of temperature. The experiments were performed in 2% H_2S-1% H_2-97% N_2. Figure 5 shows the resultant Arrhenius plots for ZnO, Z2T-a, and Z2T3-a sulfidations. The intrinsic sulfidation rate for ZnO was greater than Z2T-a and Z2T3-a at all temperatures between 400 and 800°C, while the Arrhenius plots for Z2T-a and Z2T3-a virtually overlapped. The kinetic constants obtained for ZnO and Z2T-a sulfidation are listed in Table 3. The activation energies for both ZnO and Z2T-a sulfidation were approximately the same (10.3 kcal mol and 9.3 kcal mol, respectively). The major difference was in the frequency factors. The frequency factor for ZnO sulfidation was approximately 3 times greater than that for Z2T-a. The lower frequency factor for the latter was probably caused by fewer reaction sites on the reactant surface due to the presence of titanium. The initial sulfidation rate reported in the present study for ZnO powders was approximately twice as fast as that reported by

Westmoreland *et al.* (3). This difference in sulfidation rate is believed to be due to differences in purity and crystallinity of the two solids (35).

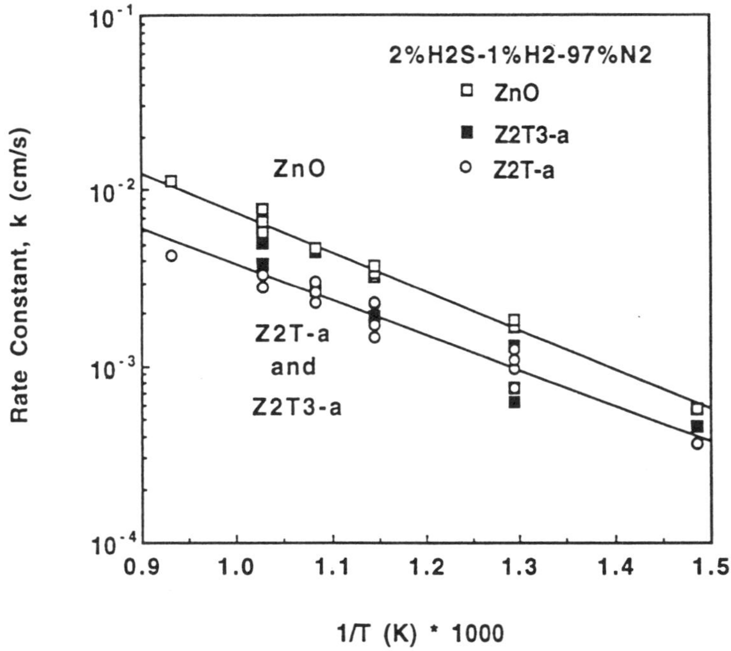

Figure 5. Comparative Arrhenius plots of the sulfidation rate constant of ZnO, Z2T-a and Z2T3-a sorbents.

Table 3. Arrhenius Constants for the Sulfidation Reactions of ZnO and Z2T-a Sorbents

reaction	Temp. (°C)	Arrhenius constants $k_{s,o}$ (cm/s)	E_a (kcal/mol)
$ZnO + H_2S \rightarrow ZnS + H_2O$	400-800	1.31	10.3
$^1/_2 Zn_2TiO_4 + H_2S \rightarrow ZnS + {}^1/_2 TiO_2 + H_2O$	400-700	0.40	9.3

Sulfidation Conversion Profiles. The sulfidation of three sorbents, ZnO, Z2T-a, and Z2T3-a, was studied in detail. Figure 6 shows the conversion profiles for these sorbents at 650°C. The fractional conversion is defined as above for reduction, only now W_f^I is the final weight at complete sulfidation assuming that the sorbent reacts completely to form ZnS and TiO_2 (for the Zn-Ti-O sorbents).

For Z2T-a and Z2T3-a, the decrease of reaction rate at high conversion is attributed primarily to nonuniformity in the grain size distribution of the solids. SEM micrographs of Z2T-a and Z2T3-a showed the presence of both small spherical grains and larger plate-like grains. In the early portion of the conversion profiles for Z2T-a and Z2T3-a (Figure

6) the small grains were mainly reacting while at higher conversion all the small grains had reacted and reaction was due only to the larger plate-like grains. The effect of nonuniform grain size was verified by sulfidation of the solid Z2T-b, which was prepared by using a 2:1 mole ratio of citric acid to metal ions instead of the usual 1:1. A solid with a more uniform, small grain size was, thus, produced. A uniform reaction rate was measured for sorbent Z2T-b up to very high conversions. The formation of a product layer of ZnS and TiO_2 around the unreacted solid core also contributes to the slower reaction rate at higher conversion (see below).

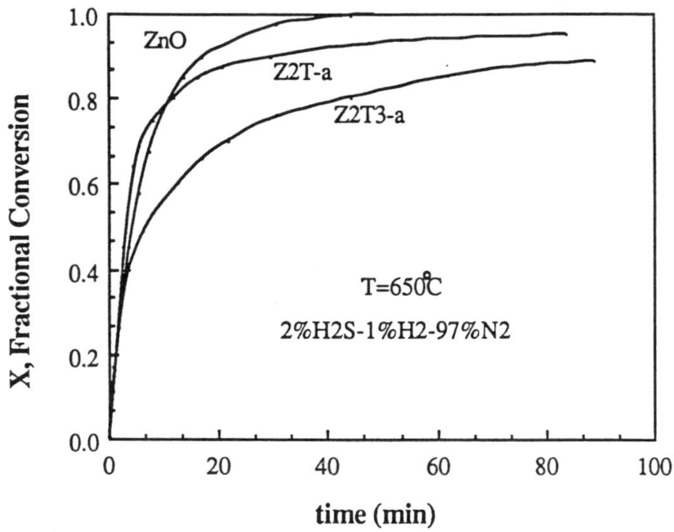

Figure 6. Comparative sulfidation conversion profiles for the sorbents ZnO, Z2T-a and Z2T3-a.

Physical Changes of Sorbents During Sulfidation. Lew, *et al.* (21, 35) described the surface area and morphological changes of Zn-Ti-O sorbents during sulfidation. Typically, an increase in surface area took place upon sulfidation of all Zn-Ti-O materials . This is attributed to roughening that takes place when a titanate of low surface area is changed to a mixture of ZnS and TiO_2. On the basis of SEM micrographs and EDS analysis of sulfided samples, it was found that sintered regions were rich in zinc sulfide. Overall, TiO_2 inhibits the sintering of zinc sulfide by acting as a physical barrier to prevent growth of ZnS particles. However, very high amounts of TiO_2 were no longer effective due to TiO_2-particle sintering. Levels up to 50% TiO_2 are most effective for preserving ZnS dispersion and total surface area in the Zn-Ti-O sorbents.

Effects of H_2. No change in the initial sulfidation rate was observed with the addition of various amounts of H_2 (1, 10 and 20%) in the gas stream at temperatures in the range of 400-800°C. In a comparison of ZnO and Z2T-a sorbents at 800°C-sulfidation, little if any zinc loss appeared to have taken place based on the final weight of the solids in the TGA. Both sorbents were completely sulfided (based on the initial sorbent weight). The absence of zinc loss is due to the particular configuration of the TGA system (i.e., a hemispherical pan with a thin layer of reactant solid). Although no weight loss was recorded, reduction did occur. Zinc vapor which formed as a result of reduction reacted with H_2S in the gas phase to form solid ZnS which was deposited on the pan. After sulfidation was completed, a white film was observed on the sample pan. Analysis of this film by atomic absorption spectroscopy identified the presence of zinc.

At high conversion, the reaction rate dropped faster when sulfidation was performed with a gas containing either 10 or 20% H_2. A possible explanation for this is pore blockage due to the deposit of solid product from the vapor-phase reaction between Zn (g) and H_2S (g). Small jumps seen occasionally in the sulfidation profiles (21, 35) are believed to be due to the formation of cracks in the product layer. As the sulfided product layer is formed, zinc and water vapor from reduction will accumulate in the particle until their pressure is sufficient to cause the product layer to crack. At that point, reaction will occur rapidly, and the conversion jump will be observed.

With the gas mixture containing 2% H_2S-20% H_2-78% N_2, large spherical agglomerates (0.7-1.3 μm radius) were formed as a result of the reaction between Zn (g) and H_2S (g) and deposited on the original surface. The agglomerates were composed of smaller crystals (~0.04 μm radius). The high zinc vapor-phase concentration during sulfidation reaction with 20% H_2 in the reactant gas is believed to lead to this enhanced sintering. Sainamthip and Amarakoon (45) have also reported enhanced grain growth in zinc vapor for manganese zinc ferrite. A lower level of hydrogen (10%) did not produce these agglomerates although the small crystals were seen. In addition, some large cracks were evident in several particles. Thus, reduction can cause not only zinc loss but also weakening of the strength of ZnO by crack formation and formation of sintered layers of ZnS on the surface.

While zinc titanates are more resistant to reduction, as discussed above, structural changes may be expected in ZnO-rich materials under high H_2-pressure. Enrichment of the subsurface with zinc and sintering of ZnS layers can create structural degradation in cyclic operation similar to that described earlier for zinc ferrite (35, 37).

Effect of H_2O. Coal-derived gas streams contain steam in various amounts depending on gasifier type and extent of quenching (3-40% H_2O). The effects of H_2O on the sulfidation kinetics of the sorbent were addressed by Lew, *et al.* (21, 35). The sulfidation rate was unaffected by the presence of H_2O. However, the sintering rate may be different (35).

Product Layer Diffusion. The sulfidation conversion-time plots, such as Figure 6, show that in addition to surface kinetics (initial linear segment) the rate is limited by diffusion at higher conversions. Diffusion limitations are more pronounced at low temperatures for both the ZnO and the Zn-Ti-O sorbents. Lew, *et al.* (35, 27) used the overlapping grain model to predict the sulfidation profiles of ZnO and Zn-Ti-O materials in the range of 400-700°C where the resistances to reaction are due to surface reaction as well as diffusion through the product layer. This model, originally developed by Sotirchos and Yu (25) allows for randomly overlapping grains in the reacting solid. For the sorbents used by Lew (35) the overlapping grain model provided a more accurate structural description than the grain model (27). On the basis of SEM micrographs of the solids, a single grain size was used for ZnO, while a discrete bimodal grain size was used for the Zn-Ti-O sorbents. Details about the model and its application to ZnO and Zn-Ti-O sulfidation can be found in Lew's thesis (35) and a previous publication (27). Here, we discuss briefly the effect of the presence of TiO_2 on the product layer diffusivity, calculated by applying the overlapping grain model to the experimental sulfidation profiles and using the experimentally determined sulfidation reaction rate constants, k_s (see above).

The calculated effective diffusion coefficient of Zn-Ti-O sorbents of various compositions indicated a dependence on the relative amount of titanium oxide. For the activated diffusion through the solid product layer, the diffusion coefficient can be represented by the Arrhenius law as follows:

$$D_e = D_{e,o} \exp(-E/RT)$$

where the preexponential factor ($D_{e,o}$) and the activation energy (E) are not temperature dependent. An Arrhenius plot of the experimentally calculated diffusion coefficients for sorbents ZnO, *Z2T-a*, and *Z2T3-a* is shown in Figure 7. Table 4 lists the calculated Arrhenius constants. The Arrhenius plot of the diffusion coefficient is shown for temperatures up to 600°C. At higher temperatures, the relative resistance of the chemical reaction becomes more dominant compared to the diffusional resistance because of the higher activation energy of the product layer diffusion coefficient (26.6 kcal/mol) compared to that of the sulfidation reaction rate constant (9-10 kcal/mol). The relative importance of the product layer and the chemical reaction resistances is a function of the product layer diffusion coefficient, reaction rate constant, temperature, and grain size. For smaller grain sizes, the product layer diffusion resistance decreases because of the smaller corresponding product layer diffusion lengths involved (35).

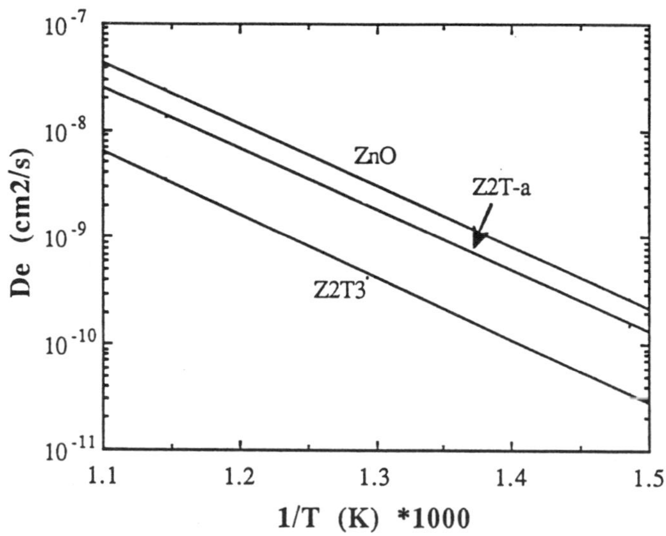

Figure 7. Arrhenius plots of the product layer diffusion coefficients of ZnO, Z2T-a and Z2T3 sulfidation sorbents.

Similar activation energies (26.6 ± 0.3 kcal/mol) were obtained for ZnO and Zn-Ti oxides. The major difference is in the pre-exponential factor. Sorbents with increasing relative concentration of titanium have correspondingly decreasing pre-exponential factor. The high activation energy is consistent with ionic diffusion through a nonporous solid. The similarity in activation energies suggests that in a product layer containing a mixture

Table 4. Arrhenius Constants for the Product Layer Diffusion Coefficient (D_e)

Sorbent	Temp. (°C)	Arrhenius Constants $D_{e,o}$ (cm^2/s)	E (kcal/mol)
ZnO	400 - 600	0.098	26.4
Z2T-a	400 -600	0.053	26.3
Z2T3-a	400 -600	0.019	27.0

of ZnS and TiO_2 diffusion occurs through ZnS and not through TiO_2. Diffusion preferentially occurs through the fastest pathway. The role of TiO_2 is to decrease the cross-sectional area available for diffusion and increase the tortuosity of the diffusion pathway. The change in the diffusion coefficient was mathematically expressed by adapting the equation for the "random pore" model proposed by Wakao and Smith (46) for diffusion in a porous solid having a bimodal pore-size distribution separated into macropores and micropores. Accordingly, the porosities were replaced by the volume fractions (v) of ZnS and TiO_2 (27). The diffusion flux in the product layer is the sum of that through ZnS, that through TiO_2, and that through ZnS and TiO_2 in series. For $D_{e,\mathrm{ZnS}} \gg D_{e,\mathrm{TiO_2}}$, the product layer diffusion coefficient is approximately equal to:

$$D_{e,ZnS+TiO_2} = v^2_{ZnS} D_{e,ZnS}$$

This expression was used to predict the product layer diffusion coefficient for various Zn-Ti oxides. The value of $D_{e,\mathrm{ZnS}}$ was obtained from ZnO sulfidation results. Good agreement between the experimental and predicted values of product layer diffusion coefficients were thus obtained for the various Zn-Ti-O sorbents (27, 35).

3.1.5 Concluding Remarks

The above case study has evaluated the addition of TiO_2 into ZnO in reductive sulfidation. The choice of zinc titanates was first made on the basis of their good sulfidation equilibria and regeneration performance. The following salient features of the kinetics comparison in reduction and sulfidation of ZnO and Zn-Ti-O sorbents have become clear from the above discussion. Addition of TiO_2 into ZnO suppresses the reducibility of the latter; all titanates were found to have lower reduction rates than ZnO. The trade-off appears in sulfidation. Sulfidation rates are also suppressed by the presence of titanium. This is true both for the intrinsic kinetics of sulfidation as well as for product layer diffusion. These factors need to be weighed properly in designing the sorbent composition, porosity, particle size and desulfurization reactor.

To avoid structural changes due to zinc migration in the sorbent, and since at high temperatures reduction and sulfidation are competing reactions, the sorbent must contain adequate amounts of TiO_2. The porosity of the sorbent is important for large particles or pellets as pore diffusion limitations may dominate the rate in the case of a low porosity material. In regards to reactor design, the fixed-bed reactor design should be avoided (33, 35) because of non-uniform gas composition along the bed length. The choice of a fluid-bed with a TiO_2-rich zinc titanate material is attractive both from the point of optimizing the ZnO performance under uniform reaction conditions as well as for providing a mechanically stronger sorbent than zinc oxide (47). However, there are issues of sorbent attrition always associated with fluid bed operation. Kinetics may be compromised too much in order to achieve acceptable attrition resistance. In such cases, the implementation of totally different designs may be recommended (33). For example, direct injection of sorbent into the gasifier exhaust duct followed by capture on a ceramic filter and regeneration has recently been proposed (48).

3.2 Comparative Study of CuO-based Sorbents

Mixed metal oxides of copper have been studied in recent years as alternatives to ZnO-based sorbents for high-temperature coal gas desulfurization. There is no volatilization issue with copper, but metallic copper has inferior sulfidation equilibria to those of ZnO. As mentioned before, uncombined copper oxide is not a suitable sorbent due to sintering of the metal and sulfide formed (15, 31). However, CuO has the best sulfidation equilibria among all oxides, e.g. at 900 K the sulfidation equilibrium constant $K_s = 6.3 \times 10^{17}$ for the reaction of $2CuO + H_2S + H_2 = Cu_2S + 2H_2O$. Over the whole range of practical temperatures, cupric (and cuprous) oxide can reduce H_2S to sub ppm levels.

Combining copper oxide with other metal oxides would be advantageous from the point of view of preserving the oxidic state of copper (+2 or +1) during all or part of the reductive sulfidation period. This approach was followed in most of our work with CuO-based sorbents (15, 16, 29, 30, 34, 36, 37, 49) as well as in work by other groups (31, 50, 51). For example, the compounds of copper oxide with the oxides of iron, aluminum and chromium in the form of the spinels $CuFe_2O_4$, $CuAl_2O_4$ and $CuCr_2O_4$, are relatively stable in reducing atmospheres considering both the phase stability diagrams and the kinetics of reduction. Of these compounds, $CuCr_2O_4$ has the highest reduction stability (30).

There are other oxides, such as MgO, TiO_2, and CeO_2, that do not form compounds or oxide solid solutions with CuO at the temperatures of interest to hot gas cleanup (<800°C). These oxides may be used as dispersants of copper. Finally, the combination of CuO with silica and zeolites has been studied by Kyotani, et al. (52, 53) who established that acidic oxides are also suitable carriers of CuO.

We have recently reported (29, 30) that the binary oxides of CuO-Cr_2O_3 and CuO-CeO_2 show high reactivity in both sulfidation and regeneration. The H_2S removal efficiency of both types of sorbents was better than 99% and complete CuO conversion was attained at 750°C.

3.2.1 Experimental Apparatus and Procedures

Sorbent preparation and characterization. The binary Cu-Cr-O and Cu-Ce-O sorbents were prepared from amorphous citrate precursors (43, 44). A similar procedure to the one described above for zinc titanate was followed (29, 30). The precursor solutions here were aq. cerium or chromium nitrate mixed with aq. solution of cupric nitrate in the desired molar ratio. The mixed solution was added dropwise to aq. citric acid solution under continuous stirring at room temperature. Typically the metals: citric acid molar ratio was 1:1. The resulting solid foam was calcined in air in a muffle furnace at 1000°C for 1 hour. After calcination, the solid was crushed and sieved to desired size. Three sorbent compositions with CuO: Cr_2O_3 or CuO: CeO_2 mole ratios of 3:1, 1:1 and 1:3 were prepared for each of the Cu-Cr-O and Cu-Ce-O sorbents. Similar analysis methods to those described above for ZnO-sorbents were used to fully characterize the sorbent samples.

TGA tests. The weight change of the sample during reduction, sulfidation and regeneration was measured in the Cahn System 113-X TGA system described above. The inlet gas flow rate was controlled by four Brooks Model 5850 mass flow controllers. Water was injected into the heated gas line by a calibrated syringe pump. A simulated coal gas mixture containing H_2S-H_2-H_2O-N_2 was used in sulfidation. Regeneration was carried out in air diluted with N_2. The total gas flow rate was 564 sccm. At this rate it was confirmed that bulk gas diffusion was eliminated.

In all TGA tests 2-3 mg of sorbent were placed in a hemispherical quartz pan suspended by a quartz hangdown wire. A thin layer of quartz wool was laid on the bottom of the quartz pan to provide a better contact between the reactant gases and the solid particles. The typical size of solid particles was 125-180 μm. After loading the sample, the furnace was heated to a set point at a rate of 22°C/min under pure N_2 and was maintained at the set value throughout the experiment. Typically, reduction lasted for 10 min, sulfidation 20 min and regeneration 10 min. An intermittent nitrogen purge of 20 min was used between sulfidation and regeneration tests.

Packed-bed microreactor tests. Sulfidation and regeneration tests were performed in a packed-bed microreactor to evaluate the overall H_2S removal efficiency and regenerability of the synthesized sorbents. The packed-bed microreactor system and experimental procedure were the same as those previously described by Lew, *et al.* (19). About 50 mg sorbent of 420-800 μm in diameter was used. The gas hourly space velocity, S.V., was 3,000 h^{-1} (NTP). H_2S and SO_2 concentration in the exit gas were measured by a HP5880A gas chromatograph (GC), equipped with a Flame Photometric Detector (FPD) and a Thermal Conductivity Detector (TCD) used, respectively, during sulfidation and regeneration.

3.2.2 Sulfidation Performance Evaluation in the Packed-Bed Microreactor

The various sorbent compositions prepared and characterized as above were evaluated in sulfidation/regeneration tests in the packed-bed reactor. Table 5 lists the properties of the fresh and sulfided sorbents. At a calcination temperature of 1000°C, the specific surface area of the two types of sorbents decreased with the CuO content. X-ray diffraction identified that cupric chromite, $CuCr_2O_4$, was formed for CuO-lean Cu-Cr-O sorbent, and Cr_2O_3 was in excess. For equimolar CuO-Cr_2O_3 material, $CuCr_2O_4$ was the only crystalline phase formed and no separate CuO or Cr_2O_3 phase was identified. Scanning electron microscopy also showed that well formed $CuCr_2O_4$ crystals were produced after calcination at 1000°C. These crystals have uniform structure and are ~0.5 μm in size. Cuprous chromite, $CuCrO_2$, and CuO phases coexisted for the CuO-rich sorbent $3CuO-Cr_2O_3$. The crystal size of this material was ~2 μm. For Cu-Ce-O sorbents, CuO and CeO_2 existed in separate phases for all the sorbent compositions. Copper and cerium oxide are immiscible oxides (54).

Dry sulfidation tests were performed with each of the six Cu-Cr-O and Cu-Ce-O sorbent compositions listed in Table 5. A gas mixture containing (mol) 2% H_2S-20% H_2-78% N_2 at S.V. = 3,000 h^{-1} (NTP) was used in these tests. The resulting performance is shown in Figure 8 in terms of H_2S breakthrough curves at 850°C. The normalized time, t/t^*, is defined as the ratio of the real reaction time, t, to the theoretical time, t^*, required for complete sulfidation of the sorbent to form Cu_2S. This assumes that neither Cr_2O_3 nor CeO_2 forms bulk sulfides under the tested conditions. The H_2S breakthrough is reported here for 100 ppmv H_2S in the effluent gas.

Table 5. Characterization of Fresh and Sulfided Sorbents

sorbent	BET surface area[1] (m^2/g)	max. theoretical sulfur loading[2] (g sulfur/100 g sorbent)	crystalline phases	
			fresh	sulfided[4]
$3CuO-Cr_2O_3$	0.6	12.2	$CuCrO_2$, CuO	Cr_2O_3, Cu_9S_5, Cu_2O
$CuO-Cr_2O_3$	2.2	6.7	$CuCr_2O_4$	Cr_2O_3, Cu_9S_5
$CuO-3Cr_2O_3$	3.7	3.0	$CuCr_2O_4$, Cr_2O_3	Cr_20_3, Cu_9S_5
$3CuO-CeO_2$	1.1	17.7	CuO, CeO_2	CeO_2, Cu_8S_5, Ce_2S_3
$CuO-CeO_2$	4.4	6.4	CuO, CeO_2	CeO_2, Cu_8S_5, Ce_2S_3
$CuO-3CeO_2$	5.8	2.7	CuO, CeO_2	CeO_2, Cu_8S_5, $Cu_{1.8}S$ Ce_2S_3

1. After calcination in air at 1000°C for 1 hr.
2. Assuming Cu_2S as the only sulfidation product.
3. Identified by X-ray diffraction analysis.
4. Sulfided under 2% H_2S, 20% H_2, bal. N_2, at 850°C .

The H_2S breakthrough curves in Fig. 8 show that complete desulfurization of the feed gas stream can be obtained for all the Cu-Cr-O and Cu-Ce-O sorbents even at the high temperature of 850°C. This was especially true for the Cu-lean sorbents $CuO-3Cr_2O_3$ and $CuO-3CeO_2$. As the CuO content was increased, H_2S breakthrough occurred earlier, with corresponding lower sulfur removal efficiency and lower CuO utilization. For example, the normalized pre-breakthrough time decreased from 1.5 to 0.65 to 0.2 as the CuO/Cr_2O_3 molar ratios changed from 1/3 to 1/1 to 3/1, respectively. A plausible explanation for this is that in higher CuO-containing sorbents, CuO exists in large crystals, which provide increased diffusion resistance for reactant gas through a shell of product Cu_2S. The sulfur loading of the sorbents at 100 ppmv H_2S breakthrough level was approximately 2.2, 3.7 and 4.3 g sulfur per 100 g of sorbent for the 3CuO-Cr_2O_3, CuO-Cr_2O_3 and CuO-3Cr_2O_3 sorbents, and 4.4, 4.2 and 2.2 g sulfur per 100 g of sorbent for the 3CuO-CeO_2, CuO-CeO_2 and CuO-3CeO_2 sorbents, respectively. Compared to the theoretical values listed in Table 5, it is obvious that the actual utilization of copper in Cu-rich sorbents is very low. For Cu-lean sorbents, however, full utilization of copper has been achieved and the actual sulfur loading exceeds the theoretical values for the CuO-3Cr_2O_3. This may be due to the formation of non-stoichiometric copper sulfides.

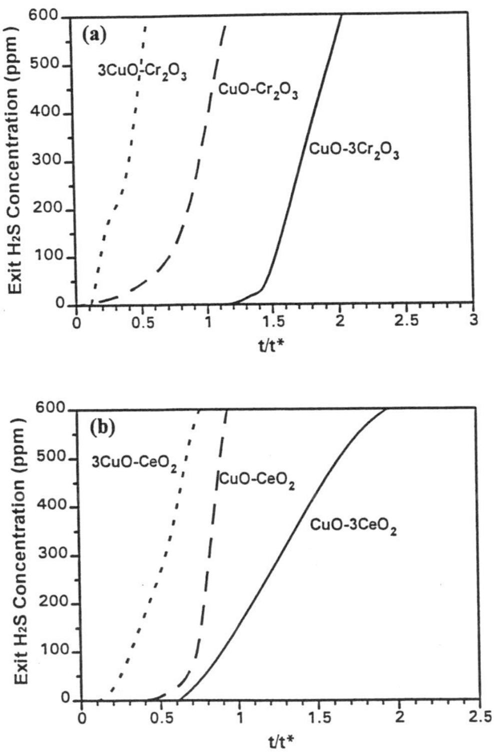

Figure 8. H_2S Breakthrough curves for (a) Cu-Cr-O and (b) Cu-Ce-O sorbents with various CuO/Cr_2O_3 or CuO/CeO_2 ratios. Sulfidation condition: T=850°C, inlet gas: 2% H_2S-20% H_2-78% N_2, S.V.=3,000 h^{-1} (NTP).

It is noteworthy that the pre-breakthrough levels of H_2S (< 5 ppm) are much lower than the equilibrium H_2S levels corresponding to reaction of metallic copper with H_2S. Under the conditions of the experiments (850°C, 20% H_2), the equilibrium H_2S level is ~300 ppmv (55). Hence, copper metal alone is not the active phase in these sorbents. Similar observations were made in the microreactor tests conducted at lower temperatures, both in this lab and at the Institute of Gas Technology using the equimolar CuO-Cr_2O_3 and CuO-CeO_2 sorbents (30b). The presence of water vapor and carbon oxides in the reactant gas did not alter the pre-breakthrough H_2S level, although it could shift the breakthrough time.

Sulfided solids from the packed-bed microreactor tests were analyzed by XRD and SEM. As listed in Table 1, non-stoichiometric Cu_9S_5 is the major phase of copper sulfides for all three compositions of Cu-Cr-O mixtures after sulfidation. Cr_2O_3 was also identified by XRD, while no chromium sulfides were found. A small amount of Cu_2O (based on signal intensity) was observed for the sulfided $3CuO\text{-}Cr_2O_3$ solid, in agreement with the low Cu utilization in the microreactor test of this composition. For the Cu-Ce-O mixtures, Cu_8S_5 was the major sulfide phase. In contrast to the Cr_2O_3, some CeO_2 also reacted with H_2S to form cerium sulfides, since the Ce_2S_3 was consistently identified in the three sulfided Cu-Ce-O sorbents. This is in agreement with the report that reduced cerium oxides, CeO_{2-x}, are good sulfidation materials at high temperatures (56).

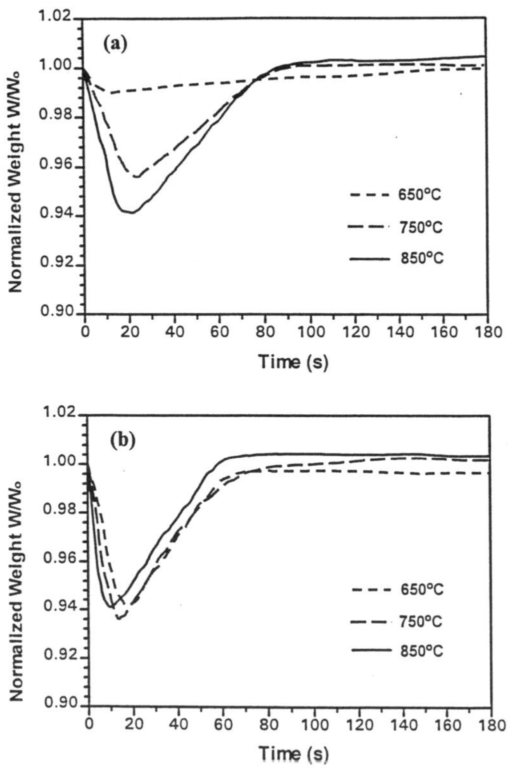

Figure 9. Normalized sorbent weight profiles during sulfidation of (a) $CuO\text{-}Cr_2O_3$ and (b) $CuO\text{-}CeO_2$ at various temperatures. Sulfidation condition: feed gas: 0.5% H_2S-10% H_2-10% H_2O-bal. N_2.

3.2.3 Comparative Kinetics of $CuO\text{-}Cr_2O_3$ and $CuO\text{-}CeO_2$ Sorbents

Kinetic evaluation in reductive sulfidation was conducted with the equimolar sorbents, $CuO\text{-}Cr_2O_3$ and $CuO\text{-}CeO_2$, in the TGA apparatus. Fig. 9 shows the weight profiles, normalized by initial sorbent weight W_0, vs. reaction time during sulfidation of the fresh $CuO\text{-}Cr_2O_3$ and $CuO\text{-}CeO_2$ sorbents at 650, 750 and 850°C in a gas mixture containing (mol) 0.5% H_2S-10% H_2-10% H_2O-bal. N_2.

From this figure it is apparent that the weight change during sulfidation consisted of two segments: an initial weight-loss followed by weight-gain. Such weight change behavior points out to two reactions taking place during reductive sulfidation: one is reduction of Cu^{2+} oxides to Cu^{1+} oxides or metal copper, with concomitant weight decrease; the second is sulfidation of metal copper or copper oxides to copper sulfides, leading to weight increase. Although both reduction and sulfidation reactions occurred simultaneously, once the sorbents were contacted with the inlet gas containing H_2 and H_2S, reduction was much faster than sulfidation and dominated the earlier part of reaction. However, as the reaction proceeded, sulfidation began to dominate the entire reaction. Thus, the reaction path for sulfidation of these copper oxide-containing sorbents appears to be $CuO \rightarrow Cu_2O/Cu \rightarrow Cu_xS$ ($x < 2$) over the tested temperature range.

Comparing Fig. 9 (a) and (b), it is seen that the reduction of CuO-Cr_2O_3 sorbent was much slower when the temperature was decreased from 850 to 650°C. However, temperature had little effect on the CuO-CeO_2 sorbent. In other words, reduction of CuO and/or CeO_2 took place to the same extent at all temperatures in the range of 650-850°C. For the CuO-Cr_2O_3 sorbent consisting of $CuCr_2O_4$ compound, the stability of $CuCr_2O_4$ suppresses reduction, even at $T \geq 750$°C (30a). Correspondingly, the H_2S removal efficiency obtained in the packed-bed microreactor was higher than that of copper metal. On the other hand, the reducibility of the Cu-Ce-O sorbents shown in Fig. 9 (b) does not match the H_2S removal efficiency shown in Fig. 8 (b). If all the copper were in metallic state, then the exit H_2S level in Fig. 2 (b) should be approximately 300 ppm as stated above (based on the equilibrium of Cu with H_2S). Reduction of cerium oxide can account in part for the observed loss of oxygen. This is facilitated by the presence of copper oxide as shown in temperature programmed-reduction (TPR) experiments with Cu-Ce-O; Liu, *et al.* (57) have found a bimodal reduction by hydrogen for 15 (at)% Cu-containing cerium oxide showed a TPR profile with a first peak at 120-145°C corresponding to CuO reduction, and a broader second peak from 450 to 850°C corresponding to CeO_2 reduction. Notably, each phase alone (CuO or CeO_2) is less reducible. Also, as indicated by the XRD data in Table 5, cerium oxide has participated in the reaction with H_2S, which may also explain the high H_2S removal efficiency observed for the Cu-Ce-O sorbents in Fig. 8 (b) (56).

The effect of H_2 concentration on sulfidation at T = 750°C is shown in Fig. 10 (a) and (b) for CuO-Cr_2O_3 and CuO-CeO_2, respectively. For both sorbents, the reduction rate and reduction conversion increased with the H_2 concentration. Above 750°C complete reduction of CuO to metal copper during sulfidation took place for CuO-Cr_2O_3 sorbents when H_2 was above 30 mol% (Fig. 10 (a)). For CuO-CeO_2 sorbents, a lower H_2 concentration of 10 mol% was sufficient to reduce the sorbent to an extent corresponding to all oxidic copper reduced to the metal state. As can be seen in Fig. 10 (a) and (b), a small weight decrease was measured even when the inlet gas did not contain H_2 ($H_2 = 0$% curves). Two factors could contribute to the weight decrease at this condition. One is the reduction of CuO by H_2 produced by the decomposition of H_2S, while the other is the oxidation of H_2S by CuO or CeO_2 to form SO_2. The former was verified by the condensation of elemental sulfur on the cooler walls of the quartz tube and in the water trap located downstream of the TGA. The equilibrium H_2 concentration for H_2S decomposition at 750°C is calculated to be ~1400 ppm in the absence of hydrogen in the inlet gas. On the other hand, the oxidation of H_2S to SO_2 has been reported to be catalyzed by ceria (58). To check for this, we analyzed the TGA off-gas in the GC-FPD. The measurement showed that approximately 50 ppm SO_2 could be obtained in the TGA outlet gas when low H_2 concentration (<10 mol%) was used. Thus, under some conditions, other reactions, such as decomposition and/or oxidation of H_2S, can also take place in addition to the main reaction: $CuO \rightarrow Cu_xS$.

Separate tests of reduction and sulfidation of reduced sorbent samples were performed to measure the kinetics of each reaction and to study parametric effects on the reduction and sulfidation performance of each sorbent type. Details about this work can be found in references 29, 30a and b. We focus here on comparison of the reaction kinetics.

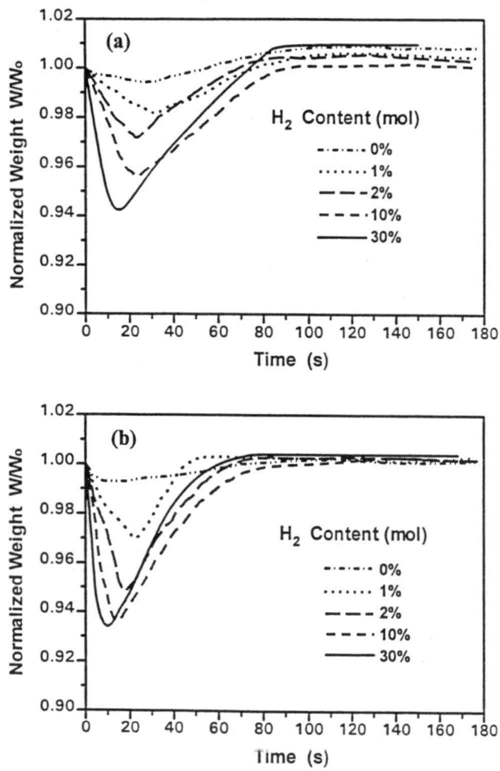

Figure 10. Normalized sorbent weight profiles during sulfidation of (a) $CuO-Cr_2O_3$ and (b) $CuO-CeO_2$ at various H_2 concentrations. Sulfidation condition: T=750°C, feed gas=0.5% H_2S-x% H_2-10% H_2O-bal. N_2.

3.2.3.1 Reduction of Cu-Cr-O and Cu-Ce-O Solids

Reduction tests were conducted over a wide temperature range from 350 to 850°C in a gas mixture of (mol) 10% H_2 and 10% H_2O in N_2. As can be seen from Fig. 11, when the temperature was decreased from 850 to 550°C, the reduction rate of $CuCr_2O_4$ decreased dramatically. But for $CuO-CeO_2$, no appreciable change of reduction rate was found even over a wider temperature range (350-850°C). Reduction of the $CuO-Cr_2O_3$ and $CuO-CeO_2$ solids under various H_2 concentrations was conducted at 750°C with $[H_2]$ = 2-40 mol%. Fig. 12 shows the profiles of the normalized weight vs. time on stream at three H_2 concentrations: 2, 10, and 40%. As can be seen from this figure, the reduction rate of both sorbents increased with the H_2 concentration in the feed stream. The $CuO-Cr_2O_3$ material showed a stronger dependence on H_2 concentration than $CuO-CeO_2$, apparently due to the stability of the $CuCr_2O_4$ compound. The kinetic parameters of reduction were measured in the TGA.

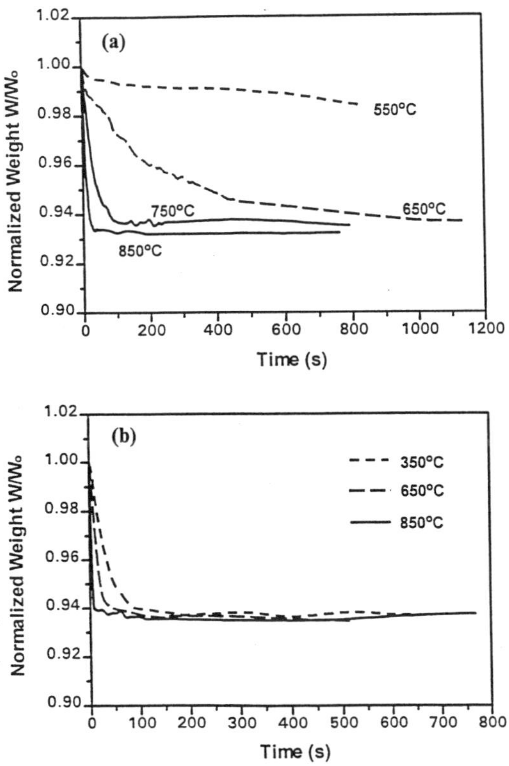

Figure 11. Normalized sorbent weight profiles during reduction of (a) $CuO-Cr_2O_3$ and (b) $CuO-CeO_2$ at various temperatures. Reduction condition: feed gas=10% H_2-10% H_2O- bal. N_2.

Initial reduction rates of $CuO-Cr_2O_3$ and $CuO-CeO_2$ in Arrhenius-type plots are shown in Figure 13. Table 6 shows the Arrhenius constants, calculated from the initial reduction rates, R_o, and for reaction orders n = 0.56 and 0.42, respectively, for $CuO-Cr_2O_3$ and $CuO-CeO_2$ at 750°C. The former had much higher activation energy for reduction, 23.1 kcal/mol, while $CuO-CeO_2$ was easier to reduce with a lower activation energy of 6.3 kcal/mol.

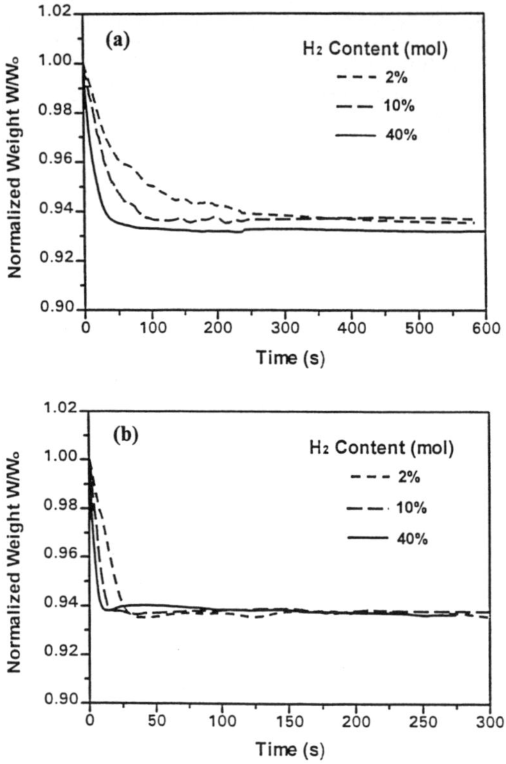

Figure 12. Normalized sorbent weight profiles during reduction of (a) $CuO\text{-}Cr_2O_3$ and (b) $CuO\text{-}CeO_2$ at various H_2 concentrations. Reduction condition: T=750°C, feed gas=x% H_2-10% H_2O-bal. N_2.

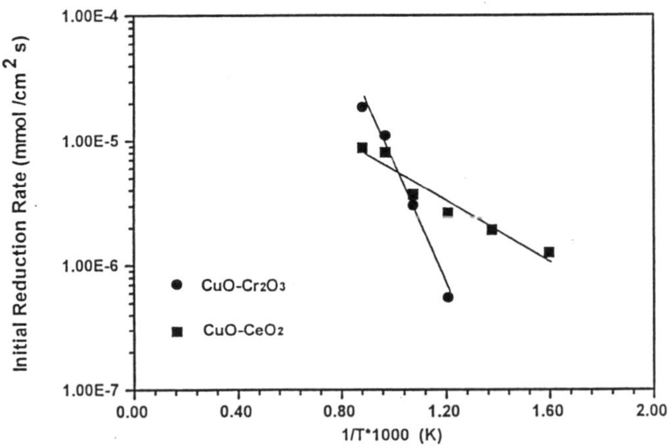

Figure 13. Arrhenius-type plot of initial reduction rate vs. temperature for $CuO\text{-}Cr_2O_3$ and $CuO\text{-}CeO_2$. Feed gas: 10% H_2-10% H_2O-80% N_2.

3.2.3.2 Sulfidation of Pre-reduced Sorbents

Sorbents were reduced at 750°C with a gas mixture containing (mol) 10% H_2-10% H_2O-80% N_2. Sulfidation kinetics of pre-reduced sorbents were measured in the TGA. The surface areas of the reduced solids were used to calculate the initial sulfidation rates. The stoichiometry of 2Cu → Cu_2S was the only one considered.

As a result of complete pre-reduction, sulfidation of Cu in both samples was fast and almost independent of temperature in the range of 650-850°C. Complete conversion of Cu to copper sulfides was achieved at all the conditions tested except at 650°C for Cu-Cr_2O_3 (29) in a gas stream containing (mol) 10% H_2-10% H_2O-0.5% H_2S-bal. N_2. Cu-CeO_{2-x} had faster overall kinetics than Cu-Cr_2O_3 as indicated by conversion-time plots (29). This difference may be caused by the different interaction of copper particles with each of the two oxides or by potential participation of cerium oxide in sulfidation.

Fig. 14 (a) and (b) show the 750°C-sulfidation profiles of pre-reduced Cu-Cr_2O_3 and Cu-CeO_{2-x} samples, respectively, at various concentrations of H_2S. Clearly, the rates increase appreciably with the H_2S concentration. With the exception of the CuO-Cr_2O_3 at 0.125% H_2S, Fig. 14 (a), the W/W_0 increased linearly with the reaction time up to complete conversion indicating that sulfidation rates are diffusion-free at this temperature.

Fig. 15 shows Arrhenius-type plots for the initial sulfidation rates of the pre-reduced solids, while Table 6 lists the corresponding parameters. The activation energy was determined to be 4.7 and 4.0 kcal/mol, respectively, for CuO-Cr_2O_3 and CuO-CeO_2. Thus, sulfidation of copper is a weak function of temperature with activation energy lower than that of reduction. Potentially, the sulfidation is limited by adsorption. The initial rates of sulfidation of the two sorbents do not differ appreciably. This is expected as Cu metal is the major active phase in the reduced sorbents. However, the kinetics of CeO_{2-x} sulfidation need to be examined in separate experiments and compared to those reported here.

Table 6. Kinetic Parameters of Reduction and Sulfidation of CuO-Cr_2O_3 and CuO-CeO_2

	reduction			sulfidation[1]		
	$k_{r,o}$ (mmol^{1-n} cm^{3n-2} s^{-1})	E_a (kcal mol^{-1})	n[2]	$k_{s,o}$ (mmol^{1-n} cm^{3n-2} s^{-1})	E_a (kcal mol^{-1})	n[2]
CuO-Cr_2O_3	1.07 x 10^3	23.1	0.58	2.73 x 10^{-2}	4.7	0.70
CuO-CeO_2	2.58 x 10^{-2}	6.3	0.42	7.50 x 10^{-2}	4.0	0.83

1: Samples used in sulfidation were pre-reduced at 750°C with 10% H_2-10% H_2O-80% N_2. The BET surface area of the pre-reduced sorbents was 7.4 and 5.0 m^2/g, respectively, for Cu-Cr_2O_3 and Cu-CeO_2.

2: Reduction and sulfidation order in H_2 and H_2S, respectively, from measurements at 750°C.

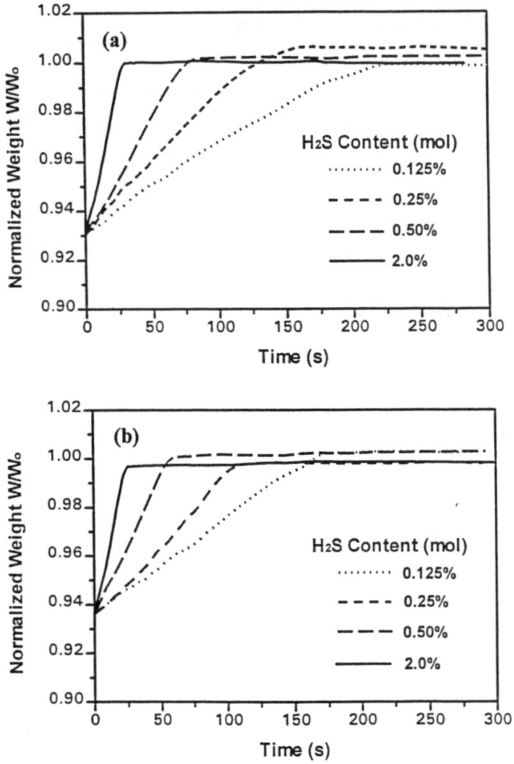

Figure 14. Normalized sorbent weight profiles during sulfidation of pre-reduced (a) Cu-Cr_2O_3 and (b) Cu-CeO_2 at various H_2S concentrations. Reduction condition: T=750°C, feed gas: 10% H_2-10% H_2O-bal. N_2. Sulfidation condition: T=750°C, feed gas: x% H_2S-10% H_2-10% H_2O-bal. N_2.

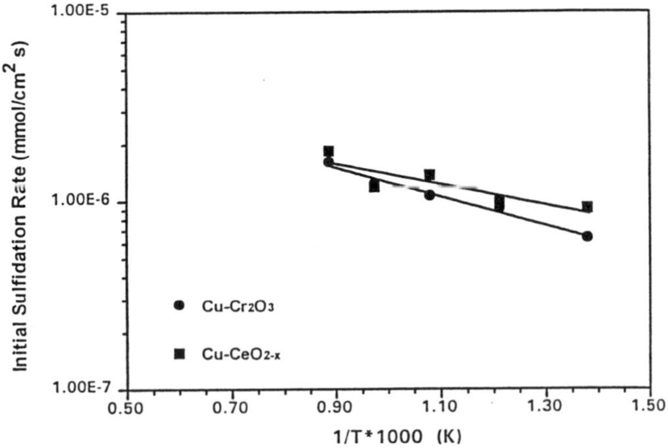

Figure 15. Arrhenius-type plot of initial sulfidation rate vs. temperature for pre-reduced Cu-Cr_2O_3 and Cu-CeO_2. Feed gas: 0.5% H_2S-10% H_2-10% H_2O-bal. N_2.

3.2.4 Cyclic Sulfidation/Regeneration Tests

Details of sorbent regeneration performance can be found elsewhere (29). The regeneration reaction paths were complex over the tested temperature range. At temperatures below 700°C, $CuSO_4$ was formed on both sorbents and the reaction path appears to be $Cu_xS \rightarrow Cu_2O \rightarrow CuSO_4/CuO \rightarrow CuO$ for $CuO\text{-}Cr_2O_3$ and Cu_2S $CuSO_4/CuO \rightarrow CuO$ for $CuO\text{-}CeO_2$. At higher temperatures a rather simple mechanism involving no intermediate sulfate is inferred from the data for $CuO\text{-}Cr_2O_3$: $Cu_2S \rightarrow Cu_2O/CuO$. However, in the case of $CuO\text{-}CeO_2$, some sulfate still persists after regeneration at 750°C with 6% O_2 in N_2. This is attributed to stable surface sulfates of cerium.

From a practical view point, both Cu-Cr-O and Cu-Ce-O sorbents should possess regenerability in cyclic operation. This was shown to be good in microreactor tests. In this work, to determine the sorbent regenerability, two and one half cycles of consecutive sulfidation-regeneration were performed with the $CuO\text{-}Cr_2O_3$ and $CuO\text{-}CeO_2$ sorbents. The results are shown in Fig. 16 (a) and (b), respectively.

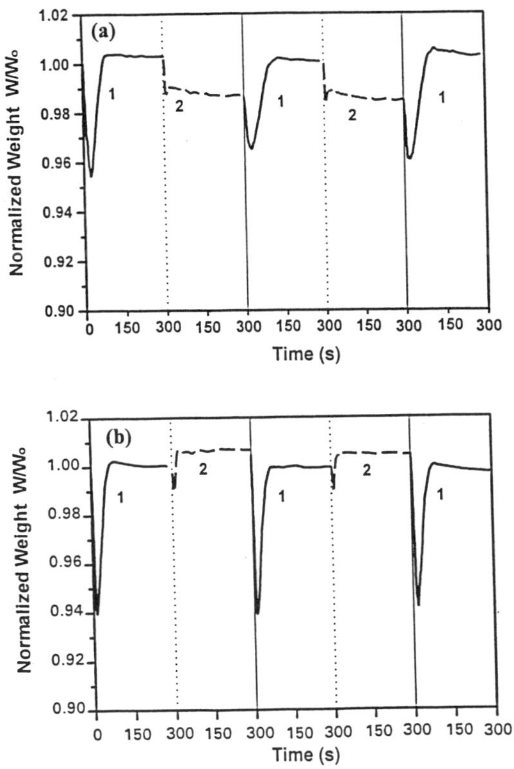

Figure 16. TGA sulfidation-regeneration cycles of (a) $CuO\text{-}Cr_2O_3$ and (b) $CuO\text{-}CeO_2$. Sulfidation condition: T=750°C, feed gas: 0.5% H_2S-10% H_2-10% H_2O-bal. N_2. Regeneration condition: T=750°C, feed gas: 6% O_2-bal. N_2.

Sulfidation was carried out with a wet gas mixture containing (mol) 0.5% H_2S-10% H_2O-10%H_2-bal.N_2 for 20 min. Sulfided solids were regenerated with a 6 mol% O_2-N_2 mixture for 10 min after nitrogen purge was switched off. All reactions were performed at 750°C. Since there was no significant weight change in the later period of sulfidation and regeneration, only the first 5 min in each of the processes are shown in Fig. 16. As can be seen from this figure, no significant changes were found, except for lower starting values of W/W_0 for CuO-Cr_2O_3 sorbent in the second and third cycles as a result of the formation of Cu_2O in the regeneration step of the previous cycles. Thus, the activities of both CuO-Cr_2O_3 and CuO-CeO_2 are stable in cyclic operation.

3.2.5 Concluding Remarks

The above case study has compared two CuO-based sorbents in terms of their desulfurization efficiency, reactivity and regenerability for application to coal gas stream cleanup at high temperatures (650-850°C). Two very different sorbent structures were selected for this study. In one, CuO-Cr_2O_3, the compound of copper chromite is formed, while in the second, CuO-CeO_2, copper oxide is kept in dispersed state. CeO_2 and CuO are immiscible oxides. Copper chromite has the lowest reducibility amongst all copper oxide-containing compounds reported in the literature. On the contrary, in the CuO-CeO_2 system, the reducibility of both the CuO and CeO_2 phases is enhanced. Yet, both sorbents can desulfurize a fuel gas to less than 1-10 ppmv in the temperature range of 650-850°C and in the presence of 20 mol% H_2-10mol% H_2O. This is well below the equilibrium H_2S level corresponding to sulfidation of metallic copper. Retention of oxidic copper (+2 or +1 oxidation state) in the Cu-Cr-O solids and reduced cerium oxide, CeO_{2-x}, in the Cu-Ce-O materials, are believed to be responsible for the excellent desulfurization efficiency of these sorbents.

The kinetics of sulfidation are fast and complete conversion of copper oxide takes place in the equimolar CuO-Cr_2O_3 and CuO-CeO_2 sorbents. The initial reaction kinetics of Cu-CeO_2 are faster than those of other metals and oxides and comparable to MnO. The overall sulfidation rate of CuO-CeO_2 is diffusion-free under the conditions examined here. Regeneration rates are fast. However, regeneration depends on the oxygen-temperature combination. Higher temperature will be required to complete regeneration of the CuO-CeO_2 material.

Overall, this work has shown that CuO-based sorbents of the type CuO-Cr_2O_3 and CuO-CeO_2 can be tailored to meet a wide range of applications in hot gas cleanup. The catalytic effects of these systems in the water gas shift reaction and other side reactions need to be properly accounted for during the sorbent/reactor design stage. In general, the desulfurization efficiency and reactivity as well as the regenerability of CuO-based sorbents can all be optimized for a particular fuel gas composition-temperature-reactor design by suitable sorbent composition and structure selection.

4 Conclusions

In this paper we have presented an overview of the important issues identified in studies of the desulfurization kinetics of mixed metal oxides. The special cases of ZnO-based and CuO-based sorbents were examined in detail to illustrate these issues and demonstrate the approaches needed to evaluate candidate sorbent compositions. In general, sulfidation kinetics analysis comprises studies of reduction kinetics (for reducible sorbents), initial sulfidation rate measurements of key sorbent component (metal or oxide) and diffusion studies (at high conversions); the latter are often dominated by product layer diffusion resistance at the low end of hot gas cleanup temperatures. Primarily this is due to sulfidation kinetics with low apparent activation energies, a common feature in all the sorbents studied to date. At the high end of the temperature range, pore diffusion may dominate. To account properly for structural changes of the sorbent during reductive sulfidation, solid analyses should be carried out to complement the kinetic studies.

The case studies provided in this paper clearly demonstrate that it is possible to design a suitable sorbent system for a given gasifier-type gas and temperature by careful choice of the sorbent type and physical properties. There is enough flexibility between compound formation and/or dispersion of the active phase to tailor a sorbent to the desired application.

5 Acknowledgments

The financial support of various parts of the work described in this paper by the U.S. Department of Energy is gratefully acknowledged. Also, the financial support of the MIT's Energy Laboratory in 1985-89 is acknowledged. The first author also wishes to thank the many students and colleagues who have participated in this work over the years, especially G.R. Gavalas, S.S. Tamhankar, M. Bagajewicz, V. Patrick, P.K. Sharma, S. Lew, K. Jothimurugesan, A.F. Sarofim, L.R. Shao, T.U. Yu, Z. Hu, L. Bo, L. Li, C. Wadia, J. Abbasian, A.H. Hill, K. Ishikawa, and her co-author Z. Li for their many valuable contributions.

6 References

1. Westmoreland, P.R. and Harrison, D.P., *Environ. Sci. Technol.* 10, 659 (1976).
2. Ruth, L.A., Squires, A.M. and Graff, R.A., *Environ. Sci. Technol.* 6, 1009-1014 (1972).
3. Westmoreland, P.R., Gibson, J.B. and Harrison, D.P., *Environ. Sci. Technol.* 11, 488-491 (1977).
4. Curran, G.P., Clensey, J.T., Pase, B., Pell, M., Rutledge, G.D. and Gorin, E.P., Final Report EPA-650/2-73-049. EPA Contract No. EHSD71-15, 1973.
5. Leurnberger, E.L., ERDA Annual Report FE-2033-17, 1976.
6. Oldaker, E.C., Poston, A.M., Jr. and Farrior, W.L., Jr., 170th National Meeting, *ACS, Div. Fuel Chem.* 20 (4), 227 (1975).
7. Oldaker, E.C. and Gillmore, D.W., 172nd National Meeting, *ACS, Div. Fuel Chem.* 21 (4), 79 (1976).
8. Tamhankar, S.S., Hasatani, M. and Wen, C.Y., *Chem. Eng. Sci.* 36, 1181-1191 (1981).

9. MERC Hot Gas Cleanup Task Force, Chemistry of Hot Gas Cleanup in Coal Gasification and Combustion. MERC/SP-78/2, February 1978.
10. Furimsky, E. and Yumura, M., *Sci. Technol.* 39, 163-172 (1986).
11. Efthimiadis, E.A. and Sotirchos, S.V., *Ind. Eng. Chem. Res.* 31, 2311-2321 (1992).
12. Krishnan, S.V. and Sotirchos, S.V., *Ind. Eng. Chem. Res.* 33, 1444-1453 (1994).
13. Brewster, L. in written communication to T. Grindley of DOE/METC, April 27, 1982.
14. Bagajewicz, M., Tamhankar, S.S., Gavalas, G.R. and Flytzani-Stephanopoulos, M., *Environ. Sci. Technol.* 22 (4), 467-470 (1988).
15. Flytzani-Stephanopoulos, M., Gavalas, G.R., Tamhankar, S.S. and Sharma, P.K., "Novel Sorbents for High-Temperature Regenerative H_2S Removal," *Final Report to DOE*, DOE/MC/20417-1898, Oct. 1985.
16. Flytzani-Stephanopoulos, M., *et al. U.S. Patent No.* 4, 729, 889 (1988).
17. Jalan, V. and Wu, D., "High Temperature Desulfurization of Fuel Gases for Molten Carbonate Fuel Cell Power Plants. National Fuel Cell Seminar, San Diego, CA, 1980; also in Giner, Inc., *Final Report to Argonne National Laboratory*, Contract No. 31-109-39-5804, Waltham, MA, 1981.
18. Grindley, T. and Steinfeld, G., "Development and Testing of Regenerable Hot Coal Gas Desulfurization Sorbents"; *Final Report DOE/*MC/16545-1125, Oct. 1981.
19. Lew, S., Jothimurugesan, K., Flytzani-Stephanopoulos, M., *Ind. Eng. Chem. Res.* 28, 535-541 (1989).
20. Lew, S., Sarofim, A.F., and Flytzani-Stephanopoulos, M., *Chem. Eng. Sci.* 47 (6), 1421-1431 (1992).
21. Lew, S., Sarofim, A.F., and Flytzani-Stephanopoulos, M., *Ind. Eng. Chem. Res.* 31, 1890-99 (1992).
22. Gibson, J.B. and Harrison, D.P., *Ind. Eng. Chem. Proc. Des. Dev.* 19, 231 (1980).
23. Ranade, P.V. and Harrison, D.P., *Chem. Eng. Sci.* 36, 1079-1089 (1981).
24. Szekely, J. and Evans, J.W., *Chem. Eng. Sci.* 25, 1091 (1970).
25. Sotirchos, S.V. and Yu, H.C., *Ind. Eng. Chem. Res.* 27, 836 (1988).
26. Yu, H.C. and Sotirchos, S.V., *AIChE J.* 33, 382-393 (1987).
27. Lew, S., Sarofim, A.F., and Flytzani-Stephanopoulos, M., *AIChE J.* 38 (8), 1161-1169 (1992).
28. Efthimiadis, E.A. and Sotirchos, S.V., *Chem. Eng. Sci.* 48 (11), 1971-1984 (1992).
29. Li, Z. and Flytzani-Stephanopoulos, M., *Ind. Eng. Chem. Res.* in press (1997).
30. a) Abbasian, J., Flytzani-Stephanopoulos, M., Hill, A.H., Bo, L. and Wangerow, J.R. *IGT-Final Technical Report* to CRSC, "Development of Novel Copper-Based Sorbents for Hot-Gas Cleanup", IGT Project No. 40330, Aug. 1992; b) Abbasian, J., Hill, A.H., Flytzani-Stephanopoulos, M. and Li, Z., *IGT-Final Technical Report to ICCI/DOE*, "Development of Novel Copper-Based Sorbents for Hot-Gas Cleanup", DE-FC22-92PC92521, Aug. 1994.
31. Sick, G., and Schwerdtfeger, K., *Metall. Trans.*, 18B, 603-609 (1987).
32. Satterfield, C.N., "Heterogeneous Catalysis in Industrial Practice", pp. 63-64, 2nd ed. McGraw-Hill, New York, 1991.
33. Flytzani-Stephanopoulos, M., "Alternate Sorbents Development", keynote lecture at DOE/METC Workshop on "Status and Direction of Research and Development for High Temperature Sulfur Removal Sorbents", Morgantown, WV, Jan. 28, 1993.
34. Flytzani-Stephanopoulos, M., Gavalas, G.R., Jothimurugesan, K., Lew, S., Sharma, P.K., Bagajewicz, M.J., Patrick, V., "Detailed Studies of Novel Sorbents for Hot Coal Gas Desulfurization", *Final Report to DOE*, DE-FC21-85-MC22193, Oct. 1987.
35. Lew, S., "The Reduction and Sulfidation of Zinc Titanate and Zinc Oxide Solids", *Ph.D Dissertation*, MIT, Cambridge, MA, 1990.

36. Tamhankar, S.S., Bagajewicz, M.J., Gavalas, G.R., Sharma, P.K. and Flytzani-Stephanopoulos, M., *Ind. Eng. Chem. Process Des. Dev.*, 25, 429-437 (1986).
37. Flytzani-Stephanopoulos, M., Yu, T.U. and Lew, S., "Development and Testing of Desulfurization Sorbents", *Topical Report to Texaco*, under subcontract, DOE Coop. Agreement No. DE-FC21-87MC23277, Dec. 1988.
38. Anderson, G.L., Berry, F.O. and Gross, M.N., "Development of a Hot Gas Cleanup System for Integrated Coal Gasification/Molten Carbonate Fuel Cell Power Plants", *Topical Report*, DOE/MC/19403-1544.
39. Jalan, V., "Studies Involving High-Temperature Desulfurization/Regeneration Reactions of Metal Oxides for Fuel Cell Development", *Final Report* *DOE*/MC/16021-1486, 1983.
40. Siriwardane, R.V. and Poston, J.A., *Appl. Surf. Sci.* 45, 131 (1990).
41. Siriwardane, R.V. and Poston, J.A., *Appl. Surf. Sci.* 68, 65 (1993).
42. Farha, F.E., Jr. and Gardner, L.E., *U.S. Patent No.* 4, 313, 820 (1982).
43. Marcilly, C., Courty, P., Delmon, B., *J. Am. Ceram., Soc.* 53 (1), 56-57 (1970).
44. Courty, P., Ajot, H., Marcilly, C. and Delmon, B., *Powder Technol.* 7, 21-38 (1973).
45. Sainamthip, P. and Amarakoon, V.R.W., *J. Am. Ceram. Soc.* 71 (8), 644-648 (1988).
46. Wakao, N. and Smith, J.M., *Chem. Eng. Sci.* 17, 825-834 (1962).
47. Flytzani-Stephanopoulos, M. and Jothimurugesan, K., *U.S. Patent No.* 4, 977, 123 (1990).
48. Flytzani-Stephanopoulos, M., *U.S. Patent No.* 5, 538, 703 (1996).
49. Patrick, V., Gavalas, G.R., Flytzani-Stephanopoulos, M. and Jothimurugesan, K., *Ind. Eng. Chem. Res.* 28, 931-940 (1989).
50. Patrick, V. and Gavalas, G.R., *J. of Am. Ceram. Soc.* 73 (2), 358-369 (1990).
51. Jalan, V. and Desai, M., "Copper-based Sorbents for Hot Gas Cleanup", DOE/METC-89/6107, Vol. 2, pp 541-547, 1989.
52. Kyotani, T., Kawashima, H., Tomita, A., Palmer, A. and Furimsky, E., *Fuel*, 68, 74-79 (1989).
53. Kyotani, T., Kawashima, H. and Tomita, A., *Envir. Sci. & Tech.* 23 (2), 218-223 (1989).
54. Liu, W., Sarofim, A.F. and Flytzani-Stephanopoulos, M., *Appl. Catal-B* 4, 167 (1994).
55. Barin, I. and Knacke, O., "Thermochemical Properties of Inorganic Substances", Springer Verlag, New York, 1973; Barin, I., Knacke, O. and Kubaschewski, O., Supplement, ibid, 1977.
56. Kay, D.A.R., *et al., U.S. Patent, No.* 4, 826, 664 (1989).
57. Liu, W., Wadia, C. and Flytzani-Stephanopoulos, M., *Catal. Today.* 28 (4), 1996.
58. Cahn, R.H. and Longo, J.M., *U.S. Patent*, No. 4, 346, 063 (1982).

Performance Analysis of ZnO-Based Sorbents in Removal of H_2S From Fuel Gas

Douglas P. Harrison

Department of Chemical Engineering, Louisiana State University, Baton Rouge, LA 70803, USA

Abstract. Processes for the high temperature desulfurization of coal-derived gases are an important component in the development of advanced electric power generation processes. Much of the desulfurization research and development effort has used zinc-based sorbents, which are quite reactive at the temperatures and pressures of interest, have acceptable sulfur capacities, and are capable of reducing H_2S concentrations to less than 20 ppmv. The sorbent must be regenerable and maintain its activity and structural integrity through many sulfidation-regeneration cycles.

Much of the early research used the mixed metal oxide, $ZnFe_2O_4$. However, a number of limitations, including the formation of Zn(g) and over-reduction of Fe_2O_3 to FeO (or even Fe), soon became apparent. Multicycle durability was not acceptable. Recently, zinc and titanium mixed metal oxides, $xZnO{\cdot}TiO_2$, as well as a proprietary sorbent, Z-Sorb, have received increased attention. Reduction of ZnO and formation of Zn(g) is retarded in the presence of titanium oxide, and reduction problems associated with iron oxide are eliminated. A major drawback, however, is the reduced capacity since TiO_2 does not react with H_2S. In general, $xZnO{\cdot}TiO_2$ and Z-Sorb have proven to be more durable, and have extended the allowable operating window to higher temperature and to coal gases having greater reducing power.

Laboratory research began with single particle kinetic studies using electrobalance reactors and integral studies using small-scale fixed-bed reactors and simulated coal gases. While convenient for laboratory use, the fixed-bed reactor is generally not suitable for commercial operation. Moving-bed and fluidized/transport reactors offer many advantages including steady-state operation, improved temperature control, and simplified materials of construction. All of these reactor types are currently included in Clean Coal Demonstration projects jointly funded by the U.S. Department of Energy and industrial organizations.

1 Introduction

Zinc oxide guard beds have long been used for the desulfurization of natural gas feed-stock for the steam reforming reaction. Sulfur is present primarily as hydrogen sulfide and the chemical reaction is

$$ZnO(s) + H_2S(g) \rightarrow ZnS(s) + H_2O(g) \qquad (1)$$

NATO ASI Series, Vol. G 42
Desulfurization of Hot Coal Gas
Edited by A. T. Atimtay and D. P. Harrison

A fixed-bed reactor is normally used, and typical operating conditions of 370°C and space velocity of 400 hr^{-1} result in an axial average sulfur loading in the bed at breakthrough of 22 to 24 wt% (Phillipson, 1970). This compares to a maximum loading of 33% corresponding to complete conversion of ZnO to ZnS. Sulfur concentrations in the natural gas are generally sufficiently small that regeneration is not necessary. That is, upon breakthrough, the sorbent bed is replaced with a fresh batch of zinc oxide and sulfur removal continues.

Because of the successful use in natural gas desulfurization, it is logical that ZnO, or zinc-based sorbents in general, would be considered for hot coal gas desulfurization. Process requirements are quite different, however, and it quickly became clear that the natural gas desulfurization process could not be simply transferred to coal gas desulfurization. Among the important differences are the need for higher operating temperature in coal gas desulfurization, the presence of steam and a host of other contaminants, and larger H_2S concentrations which necessitate that the sorbent be regenerable instead of replaced at the end of a sulfidation cycle.

The purpose of this paper is to review important steps in the development of a zinc-based sorbent desulfurization process which have occurred over the past two decades, and to summarize the current status of the technology. Only highlights are covered as a complete review would require much more space than allotted here. The sulfidation phase is emphasized in this paper, and the regeneration step is examined in more detail in a subsequent paper. Much of the discussion is based upon research carried out in the United States under the sponsorship of the U.S. Department of Energy, Morgantown Energy Technology Center.

2 The Advantages and Disadvantages of Zinc-Based Sorbents

A sorbent must meet several requirements to be considered for high temperature coal gas desulfurization. Perhaps the most important is that it be thermodynamically capable of reducing gas phase sulfur concentrations to the low levels required by downstream processes, which for an integrated gasification combined cycle (IGCC) plant is about 20 ppmv H_2S. In addition, the sorbent must have acceptable sulfur capacity (both theoretical and achievable), be regenerable, and maintain activity and capacity through a large number of sulfidation/regeneration cycles. Finally, the cost of the sorbent must be reasonable.

Within limits, zinc-based sorbents satisfy all of these requirements. The thermodynamics of H_2S removal using ZnO is superior to most other metal oxides, and H_2S concentration in the product gas can be reduced to less than 10 ppmv. Zinc oxide is more stable in the reducing coal gas atmosphere than, for example, Fe_2O_3 or CuO, although reduction of ZnO to volatile elemental zinc vapor can occur at high temperature in highly reducing atmospheres (Westmoreland and Harrison, 1976). The rate of reaction, which depends on the structural properties of the sorbent, is generally fast, and laboratory studies using spherical ZnO pellets (Gibson and Harrison, 1980) have successfully converted all of the ZnO to ZnS. Many of the problems with zinc-based sorbents relate to regenerability and multicycle durability.

3 Thermodynamic Analysis

The equilibrium concentration of H_2S (ppmv) over ZnO as a function of temperature at a pressure of 20 atm is shown in Figure 1. The results represent a global equilibrium between 1 mol of ZnO and 10 mols of coal gas whose composition (see Figure) approximates that from a Texaco gasifier. These results were obtained by free energy minimization using the CHEMQ program (Kirkpatrick and Pike, 1994). ZnO is theoretically capable of reducing the H_2S content to near 1 ppmv at 700K, to 20 ppmv at 900K, and 100 ppmv at 1050K.

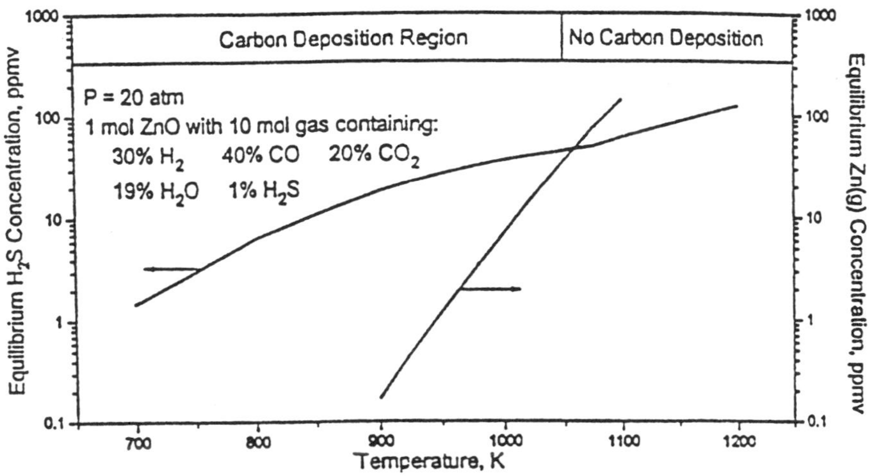

Figure 1. Thermodynamics of High Temperature Desulfurization Using ZnO: H_2S Removal, Zn(g) Formation, and Carbon Deposition Tendency.

Two ZnO problem areas are also illustrated in Figure 1. The tendency for the reduction of ZnO(s) to Zn(g) has long been recognized. The concentration of Zn(g) (ppmv) in the equilibrium gas, shown on the right-hand ordinate of Figure 1, is a stronger function of temperature than the H_2S concentration. A maximum operating temperature of approximately 900K would be required to insure that the concentration of Zn(g) remained below the 0.1 ppmv level. The second problem, which is not unique to desulfurization with ZnO, is illustrated by the bar across the top of Figure 1. Carbon deposition is thermodynamically favored in this gas at 20 atm for all temperatures below about 1060K. The carbon deposition temperature limit will vary only moderately for other metal oxide sorbents.

A number of zinc-based mixed metal oxide sorbents have also been investigated rather extensively -- $ZnFe_2O_4$, $xZnO{\cdot}TiO_2$, and Z-Sorb. Z-Sorb, from Phillips Petroleum, is composed of ZnO on a porous matrix with a nickel oxide promoter, but other details of its composition are proprietary (Gangwal et al., 1994). In the coal-gas atmosphere, $ZnFe_2O_4$ is reduced to ZnO plus Fe_3O_4 (or FeO) and its performance is thermodynamically similar to ZnO with respect to equilibrium H_2S and Zn(g) concentrations. $xZnO{\cdot}TiO_2$ is of interest because TiO_2 stabilizes the ZnO and reduces the tendency for the formation of Zn(g) (Flytzani-Stephanopoulos et al., 1986). Thermodynamic equilibrium calculations for Zn_2TiO_4 ($2\ ZnO{\cdot}TiO_2$) at conditions equivalent to those shown in Figure 1 show a moderate decrease in H_2S removal capability and significant reduction in Zn(g) concentration. At 700K and 900K, the equilibrium H_2S concentrations are about 1.5 and 18 ppmv, respectively, over ZnO, compared to 5.6 and 58 ppmv over Zn_2TiO_4. However, the temperature at which the equilibrium Zn(g) concentration reached 0.1 ppmv is increased from about 890K over ZnO to 920K over Zn_2TiO_4. It is important to recognize that TiO_2 addition also reduces the theoretical sorbent capacity, from 33% S for ZnO to 23% S for Zn_2TiO_4. In spite of the reduced capacity most recent studies have used either $xZnO{\cdot}TiO_2$ or Z-Sorb.

4 Single Particle Kinetics

An electrobalance reactor (thermogravimetric analyzer) is a convenient tool for studying the kinetics of noncatalytic gas-solid reactions between H_2S and single particles of sorbent. The progress of the reaction is followed by monitoring the change in solid mass as a function of time at constant temperature,, pressure, and gas composition.

Gibson and Harrison (1980) used an electrobalance reactor to study the reaction between H_2S and spherical pellets of commercial ZnO whose properties, both as-received and after preheating to 500°C in N_2, are presented in Table 1. All tests were at 1 atm and temperature was varied between 375 and 800°C. The reacting gas contained H_2S in the range of 1 to 6 mol%, H_2 at a constant ratio of 5 mol H_2 per mol H_2S, and balance N_2. Reaction conditions were such that mass transfer, diffusion, and surface reaction resistances were all important at some stage of the reaction. The effects of temperature and H_2S concentration on the fractional conversion of ZnO versus time are shown in Figures 2 and 3, respectively. In the lower temperature range (≤500°C) the reaction effectively stopped well before complete conversion of ZnO.

Table 1. Chemical and physical properties of ZnO spherical pellets: G-72C from the Catalysts Division of Chemetron Corp. (Gibson and Harrison, 1980)

As-Received (manufacturer's specification)		
Chemical composition	-	85 wt % ZnO (minimum)
Weight loss on ignition	-	< 10% at 1000°F
Surface area	-	16.6 - 26.0 m^2/g
Pore volume	-	0.22 cm^3/g, 0.25 - 0.035 μm
Bulk density	-	96 ± 4 lb/ft^3
After 500°C Pretreatment (measured)		
Chemical composition	-	94.5 wt% ZnO
Surface area	-	17.9 m^2/g
Pellet porosity	-	0.64 - 0.69

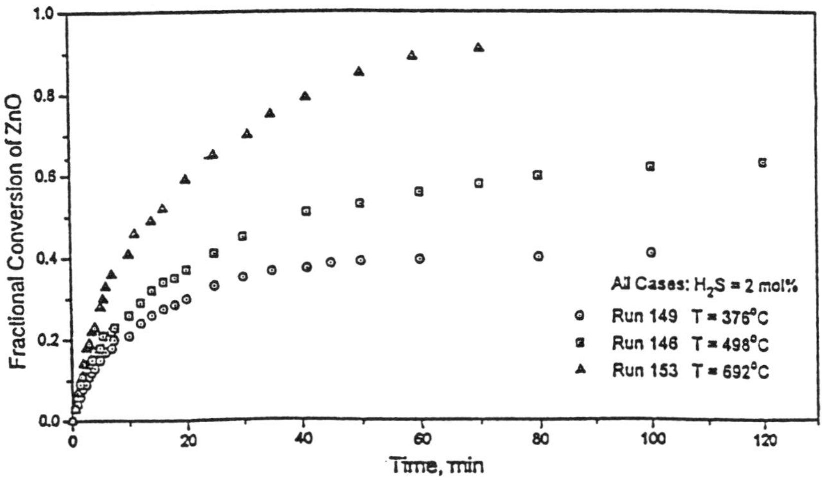

Figure 2. The Effect of Temperature on the Conversion of Spherical Pellets of ZnO to ZnS (Gibson and Harrison, 1980)

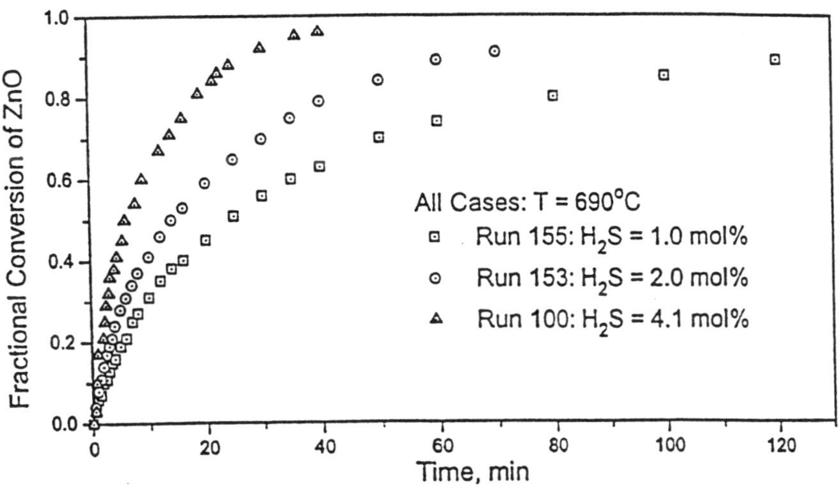

Figure 3. The Effect of H_2S Concentration on the Conversion of Spherical Pellets of ZnO to ZnS (Gibson and Harrison, 1980)

In the intermediate temperature range (600 to 700°C), effectively all ZnO was converted to ZnS in a reasonable reaction time. At higher temperature, in the vicinity of

800°C (not shown in Figure 2), the global rate again approached zero well before complete conversion of ZnO. Microscopic examination showed that an impervious layer had formed on the outside of the pellet. The high temperature data were consistent with reduction of ZnO to produce volatile metallic Zn(g), followed by diffusion of Zn(g) from the interior of the pellet toward the surface where it contacted H_2S. A gas phase reaction between Zn(g) and H_2S(g) deposited solid ZnS at the surface of the pellet which effectively blocked further reaction. The effect of H_2S concentration is shown in Figure 3. At the intermediate temperature of 690°C, the global rate increased with increasing H_2S concentration and complete conversion of ZnO was closely approached in each case. The data are consistent with a global reaction rate which is first order in H_2S.

Ranade and Harrison (1981) extended the above study by measuring the variation in pellet structure caused by sintering and reaction. Sintering studies were carried out at 1 atm in N_2 over a temperature range of 500 to 800°C. Reaction studies were limited to 1 atm, 350°C, in an atmosphere 5% H_2S - balance N_2. Sintering was negligible at 350°C so that structural changes at that temperature were due solely to the conversion of ZnO to ZnS. A correlation showing surface area as a function of temperature, time, and fractional conversion was developed, and used in a variable property grain model (Ranade and Harrison, 1979) to interpret the kinetic results of Gibson and Harrison (1980). The variable property grain model contained four resistances -- external mass transfer, pore diffusion, product layer diffusion, and surface reaction. ailure to achieve complete pellet conversion at low temperature was attributed to the solid state diffusion resistance through the impermeable product layer surrounding each grain. Diffusion through an impermeable solid is an activated process whose rate can be described by an Arrhenius equation. With the activation energy of 22,400 cal/mol used by Ranade and Harrison (1981), the solid state diffusion coefficient varied from 1.8×10^{-11} cm^2/s at 375°C to 4.8×10^{-9} cm^2/s at 690°C. The low temperature value was sufficiently small to cause the reaction rate to approach zero well before complete conversion of ZnO, while the high temperature value was sufficiently large to permit complete conversion of ZnO. The magnitudes of the solid state diffusion coefficients compared reasonably well to values reported elsewhere under both reacting (Georgakis et al., 1979) and nonreacting conditions (Barrer, 1941).

Interest in zinc-based sorbents shifted to mixed metal oxides, initially to zinc ferrite ($ZnFe_2O_4$) and later to zinc titanate ($xZnO{\cdot}TiO_2$) and Z-Sorb. Focht et al. (1988a) reported the results of atmospheric pressure electrobalance tests using cylindrical pellets of $ZnFe_2O_4$ whose properties are summarized in Table 2. The temperature range studied was from 300 to 700°C, and the reactant gas contained all major components of coal gas as shown in Table 3. In the reducing coal gas atmosphere, $ZnFe_2O_4$ breaks down into the constituent oxides and Fe_2O_3 is reduced to Fe_3O_4 or FeO. Consequently, the pellets were initially exposed to H_2S-free reducing gas to simulate the behavior of a fixed-bed reactor in which H_2S removal would first occur near the entrance leaving the majority of the sorbent exposed to sulfur-free reducing gas. After reduction, H_2S was introduced and both zinc and iron oxides were converted to the respective sulfides. Stoichiometric equations representing the reactions are shown below.

Reduction

$$3ZnFe_2O_4 + H_2 \rightarrow 3ZnO + 2Fe_3O_4 + H_2O, \text{ or} \tag{2}$$
$$ZnFe_2O_4 + H_2 \rightarrow ZnO + 2FeO + H_2O$$

Sulfidation

$$ZnO + H_2S \rightarrow ZnS + H_2O$$
$$Fe_3O_4 + 3H_2S + H_2 \rightarrow 3FeS + 4H_2O, \text{ or} \tag{3}$$
$$FeO + H_2S \rightarrow FeS + H_2$$

Complete conversion of $ZnFe_2O_4$ to ZnS and FeS corresponds to a sulfur capacity of 35%, slightly larger than the ZnO capacity.

Table 2. Structural properties of zinc ferrite pellets as reported by United Catalysts Inc . (Focht et al., 1988)

Designation	L-1442
Shape	Cylindrical extrudates
Normal size	
Diameter (in)	3/16
Length to diameter ratio	2.35
Binder	Methocel
Firing temperature (°C)	816
Specific surface area (m^2/g)	6.6
Pore volume (cm^3/g)	0.36
Average pore radius (Å)	760

Table 3. Simulated coal gas used in zinc ferrite study (Focht et al., 1988)

Component	Mol %
H_2	10
H_2O	30
CO	30
CO_2	30 - H_2S
H_2S	1-3

Results of reduction tests in terms of normalized sorbent mass as a function of temperature and time are shown in Figure 4. Complete reduction of Fe_2O_3 to Fe_3O_4 represents a 2.2% mass loss (W/W_o=0.978) while further reduction to FeO corresponds to a total mass loss of 6.6% (W/W_o = 0.934). At the end of the 15 minute reduction test, the iron components corresponded to a mixture of Fe_2O_3 and Fe_3O_4 at 500°C, essentially pure Fe_3O_4 at 600°C, and a mixture of Fe_3O_4 and FeO at 700°C and 800°C. The effect of temperature on sulfidation is shown in Figure 5 where fractional sulfidation at selected reaction times is plotted versus temperature. The kinetics are a strong function of temperature in the range of 300-500°C, and relatively independent of temperature in the range of 500 to 650°C. At higher temperatures, the conversion dropped precipitously and at 750°C, very little sulfidation occurred. The adverse effect of high temperature was attributed to "over-reduction" of Fe_3O_4 to FeO.

The lack of strong temperature dependence in the intermediate range suggests that the global rate is determined by transport resistances, while the relatively strong dependence at lower temperatures indicates that the surface reaction resistance is important, interpretations confirmed by microscopic examination of the cross-section of

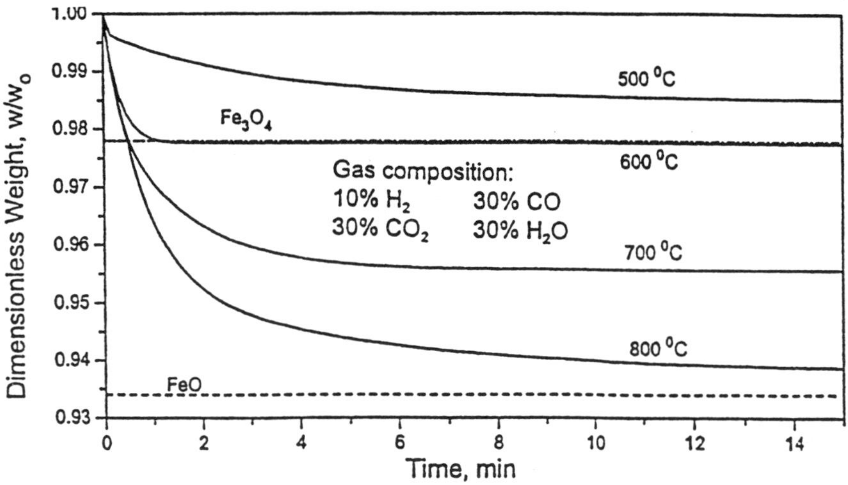

Figure 4. Reduction of Zinc Ferrite in Sulfur-Free Coal Gas as a Function of Temperature (Focht et al., 1988)

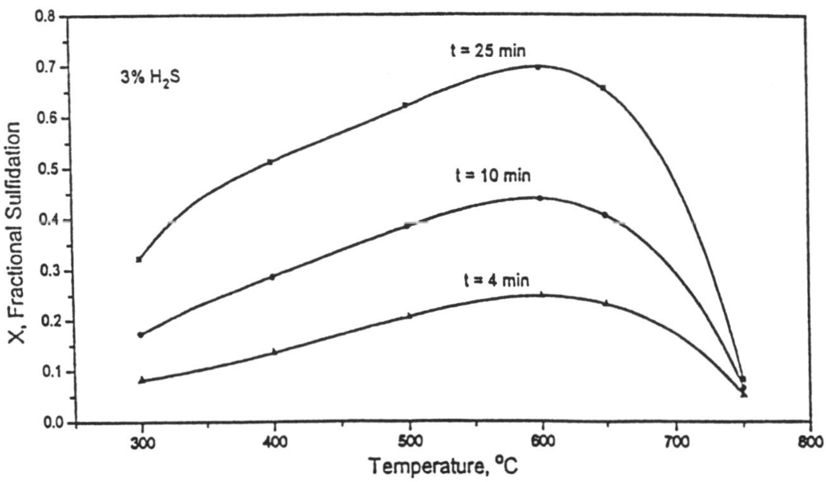

Figure 5. The Effect of Reduction and Sulfidation Temperature on the Sulfidation of Zinc Ferrite (Focht et al., 1988)

partially reacted pellets. A sharp boundary between unreacted metal oxide at the center of the pellet and product metal sulfide around the exterior of the pellet was clearly visible at 650°C, as illustrated in Figure 6. No boundary was visible at lower reaction temperatures, indicating that the reaction proceeded homogeneously throughout the cross-section.

Similar single-pellet electrobalance studies were conducted using zinc-titanate cylindrical pellets (1.5 $ZnO{\cdot}TiO_2$, designation L-3140) to confirm the expected increase in permissable operating temperature (Woods et al., 1990). A high pressure electrobalance was used, thus providing the first study of single pellet behavior above atmospheric pressure. The study also included structural property measurements as a function of temperature at various stages of sulfidation. The gas composition (see Table 4) represented the product from a partially quenched, air-blown, fluidized-bed gasifier. Each test began with a 30 minute exposure to a sulfur-free reducing gas before H_2S was added to initiate sulfidation. The effect of temperature in a series of 1 atm tests using 2.5% H_2S is shown in Figure 7. Complete conversion of ZnO to ZnS with no weight loss during reduction corresponds to a final value of W/W_o=1.116. In spite of the reduction weight loss at temperatures of 815 and 875°C, sulfidation proceeded normally with the smaller final values of W/W_o corresponding approximately to the Zn(g) lost during reduction. Sulfidation was quite slow at 260°C, and was only about 12% complete after 350 min. However, the sulfidation rate increased rapidly between 260 and 540°C, where the global

Figure 6. Photomicrograph of the Cross-Section of a Zinc Ferrite Pellet Partially Sulfided at 650 °C (Focht et al., 1988)

rate was controlled by surface reaction resistance. Primary interest in zinc-titanate sorbent is at higher temperatures, where zinc ferrite is unsuitable. Behavior in this range is represented by the 705°C test in Figure 7. Effectively no zinc was lost during reduction, and sulfidation was rapid and approached completion after 150 min. The effect of temperature on sulfidation kinetics was small in this temperature range, again suggesting that the global rate was controlled by transport resistances. The effect of pressure is illustrated in Figure 8 in a series of tests at 705°C in which the H_2S content and total pressure were varied simultaneously to keep the H_2S partial pressure constant at 0.025 atm.

Table 4. Composition of simulated coal gas used in zinc titanate study (Woods et al., 1990)

Component	Mol %
H_2	10
CO	15
CO_2	5
H_2O	15
H_2S	0.25 - 2.5
N_2	Balance

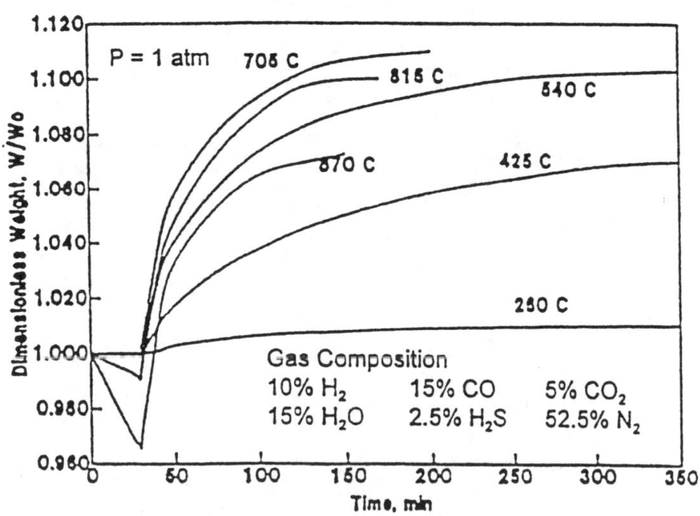

Figure 7. The Effect of Temperature on the Reduction and Sulfidation of L-3140 Zinc Titanate ($1.5ZnO.TiO_2$) (Woods et al., 1990).

The decrease in the sulfidation rate with increasing pressure was explained on the basis of a decrease in the mass transfer coefficient and effective diffusivity with increasing pressure.

The variation in structural properties with temperature at various stages of reduction and sulfidation is summarized in Table 5. Pore volume and surface area did not change appreciably during reduction. Pore volume decreased during sulfidation as the smaller oxygen atom was replaced by the larger sulfur, but was effectively independent of Sulfidation produced an increase in surface area at temperatures to 760°C, no change in surface area at 815°C, and the surface area was reduced only when the temperature reached 870°C, indicating that sintering became significant at about 800°C.

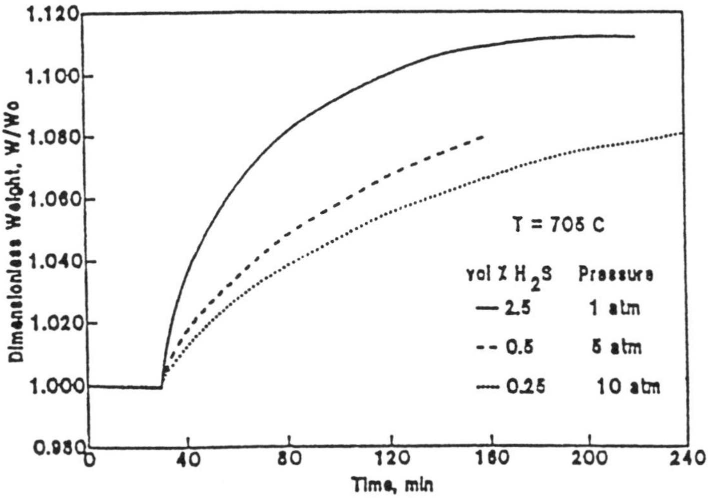

Figure 8. The Effect of Pressure on the Sulfidation Kinetics of L-3140 Zinc Titanate ($1.5ZnO.TiO_2$) at Constant H_2S Concentration (Woods et al., 1990)

5 Laboratory-Scale Integral Reactor Tests

Laboratory-scale fixed-bed and fluidized-bed reactors have been used to study the desulfurization and regeneration reactions. Both reactor types operated in a batch mode with respect to solid and with once-through gas flow. In multicycle tests, the sorbent remained in the reactor, and the feed gas alternated between the reducing coal gas and the oxidizing regeneration gas.

5.1 Fixed-Bed Reactor Studies

The performance of a fixed-bed reactor during the desulfurization cycle is shown in Figure 9. At a given time, the bed may contain three zones as shown in Figure 9a. Sorbent near the entrance has reacted completely and the H_2S concentration is constant at its inlet value. Reaction occurs in the middle zone, where the H_2S concentration is reduced from inlet to near the equilibrium value. Sorbent in the third zone has not yet been exposed to H_2S. The reaction zone, which is initially established at the reactor entrance, moves through the bed as reaction time increases, as illustrated in Figure 9b where H_2S concentration is shown as a function of bed position at selected times. H_2S concentration in the reactor product remains near its equilibrium value until the leading edge of the reaction zone reaches the exit of the bed. From that time, the H_2S concentration in the product increases to form the typical s-shaped breakthrough curve shown in Figure 9c. Equal inlet and outlet H_2S concentrations correspond to complete conversion of sorbent. The sulfidation reaction is exothermic and the movement of the reaction front may be accompanied by a temperature

Table 5. Structural property changes during the reduction and sulfidation of L-3140 zinc titanate (Woods et al., 1990)

	Pore Volume, cm^3/g	Surface Area m^2/g
As-received sorbent	0.37	3.6
After reduction at		
650°C	0.36	3.5
705°C	0.36	3.5
760°C	0.36	3.5
After ~70% sulfidation at		
650°C	0.26	5.6
705°C	0.26	5.4
760°C	0.26	4.3
After complete sulfidation		
	0.22	6.1
650°C	0.23	4.9
705°C	0.24	4.7
760°C	0.23	3.6
815°C	0.23	1.8
870°C		

front as shown in Figure 9d. Since reaction occurs within the sorbent, a temperature difference between the solid and gas phases may also be established.

Fixed-bed reactor performance is often judged on the basis of H_2S concentration prior to breakthrough, and the axial average sulfur loading at breakthrough. Performance is are determined by the combination of equilibrium and global reaction kinetics, with the latter being a function of temperature, pressure, space velocity, gas composition, and the nature and structural properties of the sorbent.

Early fixed-bed research was carried out in laboratories of the U.S. Department of

Energy. Grindley and Steinfeld (1983) compared the performance of a number of sorbents, including two commercial zinc oxide shift catalyst guard beds, the mineral franklinite (a naturally occurring zinc ferrite), a steel industry waste containing franklinite along with excess iron oxide, and a number of specially formulated zinc ferrites from United Catalysts Inc. The zinc ferrites were made using different oxide raw materials and preparation recipes. Prebreakthrough H_2S concentrations below 1 ppmv and average sulfur loadings at breakthrough as large as 17% S were obtained using zinc oxide and zinc ferrite. The low prebreakthrough concentrations were possible because there was no H_2O in the feed gas; in a typical coal gas containing H_2O such small prebreakthrough concentrations are not thermodynamically possible. Desulfurization performance of franklinite was inferior. The steel industry waste reduced prebreakthrough H_2S concentrations to low levels, but the sulfur loading decreased considerably in the second and third cycles. The overall results were judged to be favorable in that "the hydrogen sulfide levels achieved correspond quite well with those predicted by chemical thermodynamic analysis, provided that the space velocities are sufficiently low."

Two studies in which desulfurization was coupled to coal gasifiers were then undertaken. The first small-scale test used a slip-stream from the METC fixed-bed gasifier and a fixed-bed reactor containing zinc ferrite (Grindley and Steinfeld, 1984). Favorable results led to a large-scale test at the KRW Energy Systems fluidized-bed gasifier (Schmidt et al., 1987). More complete results from these studies are described in a subsequent section of this paper. Suffice it at this point to say that significant problems emerged in the KRW test. Pressure drop increased causing runs to be terminated prematurely. The increased pressure drop was attributed to the accumulation of fines carried over from the gasifier into the sorbent bed as well as to fines caused by the breakup of the zinc ferrite sorbent pellets. Because of these problems, experimental effort returned to the laboratory and smaller-scale tests under tightly controlled conditions were continued.

Gangwal et al. (1988) reported results of a fixed-bed laboratory-scale study designed to better define the cause of the KRW problems. A fifteen-cycle test in which the sorbent was mixed with 5 wt % coal fines obtained from candle filters in the KRW unit was carried out to evaluate the effect of gasifier fines carryover on desulfurization performance. The feed gas contained 1% H_2S and 15% steam and each sulfidation cycle was terminated when the product H_2S concentration reached the 10 to 20 ppmv level. Prebreakthrough H_2S concentrations were below 5 ppmv in each cycle. The average sulfur loading at breakthrough was quite variable, ranging from 5.7% S in cycle 1 to a minimum of 3.2% S in cycle 4, and a maximum 12.9% S in cycle 9. The smaller than expected sulfur loading in cycle 1 was attributed to fines causing channeling and blocking pores in the sorbent pellets. Many of the fines were burned to ash during the regeneration. No additional fines were added prior to the second desulfurization cycle and the breakthrough capacity increased to 8.3% S. Thereafter, whenever additional fines were added after regeneration, the breakthrough sulfur loading was near that measured in cycle 1. Approximately 15% of the sorbent pellets had fragmented by the end of the 15 cycles. Pellet crush strength was reduced by almost 70% with most of the loss occurring during the first five cycles. Problems were attributed to carbon deposition and subsequent carbon-sorbent interaction. A second multicycle test in which the steam concentration was reduced to 5% and no coal fines were added was then begun.

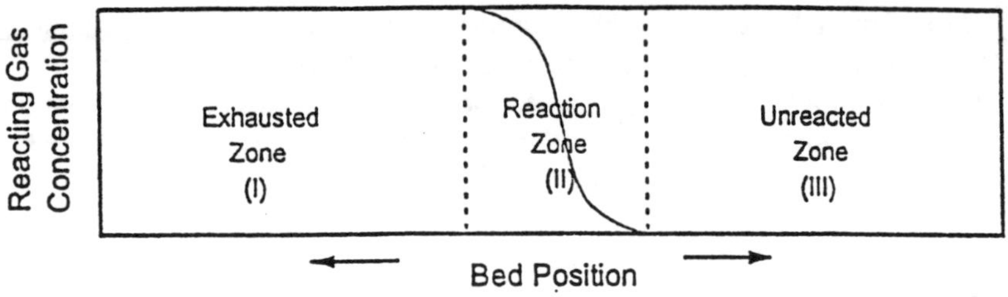

a. H_2S Concentration Profile at Time t.

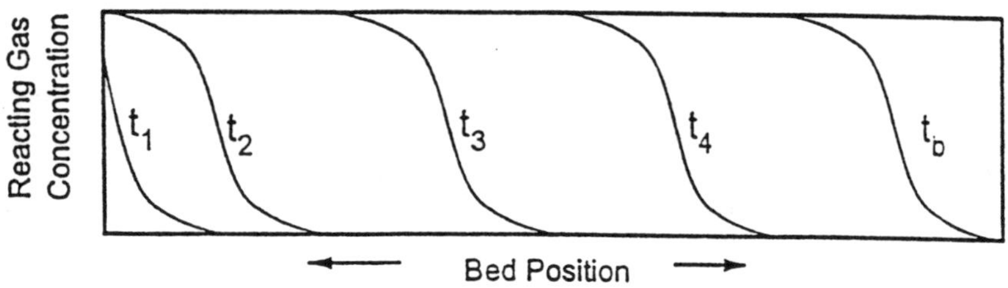

b. Movement of the H_2S Concentration Profile with Time.

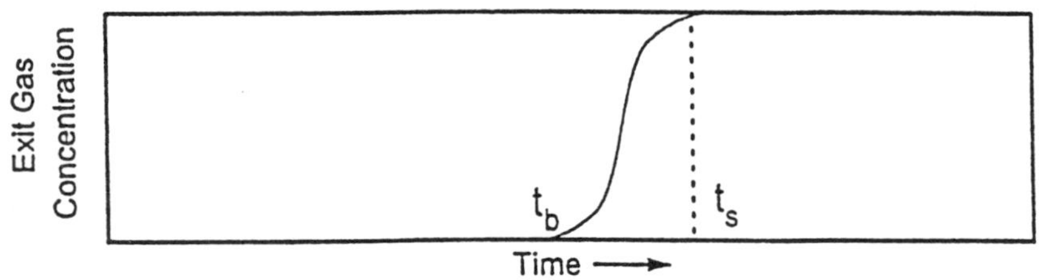

c. H_2S Breakthrough Curve.

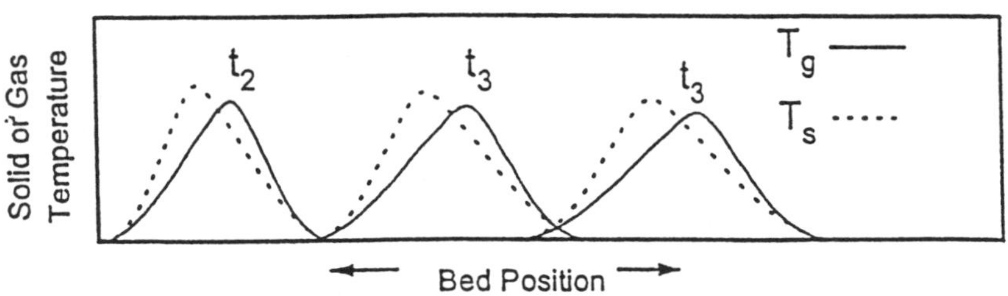

d. Movement of the Temperature Front Through the Reactor.

Figure 9. Typical Performance of a Fixed-Bed Reactor

Prebreakthrough H_2S concentrations were below 1 ppmv, and the duration of the prebreakthrough period increased with increasing cycle number. However, large quantities of carbon were deposited and the sorbent pellets were seriously fragmented after five cycles. The increased duration of the prebreakthrough period was attributed to the fact that the smaller particles resulted in decreased pellet diffusion resistance.

Fragmentation was attributed to carbon deposition and zinc sulfate formation during regeneration.

The limitations of zinc ferrite, which included formation of Zn(g) at high temperature in reducing gas, the possibility of over-reduction of Fe_2O_3 with resultant loss in activity, and the multicycle decrease in crush strength with resultant pellet break-up were fully recognized by this time. Alternate sorbent formulations including zinc titanate and copper-based sorbents were receiving greater attention.

A 0.8 $ZnO{\cdot}TiO_2$ sorbent tested through five cycles at the Research Triangle Institute (Gangwal and Harkins, 1988) exhibited improved durability and was simpler to regenerate than zinc ferrite. rebreakthrough H_2S concentrations were somewhat larger, but were below 10 ppmv to temperatures in excess of 600°C. reakthrough sulfur capacity exceeded 9% S in the first cycle (compared to theoretical capacity of 16% S) and was relatively steady between 7.5 and 8.0% S in the remaining cycles. Pellet crush strength did not deteriorate and significantly less sulfate was formed during regeneration. Zinc losses by vaporization were estimated to be less than 1% per 1000 hours of operation at 730°C, also a significant improvement over zinc ferrite.

Somewhat earlier, General Electric (Smith et al., 1987) began development of a high-temperature desulfurization process using zinc ferrite in a intermittent moving-bed reactor. orbent physical strength and attrition resistance were considered to be more important than high reactivity and sulfur capacity. Consequently, the sorbents tested by GE contained higher concentrations of inorganic binders, and were somewhat less reactive. Although the process reactor would operate in a moving-bed mode, experimental tests were carried out in laboratory-scale fixed-bed reactors. Ayala and Gal (1991) reported test results in which twenty zinc ferrite formulations were evaluated for the proper compromise between chemical reactivity and durability, and pellet strength and attrition resistance. The pellets were ellipsoidal to provide better transport in the moving-bed, and the key formulation variables were binder type and content, iron oxide source, and firing temperature. Five of the most promising formulations were subjected to eight-cycle fixed-bed tests. The desulfurization phase used a simulated air-blown fixed-bed gasifier product containing 1% H_2S at a temperature of 538°C. Regeneration was conducted in a step-wise manner beginning at 538°C with 1% O_2 in N_2 and ending at 760°C with pure air. Sulfate was formed during the early stages of regeneration, and decomposed during the latter stages. The five sorbents were evaluated on the basis of a performance index which combined mechanical strength and chemical reactivity. Results are shown in Figure 10. The "best" formulation, L-3769M, contained 88% zinc ferrite with 10% calcium sulfate and 2% bentonite binders. Although considerably less reactive than the L-3404 formulation, which contained 98% zinc ferrite and 2%bentonite, the superior strength of L-3769M was judged to provide the best overall performance for moving-bed applications.

However, attention at GE was also turning to alternate sorbents. Zinc titanate and Z-Sorb formulations were tested in a manner similar to that previously described for zinc ferrite (Ayala et al., 1992; Ayala et al., 1994). Zinc titanate formulation variables were zinc to titanium ratio (0.8 to 2.0), bentonite binder content, MoO_3 additive, organic burnout additive, and calcination temperature. Z-Sorb formulation methods were not disclosed, but both standard Z-Sorb and special formulations for moving-bed applications were tested. Sulfidation and regeneration test conditions were similar to

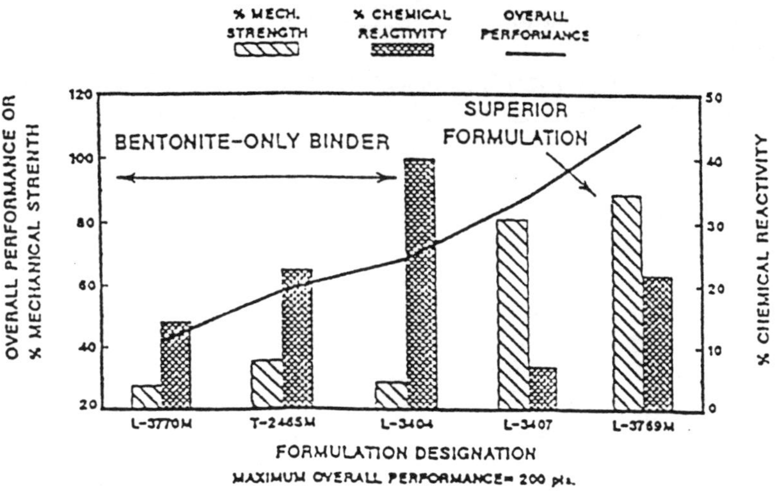

Figure 10. Comparison of the Overall Performance Index of Five Zinc Ferrite Formulations for Moving-Bed Applications (Ayala and Gal, 1991)

those described earlier for zinc ferrite except that the sulfidation gas simulated the product from an oxygen-blown gasifier. The performance of L-4103 zinc titanate and Z-Sorb B is compared in Table 6. Both were highly effective in removing H_2S, with prebreakthrough concentrations less than 20 ppmv using zinc titanate and less than 10 ppmv using Z-Sorb.

Table 6. Comparison of the desulfurization performance of zinc titanate and Z-Sorb for moving bed applications (Ayala et al., 1994)

	L-4103 B2 Zinc Titanate	Z-Sorb B
Max. sulfur capacity (% S)	25	20
Pre-breakthrough concentration	<20 ppm H_2S	<10 ppm H_2S
% Total S (5 Sulfidations)	6.03 - 8.87	10.62 - 16.17
% of Max sulfur capacity	27 - 40	58 - 88
lb sulfur/ft^3 of bed	7.3 - 10.7	7.0 - 10.7
% carbon	0.14 - 0.18	0.14 - 0.20

The sulfur loading was considerably higher with Z-Sorb on a weight basis, but, because of the lower bulk density of Z-Sorb, the sulfur loadings per unit bed volume were approximately equal. It was on the mechanical side that the Z-Sorb performance was uperior. Almost half of the zinc tinanate pellets were cracked at the conclusion of the test while almost none of the Z-Sorb pellets had cracked. The crush strength of the zinc titanate decreased while the crush strength of Z-Sorb increased. The cracking and decrease in crush strength with zinc titanate was correlated with the amount of zinc sulfate formed during regeneration. Little deterioration was noted when sulfate was limited to 2% or less, but deterioration occurred when the sulfate level increased to about 4%. In contrast, up to 6% sulfate was found in the Z-Sorb pellets, which did not crack or lose crush strength. This improved performance may be due to the larger pore volume (decreased pellet density) of the Z-sorb which allowed formation of the larger sulfate molecule without creating excessive internal stress.

Sulfate formation during regeneration was, by this time, identified as a major cause of sorbent deterioration. The GE lab-scale reactor was upgraded for pressure operation, and results using Z-Sorb were reported by Ayala et al. (1995). Sulfidation conditions were as described previously except that the pressure was 5 atm. Two regeneration schemes were tested. In "wet" regeneration the feed gas contained 2% steam while steam was omitted in "dry" regeneration. The initial regeneration temperature was 538°C, and the oxygen content was controlled between 0 and 4% (in N_2) to prevent the peak bed temperature from exceeding 732°C. At the conclusion of each regeneration cycle, the reactor was depressurized to 1 atm at 732°C under N_2 to decompose sulfate. The wet regeneration tests, originally planned for 50 cycles, were terminated after 12 cycles due to loss of sulfidation capacity, which was attributed to a combination of the steam and sulfate formation. The loss of capacity was sufficiently large that Phillips Petroleum (the Z-Sorb developers) began work on a more durable Z-Sorb formulation.

5.2 Fluidized-Bed Reactor Studies

Although fixed-bed reactors are convenient for laboratory testing, there are a number of disadvantages associated with their commercial application. Because of the unsteady-state operation, the sulfur content of the product gas will vary with time. Large valves capable of automatic operation in the high temperature, high pressure, corrosive environment are required to direct coal gas and regeneration gas to at least two reactors, with each reactor alternating between sulfidation and regeneration. Materials of construction in both reactors must be suitable to both highly reducing and highly oxidizing atmospheres. Temperature control in the fixed-bed reactor is difficult, particularly during the highly exothermic regeneration cycle.

Many of these disadvantages are avoided, or at least decreased in severity, in fluidized/transport or moving-bed reactors. At least two reactors are required as before, but each is exposed only to reducing coal gas or oxidizing regeneration gas. Steady-state operation is feasible, and in fluidized reactors the intense solid mixing permits effectively isothermal operation. Valve problems are reduced since it is not necessary to switch gas flow paths. Sorbent attrition resistance replaces pellet crush strength as an important durability parameter.

Grindley (1989) and Gansley and Lee (1990) studied the performance of zinc ferrite in semi-batch fluidized-bed reactors operating at both atmospheric and elevated pressure and using both simulated and actual coal gas. The semi-batch fluidized-bed reactor is an unsteady-state system in which H_2S breakthrough curves, similar to the those from fixed-bed reactors, are produced. Greater than 99% H_2S removal was achieved during the prebreakthrough period with sorbent loadings at breakthrough as large as 19% S.

Gangwal (1990) conducted multi-cycle semi-batch fluidized-bed tests using zinc ferrite and simulated coal gas feed. Particles having a mean diameter of 192 μm were obtained by crushing and screening zinc ferrite pellets. A 100-cycle test at a nominal sulfidation temperature of 625°C and pressure of 15 atm was carried out. Regeneration conditions were 760°C and 15 atm using 2% O_2 in N_2. The high regeneration temperature was chosen to eliminate sulfate formation. H_2S concentration was reduced from 4500 ppmv to less than 20 ppmv during the early stages of each sulfidation cycle. Sulfidation was terminated at H_2S breakthrough, defined as 500 ppmv H_2S in the product gas. Each regeneration cycle was continued until the SO_2 concentration dropped below 200 ppmv. The sorbent capacity at breakthrough dropped rapidly, from 16% S to 5% S the first 20 cycles, and then slowly to 3% S over the remaining 80 cycles. Particle break-up due to mechanical and/or chemical attrition with subsequent fines carryover caused a loss of 42% of the sorbent from the bed during the 100 cycles. Zinc loss due to reduction and vaporization was approximately 8%. Much of the particle break-up and fines carryover was attributed to mechanical attrition associated with the rough edges produced during pellet crushing. As the particles became smoother, the primary cause of break-up and fines carryover shifted to chemical attrition, which was attributed to over-reduction of iron to FeO (and perhaps formation of FeC) and reduction and vaporization of ZnO to Zn(g). The results again confirmed that zinc ferrite was unsuited for use at temperatures exceeding about 550°C in highly reducing gas.

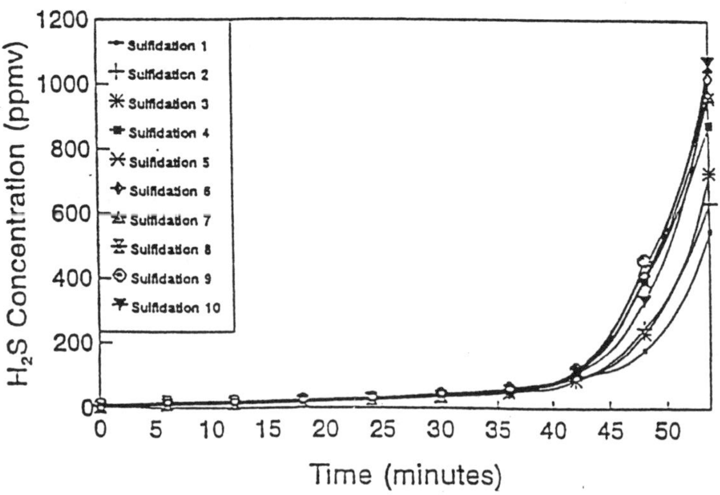

Figure 11. H_2S Breakthrough in a Ten-Cycle Fluidized-Bed Test Using ZT-4 Zinc Titanate ($1.5ZnO.TiO_2$) in a Simulated Texaco Gas (Gangwal and Gupta, 1991)

In a subsequent paper, Gangwal and Gupta (1991) reported results of semi-batch fluidized-bed ten-cycle tests using zinc titanate and simulated coal gas. The sorbent, prepared by a proprietary granulation technique, exhibited superior mechanical attrition resistance. Sulfidation temperatures of 650°C and 750°C, well beyond the maximum temperature suitable for zinc ferrite, were studied. Other reaction conditions were similar to those used in the previous zinc ferrite tests. A zinc titanate having a composition of $1.5ZnO{\cdot}TiO_2$ performed well at 750°C and 15 atm in a simulated Texaco gas. H_2S concentrations were reduced to less than 20 ppmv during the early stages of each cycle. The breakthrough curves were effectively the same in each of the 10 cycles as shown in Figure 11, and the capacity at breakthrough was approximately 12% S in each cycle. The attrition rate, defined as the rate of weight loss due to particle break-up and fines carryover in a standard test, was about 1% per hour, significantly lower than the attrition rate for the zinc ferrite fluidized particles reported earlier.

In a follow-up 100-cycle test under similar reaction conditions (Gupta and Gangwal, 1994), the H_2S content was reduced to less than 20 ppmv in the early stages of each cycle. Sorbent capacity at breakthrough dropped from 12.6% S in cycle 1 to 5.8% S in cycle 100, and averaged about 9% S over the 100 cycles. The majority of the capacity decrease occurred during the first 50 cycles. The attrition rate of the sorbent after 100 cycles was considerably less than that of the fresh sorbent, and the total sorbent loss from the reactor due to particle break-up and fines carryover was less than 2%. Zinc loss was negligible.

Phillips Petroleum developed a fluidizable version of Z-Sorb and initial test results were presented by Gangwal et al. (1994). The sorbent had a range of particle diameters between 50 and 300 μm. Fifty sulfidation-regeneration cycles were carried out in two 25-cycle segments using a semi-batch fluid-bed reactor. Sulfidation conditions were 538°C, 20 atm, using a simulated U-gas feed containing 5000 ppmv H_2S. Regeneration was carried out at 20 atm and a typical temperature range of 649 to 690°C (with periodic excursions to a low of 580°C and a high of 780°C) with the regeneration gas containing 2.0 to 2.5% O_2 in N_2. Prebreakthrough removal of H_2S approached 100% in each cycle, with the sulfur loading at breakthrough shown in Figure 12. Capacity during the first ten cycles was effectively constant at 20.2% S, quite close to the theoretical capacity. An unexplained temperature excursion to 760°C occurred during the tenth regeneration cycle, which seemed to trigger the capacity decrease in subsequent cycles. The breakthrough time decreased by about 30% over the first 25 cycles, dropping the sorbent capacity to about 16% S. A slight improvement in performance was noted in cycle 26, immediately following a several week interruption. Capacity continued to decline slowly thereafter to about 10% S after 50 cycles. Only about 2.5% of the sorbent mass was lost through attrition over the fifty cycles.

Greenwood et al. (1995) reported results of an eight-cycle fluidized-bed test using Z-Sorb which is of particular interest because the sulfidation feed simulated the product from a highly reducing Shell gasifier. The composition of the Shell gas along with key reaction conditions are presented in Table 7; the composition of the regeneration gas was not reported. Similarly, there was no mention of prebreakthrough H_2S concentrations or of sorbent attrition losses. However, the near-constant sulfur loading, which exceeded 15% S during each cycle, shows that the sorbent was quite stable even in the highly reducing Shell gas.

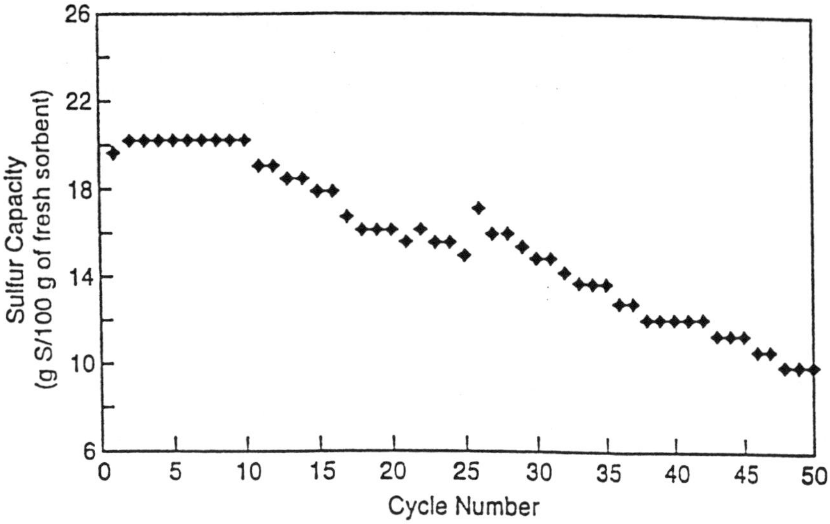

Figure 12. Sulfur Loadings at Breakthrough in a Fifty-Cycle Fluidized-Bed Test Using Z-Sorb III in Simulated U-Gas (Gangwal et al., 1994)

Table 7. Operating conditions associated with the eight-cycle fluidized-bed test of Z-Sorb (Greenwood et al., 1995)

Sulfidation gas feed composition (mol%) (simulated Shell gas)			
H_2	28.5	H_2O	5.0
CO	58.9	N_2	3.3
CO_2	2.2	H_2S	1.0
Reaction conditions	**Sulfidation**	**Regeneration**	
Temperature (°C)	400	600	
Pressure (atm)	20	20	

6 Integrated Operation

A more severe test of the high temperature desulfurization process arises when desulfurization is coupled with an actual gasifier. The presence of contaminants in the coal gas, including particulates, tars, ammonia, chlorides, and trace metals, may impact the operation of the sulfur removal units. In addition, the composition, flow rate, temperature, and pressure of the coal gas will vary in response to changes in gasifier operating conditions.

6.1 Fixed-Bed Tests

Perhaps the earliest tests involving actual coal gas were reported by Grindley and Steinfeld (1984). A slip-stream from the METC fixed-bed gasifier was fed to a fixed-bed desulfurization reactor containing zinc ferrite. Operating temperatures and pressures during sulfidation and regeneration ranged from 540 to 650°C and 9 to 17 atm. Three different coals were gasified and five sorbents were subjected to a total of 13 sulfidation and 9 regeneration cycles. Prebreakthrough H_2S concentrations were generally less than 5 ppmv (dry basis), and axial average sulfur loadings at breakthrough were in the range of 10 to 15% S. The favorable results led to the large-scale test at the KRW Energy Systems fluidized-bed gasifier (Schmidt, 1987; Schmidt, 1988). The gasifier could be operated with limestone injection to provide in-bed sulfur removal, in which case the zinc ferrite operated in a polishing mode. Without limestone injection, the zinc ferrite bed operated in the bulk desulfurization mode to remove essentially all of the H_2S. Two fixed-bed reactors were tested, with the larger being 4.5 ft inside diameter by 6 ft long, and the smaller having an inside diameter of 2 ft and a length of 6 ft. In the most successful desulfurization test, which used the larger reactor and lasted 116 hours, the H_2S content in the product gas was consistently in the 20-40 ppmv range. Operating conditions were 15 atm, 650°C, and space velocity of 794 hr^{-1} (stp). The zinc ferrite sorbent had been subjected to two previous sulfidation and regeneration cycles. The most significant problem was caused by a continuous increase in reactor pressure drop, from 2.94 psi at the beginning of the test to 6.5 psi at the end. There was also evidence of significant channeling of gas along the reactor walls. In subsequent runs using the smaller reactor in a sulfur polishing mode, the pressure drop increased at an even faster rate. The pressure drop increase was caused by a combination of fines carried over from the gasifier and high efficiency cyclones and fines caused by the break-up of zinc ferrite sorbent particles. A laboratory study was then initiated to better define the cause of the problems, and a brief description of key results has already been presented (Gangwal et al., 1988).

Leininger et al. (1991, 1992) tested five zinc-based sorbents -- two zinc ferrites, two zinc titanates, and a zinc copper ferrite -- using a slip stream from the Texaco entrained-bed process development gasifier. Small fixed-bed desulfurization reactors were used at a space velocity of 2000 hr^{-1} and inlet temperatures between 500 and 640°C, depending on gasifier operating conditions. The gasifier was operated in both oxygen-blown and air-blown modes. All sorbents provided greater than 99% H_2S removal prior to breakthrough, but the ferrite sorbents suffered a significant decrease in crush strength in the oxygen-blown mode and also catalyzed methanation of the coal gas. Neither of these problems was significant in the air-blown operating mode. The performance of zinc titanate sorbents, which retained most of their strength and did not catalyze the methanation reaction, was clearly superior to that of the zinc ferrites. Multicycle tests to evaluate the durability of four zinc titanate formulations were then carried out. Significant cracking and spalling occurred with each formulation after 4½ cycles (five sulfidations and four regenerations). Although no reasons for sorbent deterioration were presented, one can speculate that the poor performance of the zinc ferrites was due to over-reduction of Fe_2O_3 in the high temperature, highly reducing atmosphere, coupled with zinc sulfate formation during regeneration. The multicycle deterioration of the zinc titanates was likely associated with sulfate formation.

Full integration of hot gas desulfurization with gasification requires that both sulfidation and regeneration cycles be utilized in a manner to provide continuous operation. Those tests previously described were integrated only in the sense that the desulfurization feed gas was obtained directly from the gasifier.

6.2 Moving-Bed Tests

In the mid-1980s, work began at General Electric on facilities to integrate GE's moving-bed high temperature desulfurization process with an existing fixed-bed, air-blown, dry-ash gasifier (Smith et al., 1987). The gasifier was capable of treating about 2000 lb/hr of coal to produce coal gas at 20 atm and 540 to 650°C. Design sulfur content of the coal gas was in the 3000 to 5000 ppmv range.

A schematic of the moving-bed sulfur removal process is shown in Figure 13. Coal gas enters at the bottom for the absorber (desulfurizer) and flows countercurrent to intermittently moving sorbent. Particulate from the gasifier is removed in high efficiency cyclones. Partially sulfided sorbent exits from the bottom of the absorber into a lock hopper and then into the regeneration reactor. The regeneration reactor shown contains three stages although numerous changes have been made since the original design, and currently regeneration is carried out in two stages. The first stage is designed for co-current flow of sorbent and regeneration gas containing 1.0 to 2.5% O_2, 10 to 12% SO_2 and balance nitrogen. Only about 30% of the regeneration occurs in this stage and some sulfate formation is expected. The majority of the regeneration occurs in the second stage where additional oxygen is added. The design outlet temperature from stage 2 is 790°C, sufficiently high to decompose most of the sulfate. Dry air is fed to the third stage at 315 to 425°C to complete the regeneration, purge SO_2 rich gas, and to cool the sorbent. Regenerated sorbent exits into a lock hopper, and is transferred by bucket elevator to the lock hopper above the desulfurization reactor. The regeneration gas recycle section operates with dry air input and a product gas containing about 13% SO_2, which is a suitable feed to a sulfuric acid plant. No external diluents such as steam or nitrogen are required.

Preliminary integrated tests were carried out in 1989 and 1990 using zinc ferrite sorbent. Cook et al. (1991) reported results of a 60-hour integrated run in early 1991, which included 30 sorbent movement cycles. Exit gas H_2S concentrations were in the 10 to 20 ppmv range when fully regenerated sorbent was added to the absorber. Reactor pressure drop increased between each sorbent movement but returned to near the original value when regenerated sorbent was added and spent sorbent removed. Sorbent transfer provided a self-cleaning mechanism for removal of fines carried over from the gasifier and formed by sorbent attrition. SO_2 concentrations in the regenerator product gas as large as 11% were measured during portions of the test. However, temperature excursions in the regenerator led to the formation of sintered clumps of zinc ferrite

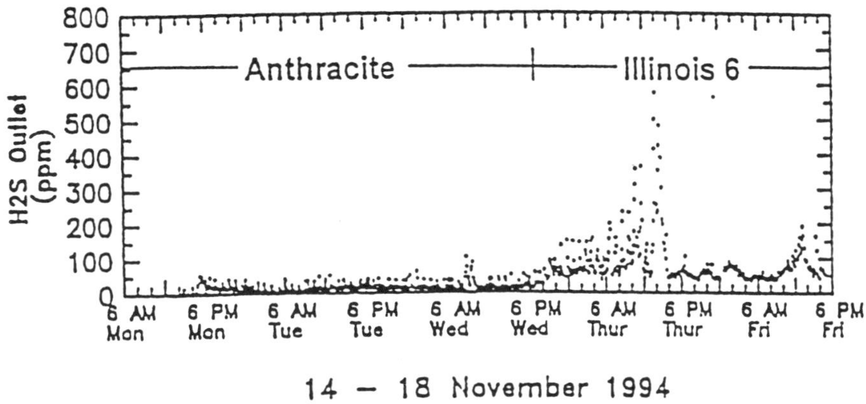

Figure 13. H_2S Concentrations in the Product Gas During a 100-Hour Integrated Test of the GE Moving-Bed Process (Bevan et al., 1995)

which prevented sorbent transfer. Problems associated with the interaction of chlorine compounds in the coal gas and the sorbent were recognized at this time, and a nahcolite chloride removal system was installed.

Results of moving-bed tests using zinc titanate were reported in 1992 (Cook et al., 1992). Modifications in the regeneration reactor were made during this period. The most successful zinc titanate test was reported by Bevan et al. (1994) in which 65 hours of integrated gasifier-hot gas desulfurization operation were achieved. Inlet H_2S concentrations of 4000 to 6000 ppmv were reduced to 50 to 500 ppmv, with the higher concentrations attributed to incomplete regeneration. The first test using Z-Sorb was also reported by Bevan et al. (1994). In an 80-hour integrated test, the exit H_2S concentration was less than 50 ppmv for extended periods. Regeneration performance was improved but incomplete regeneration still led to spikes in the exit H_2S concentration. Sorbent particles were subjected to eight sulfidation-regeneration cycles during this test with only about 0.4% attrition loss.

A total of 100 hours of uninterrupted integrated operation was achieved in a November 1994 test using Z-Sorb (Bevan et al., 1995). Anthracite was fed during the first half of the run while Illinois No. 6 coal was gasified during the second half of the run. H_2S concentration in the product gas as a function of time is shown in Figure 14. Input concentrations were about 1000 pmv H_2S during the anthracite portion of the run, and product H_2S concentrations were reasonably steady at about 20 ppmv. Product H_2S concentrations increased during the first 24 hours that Illinois No. 6 was fed, but subsequently dropped to a reasonably constant 50 ppmv. The H_2S increase following the coal switch was caused by a lag in increasing the sorbent transfer rate to compensate for the increased H_2S feed rate.

6.3 Fluidized-Bed Tests

Fully integrated testing of fluidized-bed desulfurization has been carried out at the Enviropower facility in Finland (Salo et al., 1994; Ghazanfari et al., 1995). An air-blown, fluidized-bed gasifier having a nominal capacity of 35 tons of coal per day and capable of producing coal gas at 30 bar and 500°C has been fitted with dual circulating fluidized-bed desulfurization-regeneration reactors. Two zinc titanate sorbent formulations, ZT-4-L from Research Triangle Institute and UCI-5 from United Catalysts, have been tested in three runs each of 5 to 6 days duration. The gasifier was fed with Polish bituminous coal and the coal gas contained about 800 ppmv total reduced sulfur. UCI-5 sorbent was used in two of the tests, one of which used N_2 dilution to produce a regenerator feed gas containing 2 to 4% O_2.Steam dilution was used to achieve the same O_2 concentration in the other test. N_2 dilution was used in the ZT-4-L test. Steam did not produce any detrimental effect on sorbent performance. The sulfur removal efficiency using UCI-5 was in the range of 90 to 95%, and the H_2S content of the product gas was less than 100 ppmv. SO_2 content of the regenerator off-gas averaged about 2% with the reactor at 30 bars and 600 to 700°C. ZT-4-L sorbent was somewhat more effective with a sulfur removal efficiency of about 98% to produce from 10 to 20 ppmv H_2S in the product gas. Attrition losses were said to be satisfactory and changes in properties of the sorbent were essentially the same as measured in laboratory tests using simulated coal gas. No evidence of zinc sulfate formation was found.

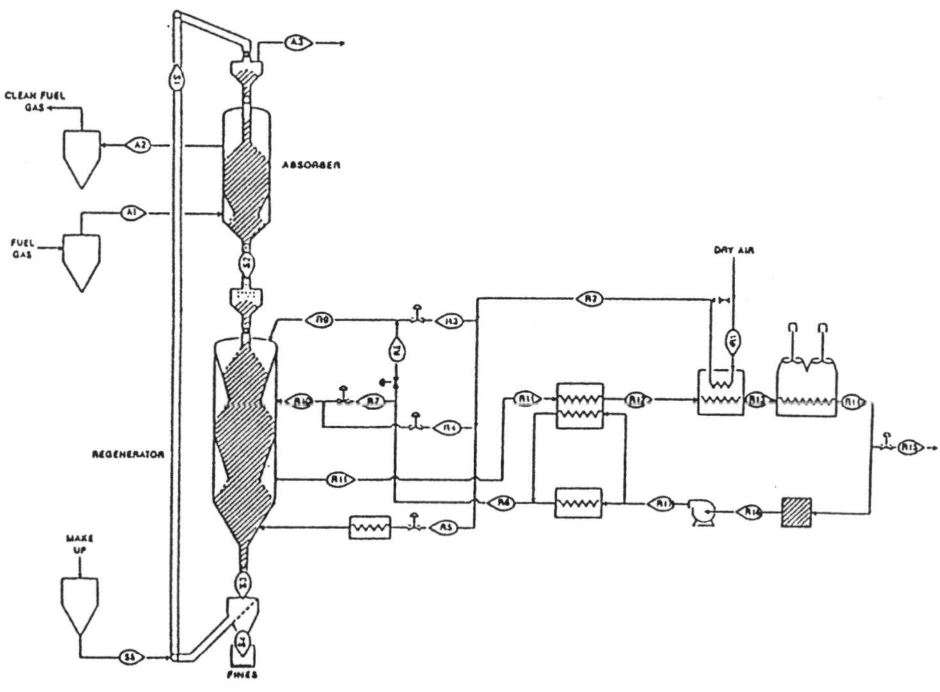

Figure 14. Schematic of the GE Moving-Bed Desulfurization Process (Smith et al., 1987)

7 Other Contaminants

A number of additional contaminants in the coal gas may require control, either to meet environmental standards or for process reasons. These include particulates, tars, nitrogen compounds, chlorides, alkalis, and volatile metals. Some of these contaminants, especially particulates, tars, and chlorides, may impact the performance of the desulfurization step.

Particulate carryover from the gasifier and particulate control devices may create a of problems in the desulfurization system, including contributing to pressure drop increase, channeling, and blockage of sorbent pores. In addition, the particulate may react chemically carbon and metal silicates from ash components are of concern. Problems reported with the KRW Energy Systems tests (Schmidt, 1987; Schmidt, 1988) were attributed to particulate carryover. Moving-bed and fluidized-bed reactors provide an opportunity for separating particulate matter and sorbent fines between the desulfurization and regeneration reactors. The General Electric moving-bed process incorporates a fines separator at the bottom of the regeneration reactor. Cook et al. (1991) reported a gradual increase in pressure drop during periods when the bed was stationary, but recovery following each bed movement. Carryover of tar and carbon (coke) from the gasifier creates problems similar to those caused by inorganic ash in the desulfurization step. Additional problems are created during regeneration where oxidation of the tar and carbon increases the rate of heat release, and alters the quantity of oxygen available for sorbent regeneration.

Chlorine in the coal is converted in the gasifier to HCl with estimated coal gas concentrations as large as 500 ppmv, depending upon the composition of the coal and the gasification process. Grindley (1990) first identified the harmful effect of HCl in laboratory fixed-bed tests using zinc ferrite sorbent. Without HCl in the feed gas, the breakthrough time decreased by 15% during a 10-cycle test. When 1000 ppmv of HCl was added, the breakthrough time decreased by 60% after 10 cycles at the same reaction conditions. Both zinc and iron were lost due to the formation of volatile zinc and iron chlorides. General Electric has been forced to terminate prematurely an integrated test when heat exchangers in the regenerator recycle gas loop became plugged by deposits of zinc sulfate and zinc chloride (Bevan et al., 1994). Krishnan et al. (1994) have evaluated sodium-based minerals for chloride removal. The research concentrated on naturally occurring, inexpensive materials since the spent sorbent is nonregenerable. Nahcolite was effective in reducing HCl concentration to less than 1 ppmv at 650°C, and the conversion of sodium to NaCl approached 100%. General Electric has subsequently modified their system to include a chloride guard bed.

8 Commercial Development

The Clean Coal Technology Demonstration program instituted by the U.S. Department of Energy in 1985 has provided $2.42 billion in government funds matched by $4.72 billion from the private sector to support forty-five projects aimed at demonstrating on a commercial scale a new generation of advanced coal-based technologies (U.S. Department of Energy, 1995). Three of these projects involve integrated gasification combined cycle (IGCC) technology with hot gas desulfurization for electric power generation.

8.1 Piñon Pine Project.

Jointly funded by the U.S. Department of Energy, Sierra Pacific Power Company, Foster Wheeler Corporation, and The M.W. Kellogg Company, this project incorporates a KRW air-blown, pressurized, fluidized-bed gasifier, hot gas desulfurization, and gas and steam power cycles to produce 99 MWe (net) of electric power with a bus-bar efficiency of 43.7%.

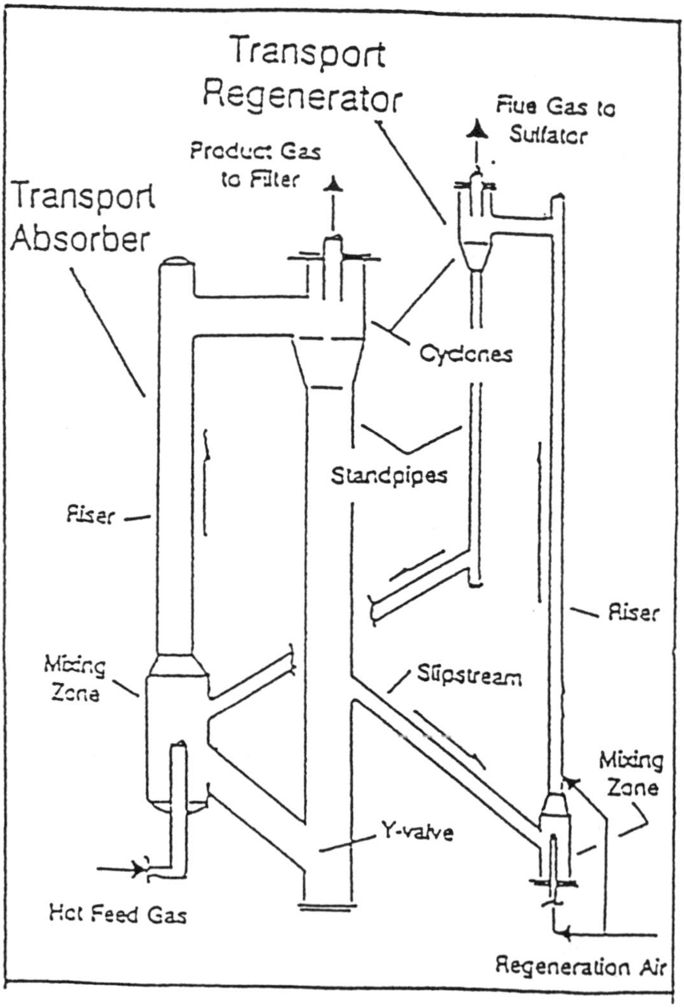

Figure 15. Schematic of the Transport Reactors Used in the Pinon Pine Project (Campbell et al., 1995)

Bulk desulfurization is accomplished by injection of crushed limestone into the gasifier with remaining sulfur compounds removed using a zinc-based sorbent in a transport reactor. Regenerator off-gas is to be recycled to the gasifier so that effectively all sulfur is ultimately captured by the limestone. The transport reactor, shown in Figure 15 (Campbell et al., 1995), differs from the standard fluidized-bed reactor in that gas velocities are sufficient to entrain the sorbent particles. Short contact times in the sulfidation reactor avoid the large sulfur loadings which are thought to contribute to sorbent deterioration. Regeneration is controlled by the short contact time and the use of sub-stoichiometric oxidant to limit the formation of zinc sulfate and control temperature by limiting the oxidation exotherm.

Both Western bituminous coal containing 0.5 to 0.9% sulfur and West Virginia or Pennsylvania coal containing 2 to 3% sulfur are to be processed in the demonstration plant. Design specifications call for SO_2 emissions to be below 0.045 lb/10^6 Btu, corresponding to 98 to 99% reduction for high sulfur coals. The primary drawback is the large amount of solid waste associated with limestone injection

8.2 Tampa Electric Project

This project couples a Texaco pressurized, oxygen-blown, entrained flow gasifier with General Electric's moving-bed high temperature desulfurization process. Tampa Electric, Texaco, General Electric, and Bechtel Power Company are cost sharing with the Department of Energy. The process differs from the Piñon Pine project in several important aspects. There is no limestone injection into the gasifier, and effectively all of the sulfur in the coal emerges in the coal gas. Both high and low temperature options for sulfur
removal are provided, with the units to operate in parallel. The low temperature unit will utilize conventional wet scrubbing technology. The regenerator off-gases will be suitable for sulfuric acid production.

The project will generate 250 MWe (net) at an expected heat rate of 8600 Btu/kWh, corresponding to a 40% overall station efficiency. Both Illinois 6 and Pittsburgh 8 bituminous coals are to be used in the demonstration. NO_x emission reductions of about 90% are expected based on the use of an advanced gas turbine with nitrogen injection. The project, with the exception of hot gas desulfurization and other advances, represents a scale-up the successful 100 MWe Cool Water demonstration.

8.3 Tom's Creek Project

The Tom's Creek project, jointly funded by Tampella Power, Coastal Power Production Co., and the Department of Energy was to have been the third project involving hot gas desulfurization. However, the project is now on hold. The 190 MWe plant was to utilize a U-Gas pressurized, air-blown, fluidized-bed gasifier with in-bed injection of calcium-based sorbent for bulk sulfur removal. Final sulfur removal was to be accomplished in recirculating fluidized-bed sulfidation-regeneration reactors using zinc-based sorbent. Regenerator off-gases were to be recirculated to the gasifier so that all sulfur would be eventually removed as $CaSO_4$ and disposed of by landfill. The design heat rate of the demonstration facility was 8,720 Btu/kWh, corresponding to 39% efficiency. In a larger, fully commercial unit, the heat rate was estimated to be 7,750 Btu/kWh (44% efficiency). SO_2 emission reductions of 99% were expected.

9 References

Ayala, R.E., and Gal, E., 1991, Enhanced Durability of High-Temperature Desulfurization Sorbents for Moving-Bed Applications, Proceedings of the Eleventh Annual Gasification and Gas Stream Cleanup Systems Review Meeting, DOE/METC-91/6123, Vol. 2, p. 377.

Ayala, R.E. et al., 1992, Enhanced Durability of High-Temperature Desulfurization Sorbents for Moving-Bed Applications, Proceedings of the Twelfth Annual Gasification and Gas Stream Cleanup Systems Review Meeting, DOE/METC-92/6128, Vol. 1, p. 230.

Ayala, R.E., et al., 1994, Development of Sorbents for High-Temperature Desulfurization in Moving-Bed Systems, Proceedings of the Coal-Fired Power Systems 94 -- Advances in IGCC and PFBC Review Meeting, DOE/METC-94/1008, Vol. 2, p. 637.

Ayala, R.E., et al., 1995, Moving-Bed Sorbents, Proceedings of the Advanced Coal-Fired Power Systems '95 Review Meeting, DOE/METC-95/1018, Vol. 2, p. 591.

Barrer, R.M., 1941, Diffusion In and Through Solids, Macmillan, New York.

Bevan, S., et al., 1994, Integrated Operation of a Pressurized Gasifier, Hot Gas Desulfurization System and Turbine Simulator, Proceedings of the Coal-Fired Power Systems 94 -- Advances in IGCC and PFBC Review Meeting, DOE/METC-94/1008, Vol. 1, p. 222.

Bevan, S., et al., 1995, Integrated Testing of a Pressurized Fixed-Bed Gasifier, Hot Gas Desulfurization System, and Turbine Simulator, Proceeding of the Advanced Coal-Fired Power Systems '95 Review Meeting, DOE/METC-95/1018, Vol. 1, p. 187.

Campbell, W.K. et al., 1995, Zinc Titanate Tests in Transport Reactor, Proceedings of the Advanced Coal-Fired Power Systems '95 Review Meeting, DOE/METC-95/1018, Vol. 2, p. 215.

Cook, C.R., et al., 1991, Integrated Operation of a Pressurized Fixed-Bed Gasifier and Hot Gas Desulfurization System, Proceedings of the Eleventh Annual Gasification and Gas Stream Cleanup Systems Review Meeting, DOE/METC-91/6123, Vol. 1, p. 45.

Cook, C.R., et al., 1992, Integrated Operation of a Pressurized Fixed-Bed Gasifier and Hot Gas Desulfurization System, Proceedings of the Twelfth Annual Gasification and Gas Stream Cleanup Systems Review Meeting, DOE/METC-92/6128, Vol. 1, p. 84.

Flytzani-Stephanopoulos, M., et al., 1986, Detailed Studies of Novel Regenerable Sorbents for High Temperature Coal-Gas Desulfurization, Proceedings of the Sixth Annual Meeting on Contaminant Control in Coal-Derived Gas Streams, DOE/METC-86/6042, p. 257.

Focht, G.D., et al., 1988, High Temperature Desulfurization using Zinc Ferrite:

Reduction and Sulfidation Kinetics, Chem. Engng. Sci., 43, 3005.
Gangwal, S.K., et al., 1988, Multicycle Testing of Zinc Ferrite, DOE/MC/23126-2644.
Gangwal, S.K., and Harkins, S.M., 1988b, Bench-Scale Testing of Novel High-Temperature Desulfurization Sorbents, in Gas Stream Cleanup Papers from DOE/METC Sponsored Contractors Review Meetings in 1988, DOE/METC-89/6099, p. 108.
Gangwal, S.K., 1990, Enhanced Durability of Desulfurization Sorbents for Fluidized-Bed Applications, Proceedings of the Tenth Annual Gasification and Gas Stream Cleanup Systems Review Meeting, DOE/METC-90/6115, Vol. 1, p. 102.
Gangwal, S.K., and Gupta, R.P., 1991, Enhanced Durability of Desulfurization Sorbents for Fluidized-Bed Applications, Proceedings of the Eleventh Annual Gasification and Gas Stream Cleanup Systems Review Meeting, DOE/METC-91/6123, Vol. 2, p. 386.
Gangwal, S.K., et al., 1994, Fluidization Studies Using Phillips Z-SORB Sorbent, Proceedings of the Coal-Fired Power Systems 94 -- Advances in IGCC and PFBC Review Meeting, DOE/METC-94/1008, Vol. 2, p. 654.
Gangwal, S.K., and Gupta, R.P., 1994, Fluidized-Bed Sorbents, Proceedings of the Coal-Fired Power Systems 94 -- Advances in IGCC and PFBC Review Meeting, DOE/METC-94/1008, Vol. 2, p. 646.
Gansley, R.R., and Lee, G.T., 1990, Zinc Ferrite Desulfurization Studies in Fluid-Bed Reactors, Proceedings of the Tenth Annual Gasification and Gas Stream Cleanup Systems Review Meeting, DOE/METC-90/6115, Vol. 1, p. 83.
Georgakis, C., et al., 1979, A Changing Grain Size Model for Gas-Solid Reaction, Chem. Engng. Sci., 34, 1072.
Ghazanfari, R., et al., 1995, Pilot Scale Experience on IGCC Hot Gas Cleanup, Proceedings of the Advanced Coal-Fired Power Systems '95 Review Meeting, DOE/METC-95/1018, Vol. 1, p. 202.
Gibson, J.B., and Harrison, D.P., 1980, The Reaction Between Hydrogen Sulfide and Single Spherical Pellets of Zinc Oxide, IEC, Proc. Des. Devel. 19, 231.
Greenwood, G.J., et al., 1995, Pilot Plant Test of Z-SORB Sorbent, Proceedings of the Advanced Coal-Fired Power Systems '95 Review Meeting, DOE/METC-95/1018, Vol. 1, p. 241.
Grindley, T., and Steinfeld, G., 1983, Zinc Ferrite Hydrogen Sulfide Absorbent, Proceedings of the Third Annual Contaminant Control in Hot Coal Derived Gas Streams Contractors' Meeting, Morgantown, WV, DE84000216, p. 145.
Grindley, T., and Steinfeld, G., 1984, Testing of Zinc Ferrite Hydrogen Sulfide Absorbent in a Coal Gasifier Sidestream, Proceedings of the Fourth Annual Contractor's Meeting on Contaminant Control in Hot Coal-Derived Gas Streams, DOE/METC-85/3, p. 314.
Grindley, T., 1989, Study of Fluidized-Bed Desulfurization With Zinc Ferrite, Proceedings of the Ninth Annual Gasification and Gas Stream Cleanup Systems Review Meeting, DOE/METC-89/6107, Vol. 2, p. 548.
Grindley, T., 1990, Effect of Chlorine on Hot Gas Desulfurization Sorbents, Proceedings of the Tenth Annual Gasification and Gas Stream Cleanup Systems Review Meeting, DOE/METC-90/6115, Vol. 1, p. 215.

Kirkpatrick, M., and Pike, R., 1994, Linking Database Management Systems and Chemical Engineering Application Programs - Prediction of Chemical Equilibrium, AIChE Symposium Series, Vol. 90, No. 298, p. 174.

Krishnan, G.P., et al., 1994, Development of Disposable Sorbents for Chloride Removal from High-Temperature Coal-Derived Gases, Proceedings of the Coal-Fired Power Systems 94 -- Advances in IGCC and PFBC Review Meeting, DOE/METC-94/1008, Vol. 1, p. 478.

Leininger, T.F., et al., 1991, Integration and Testing of Hot Desulfurization and Entrained-Flow Gasification for Power Generating Systems, Proceedings of the Eleventh Annual Gasification and Gas Stream Cleanup Systems Review Meeting, DOE/METC-91/6123, Vol. 1, p. 56.

Leininger, T.F., et al., 1992, Integration and Testing of Hot Desulfurization and Entrained Flow Gasification for Power Generating Systems, Proceedings of the Twelfth Annual Gasification and Gas Stream Cleanup Systems Review Meeting, DOE/METC-92/6128, Vol. 1, p. 95.

Phillipson, J.J., 1970, Desulfurization, in Catalyst Handbook, Wolfe Scientific Books, London, p. 46.

Ranade, P.V., and Harrison, D.P., 1979, The Grain Model Applied to Solids With Varying Structural Properties, Chem. Engng.. Sci., 34, 427.

Ranade, P.V., and Harrison, D.P., 1981, The Variable Property Grain Model Applied to the Zinc Oxide-Hydrogen Sulfide Reaction, Chem. Engng. Sci., 36, 1079.

Salo, K., et al., 1994, Enviropower Hot Gas Desulfurization Pilot, Proceedings of the Coal-Fired Power Systems 94 -- Advances in IGCC and PFBC Review Meeting, DOE/METC-94/1008, Vol. 1., p. 236.

Schmidt, D.K., et al., 1987, KRW Process Development Coal Gasification/Hot Gas Cleanup, Proceedings of the Seventh Annual Gasification and Gas Stream Cleanup Systems Review Meeting, DOE/METC-87/6079, Vol. 2, p. 668.

Schmidt, D.K., et al., 1988, Fluidized-Bed Gasification With Hot Gas Cleanup, Proceedings of the Eighth Annual Gasification and Gas Stream Cleanup Systems Review Meeting, DOE/METC-88/6092, Vol. 1, p. 32.

Smith, D.P. et al., 1987, Design Studies for Hot Gas Cleanup System in Load Following Mode, Proceedings of the Seventh Annual Gasification and Gas Stream Cleanup Systems Review Meeting, DOE/METC-87/6079, Vol. 2., p. 678.

U.S. Department of Energy, 1995, Clean Coal Technology Demonstration Program - Program Update 1994, DOE/FE-0330.

Westmoreland, P.R., and Harrison, D.P., 1976, Evaluation of Candidate Solids for High-Temperature Desulfurization of Low-Btu Gases, Env. Sci. Tech., 10, 659.

Woods, M.C., et al., 1990, Reaction Between H_2S and Zinc Oxide-Titanium Oxide Sorbents. 1. Single-Pellet Kinetic Studies, IEC Research, 29, 1160.

Performance of Manganese-based Sorbents in High-Temperature Coal Gas Desulfurization

W.J.W. Bakker, J.C.P. van Rossen, J.M. Mugge, F. Kapteijn, and J.A. Moulijn

Delft University of Technology, Department of Chemical Engineering,
Julianalaan 136, 2624 RT Delft

Abstract. In this chapter the structure and properties of (supported) manganese oxides and manganese aluminate is discussed. The thermodynamics of sulfidation, the regeneration with different reactants and the influence of other components on the (thermodynamics of) sulfidation of manganese based sorbents are given.

The applicability of a manganese based sorbent for high temperature desulfurization of coal gas has been studied. It appears that a "duo" sorbent consisting of bulk $MnAl_2O_4$ and disperse MnO performs best in the desulfurization of "dry" coal gas. The acceptor capacity is high, up to 20 wt.% sulfur and the sample appears to be stable during 110 sulfidation and regeneration cycles. The sample can not only be regenerated with steam under the formation of H_2S but also with SO_2 under the formation of elemental sulfur. The sorbent can be used in the temperature range between 673 to 1273 K. The optimal performance temperature is around 1123 K.

Additionally, a short comparison is given of some process configurations for high temperature desulfurization. The application of a rotating monolith reactor for continuous high temperature desulfurization is feasible.

1 Structure and Properties

1.1 Thermodynamic Stability of Manganese oxides, Aluminates and Sulfides

Many different types of manganese oxides exist. The majority of literature on manganese oxides applies to oxides with manganese in the oxidation states 2, 3, 4, and 7. Most manganese oxides can be formed out of another manganese oxide, depending on the temperature and atmosphere.

In Fig. 1.1 the stability of manganese oxides as function of temperature is shown. MnO_2 is the most stable form until up to about 770 K. Above this temperature it decomposes towards Mn_2O_3 which is stable up to about 1250 K. Hereafter it is further reduced towards Mn_3O_4. Above 1900 K MnO is the most stable phase. Further reduction to metallic manganese is neither possible with hydrogen nor with carbon monoxide as reducing agents. At room temperature MnO is oxidized in air to an

NATO ASI Series, Vol. G 42
Desulfurization of Hot Coal Gas
Edited by A.T. Atimtay and D. P. Harrison

extent depending on the particle size. Usually only the surface of MnO particles will be oxidized.

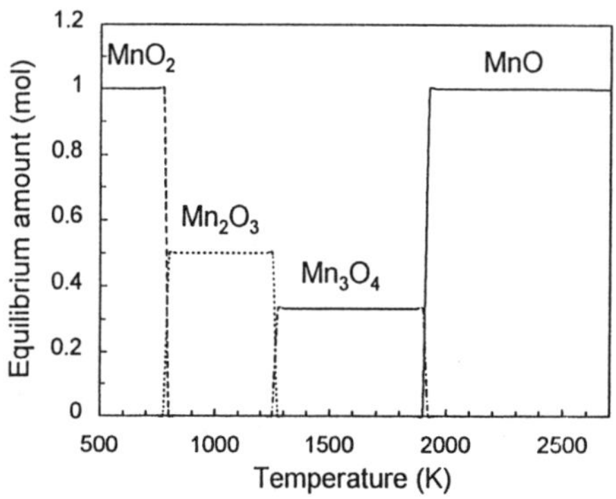

Fig 1.1 Stability diagram for the Mn-O system. Equilibrium amounts for reaction of 1 mol O_2 and 1 mol Mn as a function of the temperature. Calculated with "Chemsage 3.0" and "Microterm" database.

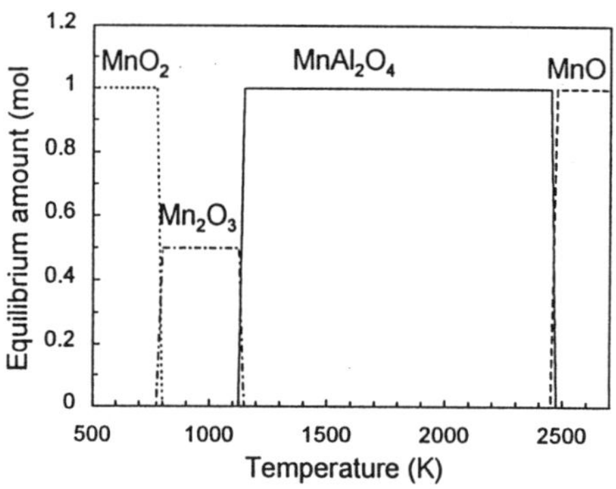

Fig 1.2 Stability diagram for the Mn-O-Al system. Equilibrium amounts for reaction of 1 mol O_2, 1 mol Mn and 1 mol Al_2O_3 as a function of the temperature. Calculated with "Chemsage 3.0" and "Microterm" database.

From Fig. 1.2 it arises that if Al_2O_3 is present $MnAl_2O_4$ (manganese aluminate) is the most stable phase between 1100 and 2450 K. Mn_3O_4 is not stable at any temperature.

The stability diagram for manganese sulfide is given in Fig. 1.3. Up to about 625 K MnS_2 is the most stable phase. At higher temperatures manganese is reduced from Mn^{4+} to Mn^{2+} under the formation of sulfur and MnS.

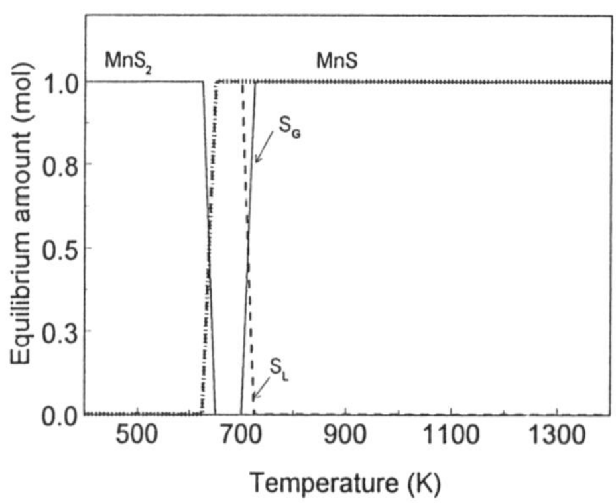

Fig. 1.3 Stability diagram for the Mn-S system. Equilibrium amounts for reaction of 1 mol S_2, and 1 mol Mn as a function of the temperature. Calculated with "Chemsage 3.0" and "Microtherm" Database.

1.2 Lattice Structures and Colours [1-3]

The unit cell of a spinel $MnAl_2O_4$ contains 32 oxygen ions in cubic closest packing and corresponds to the formula $Mn_8Al_{16}O_{32}$. It is depicted in Fig. 1.5 This unit cell contains 32 octahedral holes and 64 tetrahedral holes, in which 8 Mn^{2+} ions and 16 Al^{3+} ions are arranged. 8 Out of the 64 tetrahedral holes are occupied by Mn^{2+} ions and 16 out of the 32 octahedral holes are occupied by Al^{3+} ions, which is normally the case for spinel structures. The number of unoccupied sites is large and therefore metal ions are relatively mobile in this lattice.

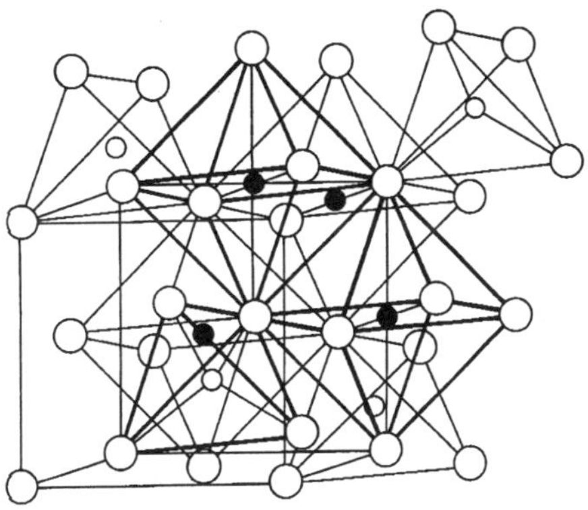

Fig 1.4 The lattice structure of spinels. The large circles represent oxygen ions. Small circles represent octahedral cations (black) and tetrahedral cations (white), respectively. $MnAl_2O_4$ has a spinel structure.

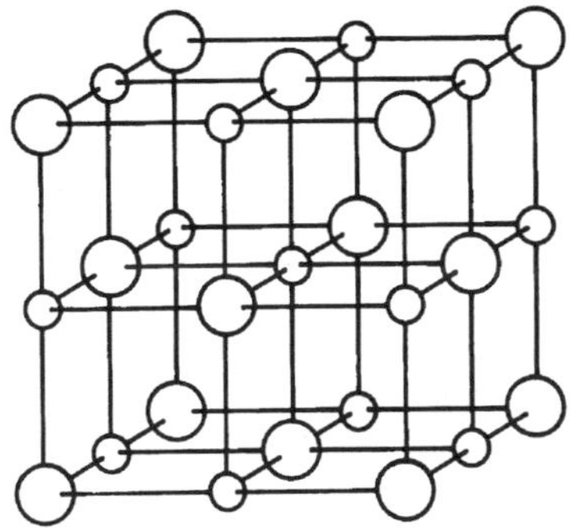

Fig 1.5 The rock salt crystal structure. Small circles represent metal ions, large circles oxygen ions. MnO and MnS have this structure.

The position of the Mn^{2+} ion in the aluminate spinel is different from that in its simple oxide. In MnO all the M^{2+} ions occupy octahedral holes. MnO and MnS have a rock salt structure and are shown in Fig. 1.5.

γ-Al_2O_3 is an important alumina support, with a high surface area. This material is relatively stable up to 1073 K. The structure of γ-Al_2O_3 is very similar to the spinel structure discussed above. Compared to a spinel however, one-ninth of the positions normally occupied by cations are unoccupied. These cation vacancies may be distributed in different ways among the tetrahedral and octahedral holes. An increased stability and mechanical strength can be obtained by addition of a small amount of divalent metal ions.

The color of manganese oxides is dark brown to black, manganese aluminate is beige and MnS green. Amorphous MnS is pink [3].

2 Reactions and Thermodynamics during Sulfiding of MnO and $MnAl_2O_4$ and Regeneration of MnS

In this chapter the thermodynamics of the reactions during sulfidation of a manganese based sorbent in a coal gas environment are given. Also sorbent regeneration with different reactants is discussed.

Equilibrium constants or specific concentrations for the reactions are calculated as function of temperature. In most cases also the heat of reaction is calculated. The thermodynamic data are not specified for each crystallographic structure separately and may, therefore, be not complete.

The regenerative desulfurization process is a gas-solid reaction where the gas is passing the solid acceptor material. For the calculations made, the solid material and the gas are supposed to be in a closed box. The equilibrium compositions of the reaction are calculated by minimizing the Gibbs Energy. Hereby, the program "Chemsage 3.0" with "Microterm" database is used.

When the solid phase is in excess one can assume that the solid "activity" is constant and the equilibrium constant can be expressed as the ratio of the gas phase concentrations (Eq. 1). The heat of reaction is calculated from the sum of the heat of formation (Eq. 2).

$$K_{eq} = [H_2O]/[H_2S] \quad (1)$$

$$\Delta H_{react} = \Sigma \nu_i \bullet \Delta H_{fi} \quad (2)$$

In previous work [4] mainly surface sites were used as acceptor sites during the sulfidation. In analogy to heterogeneous catalysis one can argue that surface site concentrations are not constant and should be included in Eq. 1.

$$K^i_{eq} = \frac{[H_2O]}{[H_2S]} \cdot \frac{[\text{sulfided surface sites}]}{[\text{free surface sites}]} \quad (1a)$$

2.1 Sulfiding of MnO and $MnAl_2O_4$

During sulfiding of metal oxides, oxygen is exchanged with sulfur, according to Eqs. 3 and 4. The reaction for the sulfiding of MnO with H_2S is:

$$MnO + H_2S \Leftrightarrow MnS + H_2O \quad (3)$$

For $MnAl_2O_4$ the reaction is given by:

$$MnAl_2O_4 + H_2S \Leftrightarrow MnS + \gamma\text{-}Al_2O_3 + H_2O \quad (4)$$

The equilibrium constants of the reaction of $MnAl_2O_4$, MnO and some other manganese oxides with H_2S are presented in Fig. 2.1. Sulfiding is thermodynamically possible over the whole temperature range from 400 to 1400 K. For $MnAl_2O_4$ it is however less favorable than for MnO. At 600 K the equilibrium constant for sulfiding of MnO is 30000 while for $MnAl_2O_4$ it is only 25. For the other manganese oxides the difference is even larger. The heats of reaction of sulfiding, which are given in Table 2.1, are about constant with increasing temperature and amounts to -10 kJ/mol for $MnAl_2O_4$ and -50 for kJ/mol for MnO. The reaction heat for the other components is high and amounts to about 150 kJ/mol.

In coal gas part of the sulfur is present as COS; typically 150 ppm. Both MnO and $MnAl_2O_4$ effectively remove COS from the coal gas, although MnO to a lower level than $MnAl_2O_4$; see Table 2.1. The reaction for the sulfiding of MnO with COS is:

$$MnO + COS \Leftrightarrow MnS + CO_2 \quad (5)$$

For $MnAl_2O_4$ the reaction is given by:

$$MnAl_2O_4 + COS \Leftrightarrow MnS + \gamma\text{-}Al_2O_3 + CO_2 \quad (6)$$

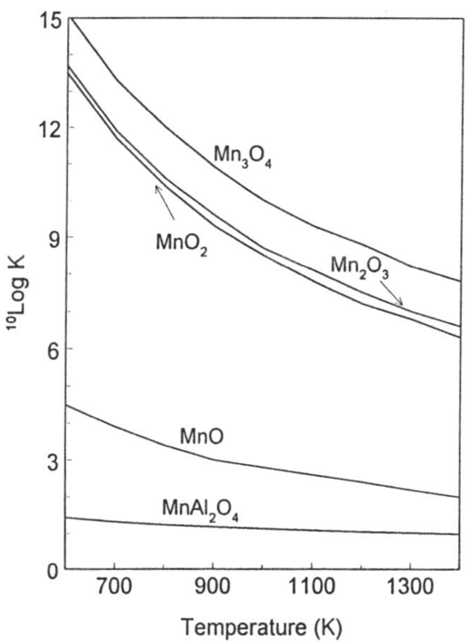

Fig 2.1 Equilibrium constants of different manganese oxides and $MnAl_2O_4$ for reaction with H_2S (Eqs. 3, 4, 22, 23, 24).

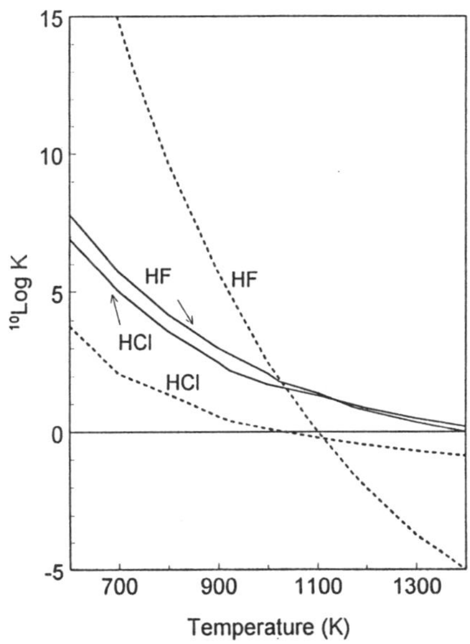

Fig. 2.2 Equilibrium constants for reaction of MnO (solid lines) and $MnAl_2O_4$ (dotted lines) with HF and HCl (Eqs. 24-27).

2.2 Influence of HCl, HF and other Impurities on the Sulfidation Performance

The influence of other impurities than H_2S and COS on the performance of the acceptor is a point of major concern. If these impurities react with the acceptor and this reaction is irreversible the acceptor capacity may be reduced during prolonged operation. Also the formation of coke may decrease the acceptor performance.

Thermodynamic calculations showed that HCl and HF are also absorbed by the manganese based sorbent, see Fig. 2.2, while NH_3 and HCN do not react with the system. The reaction equilibrium constant at high temperatures for HF and HCl is lower than for H_2S and COS. This means that H_2S and COS are preferably adsorbed. During sulfiding not the whole acceptor capacity is used by H_2S and COS, so HCl and HF will also be adsorbed on the acceptor. As the reaction is reversible HCl and HF will be removed during the regeneration with steam. Metal chlorides are relatively volatile and therefore some loss of manganese may occur in case the sorbent is applied at very high temperatures. (T_{boil} $MnCl_2$ = 1503 K). For metals other than manganese the boiling point of their chlorides is significantly lower which makes them unusable for application at high temperatures in case the coal gas contains a considerable amount of HCl.

Manganese is not a catalyst for coke formation contrary to *e.g.* iron. Further, when a minor amount of coke is formed, it is easily removed during regeneration with steam under the formation of H_2 and CO.

2.3 Other Reactions during Sulfidation

Coal gas contains various components and many reactions occur in the coal gas [6]. Some of these reactions occur in the gas phase but are accelerated by the presence of a sorbent. The most important reactions for the manganese based sorbent are summarized below.

$$H_2S \Leftrightarrow 0.5\ S_2 + H_2 \qquad (7)$$
$$H_2S + 2\ H_2O \Leftrightarrow 3\ H_2 + SO_2 \qquad (8)$$

Water Gas Shift Reaction

$$CO + H_2O \Leftrightarrow H_2 + CO_2 \qquad (9)$$

"Sour-gas Shift Ractions"

$$CO + H_2S \Leftrightarrow COS + H_2 \qquad (10)$$
$$COS + H_2O \Leftrightarrow CO_2 + H_2S \qquad (11)$$

2.4 Regeneration

Sorbent regeneration is the exchange of a sulfur ion for an oxygen ion. Generally oxygen is applied for regeneration. Without dilution to a few % oxygen the regeneration process is very difficult to control because of its highly exothermal nature. Moreover the regeneration product, diluted SO_2, is not very attractive. Alternative regeneration methods are the use of H_2O, SO_2 or CO_2 to remove sulfur from the acceptor bed. Even exchange of a sulfur ion for another ion than oxygen is a possibility although that is not (yet) a realistic one. *E.g.* regeneration with OF_2 leads to the direct production of commercially attractive SF_6. A number of possible regeneration reactions are summarized below and in Table 2.1. Reaction equilibrium constants as function of temperature for MnS and MnS on γ-Al_2O_3 are depicted in Figs. 2.3 and 2.4 and an overview of heats of reaction is given in Table 2.1.

Regeneration of Manganese Sulfide

$MnS + 1.5\ O_2$	$\Leftrightarrow$	$MnO + SO_2$	(12)
$MnS + H_2O$	$\Leftrightarrow$	$MnO + H_2S$	(13)
$MnS + 0.5\ SO_2$	$\Leftrightarrow$	$MnO + 0.75\ S_2$	(14)
$MnS + CO_2$	$\Leftrightarrow$	$MnO + CO + 0.5\ S_2$	(15)
$MnS + 3\ CO_2$	$\Leftrightarrow$	$MnO + SO_2 + 3\ CO$	(16)

Regeneration of Manganese Sulfide on Alumina

$MnS + \gamma\text{-}Al_2O_3 + O_2$	$\Leftrightarrow$	$MnAl_2O_4 + SO_2$	(17)
$MnS + \gamma\text{-}Al_2O_3 + H_2O$	$\Leftrightarrow$	$MnAl_2O_4 + H_2S$	(18)
$MnS + \gamma\text{-}Al_2O_3 + 0.5\ SO_2$	$\Leftrightarrow$	$MnAl_2O_4 + 0.75\ S_2$	(19)
$MnS + \gamma\text{-}Al_2O_3 + CO_2$	$\Leftrightarrow$	$MnAl_2O_4 + CO + 0.5\ S_2$	(20)
$MnS + \gamma\text{-}Al_2O_3 + 3\ CO_2$	$\Leftrightarrow$	$MnAl_2O_4 + SO_2 + 3\ CO$	(21)

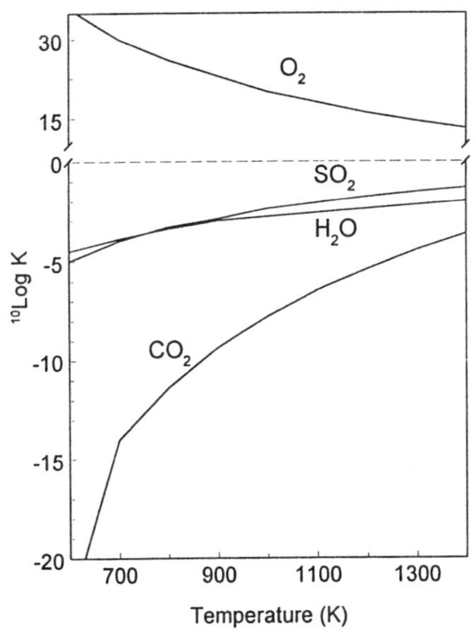

Fig. 2.3 Thermodynamic equilibrium constants for reaction of MnS with different regenerants (Eqs. 12-16).

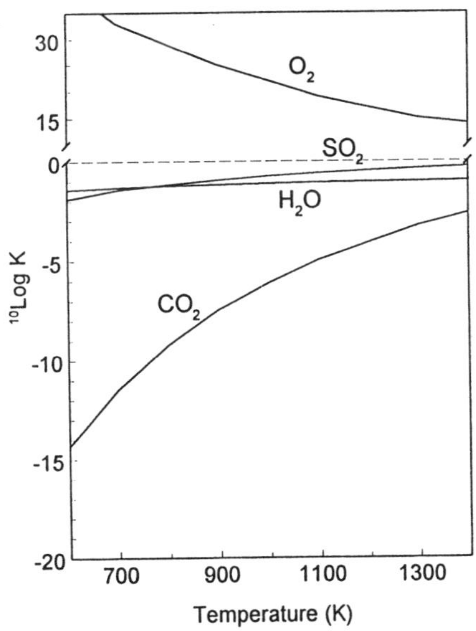

Fig. 2.4 Thermodynamic equilibrium constants for reaction of MnS on Al_2O_3 with different regenerants (Eq. 17-21).

Steam-regeneration is simply the reverse of the original sulfiding and H_2S is formed. The regeneration product which is formed with CO_2 are S_2 and SO_2, while regeneration with SO_2 produces mainly elemental sulfur. Of course the latter is in most cases desirable.

Table 2.1 Equilibrium constants and heats of reaction for several reactions with manganese based compounds. Most important reactions are written in italics.

Nr.	Reaction			$^{10}\log K_{eq}$		Heat of reaction	kJ/mol
			Temperature (K)	600	1400	600	1400
22	$MnO_2 + 2\ H_2S$	⇔	$MnS_2 + 2\ H_2O$	13.5	6.3	-144	-143
23	$Mn_2O_3 + 3\ H_2S$	⇔	$MnS_2 + MnS + 3\ H_2O$	13.7	6.6	-145	-148
14	$Mn_3O_4 + 4\ H_2S$	⇔	$MnS_2 + 2\ MnS + 4\ H_2O$	15.1	7.8	-149	-152.7
3	*$MnO + H_2S$*	⇔	*$MnS + H_2O$*	*4.5*	*2*	*-50*	*-51*
4	*$MnAl_2O_4 + H_2S$*	⇔	*$MnS + Al_2O_3$*	*1.4*	*1*	*-10*	*-10*
5	$MnO + COS$	⇔	$MnS + CO_2$	7.1	3.1	-81	-82
6	$MnAl_2O_4 + COS$	⇔	$MnS + Al_2O_3 + CO_2$	4.0	2.0	-40	-39
7	$MnS + 1.5\ O_2$	⇔	$SO_2 + MnO$	36	13.2	-469	-469
8	*$MnS + H_2O$*	⇔	*$MnO + H_2S$*	*-4.5*	*-2*	*49*	*52*
9	*$MnS + 0.5\ SO_2$*	⇔	*$MnO + 0.75\ S_2$*	*-5*	*-1.3*	*73*	*73*
10	$MnS + CO_2$	⇔	$MnO + 0.5\ S_2 + CO$	-11.2	-2.5	176	174
9	$MnS + 3\ CO_2$	⇔	$MnO + SO_2 + 3\ CO$	-34	-4.8	381	374
11	$MnS + Al_2O_3 + 1.5\ O_2$	⇔	$MnAl_2O_4 + SO_2$	40	14	-509	-511
13	*$MnS + Al_2O_3 + H_2O$*	⇔	*$MnAl_2O_4 + H_2S$*	*-1.44*	*-0.99*	*9.2*	*9.5*
14	*$MnS + Al_2O_3 + 0.5SO_2$*	⇔	*$MnAl_2O_4 + 0.75\ S_2$*	*-1.9*	*-0.29*	*33.4*	*31.1*
15	$MnS + Al_2O_3 + CO_2$	⇔	$MnAl_2O_4 + 0.5\ S_2 + CO$	-8.1	-1.5	135	131
16	$MnS + Al_2O_3 + 3\ CO_2$	⇔	$MnAl_2O_4 + SO_2 + 3\ CO$	-20.6	-3.8	341	331
20	$MnO + 2\ HCl$	⇔	$MnCl_2 + H_2O$	6.9	0.2	-152	-103
21	$MnO + 2\ HF$	⇔	$MnF_2 + H_2O$	7.8	0.001	-164	-119
22	$MnAl_2O_4 + HCl$	⇔	$MnCl_2 + Al_2O_3$	3.8	-0.86	-115	-60
23	$MnAl_2O_4 + 8\ HF$	⇔	$MnF_2 + 2\ AlF_3 + 4\ H_2O$	21.7	-5	-553	-487

As follows from Figs. 2.3 and 2.4, the regeneration of MnS with H_2O is thermodynamically difficult. The same holds for regeneration with SO_2 and CO_2. The regeneration of MnS on alumina is easier (with all regenerants) as the thermodynamically more stable $MnAl_2O_4$ is formed instead of MnO. SO_2 is better than H_2O above 800 K.

Application of SO_2 for the regeneration is for most metal oxides encountered with undesirable sulfate formation. For MnS on alumina, however, direct production with SO_2 is possible without sulfate formation above 700 K; see Fig. 2.5.

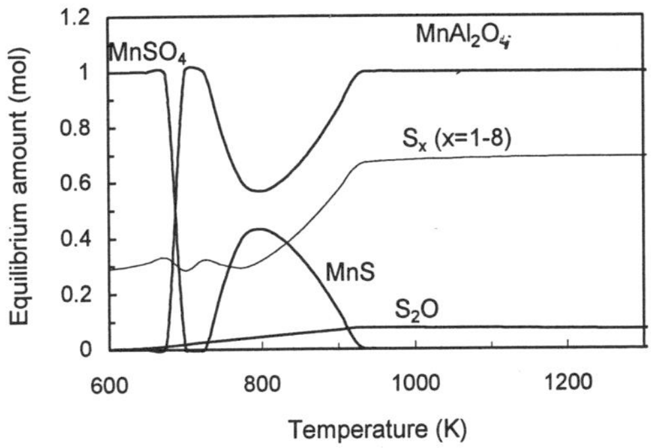

Fig. 2.5 Thermodynamic calculations for regeneration of 1 mol MnS on Al_2O_3 with 10 mol SO_2.

3 Experimental

For the desulfurization of hot dry coal gas a manganese based acceptor has been developed. Both particles and monolith shaped sorbents are being used. Particles were tested for application in a fixed bed and fluidized bed reactor and the monolith shaped sorbent in a fixed bed or a rotating monolith reactor.

3.1 Acceptor Preparation

The acceptor material was prepared by wet impregnation of the carrier material. Three types were used: pure γ-Al_2O_3 monoliths (22 cells/cm^2), cordierite monoliths (22 cells/cm^2) with about 25 wt% γ-Al_2O_3 washcoat layer, or particles in different sizes ranging from 50 μm, for fluidized bed application, to 0.7 mm for fixed bed application. The material was contacted with a 1-2 mol/l aqueous solution of $MnAc_2.4H_2O$, which was continuously circulated through the monolith channels or around the particles during impregnation. The impregnation time varied between 4-8 hours. After impregnation the excess solution was removed from the channels with pressurised air (monoliths) or by centrifugation (particles). The samples were dried for 6 minutes in a microwave oven (115 W) and 12 hours in a conventional oven at 353 K. Calcination was performed in static air at 873-973 K for at least 6 hours. The γ-Al_2O_3 monoliths/ particles and the washcoated monoliths had a 6 and 1.5 wt% manganese loading respectively, after a single impregnation with a 1.5 mol/l solution. To obtain higher manganese loadings the procedure was repeated.

A series of acceptor material was prepared with manganese loadings ranging from 1.4 to 24 wt% for washcoated monoliths and 6 to 37 wt% for γ-Al_2O_3 monoliths and particles. The colour of the samples after impregnation and calcination ranged from light brown to very dark brown/black going from low to high loadings. XRD did reveal crystalline Mn_3O_4 (PDF 16-154, international centre for diffraction data) for samples with high loadings, but not for samples with low loading (1.4 wt% washcoated support,

6 wt% pure γ-Al_2O_3 support). The surface area after preparation was depending on the manganese loading and the type of monolith and decreased about one third for low loading samples to almost two third for samples with a very high loading (>35 wt% Mn).

Manganese Content in Pure γ-Al_2O_3 and Washcoated Monoliths
The samples used are labeled by their manganese content in wt% Mn. Both pure γ-Al_2O_3 alumina support and cordierite monolith samples with a 25 wt% γ- Al_2O_3 washcoat layer are used. In the latter case, the amount of manganese relative to the alumina present is much higher. Normalized to the relative amount of washcoat, the manganese loading of washcoated cordierite samples is 33.1, 44.3 and 50.7 wt% Mn on alumina for 12.2, 19.4 and 24.9 wt% Mn on washcoated cordierite respectively. It should be noted that at 31.8 wt% Mn on alumina, the stoichiometric Mn/Al ratio for the spinel $MnAl_2O_4$ is reached. Above this percentage the sample will also contain a MnO phase.

3.2 Sulfiding and Regeneration Experiments

Absorption and regeneration experiments were carried out in a temperature programmed sulfiding apparatus (own design, see Fig. 3.1 in [7]), equipped with a UV-Vis spectro photometer, a thermal conductivity detector (TCD) and a mass spectrometer (MS). Water and H_2S were removed, after passing the UV spectrophotometer, with a tubular membrane dryer and zinc oxide bed at 723 K. In case the MS was used the zinc oxide bed and the TCD were bypassed. Gases were controlled with thermal mass flow controllers and led to the quartz reactor tube. For regeneration with steam and a sulfiding gas with water, steam was added to the feed by passing argon through a saturator. For some experiments SO_2 was used during regeneration which was dosed via a thermal mass flow controller. Both temperature programmed and isothermal experiments were performed. Isothermal experiments involved a number of sulfiding/regeneration cycles. A cycle consisted of sulfiding, flushing, regeneration, and flushing of a sample. During sulfiding, the feed contained 0.2-3 % H_2S, 15-50 % hydrogen, 0-50 % CO and 0-12 % H_2O, balanced with argon. For regeneration with steam the feed contained 15-60 % H_2O, 0-50 % hydrogen, balanced with argon. During regeneration with SO_2 the feed contained 50 % SO_2 balanced with Ar. Argon was used for flushing. The total flow rate during sulfiding and regeneration varied between 1.35 and 4 mmol/min.

Tests were carried out between 300 and 1273 K using small (height 2 cm, diameter 1 cm) monoliths (washcoated and pure alumina) and particles (50 μm - 0.7 mm) samples with different manganese loadings. In some cases the monoliths were crushed to fit in a reactor tube with an inner diameter of 0.6 cm. Most isothermal experiments were performed at 1123 K. The exact temperatures and feed compositions can be found in the legends of the Figures or in the text of the "Results" chapter. The experiments focused on:

- the optimal metal loading,
- optimal operating temperature for absorption and regeneration,

- the desulfurization depth of the acceptor
- regeneration with H_2O and SO_2
- the influence of feed composition on the sulfidation behaviour. Hereby special attention was given to H_2O.
- the stability of the sample

3.3 Characterization

The acceptors were characterised with a number of techniques. XRD and BET surface area analyses of some samples were performed after impregnation and calcination and after sulfiding or regeneration. For XRD of the washcoated monolith, the washcoat layer was mechanically removed from the washcoated monoliths and carefully ground. The γ-Al_2O_3 based samples were ground as such. XRD analyses was performed using a Siemens d.v. 100 diffractometer with diffracted beam monochromator Cu-K-radiation. BET surface analyses, by adsorption of N_2 at 77 K, were performed with a Micromeretics Digisorp 2600 and a Quantachrome Autosorp 6B.

Also temperature programmed sulfidation (TPS) and regeneration (TPR) with steam were performed with the above described experimental set-up. Hereby the sulfiding or regeneration consists of an isothermal stage, followed by the actual temperature programmed sulfiding or regeneration stage and an isothermal stage at the end of the temperature program.

To determine the metal loading of the samples the manganese was dissolved with a concentrated HCl solution. ICP-AES or flame atomic absorption spectrometry was performed to determine the metal loading of the samples.

4 Results

4.1 Characterization

A typical TPS and a TPR pattern of a sorbent after several isothermal sulfiding-regeneration cycles at 1123 K are presented in Figure 4.1 and 4.2, respectively. At the start of the experiment (isothermal adsorption at 300 K) some H_2S is adsorbed, from which most is desorbed at the beginning of the temperature program. The major uptake

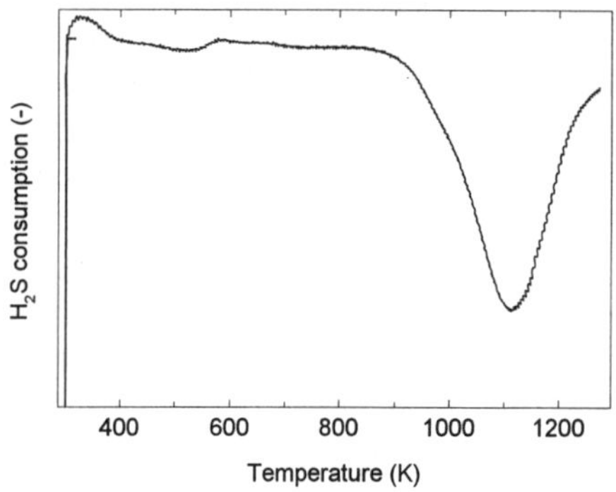

Fig 4.1 Temperature programmed sulfiding of $MnAl_2O_4$ after several cycles at 1123 K (washcoated monolith; 11 wt% Mn based on washcoat).

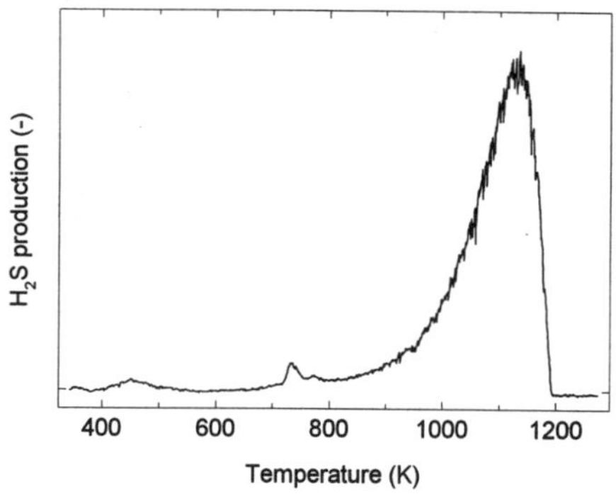

Fig 4.2 Temperature programmed regeneration of MnS on γ-Al_2O_3 after several cycles at 1123 K (17 wt% Mn on particles).

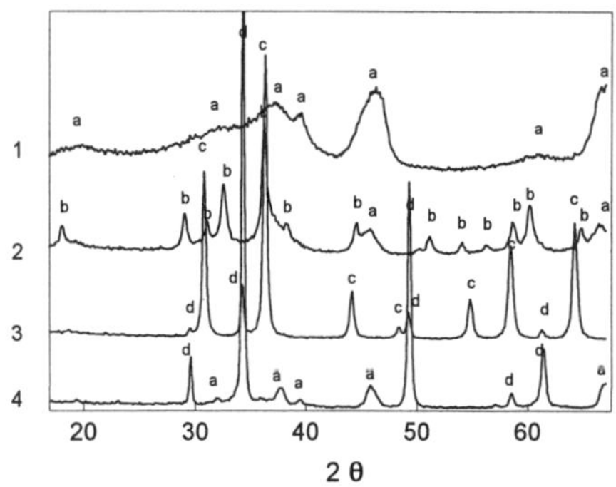

Fig. 4.3 XRD patterns for: 1) γ-Al_2O_3; 2) Mn_3O_4 on γ-Al_2O_3; 3) $MnAl_2O_4$; 4) MnS on γ-Al_2O_3. a = γ-Al_2O_3; b = Mn_3O_4; c = $MnAl_2O_4$; d = MnS. Sample contains 32 wt % Mn.
$T_{calcination}$ = 873 K (2), $T_{sulf.\ and\ reg.}$ = 1123 K (3 + 4).

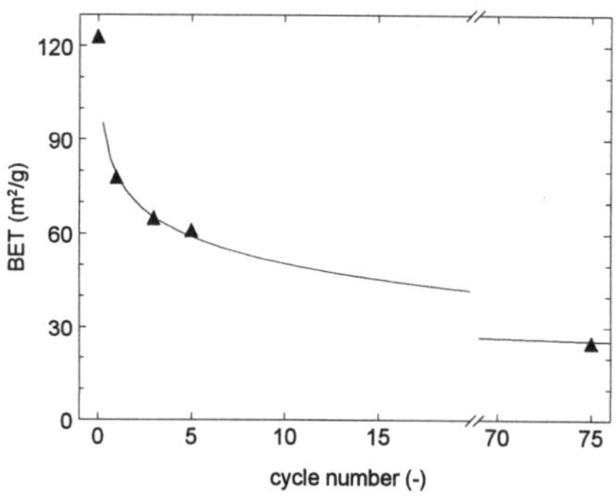

Fig 4.4 BET surface area as a function of the number of isothermal sulfiding and regeneration cycles. 21 wt% Mn on γ-Al_2O_3 acceptor.

is observed between 900-1150 K. Some absorption is observed between 400 and 550 K. During regeneration at 450 K and 730 K two minor production peaks can be observed. The major H_2S production occurs between 1020 and 1170 K. For details about TPS and TPR (with steam) one is referred to [7].

XRD patterns of an unloaded γ-Al_2O_3 monolith, a fresh impregnated and calcined sample, a regenerated and a sulfided acceptor are depicted in Figure 4.3. Pattern (1) shows the diffraction pattern of a disperse γ-Al_2O_3 phase (PDF 10-425). In pattern (2) the diffraction pattern of Mn_3O_4 (PDF 16-154) is superimposed on the pattern of γ-Al_2O_3. $MnAl_2O_4$ (PDF 29-880) is the main species in pattern (3) but also some MnS (PDF 6-518) is present. In pattern (4) MnS and γ-Al_2O_3 are present. The particle size of the manganese species in pattern 2-4 was estimated with the Scherrer formula [8] to be in between 10 and 20 nm.

Samples with lower loadings showed the same manganese phases as described above, but the peaks broaden with decreasing manganese loading. For washcoated samples and γ-Al_2O_3 samples with a manganese loading of, respectively, $\leq$ 1.5 wt% and $\leq$ 6 wt% XRD did not reveal any crystalline manganese oxide or sulfide phases. In samples with a Mn loading above the stoichiometric amount of 32 wt% Mn, present in $MnAl_2O_4$, the formation of a MnO like phase occurs. At manganese loadings $\geq$ 35 wt% Mn, MnO could be detected by XRD.

The observed sample colours were green after sulfidation and beige to grey after regeneration. For some washcoated high loading samples green or pink spots after regeneration were observed.

In Figure 4.4., it is shown that the surface area decreases during cyclic operation at 1123 K from 122 towards 25 m^2/g after 75 cycles. Surface area decease occurs mainly during the first cycles.

4.2 Isothermal Cycles

In Figure 4.5 typical curves for isothermal sulfiding (dry; 0% H_2O) and regeneration with steam at 1123 K are presented. During sulfiding the sample initially adsorbs all H_2S and no H_2S is detected in the exit ($[H_2S] < 20$ ppm). After a certain time breakthrough occurs and the signal stabilises at the feed level. The regeneration starts with a very sharp and high H_2S production peak, whereafter the production tails off until all sulfur is removed. Some elemental sulfur is observed behind the reactor after the first sulfidation of the sorbent.

The S/Mn ratio, representing the total sulfur uptake capacity, for isothermal absorption from 673 K to 1223 K and back to 673 K is depicted in Figure 4.6. Hereby, the acceptor capacity is based on the total amount of H_2S adsorbed and not on a certain breakthrough criterion. The acceptor capacity increases strongly from 873 to 1123 K, and stabilises. At the maximum of the curve the S/Mn ratio is around one. (If a breakthrough criterion is used the breakthrough capacity is not only determined by the sorbent properties but also by integral bed behavior).

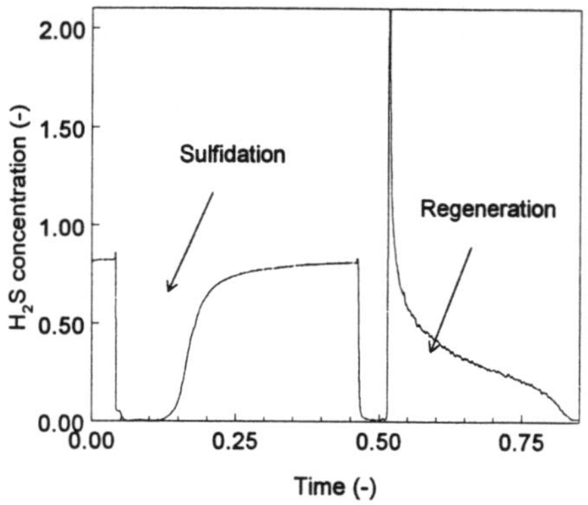

Fig. 4.5 Typical curves for isothermal sulfidation and regeneration at 1123 K in a fixed bed reactor. Washcoated cordierite sample 12. wt% Mn. (Feed sulfidation 3.3% H_2S, 40 % H_2, balance gas Ar. regeneration 15% H_2O). Signal at t = 0 is the feed signal.

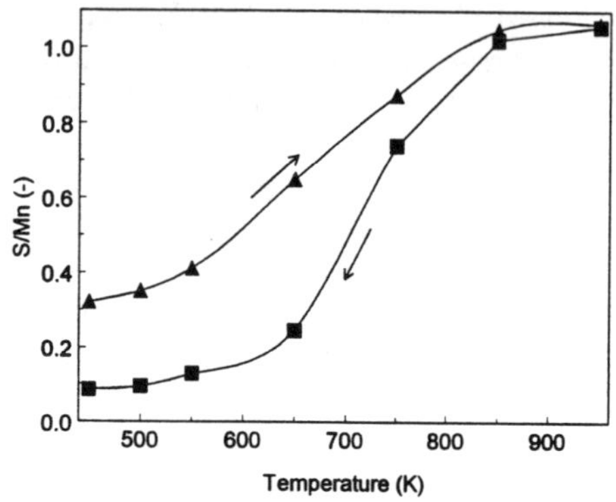

Fig 4.6 Molar S/Mn-ratio for 12 wt% Mn on alumina sample. Isothermal cycles from 723 K to 1223 K and back (■) to 723 K. Feed 3% H_2S, 50% H_2, Reg. 30%, H_2O; balance Ar.

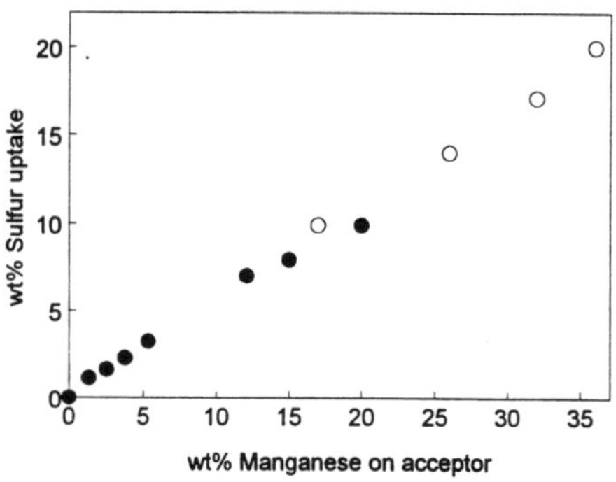

Fig 4.7. Capacity in wt% sulfur uptake for different manganese oxide samples. • = washcoated cordierite. o = γ-alumina samples. Sulf. 1-3% H_2S, 50% H_2, balance Ar. Regeneration 15 % H_2O in H_2/Ar (70/30) mixture. 15, 17 and 37 wt% Mn 30 % H_2O in argon.

The acceptor capacity at 1123 K as function of the manganese loading is given in Figure 4.7 The capacity has been measured after at least four absorption-regeneration cycles at 1123 K. The acceptor capacity increases almost linearly with the manganese loading and amounts to 1.1 to 20 wt% sulfur for the samples with manganese loadings ranging from 1.4 to 37 wt% manganese. The manganese utilisation initially decreases, starting from S/Mn = 1.2 for the lowest loading towards somewhat below one for the highest loadings.

Simulated Coal Gas

Tests with simulated coal gas ("dry" *e.g.* Shell type; 35% CO, 17% H_2, 2.1 % H_2O and 0.64% H_2S in Ar) and a washcoated cordierite sample with 19 wt% Mn showed a effective sulfur removal to a level of below 40 ppm. No coke formation was observed.

4.3 Influence of Feed Composition and Mn Loading on the Sulfidation Behavior

H_2O

In Fig. 4.8 the effect of H_2O is illustrated with a low loading sample (16 wt% Mn; a) and a high loading sample (35 wt% Mn; b). a1 and b1 are the sulfidation curves for a feed without water and a2 and b2 the curves for a feed with water. The curves for "wet" sulfidation are very different from each other.

The 16 wt% sample displays a fast breakthrough to a relatively high H_2S concentration followed by a slow tailing towards the feed concentration. The K_{eq} as calculated from this breakthrough level and Eq. 1 is 17. The 35 wt% sample shows two breakthrough levels, one low H_2S level and a second level comparable to the 16 wt% sample. Calculated K_{eg} 's are respectively 10 and 220. In Figure 4.9 the calculated K_{eq} for some other samples are depicted. The samples up to 23 wt% have a K_{eq}-value varying from 5 (6 wt% Mn) to 17 (23 wt%) Mn.

Two breakthrough levels are observed for Mn loadings somewhat above 32 wt% Mn. For samples with very high loadings, > 40 wt%, only the low breakthrough level is observed. Hereby the K_{eq} values vary between 200 and 275.

Water gas shift (WGS), COS, H_2S and H_2

The combined influence of CO and hydrogen on the sulfiding is shown in Figure 4.10. An increasing CO/H_2 ratio results in a lower breakthrough level and longer breakthrough time. It was found that the presence of CO at 1123 K enables to shift away up to 30 % of the H_2O. Other effects of CO include the formation of COS which was detected with a mass spectrometer after breakthrough occurred. The

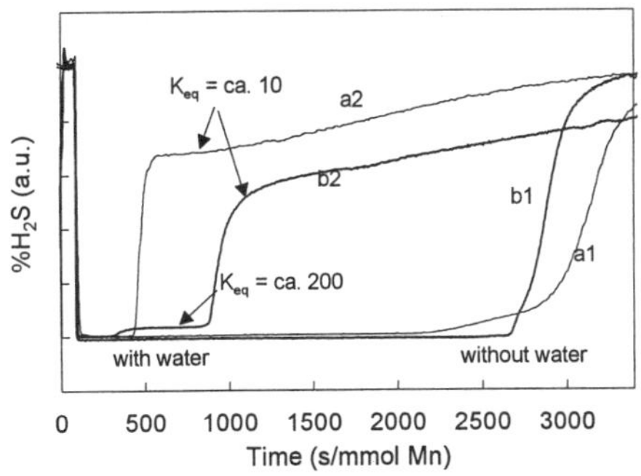

Fig 4.8 Sulfiding breakthrough profiles for two different Mn-loaded alumina samples. a = 16 wt% Mn, b = 35 wt% Mn, 1 and 2 are profiles without and with water, 10 and 6% H_2O, respectively.

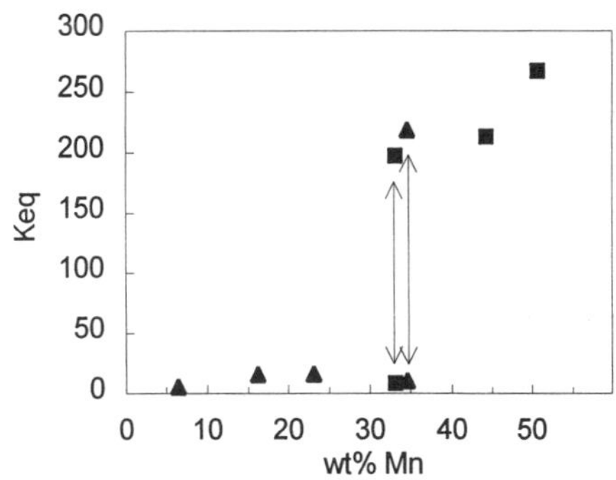

Fig 4.9 $[H_2O]/[H_2S]$ (K_{eq}; Eq. 1) as function of manganese content in alumina at 1123 K. ▲ = alumina samples, ■ = alumina-washcoated cordierite samples. Calculated from average values (feed 0-12 % H_2O). For washcoated samples the wt% Mn is based on the relative amount Mn.

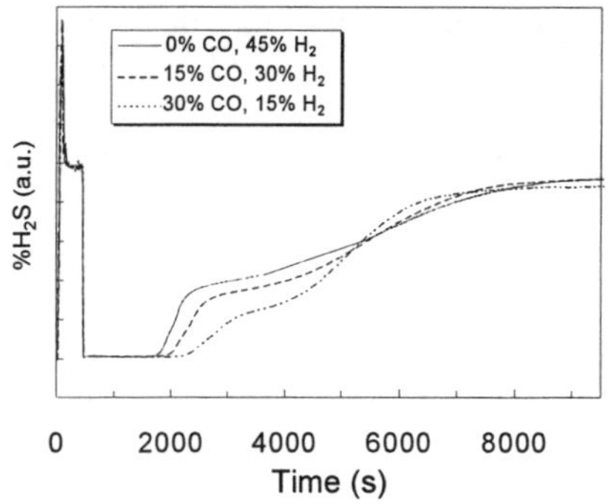

Fig 4.10. The influence of CO and H_2 concentration during wet sulfiding at 1123 K. Washcoated monolith, 12.2 wt% Mn)

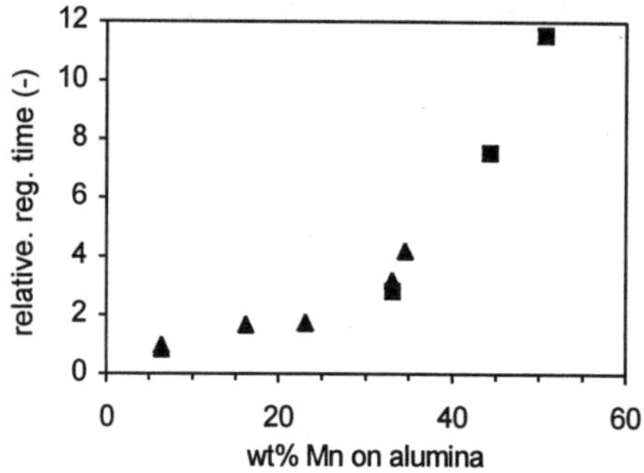

Fig. 4.11 Relative regeneration time as function of the manganese loading at 1123 K. ▲ particles and ■ washcoated monoliths (%Mn relative to washcoat).

influence of hydrogen other than via the water gas shift reaction is small although the presence of a minor amount, < 5 %, is favourable in the H_2S absorption process. Also the influence of the H_2S concentration is small. With increasing H_2S concentration, from 0.2 to 3% the capacity decreases a little (< 5%).

4.4 Regeneration

Most experiments described in this article were performed with H_2O as regeneration gas. During some experiments SO_2 was used as regeneration gas. Isothermal cycles at different temperatures were performed with both H_2S and SO_2. The results of the cycles regenerated with H_2O are depicted in Fig. 4.6 . The profile which was obtained with SO_2 was comparable to the profile in this Figure.

During the regeneration with steam the main product was H_2S. Above about 1000 K also some SO_2 and S was detected if no H_2 was present in the regeneration gas. The average H_2S concentration at 1123 K in the regeneration gas was about 10 % of the initial steam concentration. At low manganese loadings (6 wt %) this value was, however, higher, namely 20-30 % of the initial steam concentration. At high Mn loadings (> 32 wt%) it was lower. The latter is also illustrated in Figure 4.11 which shows that above 30 wt% manganese the regeneration time increases considerably which consequently lowers the H_2S concentration in the regeneration off gas.

During the regeneration with SO_2 elemental sulfur was formed as regeneration product but below 823 K also sulfate formation on the acceptor occurred. This resulted in SO_2 production during the next sulfidation. The mass balance indicated that the regeneration product at 1123 K was solely elemental sulfur. The regeneration with SO_2 at 1123 K was relatively fast. Only at very high manganese loading (35 wt%) it was difficult to regenerate the sample completely.

4.5 Deactivation

The acceptor capacity as function of the number of cycles is illustrated in Figure 4.12. During the first cycles the acceptor capacity decreased from 1.2 to 1. After this initial deactivation no further decline in acceptor capacity was observed during the following 100 cycles. Deactivation tests (20 cycles) with simulated (35.4% CO, 16.7% H_2, 2.1% H_2O and 0.64% H_2S in Ar) "Shell" coal gas for a sample with 19.4 wt% Mn on washcoated cordierite showed similar results. Moreover, during these tests sorbent effectively removes H_2S to levels below 40 ppmv and no coke formation was observed. The S/Mn ratio during this test was 0.75.

No mechanical damage was observed for all tested samples after prolonged operation. Even the washcoat layer on the washcoated samples was still intact. Generally, the mechanical strength of washcoated samples, both before and after isothermal cyles at 1123 K, was higher than for the γ- alumina monoliths.

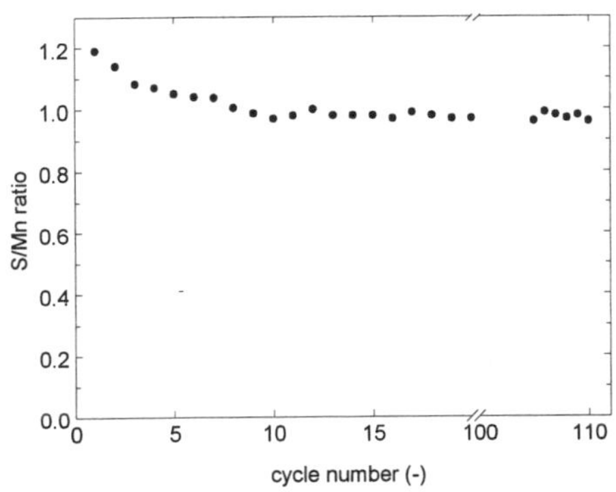

Fig 4.12 Deactivation, expressed by molar S/Mn-ratio for alumina washcoated cordierite sample (12 wt% Mn) for 110 isothermal cycles at 1123 K. Sulfidation 1% H_2S, 50 % H_2, balance Ar. Regeneration: 30 % H_2O in Ar.

5 Discussion

5.1 Characterization

The impregnation of γ-Al_2O_3 via (wet) impregnation with a manganese acetate solution appeared to be a good method for the preparation of a high capacity sorbent with a relatively high, compared to other bulk sorbents, surface area.

A small crystal size after impregnation and calcination is essential for the formation of the main active phase, $MnAl_2O_4$, during the first regeneration. According to Kapteijn *et al.* [10] the preparation of alumina-supported manganese oxide catalysts by using manganese acetate as precursor results in a highly dispersed oxide phase homogeneously distributed throughout the alumina particles. This is due to the ionic interactions with acidic (up to 2 wt% manganese) and basic (above 2 wt% manganese) surface -OH groups of the alumina. For low manganese loadings XRD could not detect micro crystalline manganese oxide which proves the presence of a highly dispersed phase. This was confirmed by XPS. With increasing loading the size of the manganese crystals increases, as for higher loading samples Mn_3O_4 was detected with XRD. For a sample with a 32 wt% Mn loading the average crystal size for a fresh, sulfided and regenerated sample was found to be between 10 and 20 nm, which is still rather small for such a high loading.

After the first sulfiding the S/Mn ratio in a sample is about 1.3 which decreases to around 1 after several cycles. Sulfidation of Mn_3O_4 leads at temperatures below 625 K to the formation of both MnS and MnS_2 (Fig. 1.3). At high temperature MnS_2 will be formed but decomposes to MnS under the formation of S and H_2S [7]. This might explain the observed formation of some elemental sulfur after the first sulfidation. Possibly also some amorphous overstoichiometric manganese sulfide with a S/Mn ratio > 1 was formed [9].

As expected from thermodynamic calculations (Fig. 2.4.) regeneration of MnS with H_2O and SO_2 is possible in the presence of γ-Al_2O_3. Without γ-Al_2O_3 this is very difficult (Fig. 2.3.) XRD shows the formation of bulk manganese aluminate ($MnAl_2O_4$; Fig. 4.3d). As the Mn concentration exceeds the stoichiometric amount of Mn in $MnAl_2O_4$ (> 32 wt% Mn) also some MnO is formed. XRD showed that this MnO is very disperse if only a few wt% MnO is present.

The presence of MnS in some high loading samples after regeneration may be attributed to a minor amount of large MnS crystals which are very difficult to regenerate. These large MnS crystals may originate from large MnO crystals formed after improper impregnation (*e.g.* because the excess solution was not removed effectively).

During sulfiding $MnAl_2O_4$ is converted to MnS and γ-Al_2O_3 (Eq. 4). For effective sulfiding and regeneration of the acceptor fast solid state diffusion has to take place which, as shown by temperature programmed sulfidation and regeneration, only occurs at significant rates at temperatures above 1000 K.

The observed sample colours after sulfiding (green) and regeneration (beige) cycles support the XRD data. Crystalline MnS is green and manganese aluminate beige [3].

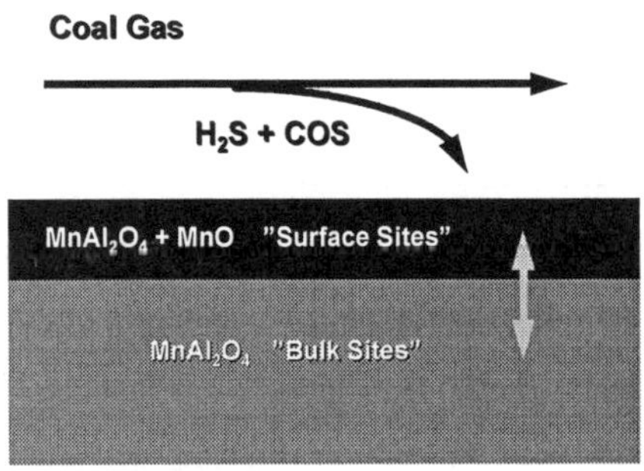

Fig. 5.1 Schematic representation of desulfurization with the manganese based sorbent.

Surface and Bulk Sites

The observed transient sulfiding and regeneration behavior can be explained by a two site (surface and bulk) model during both sulfiding and regeneration (Fig. 5.1). The behavior of surface sites can be described with a model which is also applied to describe ion exchange. Hereby the surface sites are included in the equilibrium constants (Eq. 1a). The sulfidation behavior of bulk Mn components at high temperatures (*e.g.* 1123 K) is determined by (bulk) thermodynamics (Eq. 1). Hereby, only the gas phase activities are important. At lower temperatures also solid state diffusion limitation should be encountered. At temperatures < 775 K only surface sites react because bulk diffusion in the solid is too slow [5]. The breakthrough during sulfiding starts after sulfiding of almost all (available) sites has taken place. Surface sites produce the sharp H_2S production peak during regeneration (see Fig. 4.5). Modelling of sulfidation and regeneration behavior is still in progress.

5.2 Acceptor Capacity and Reaction Conditions

From the presented absorption and regeneration experiments on the manganese based sorbent it arises that it is a good acceptor for hot dry gas desulfurization. The sorbent reduces the amount of H_2S to a level below 20 ppm if no H_2O is present (Fig. 4.5). Experiments with a simulated coal gas, with 2.1 % H_2O, proved that H_2S is effectively removed to below a H_2S level of 40 ppm. This is better than expected from thermodynamic calculations. Via the water gas shift reaction (Eq. 9) part of the H_2O will be shifted away and probably part of the H_2S is converted to COS (Eq. 11) which is absorbed more strongly than H_2S (Table 2.1).

Up to a manganese loading of 37 wt% an increase in acceptor capacity is observed which results in a maximum sulfur absorption capacity of 20 wt% sulfur for a 37 wt% manganese loading at 1123 K. This high capacity could be obtained between 1123 and 1223 K.

According to previously published work [4, 5] the optimum performance is reached with a manganese loading of 8 wt% at 873 K. This resulted in a sulfur uptake capacity between 0.5 and 1.5 wt% S. Higher loadings did not result in a higher capacity at this temperature and it was believed that, although higher temperatures increased the S/Mn ratio, excessive sintering during prolonged operation would destroy the acceptor capacity. Indeed, severe sintering was observed at high temperatures. Above 1100 K, however, the acceptor capacity remained high during subsequent cycles. Only below 1100 K a decrease in surface area causes a decrease in sulfur uptake capacity (Fig. 4.7). Apparently at high temperatures (>1100 K) the surface area is less important. As bulk diffusion is a strongly activated process (Figs. 4.1 and 4.2) only at high temperatures bulk diffusion occurs at sufficiently high rates to accomplish complete and fast sulfidation and regeneration of the whole sorbent. At low temperatures the surface-site reactions are important for the total uptake capacity and in that case sintering will cause deactivation of the sorbent.

5.3 Influence of the Feed Composition and the Manganese Loading on Sulfidation Behavior

The results in Fig. 4.8 show that the amount of water present in the feed and the manganese loading of the sorbent are key factors in the sulfidation behavior of the sorbent. The direct influence on the sulfidation behavior of other components, hydrogen, carbon monoxide, and H_2S is little, but as they influence the amount of water in the feed they can influence the overall behavior indirectly.

As has been stated above a sorbent can contain different sites (surface and bulk) and two kinds of active material ($MnAl_2O_4$ and MnO). The amount of surface sites is correlated to the surface area, so the higher the surface area the more surface sites are available. These surface sites can remove H_2S and COS to very low levels even in the presence of water [4]. The capacity of these sites is, however, limited (< 1.5% S) and decreases with increasing water amount [5]. The behavior of these sites can be compared with the sorption behavior of an ion exchange column. If the column is fully regenerated always some ions can be exchanged. Empty surface sites play a role in the thermodynamic equilibrium [Eq.1a]. For bulk sites the system consist of a gas phase and a solid phase, leading to (Eq.1) absorption of H_2S by the bulk sites when the ratio of $[H_2O]/[H_2S]$ is lower than the K_{eq}-value.

On low loading samples (< 30 wt% Mn) only a $MnAl_2O_4$ phase is present. The observed K_{eq} for these samples were close to the values obtained via thermodynamics; 5-17 compared to 12. The variation in K_{eq} can be explained by the fact that with changing manganese loading the interaction of Mn with the alumina changes somewhat and therefore also K_{eq}. Apparently at low loading (*e.g.* 6 wt% Mn) the manganese-alumina compound is more stable than at higher loadings (*e.g.* 23 %).

At very high manganese loadings also some MnO like phase appears. If the amount of Mn is only a few percent above the stoichiometric point of $MnAl_2O_4$ the MnO phase is disperse (or amorphous) and cannot be detected with XRD. The K_{eq} for MnO is much higher than for $MnAl_2O_4$ and therefore for high loading samples also high K_{eg} 's are observed (200-275) near the (calculated) K_{eq} of MnO (375). The K_{eq} increases with increasing Mn loading towards the K_{eq} of pure MnO as with increasing crystal size the interaction with alumina decreases. With very high MnO loading only breakthrough of MnO was observed which can be explained by the occurrence of integral bed behavior. All H_2S which is not taken up by $MnAl_2O_4$ is taken up by MnO and the breakthrough of $MnAl_2O_4$ is not observed separately anymore.

The water gas shift reaction (Eq. 9) is an important reaction during sulfiding. It has been shown (Fig. 4.10) that up to 30% of the water can be shifted away by this reaction. From thermodynamic calculations it is known that at 1123 K, the water gas shift reaction proceeds to an equilibrium with $K_{eq} \approx 1$.

COS is formed (Eq.10) but is very well adsorbed by the sorbent. Only when breakthrough occurs, COS could be detected with MS.

The reaction with H_2S produces some H_2O, therefore with increasing H_2S concentration the capacity of the sorbent decreases somewhat.

5.4 Regeneration

Most regenerations were performed with H_2O resulting in the production of diluted H_2S; typically 10 % of the initial steam concentration in the regeneration gas. The K_{eq} for low loading samples is relatively low (Fig. 4.11). Low loading samples are therefore more easy to regenerate, which result in a higher H_2S concentration in the regeneration off-gas. Moreover, the relatively high amount of surface sites on low loading samples also increases the average H_2S concentration in the regeneration off-gas. The formation of some S_2 and SO_2 during regeneration at T > 1000 K can be explained by the occurrence of the reactions in Eqs. 7 and 8.

Regeneration of samples with high manganese loadings (> 30 wt% manganese) required more steam because of the presence of MnO or MnO like phases which has less favourable thermodynamics and thus requires more regeneration gas. Compared to "unsupported" MnS [7], which can not be fully regenerated with steam at 1123 K the regeneration of the high loading samples is relatively easy. This may be explained by the absence of severe diffusion limitation in disperse MnS.

Regeneration with SO_2 was successful at temperatures above 823 K. Direct formation of sulfur was observed and the mass balance indicated that the conversion to S_2 was near to 100%. It was estimated that at 1123 K the regeneration gas contained about 10 % S_2, conform predicted values from thermodynamic calculation (Fig. 2.4.) Samples with a very high loading are difficult to regenerate completely, similar as in H_2O regeneration.

As already explained in chapter 2.4. the direct production of elemental sulfur during regeneration with SO_2 is in most cases preferred over the production of SO_2 or H_2S. Regeneration of MnS to MnO in high loading samples with SO_2 or H_2O is difficult and requires large amounts of regeneration gas. Therefore, in practical applications, addition of some O_2 (typically 1-3 %) will be favourable for the regeneration of high loading samples (Mn > 30%). Moreover the exothermal reaction with oxygen can compensate the endothermal regeneration with SO_2 (Table 2.1). High oxygen concentrations lead to unacceptable temperature rise in the acceptor bed and require additional cooling [11]. In Fig. 5.1 a possible process scheme for regeneration with SO_2/O_2 is depicted.

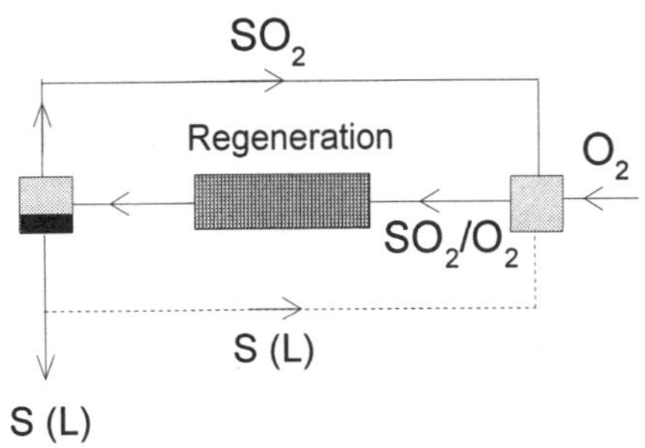

Fig. 5.1
Scheme of the regeneration with SO_2/O_2. The sulfur (gas) and the excess SO_2 are cooled to below the condensation point of S. The liquid sulfur is captured and the remaining SO_2 is recirculated to the bed. Some of the sulfur is burned with a surplus oxygen to provide the SO_2 and O_2 necessary for the regeneration.

5.5 Practical Capacity, Stability and Influence of Impurities

This sorbent has been designed for desulfurization of gas streams with little or no H_2O in the feed (*e.g.* Shell type coal gas) down to a level of below 100 ppm H_2S. $MnAl_2O_4$ can remove the bulk, say > 90%, of the H_2S and COS from the feed stream if 0-2% H_2O is present in the feed. MnO and the surface sites are needed to remove the remaining H_2S. The amount of MnO should, however, be minimized as MnS without Al_2O_3 is difficult to regenerate and thus requires large amounts of steam, SO_2 or extra addition of oxygen. The total amount of MnO should be adapted to the remaining amount of H_2S which has to be removed after absorption of H_2S by $MnAl_2O_4$. The amount of surface sites is more or less constant if the surface area remains constant. At most 1.5 % S can be captured with the surface sites if no water is present in the feed [4, 5]. With an equilibrium constant of the MnO sites of around 250 the maximum amount of water in the feed (after the WGS reaction) is about 2.5%. As a Shell type coal gas contains 0.4-2 % H_2O the sorbent is suitable for desulfurization of this gas. This was confirmed by tests with simulated coal gas with 2.1. % H_2O. H_2S was removed down to a level of less than 40 ppm. Because water was present in the feed a breakthrough capacity was obtained which was 0.75 times lower than theoretical maximum capacity obtained with a S/Mn ratio of 1.

The sorbent has a good stability as operation during 110 cycles shows only some deactivation (< 20%) during the first cycles. Also during operation with a simulated coal gas no deactivation was observed during a test of 20 cycles. In the past deactivation tests of 50 cycles at 873 K were performed with a low loading sample (8 wt% Mn) behind a pilot coal gasifier of ECN in the Netherlands [5]. Here, also only an initial deactivation was observed followed by a stable performance. The high capacity sorbent should also be tested behind a gasifier in the future.

The high mechanical strength of the washcoated cordierite monolith is not surprising, because it is designed for application in car exhaust systems in which it should be able to withstand high thermal stress.

Besides H_2S the coal gas contains other impurities like HCl and HF and alkali metals. As shown in Fig. 2.2 HCl and HF are also adsorbed by the Mn-based sorbent. These components are not or very difficult to remove with SO_2 or oxygen but they can be removed relatively easy with H_2O. In a practical application the sorbent should be treated with steam after a number of cycles to remove the absorbed HCl and HF. Possibly some deactivation might occur by volatilisation of $MnCl_2$ at 1123K. The amounts are, likely, very small as the boiling point of $MnCl_2$ is 1503 K. Removal of alkali metals at high temperatures can be performed with an other sorbent *e.g.* with aluminosilicate compounds as described in [12].

6 Continuous High Temperature Desulfurization

As has been shown above, the application of the above described manganese based sorbent is feasible for desulfurization of dry coal gas (0-2% H_2O). Several reactor/process options exist in which this sorbent can be applied. We will discuss these options very briefly as they are discussed elsewhere in this book by van Dam *et al.*

For desulfurization of a coal gas from a gasifier the following is preferred: continuous operation, low pressure drop, low dust sensitivity, little attrition of the sorbent, and a constant gas composition of the gas from the desulfurization reactor. In case the above described manganese based sorbent is used, regeneration of the sorbent under the direct production of sulfur is preferred.

The most simple reactor is a fixed bed operation in swing mode with two or three reactors. In such a set-up the reactor is subsequently used for regeneration and sulfidation. Either pellets or more structured acceptors like monoliths can be used. In treating large volumina of gas monolith shaped sorbents are preferred over conventional particles. Not only because of the low pressure drop of these structured sorbents. material but also because less maldistributions occur and the insensitivity of monoliths to dust. Moreover the gas diffusion length in particles is a factor 10-50 longer than in the monoliths and monoliths are relatively incentive to dust

For a 250 MW_e power plant based on the Shell gasification technology, tree reactors with each containing about 20 m^3 monolith acceptor (35 wt% Mn, 0.5% H_2S, 75% of sorbent used) are needed when a switching time of 3 hours is applied. Switching valves in these large systems lead to instabilities in the system of typically 15 minutes. To avoid unacceptable system instabilities 2-4 hours is a preferred minimum switching time. This leads to large reactor volumina. The size of the reactors and the fact that the composition of the gas the reactor changes in time are disadvantages of a periodically operated absorber. Moreover, the construction of corrosion resistant valves for application at high temperatures and high gas volumes is extremely difficult.

To overcome the disadvantages of the discontinuous operation of a fixed bed a fluidized bed reactor, a moving bed reactor or a rotating reactor can be used [14]. The sulfur removal efficiency in a fluidized bed is lower than that of a fixed bed system because of mixing of the acceptor bed and bypassing of acceptor particles and bubbles. Staging of a fluidized bed can overcome this problem. The use of a moving bed reactor is also a solution to this problem. An advantage of a moving bed is that sorbent attrition is less than in a fluidized bed. Sorbent attrition can also be reduced, compared with a fluidized bed with pneumatic sorbent transport, by application of an interconnected fluidized bed [15].

A rotating reactor is another promising reactor type to perform cyclic processes continuously. Also in this reactor sorbent material is subsequently exposed to a "sulfidation" and "regeneration" gas but instead of moving the sorbent, which occurs in a moving or fluidized bed, the reactor itself is rotating (see Fig. 6.1). In this way adsorption and regeneration takes place simultaneously in the same reactor and minor attrition of the sorbent and reactor will occur. If applied after a 250 MWe coal gasifier

(Shell type, 0.5% H_2S, 35 wt% Mn) only 2 m^3 sorbent material would be needed when a rotation speed of 10 rotations/ hour would be applied.

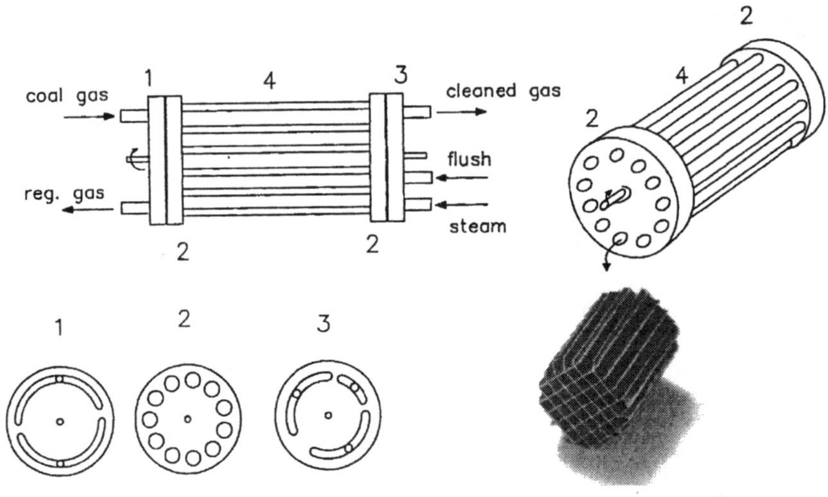

Fig. 6.1 H_2S removal by means of a rotating monolith reactor. Patent pending [16].

At the Delft University a bench-scale (max. 8 m^3_{gas}/h) rotating (monolith) reactor has been designed and constructed for continuous high temperature (600-1200 K) desulfurization of coal gas. The monolith shaped acceptor material is mounted in a housing rotating during operation. In Fig. 6.1 the reactor is shown schematically. The reactor has been operated up to 923 K without serious problems. Application of this reactor is not limited to hot gas cleaning. The reactor principle can be used for many other cyclic processes, like selective oxidation, FCC, *etc.* and is therefore very promising [16].

7 Conclusions

A stable high capacity manganese based sorbent for high temperature desulfurization of dry coal gas has been developed. Both particles and monoliths are suitable as support, and regeneration with SO_2 under the direct production of elemental sulfur is possible. The active material consists of two manganese phases: bulk manganese aluminate which provides the high sulfur uptake capacity and a disperse manganese oxide phase which provides the desulfurization to a low level. The amount of water in the feed gas and the Mn loading are the key elements in the sulfiding behavior.

Thermodynamic calculations appeared to be a very useful tool to estimate gas phase and solid phase compositions during sulfiding and regeneration.

The concept of a rotating reactor as application in high temperature desulfurization of coal gas is in principle feasible.

Acknowledgments

The investigations has been carried out in the framework of the Joule 2 program of the European Union, Jou2-CT93-0431. The KEMA (the Netherlands) and S.B.T. Ris are acknowledged for their contribution to the experiments with SO_2 as regeneration gas.

8 Literature

1 Haber J., Crystallography of Catalyst Types, in J.R. Anderson and M. Boudart (eds), "Catalysis : Science and Technology, Springer Verlag, Berlin, 1981, Vol. **2**, Chapter 2.

2 Gray T.J., Oxide spinels, in A.M. Alper, "high temperature oxides", Vol. **4**, Academic Press, New York, 1971, pp 77-107.

3 Weast, R.C., Handbook of Chemistry and Physics, 66th edition, CRC Press, Baca Raton, FL (1985), B113-B114.

4 Wakker J.P., A.W. Gerritsen and J.A. Moulijn, "High Temperature H_2S and COS Removal with MnO and FeO on γ-Al_2O_3 Acceptors", Ind. Eng. Chem. Res, **32**, 139-149 (1993).

5 Wakker J.P., high temperature desulfurization, PhD thesis TU-Delft, 1992.

6 Moulijn J.A. and Diepen, Effect of process conditions on the thermodynamics of gasification, this book.

7 Bakker W.J.W, J.C.P. van Rossen, J.P. Janssens and J.A. Moulijn, Hot gas cleaning; Sulfiding mechanisms in H_2S removal by solids, this book.

8 Guinier, A. X-ray diffraction in crystals, imperfect crystals and amorphous bodies, Freeman, San Fancisco, 1963.

9 Mangnus P.J., Characterization of hydrotreating catalysts with temperature programmed techniques, A search for improved catalysts, Phd. Thesis, University of Amsterdam, 1991.

10 Kapteijn, F. , A.D. van Langeveld, J.A. Moulijn, A. Andreini, M.A. Vuurman, A.M. Turek, J.Jehng and I.E.Wachs, "Al_2O_3-Supported Manganese Oxide Catalysts", J. Catal. **150**, 94-104 (1994).

11 Slimane, R.B. and M.T. Hepworth, "Desulfurization of Hot Coal-derived Fuel Gases with Manganese Based Regenerable Sorbents, Energy and Fuels, **8**, 1184-1191 (1994).

12 Schulz R.A., N.F. Kirkby and D. Chrysohoidou, Fixed bed sorption studies of alkali vapour by aluminosilicates at 600-950 °C, Pittsburgh coal conference proceedings, 1995, 1213-1218.

13 Cybulski, A. and J.A. Moulijn, "Monoliths in Heterogeneous Catalysis", Cat.Rev.-Sci.Eng, **36**, 1994, 179-270.

14 Bakker W.J.W, M. Vriezendorp, F. Kapteijn, J.A. Moulijn, Sorbent development for continuous regenerative H_2S removal in a rotating monolith reactor, Can. J. Chem. Eng. **74,** 1996, 1-6.

15 Korbee, Regenerative desulfurization in a interconnected fluidized bed system, PhD thesis, Delft University of Technology, 1995.

16 Bakker W.J.W. and J.A. Moulijn, roterende monoliet reactor, Dutch patent application, 1996.

IN-SITU DESULFURIZATION OF COAL GAS WITH CaO-BASED SORBENTS

Javad Abbasian, Institute of Gas Technology, Des Plaines, Illinois 60018, USA

ABSTRACT. Calcium-based sorbents, such as limestone and dolomite, are viable candidates as sulfur-capturing agents in an *in-situ* coal asification/desulfurization process. The effect of limestone addition on the hydrodynamics of a fluidized-bed gasifier and desulfurization of the product gas have been studied. The hydrodynamic characteristics of coal char/limestone mixtures, such as solid mixing and fines retention, have been studied in a 0.2-m diameter fluidization column. The results indicate that a superficial gas velocity of 0.6 cm/s is sufficient to produce a well-mixed bed of limestone/char mixture in the gasifier. The presence of limestone does not appear to affect the fines retention while the bed height has a significant effect on the fines elutriation from the fluidized bed gasifier. Reaction rate data pertaining to the reaction between calcium-based sorbents and hydrogen sulfide have been obtained by a thermogravimetric technique in the temperature range of 650° to 1050°C. The results indicate that calcium-based sorbents can be used effectively for capturing sulfur during the gasification of high sulfur coals.

Keywords. Sorbents, Coal Gas, Desulfurization, *In-Situ*, Limestone, Dolomite

1 INTRODUCTION

Coal is mainly used for power generation. More than 50% of the electricity used in the United States is generated by coal-fired power plants. Because of the abundance of coal supplies in the US, depleting petroleum reservoirs and public concerns about the safety issues of nuclear power plants, coal will continue to remain as the most important feedstock for the power industry.

Conventional pulverized coal-fired power plants are inherently 30% to 35% efficient. Because of the ever-increasing and more stringent government-imposed limitations on sulfur emission, Flue Gas Desulfurization (FGD) has been commonly practiced to meet the air pollution standards. Implementation of FGD systems, has

NATO ASI Series, Vol. G 42
Desulfurization of Hot Coal Gas
Edited by A.T. Atimtay and D. P. Harrison

resulted in higher capital and operating costs, and lower overall plant efficiency, leading to a higher cost of electricity.

Emerging technologies, such as the Integrated Gasification Combined Cycle (IGCC) and Integrated Gasification Fuel Cell (IGFC) are among the leading contenders for power generation in the future. These technologies are based on coal gasification rather than combustion and can reach overall plant efficiencies in the range of 40% to 45%. These emerging technologies, because they essentially eliminate the need for coal preparation to reduce its sulfur content, are creating a renewed interest in high sulfur coals. Several IGCC-based processes are currently in the demonstration stage under the sponsorship of the U. S. Department of Energy's Clean Coal Technology (CCT) Demonstration Program.

In these advanced coal based power generation processes, calcium-based sorbents such as limestone and dolomite are added to the gasifier to capture hydrogen sulfide as calcium sulfide (CaS). This is necessary to comply with sulfur emission standards and to prevent the H_2S from damaging turbine blades and other downstream surfaces.

Researchers[1-6] in the field of chemical reaction have already verified the potential use of these sorbents for sulfur capture. The reaction of hydrogen sulfide with calcined limestone/dolomite is very rapid, and the reaction almost approaches equilibrium. On that basis, it is possible to capture substantial quantities of sulfur and discharge it with ash. Based on the equilibrium considerations, it is feasible to remove up to 90% sulfur by using this process. A limited number of tests conducted at pilot-plant scale[7-10] have verified the feasibility of this sulfur-removal method.

This paper addresses the results obtained in a study geared toward obtaining engineering data on fluidization characteristics of char/sorbent mixtures, as well as sulfidation reactions related to the gasification process with in-situ desulfurization with calcium-based sorbents.

2 EXPERIMENTAL

2.1 Hydrodynamics

The objectives of the hydrodynamic study was to determine the mixing characteristics of char and sorbent of various sizes in the fluidized bed and to determine whether the sorbent addition to the fluidized bed of char has any noticeable influence on the rate of fines carry-over from the fluidized bed.

2.2 Material

A batch of limestone (size -12+200 mesh) was obtained and screened into three size ranges: +40 mesh; -40+80 mesh; and -80 mesh. The first two batches were mixed with a coke breeze (-12 mesh) at different proportions to produce fluidized-bed material. The size distributions of the coke breeze and batches No. 1 and 2 of the limestone are given in Tables 1 and 2, respectively. The average particle size of the limestone in batch No. 1 was 640 μm (30 mesh) and that of the second batch was 210 μm (70 mesh). The average particle size of the coke breeze was 425 μm (40 mesh). The bulk densities of the coke and the limestone were 750 and 1550 kg/m^3, respectively.

Table 1. Size Distribution of the Coke Breeze

Size(Micron)	Sieve (Mesh)	wt.% Retained	wt.% Cum.
3360	6	--	--
1680	12	12.5	12.5
850	20	18.1	30.6
425	40	22.7	53.5
212	70	25.5	78.8
106	140	14.1	92.9
53	270	5.5	98.4
	Pan	1.6	100.0
		100.0	
d_p, average		425 μm	

Table 2. Size Distribution of the Limestone

Size		Batch No. 1		Batch No. 2	
Microns	Mesh	wt.%	wt.% Cm.	wt.%	wt.% Cum.
1400	14	2.7	3.7	--	--
850	20	32.6	36.3	--	--
600	30	29.8	66.1	0.3	0.3
425	40	21.3	87.4	0.3	0.6
300	50	12.0	99.4	17.2	17.8
250	60	0.3	99.7	16.0	33.8
212	70	0.1	99.8	15.6	49.4
180	80	0.1	99.9	14.1	64.5
150	100	0.1	100	19.4	82.9
	Pan	0.0	--	17.1	100
		100.0		100.0	
dp, average		640 μm		210 μm	

The third batch of limestone (-80 mesh) was fluidized at a superficial gas velocity of 0.6 m/s (the velocity used in all experiments) in a 0.2-m diameter column (described later), and the elutriated fines were collected. The elutriated fines were divided into 2.2 kg batches to be used for injection into the fluidized bed in the fines retention tests. The size distribution of fines is presented in Table 3.

Table 3. Size Distribution of Fines Injected

Sieve (Mesh)	Size(Micron	wt.% Retained	wt.% Cum.
100	150	--	--
140	106	23.3	23.3
200	75	44.3	67.6
270	53	27.0	94.6
Pan	53	5.4	
		------	100.0
		100.0	

2.3 Fluidization Characteristics

Fluidization tests were conducted on a sample of the coke breeze, the coarse (dp=640 μm) and fine (dp=210 μm) samples of the limestone, and 50:50 mixtures of coke and two sizes of the limestone. The tests were conducted in a 0.1 m fluidized-bed column with a flat porous plate gas distributor. The aspect ratio (L/D) was about 1.5. The temperature and pressure were ambient; the fluidizing gas was nitrogen. Table 4 presents experimental values of the minimum and complete fluidization velocities.

Table 4. Fluidization Velocities of coke/Limestone Mixtures

Microns	Sieve (Mesh)	wt.%	wt.%, cum.
100	150	--	--
140	106	23.3	23.3
200	75	44.3	67.6
270	53	27.0	94.6
Pan		5.4	100.0
		100.0	

2.4 Solids Mixing and Fines Retention

The mixing tests and the fines retention tests were conducted in an 0.2-m diameter fluidization column with an expanded section and sloping grid. To avoid segregation either by size or by density in these tests, they were conducted at a velocity of 0.6 m/s, exceeding the complete fluidization velocity of the coke breeze, which had the higher value.

A schematic diagram of the fluidization test unit is presented in Figure 1. The unit essentially consists of a fluidization column, equipment for feeding and measuring the fluidization gas (air) flow rates in different streams, and pressure differential cells for measuring bed density. The fluidization column is made of transparent Plexiglass. The column consists of an 0.2-m diameter lower section containing a discharge nozzle and an 0.3-m diameter upper section with a 45° sloped transition. The distributor has a 45° slope and consists of 128 uniformly spaced holes. Air is used as the fluidization medium. Solids are fed through one of several feeding ports. The lower section of the column is 0.4-m long.

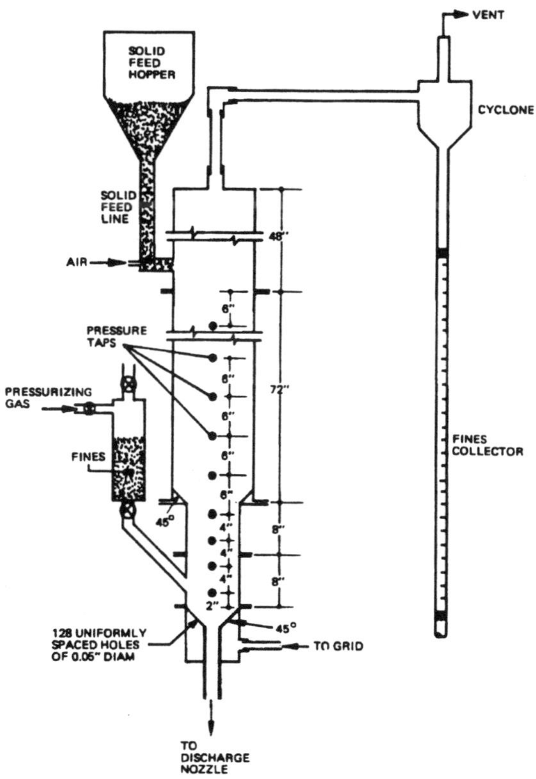

Figure 1. Schematic Diagram of the Fluidized Bed Unit

During a test, the flow through the grid and discharge nozzle was set at a minimum rate to prevent plugging, and the solids were fed from a feed hopper through one of the feed ports to a predetermined bed height. The static-bed height was generally chosen to be about 75% of the desired fluidized-bed height. The flow rates were then set at the desired values, and the fluidization behavior of the bed was observed and recorded. The pressure drop across the bed was measured by pressure transmitters and recorded. After the column was filled with the appropriate bed material, the velocity at the grid was increased to 0.375 m/s and the velocity in the discharge nozzle was increased to 7 m/s. The total air flow yielded an overall superficial velocity of 0.6 m/s in the lower section (0.2 m) and approximately 0.3 m/s in the freeboard.

2.5 Solids Mixing

In the mixing tests, 50:50 mixtures of coke/limestone of both sizes (i.e., d_p=210 μm and d_p=640 μm were fluidized. The fluidized-bed height was maintained at 0.8 m. After 30 minutes of fluidization, the flow of gases was abruptly shut off. Samples were taken to be representative of the fluidized-bed material of three sections of equal height (approximately 0.27- m each). The concentrations of limestone were determined by measuring the fraction of calcium in the mixture using the atomic absorption technique. The results, presented in Table 5, show that the concentration of limestone in all three sections was about 50%, indicating that the velocity of 0.6 m/s is sufficient for either 70 mesh (210 μm) or 30 mesh (640 μm) limestone to produce a well-mixed bed of char and limestone.

2.6 Fines Retention

Prior to the fines retention tests, the mixture of coke and limestone was fluidized at 0.6 m/s, and elutriated fines were collected in the calibrated transparent column installed at the bottom of the cyclone. The volume of fines collected was recorded at predetermined periods. About 1 hour after initial fluidization of the coke/limestone bed, the rate of collection of elutriated fines reached a relatively constant value, which corresponded to the attrition rate of the bed material. After a steady value for the attrition was attained, fines (shown in Table 3) were injected into the fluidized bed from the pressurized container (0.2 MPa). The injection point was 0.1 m above the distributor. In a typical experiment, about 2.2 kg of fines were injected. Measurement of the volume of collected fines continued until the rate approached that of the attrition rate. The bed material and fines were weighed before and after each run to determine the material balance.

To determine the effect of limestone size and concentration on fines retention, tests were conducted with bed material containing 0, 25, 50, and 75 wt.% limestone using both 30 mesh (640 μm) and 70 mesh (210 μm) limestone. The bed height in all of these runs was maintained at about 0.8 m, and the superficial gas velocity was 0.6 m/s in the lower section and 0.3 m/s in the free-board. The rate of elutriation of fines for the bed containing 100% coke by weight (or 0 wt.% limestone) is shown in Figure 2. The data for 25, 50, and 75 wt.% mixtures of limestone with coke for two different limestone sizes are shown in Figure 3. The line in Figure 3 represents the average rate for the bed containing no limestone, obtained from Figure 2. These data indicate that the deviation of the rate of fines elutriation from the bed containing limestone from that of the bed containing no limestone is statistically insignificant. Therefore, it can be concluded that the addition of limestone to a bed of coke does not alter the rate of (limestone) fines elutriation. Similarly, the presence of limestone does not affect the fines retention in the fluidized bed.

Table 5. Fluidized Bed Composition at Superficial Velocity of 0.6 m/s

Bed Section	Percent Limestone	
	d_p=210 μm	d_p=210 μm
Upper	48	48
Middle	44	55
Lower	48	51

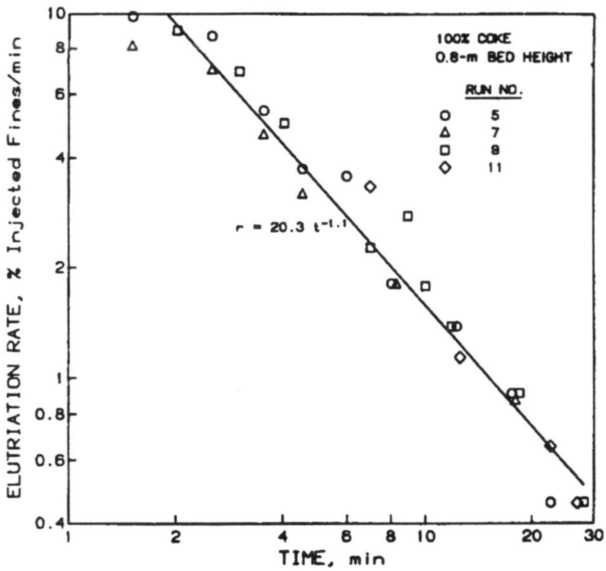

Figure 2. Elutriation Rate of Coke Breeze in 0.8 m Bed

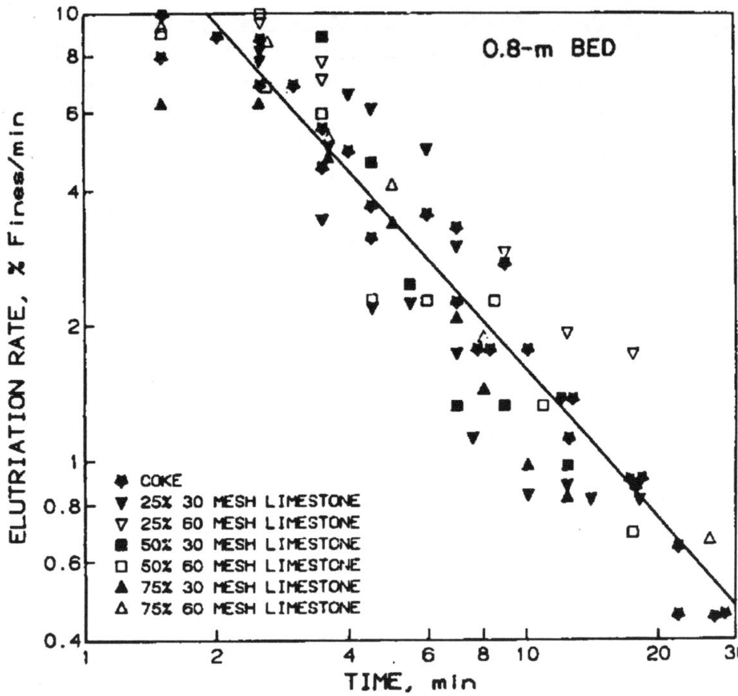

Figure 3. Elutriation Rate of Coke/Limestone Mixtures in 0.8 m Bed

Additional runs were conducted to determine the effect of bed height on fines retention. The fluidized-bed height in these runs was increased to 1.1 m. In the first run, the bed material consisted of coke only; in the second run, the material consisted of a 50:50 mixture of coke/30 mesh limestone. The results are presented in Figure 4, with the corresponding rates from a 0.8 m bed. Even for a 1.1-m bed, the fines elutriation rate was not affected by the presence of limestone. However, there was a distinct difference in the elutriation rate observed when the rate for a 32-inch (0.9-m) bed was compared with that for a 1.1-m bed. The rate of elutriation from a 1.1-m bed was about two-thirds of the rate obtained from a 0.8-m bed. These results indicate that bed height has a significant effect on the fines elutriation rate from a fluidized bed.

Fluidization tests were also attempted at greater superficial gas velocities (higher than 0.8 m/s). At these velocities, however, the attrition of the bed material was significantly higher, and accurate data on the attrition rate could not be obtained.

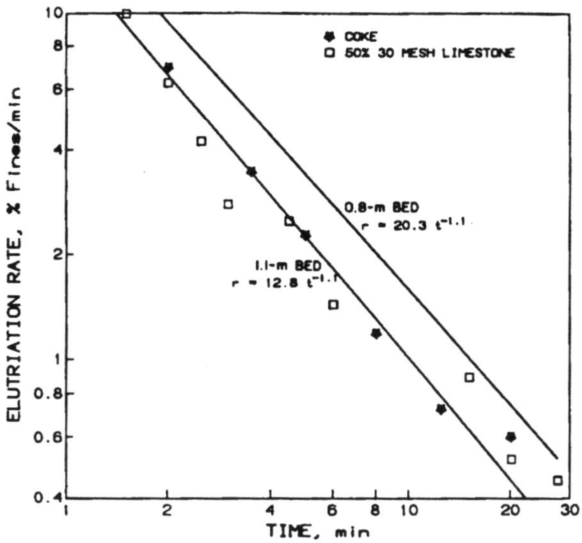

Figure 4. Elutriation Rate of Coke and Limestone Mixtures in 0.8 m Bed

2.7 Gas/Solid Reactions

In the process, the primary desulfurization reactions (sulfidation reactions) are confined mainly within the fluidized-bed gasifier under reducing conditions. The primary gas/solid sulfidation reactions, taking place in the gasifier upon the introduction of limestone and/or dolomite, are as follows:

$$CaCO_3 = CaO + CO_2 \quad (A)$$
$$CaO + H_2S = CaS + H_2O \quad (B)$$

When dolomite is introduced into the gasifier, it calcines to produce CaO and MgO. Calcium oxide and the resulting CaS react with H_2S according to Reaction (B), whereas MgO passes through the system as an inert.

2.8 Sulfidation Reaction

The sulfidation reaction tests involving the reaction between CaO and H_2S were conducted in a quartz thermobalance at atmospheric pressure. The effects of sample weight, gas flow rate, particle diameter, H_2S concentration, and temperature on the reaction rate were determined. The thermobalance apparatus and the test operating procedure have been discussed elsewhere.[11]

The sorbent chosen for the complete kinetic study in this program was obtained from New Enterprise Stone & Lime Co., Inc., located in Pennsylvania. This material has already been used by the Institute of Gas Technology (IGT) in fluidized-bed gasification tests. Elemental analysis were conducted on this sorbent, the results of which are given in Table 6. The sorbent (dolomite) was screened and divided into four batches of different particle size range. The average particle diameter for the four batches was .014, .032, .065, and .129 cm.

Table 6. Chemical Composition of the Sorbent

Analysis	wt %
Calcium	28.5
Magnesium	4.79
Silicon	3.11
Potassium	0.46
Iron	0.34
Strontium	0.10
Carbon Dioxide	40.84

A previous investigation[10] had indicated that the reaction of CaO with H_2S does not depend on the gas composition as long as a constant level of H_2S is maintained. Therefore, in all the tests conducted in this study, the reactant gas consisted of $H_2S/H_2/He$ mixtures. The ratio of hydrogen to hydrogen sulfide was maintained at about 3:1 to prevent dissociation of hydrogen sulfide to elemental sulfur.

To determine the effect of hydrogen sulfide on the reaction rate, and the order of the reaction with respect to H_2S concentration, a series of four tests were conducted with a sample from Batch No. 1 (-80 mesh) at 980°C. The concentration of H_2S in this series of runs was 0.8%, 1.6%, 2.5%, and 3.75%. A plot of CaO conversion versus time is presented in Figure 5.

The tests for the effect of particle size were conducted at 980°C with a reactant gas containing 3.75% hydrogen sulfide. Figure 6 shows the conversion of calcium oxide versus time as a function of particle size. The conversion of calcium oxide in the sulfidation reaction is calculated using the following formula:

$$\% \text{ Conversion} = \frac{56 \times (\text{weight gained})}{16 \times (\text{sample weight}) \times (\% \text{ CaO in the sample})} \qquad (1)$$

where 16 is the difference in the atomic weight of gained sulfur and lost oxygen and 56 is the molecular weight of calcium oxide.

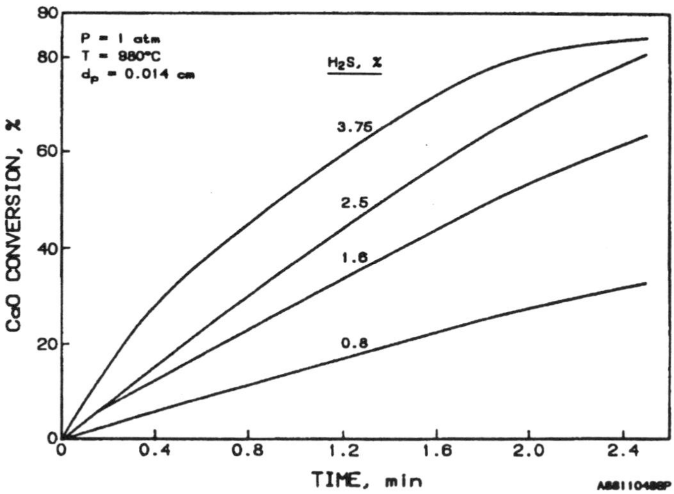

Figure 5. Effect of H_2S Concentration on the Sulfidation Reaction Rate

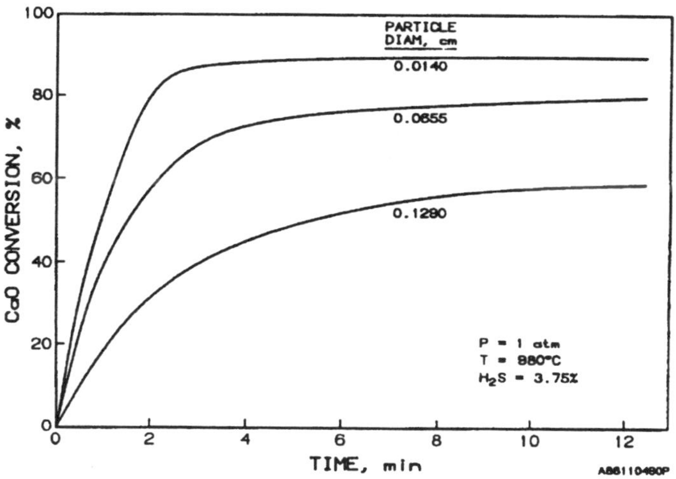

Figure 6. Effect of Particle Size on the Sulfidation Reaction Rate

The rate expression for Reaction B is --

$$-r_{CaO} = K_s[H_2S]^n \tag{2}$$

where r_{CaO} is the rate of the reaction, $[H_2S]$ is the concentration of hydrogen sulfide, and K_s is the overall reaction rate constant which can be expressed as --

$$1/K_s = 1/K_d + 1/K_s \quad (3)$$

K_s is the intrinsic reaction rate constant and K_d is the measure of the conductance due to diffusion. The intrinsic reaction rate constant is usually expressed in Arrhenius form --

$$K_s = K_s\, Exp(-E_a/R_g T) \quad (4)$$

where E_a is the activation energy for the reaction, T is the absolute temperature, and R_g is the ideal gas law constant. The diffusion conductance is usually expressed as--

$$K_d = D_{eff}/L \quad (5)$$

In equation 5, D_{eff} represents the effective diffusivity and L is the average thickness of the diffusion layer. Integrating Equation 3, it can be shown that --

$$1 - (1 - X_{CaO})^{1/3} = t/\tau \quad (6)$$

where

$$\tau = \rho_{CaO} \bullet R/(K_s \bullet [H_2S]^n) \quad (7)$$

In the above equations τ is the time required for complete reaction of a particle, X_{CaO} is the conversion of calcium oxide, ρ_{CaO} is the molar density of the sorbent, R is the average radius of the solid particles, and t is the time of reaction. It follows that [$1-(1 - X_{CaO})^{1/3}$] should be a linear function of time and the slope of such linear plot will be $1/\tau$. A plot [$1-(1 - X_{CaO})^{1/3}$] versus time is presented in Figure 7. The order of reaction with respect to hydrogen sulfide concentration was obtained by determining the slope of the line when $\ln(\tau)$ was plotted against $\ln([H_2S])$, shown in Figure 8, indicating that the reaction is first order with respect to hydrogen sulfide concentration (that is, n=1).

From the experimental data of this study, the value of intrinsic rate constant was found to be 2.99 cm/s, while the effective diffusivity within the solid particle was determined to be 0.692 cm^2/s. The diffusivity of hydrogen sulfide in a binary mixture of H_2S/He was calculated using the correlation by Fuller, *et al*[12] to be 7.97 cm^2/s. The ratio of the effective diffusivity of H_2S within the solid particle to the open space diffusivity is 0.086. This ratio was reported by Pell[13] to be 0.056.

Tests for the effect of temperature on the sulfidation reaction rate were conducted in the temperature range of 650° to 1040°C. Figure 9 shows the Arrhenius plot of the intrinsic rate constant versus reciprocal of absolute temperature. The slope of this curve indicates that the activation energy for this material in the temperature range of 650° to 1040°C is about 7.7 Kcal/g mole.

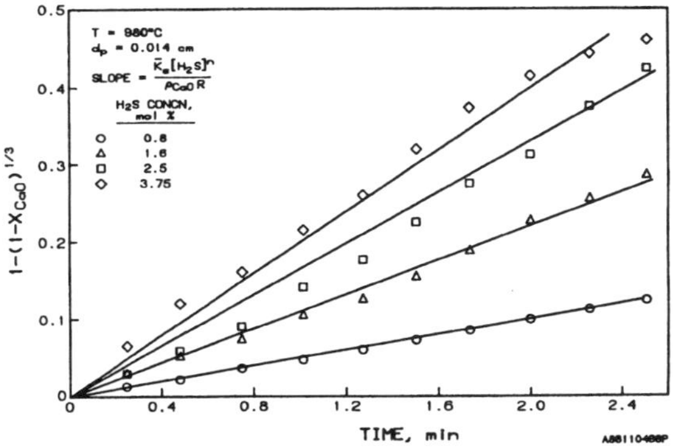

Figure 7. Plot of [$1-(1-X_{CaO})^{1/3}$] Versus Time

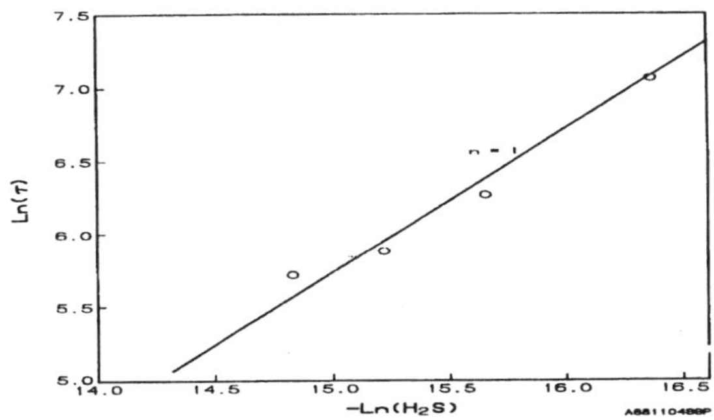

Figure 8. Determination of the Order of Sulfidation Reaction with Respect to H_2S

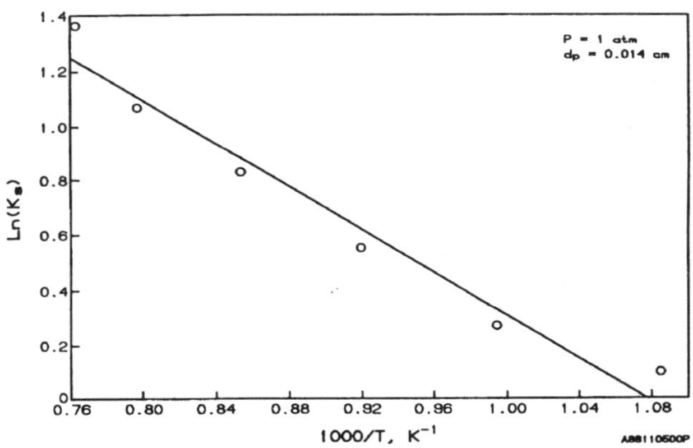

Figure 9. Arrhenius Plot of Sulfidation Rate Constant

3 CONCLUSIONS

The results obtained in this study indicate that calcium-based sorbent can be effectively used as an in-situ sorbent to remove H_2S from the gasifier coal gas at elevated temperatures. The addition of sorbent does not adversely appear to affect the fluidization characteristics of the gasifier.

4 REFERENCES

1. Squires, A. M., Graff, R. A. and Pell. M., "Desulfurization of Fuels With Calcined Dolomite." Chem. Eng. Progr. Symp. Ser. 67, 23-24, 1971.
2. Ruth, L. A., Squires, A. M. and Graff, R. A., "Desulfurization of Fuels With Half-Calcined Dolomite: First Kinetic Data." Environ. Sci. Technol. 6, (12) 1009-1014, 1972.
3. Borgwardt, R. H., Roache, N. F. and Bruce, K. R., "Surface Area of CaO and Kinetics of CaS Formation." Environ. Progr. 3, (2) 129, 1984.
4. Borgwardt, R. H. and Roache, N. F., "Reaction of H2S and Sulfur With Limestone Particles." Ind. Eng. Chem. Process. Des. Dev. 23, 742-748, 1984.
5. Kamath, V. S. and Petrie, T. W., "Rate of Reaction of Hydrogen Sulfide — Carbonyl Sulfide Mixture With Fully Calcined Dolomite." Environ. Sci. Technol. 15, 966-968, 1981.
6. Freund, H., "Intrinsic Global Rate Constant for the High-Temperature Research of CaO With H_2S." Ind. Eng. Chem. Fundam., 23, 338-341, 1984.
7. Jones, F. L. and Patel, J. G., "Performance of Utah Bituminous Coal in the U-GAS Gasifier." Paper presented at the Fifth EPRI Contractor's Conference on Coal Gasification, EPRI, Palo Alto, California, 1985.
8. Weldon, J., Haldipur, G. B., Lewandowski, D. A. and Smith, K. J., "Advanced Coal Gasification and Desulfurization With Calcium Based Sorbents." Am. Chem. Div. Fuel Chem. Prepr., 33, 3, 1986.
9. Keairns, D., Newby, R. A., O'Neill, E. P. and Archer, D. H., "High Temperature Sulfur Removal System Development for Westinghouse Fluidized Bed Coal Gasification Process." Am. Chem. Soc. Div. Fuel Chem. Prepr., 21, 91, 1976.
10. Abbasian, M. J., Rehmat, A., Leppin, D. and Banerjee, D. D., "Desulfurization of Fuel Gas With Calcium-Based Sorbents." Fuel Proc. Tech. 25, 1990
11. Roache, N. F., "Reaction of H_2S and Sulfur With Limestone Particles." Ind. Eng. Chem. Process Des. Dev., 23, 1984.
12. Fuller, E. N., Schettler, P. D. and Giddings, J. C., "A New Method for Prediction of Binary Gas-Phase Diffusion Coefficient." Ind. Eng. Chem., 58, (5) 18, 1966.
13. Pell, M., "Reaction of Hydrogen Sulfide With Fully Calcined Dolomite." Ph.D. Thesis, City University of New York, 1971.

STABILIZATION AND REGENERATION OF SPENT SORBENTS

Javad Abbasian, Institute of Gas Technology, Des Plaines, Illinois 60018, USA

ABSTRACT. The sulfur compounds present in coal are converted to hydrogen sulfide when coal is gasified. To comply with environmental regulations, a high fraction of the sulfur must be removed from the product gas stream. Calcium-based sorbents are the prime candidates for in-bed capture of sulfur.

The removal of sulfur takes place through the reaction of hydrogen sulfide with calcium-based sorbent and produces calcium sulfide, which is not stable and therefore not suitable for disposal. Calcium sulfide, however, can be either reacted with oxygen to produce stable and environmentally acceptable compounds for disposal, or regenerated with water and carbon dioxide.

This paper addresses the effects of operating variables on the successful stabilization and regeneration of sulfide-containing solid wastes from coal gasifiers.

Keywords. Stabilization, Sulfation, Oxidation, Calcium Sulfide, Solid Waste

1 INTRODUCTION

During gasification, coal is reacted with steam/oxygen or steam/air at a temperature of 1700° to 2000°F (925° to 1090°C) and pressure of 50 to 450 psig (0.4 to 3.2 MPa) to produce a gas mixture that contains carbon monoxide, hydrogen, methane and carbon dioxide (and also nitrogen if air is used). The sulfur compounds in coal, because of the prevailing reducing environment of the gasifier, are converted to hydrogen sulfide. To comply with the New Source Performance Standards (NSPS), a high fraction of sulfur must be removed from the gas stream. A number of technologies are available for gas cleanup at low temperatures. However, there are enormous cost disadvantages associated with such processes.

High efficiency advanced electric power generation systems using Integrated Coal Gasification Combined Cycle (IGCC) require very low levels of H_2S in fuel gases. Calcium-based sorbents, such as limestone and dolomite, are among the prime candidates for in-bed capture of sulfur from the fuel gases. More than 90%

NATO ASI Series, Vol. G 42
Desulfurization of Hot Coal Gas
Edited by A.T. Atimtay and D. P. Harrison

of the H_2S in the fuel gas can be removed in the gasifier by these calcium-based sorbents. An external hot gas cleanup system using regenerable mixed oxide sorbents such as zinc ferrite or zinc titanate is then used to further reduce the H_2S content of the fuel gas to acceptable levels for IGCC application.

The removal of sulfur takes place through the reaction of hydrogen sulfide with calcium oxides, calcium carbonate, or calcium hydroxide and produces calcium sulfide. The solid wastes produced from the reactions of calcium-based sorbents with hydrogen sulfide contain calcium sulfide, which is unstable and has the tendency to decompose when it contacts moisture in the air, releasing hydrogen sulfide to the atmosphere. To dispose of the solid waste products in an environmentally acceptable manner, calcium sulfide can be converted to stable calcium sulfate through reaction with oxygen --

$$CaS + 2O_2 \rightarrow CaSO_4 \qquad \text{(A)}$$

The limited experimental data on the sulfation of calcium sulfide that have been reported in the literature[1-2] indicates that calcium sulfide in partially sulfided dolomite can be completely sulfated, whereas in the case of limestone, only partial sulfation (less than 30%) can be achieved. The use of limestone rather than dolomite is generally desired because of its higher fraction of calcium, compared to dolomite, will require utilization of a smaller quantity of the sorbent for fuel gas desulfurization and will result in a smaller quantity of solid wastes for disposal.

The calcium sulfate produced by Reaction A, at temperatures of interest to the overall process (800° to 1000°C), can also react with the calcium sulfide in an apparent solid-solid reaction to release SO_2[3-9]--

$$CaS + 3CaSO_4 = 4CaO + 4SO_2 \qquad \text{(B)}$$

Reaction A is a highly exothermic reaction (ΔH_{rxn} > 350 Kcal/g-mol), which results in a significant increase in the particle temperature. Reaction B becomes more favorable as the particle temperature increases, which results in the release of SO_2, thereby reducing overall sorbent utilization. The highly exothermic nature of Reaction A and its effect on the reactor temperature significantly affects the final distribution of the products.

The overall reaction for release of SO_2 can be obtained by combining Reactions A and B, that is, [3(A) + (B) $\leftrightarrow$ 4(C)]--

$$CaS + 3/2O_2 = CaO + SO_2 \qquad \text{(C)}$$

The familiar Reaction D for sulfation of calcium oxide--

$$CaO + SO_2 + 1/2O_2 = CaSO_4 \qquad \text{(D)}$$

is also a dependent reaction with regard to independent Reactions A and B--

$$(A) - (B) = 4(C)$$

Oxidation of calcium sulfide through Reaction B decreases the sulfur content of the sorbents. In cases where both Reactions A and B are taking place, it is necessary to know the extent of conversion in competing reactions to determine the suitability of the solid products for disposal.

Another common stabilization process used to recover elemental sulfur from the spent sorbent involves quenching the hot ash discharge and sorbent residue in a volume of water maintained at atmospheric conditions. Under these conditions, the water reacts with the CaS to form $Ca(OH)_2$, and H_2S. This process, however, is generally very slow and incomplete, and residues may retain considerable amounts of sulfur, thus causing serious disposal problems. Turkdogan(10) found that wet steam facilitated sulfur removal from CaS to a much greater extent than dry steam. Experiments performed in boiling water yielded high levels of sulfur removal, typically ranging from 91% to over 98% for temperatures at 212° to 338°F. Particle size reduction was also shown to have a significant beneficial effect on the rate of desulfurization in wet steam.

In a related low-temperature method, Tsao and Kettle(11) investigated the regeneration of spent flux, containing both CaS and CaO from an iron ore process. The results of this work indicate that at pH level below 8, CaS can react with H_2O and CO_2 to regenerate the sorbent --

$$CaS + H_2O + CO_2 = CaCO_3 + H_2S \qquad (E)$$

The feasibility of different methods for stabilization and/or regeneration of calcium sulfide containing water materials is discussed in this paper.

2 EXPERIMENTAL

2.1 Sorbent Selection

Three calcium-based sorbents were selected for this study. The selection of these sorbents was primarily based on their calcium carbonate content. These sorbents are classified as limestone, dolomite, and dolomitic limestone. The sorbents were crushed and screened into narrow particle size ranges. Two cuts from each sorbent were selected for testing in this study. These selected cuts included coarse particles (25 mesh, dp = 7×10^{-4} m) and fine particles (70 mesh, dp = 2×10^{-4} m). Elemental analyses for both sorbents are given in Table 1. The calcium content of the limestone corresponds to 98% calcium carbonate and the molar ratio of calcium to magnesium in dolomite is 0.98, indicating that both the limestone and dolomite are reasonably pure. The dolomitic limestone contained about 85% calcium carbonate and 4% magnesium which corresponds to 66% limestone and 34% dolomite.

Table 1. Chemical Analyses of Sorbents

Analyses, wt %	Limestone	Dolomite	Dolomitic-Limestone
Calcium	39.20	22.15	3.95
Magnesium	0.56	13.65	3.73
Potassium	0.50	0.5	0.33
Iron	0.09	0.135	0.16
Aluminum	0.05	0.07	0.06
Silicon	0.10	0.275	0.78
Strontium	0.16	0.005	0.02
Carbon Dioxide	44.60	48.1	43.55
Oxygen(by Diff.)	14.74	15.11	17.42
Totals	100.00	100.00	100.00

2.2 Preparation of Sulfided Sorbents

Samples of both fine and coarse particles from each sorbent were calcined and then reacted with hydrogen sulfide in a ¾-inch-diameter (0.02 m) quartz fluidized-bed reactor.

$$CaCO_3 = CaO + CO_2 \tag{F}$$

$$CaO + H_2S = CaS + H_2O \tag{G}$$

A schematic diagram of the unit is presented in Figure 1. The unit essentially consists of a quartz fluidized-bed reactor that is externally heated by two electrical furnaces; equipment for feeding and measuring the flow rates of the gases, measuring and controlling the bed temperature and monitoring the fluidized-bed pressure. The fluidized-bed distributor was made of a quartz fritted disk. This unit was used for the preparation of the sulfided sorbents as well as the stabilization of the sulfided sorbents.

The sorbents were calcined during the heat-up period at 70° to 1800°F (20° to 980°C) in a nitrogen atmosphere and then reacted with hydrogen sulfide at 1800°F (980°C) for a predetermined period to achieve the desired sulfidation levels of 33%,

67%, and 100%, which corresponds to a Ca/S ratio of 3, 1.5, and 1 in the gasifier feed.

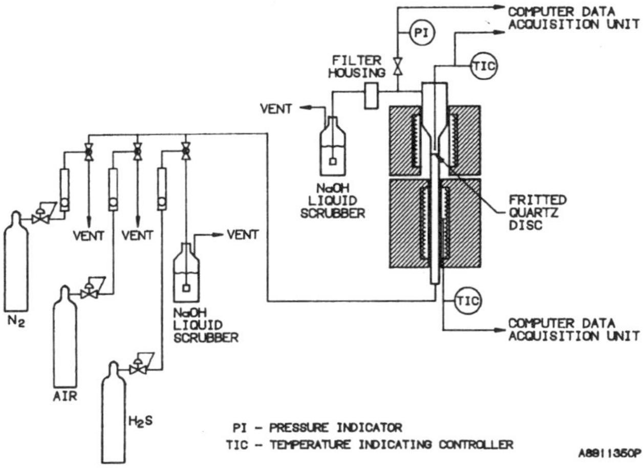

Figure 1. Process Flow Diagram of Quartz Fluidized-Bed Unit

2.3 Sulfation of Sulfided Sorbents

Tests were conducted in an ambient pressure thermogravimetric analyzer (TGA) reactor unit to determine the optimum temperature for maximum conversion of calcium sulfide to stable calcium sulfate and to determine the rate and the extent of the sulfation reaction.

$$CaS + 2O_2 = CaSO_4 \qquad \text{(A)}$$

A schematic diagram of the TGA unit used in this study is shown in Figure 2. The experimental procedure for TGA tests have been described in an earlier study.[1] The procedure for these tests included heating the sorbent in a nitrogen atmosphere to a predetermined temperature. At that point, the gas mixture containing oxygen was allowed to flow past the sorbent while the change in the sample weight was continuously monitored. The sample was exposed to the reactant gas containing oxygen until the sulfided sorbent was converted to its peak value.

The weight-versus-time data were translated into conversion-versus-time. Conversion of calcium sulfide to calcium sulfate (Reaction 3) is calculated using the following formula:

$$\% \text{ CaS Conversion} = \frac{(\text{Sample weight gain}) \times 32 \times 100}{(\text{Initial sample weight}) \times (\text{fraction of sulfur in sample}) \times (64)}$$

where 32 is the atomic weight of sulfur, 64 is the weight of 4 oxygen atoms added to the sample during the sulfation reaction

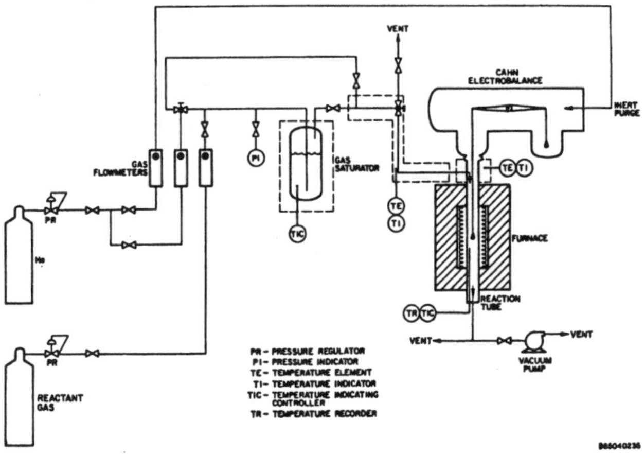

Figure 2. Process Flow Diagram of the Ambient-Pressure TGA Reactor Unit

The sulfided sorbents were also reacted with oxygen in the ¾-inch-diameter (0.02 m) fluidized-bed reactor at similar conditions. The sulfated samples were analyzed to make a comparison between the TGA and the fluidized-bed results.

2.4 Regeneration of Sulfided Sorbents

A series of tests were conducted in a Liquid-Bed Reactor (LBR) to regenerate samples of sulfided dolomite with water and carbon dioxide to determine the optimum operating condition. A sample of sulfided dolomite was regenerated at the optimum condition over four successive cycles with and four cycles without, adding binder during pelletization. The regenerated sorbents were analyzed after each test and the extent of regeneration was determined.

3 RESULTS AND DISCUSSION

A series of tests was conducted with fine dolomite particles to determine the optimum operating temperature for the sulfation reaction (stabilization). The range of temperatures in this series of tests was 1200° to 1800°F (650° to 980°C). The level of sulfation of the samples used in this series of tests was 43%. The results, as presented in Figures 3 and 4, indicate that the optimum temperature for maximum sulfation is in the range of 1500° to 1650°F (815° to 879°C). Lower reaction rates at lower temperatures are expected because the reaction rate constant is lower at lower temperatures. The lower maximum levels of conversion at higher temperatures are caused by local sintering at higher temperatures, as well as SO_2 evolution in the reactions.

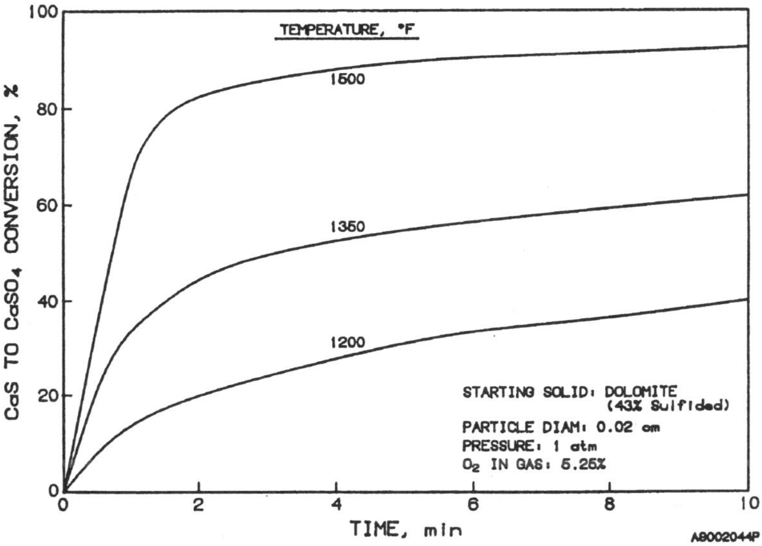

Figure 3. Effect of Temperature on Stabilization of Sulfided Dolomite

Figure 5 shows the effect of the level of sulfidation on the sulfation (stabilization) of the sulfided sample. The dolomite samples with lower sulfidation levels react faster and are sulfated to a higher extent compared to samples with higher levels of sulfidation. Lower levels of sulfidation correspond to thinner layers of calcium sulfide. Because Reaction (3) is probably diffusion limited, thinner layers of calcium sulfide should result in a higher reaction rate, which is consistent with the results seen in Figure 4. Similar results were observed for the limestone and dolomitic limestone that are presented in Figures 6 and 7, respectively. The results in Figure 6 indicate that high levels of sulfation should not be expected for limestone particles that have sulfidation levels over 37% (Ca/S <3). The extent of sulfation with the dolomitic limestone is between those of limestone and dolomite. A similar trend was observed for coarse particles.

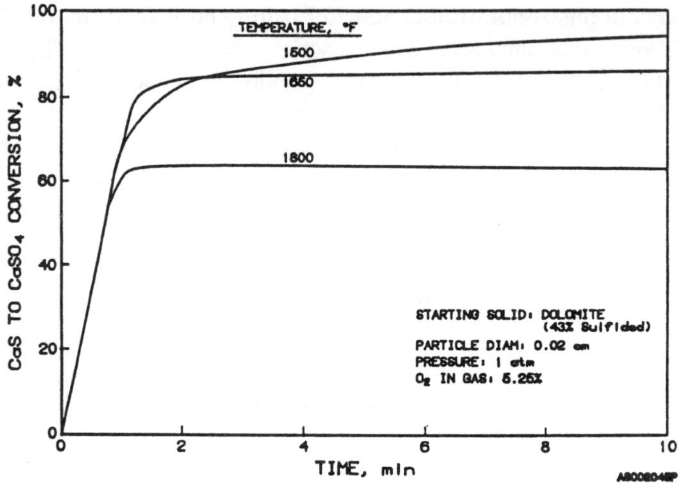

Figure 4. Effect of Temperature on Stabilization of Sulfided Dolomite

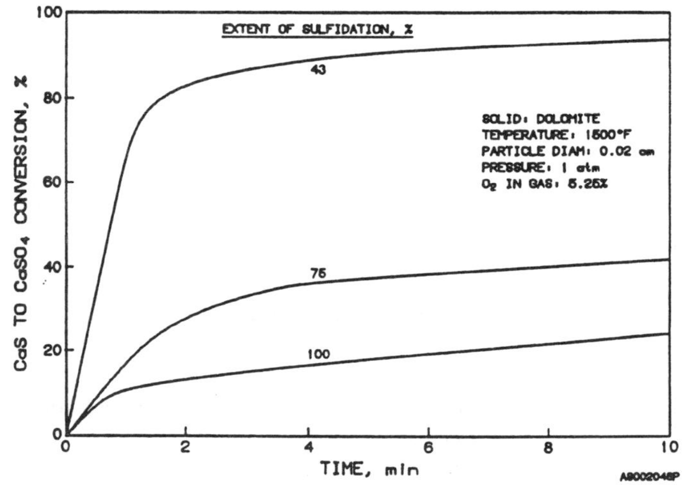

Figure 5. Effect of the Level of Sulfidation on Stabilization of Sulfided Dolomite

The sulfided samples of the three sorbents in two particle sizes were also reacted with oxygen in the ambient-pressure fluidized-bed unit. The solid products from these tests were analyzed and the extent of sulfation were determined. A comparison of the maximum level of sulfation achieved in the two units with different samples, presented in Figure 8, indicates that the results obtained in fluidized-bed experiments are in very good agreement with the TGA test results. The maximum extent of sulfation as a function of the level of sulfidation (Ca/S ratio in the gasifier), sorbent type (magnesium content), for the coarse particle and the fine particles are present in Figures 9 and 10, respectively. A sample of solid wastes from a U-GAS pilot plant gasification run that used limestone as sulfur capturing medium was also oxidized to convert calcium sulfide to calcium sulfate, and the result is also presented in Figure 9, which indicates that the level of sulfidation achieved with U-GAS solid discharge is consistent with those sorbents sulfided in the laboratory.

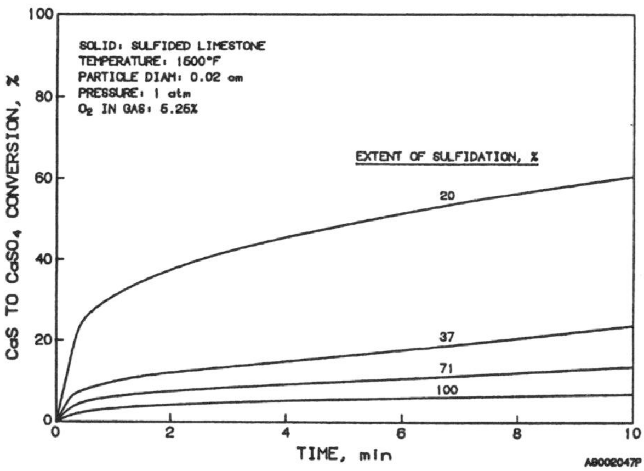

Figure 6. Effect of the Level of Sulfidation on Stabilization of Sulfided Limestone

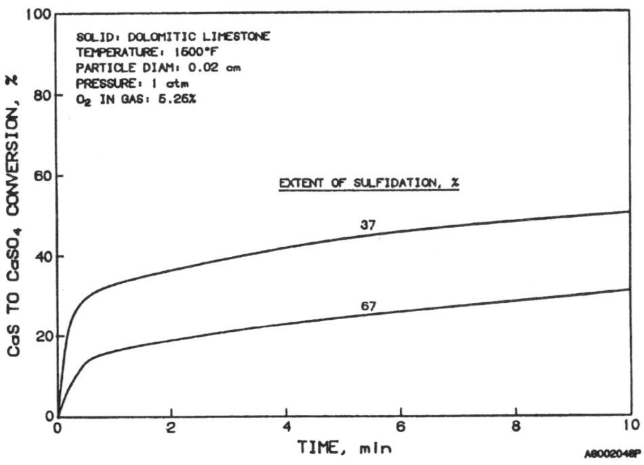

Figure 7. Effect of Sulfidation Level on Stabilization of Sulfided Dolomitic Limestone

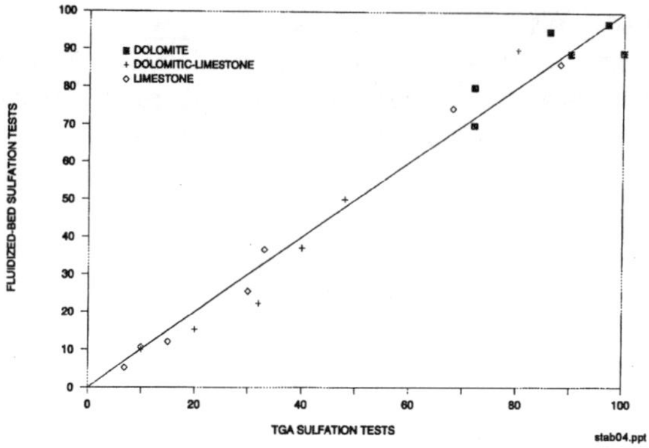

Figure 8. Comparison of the Maximum Level of Sulfation in TGA and Fluidized-Bed Units

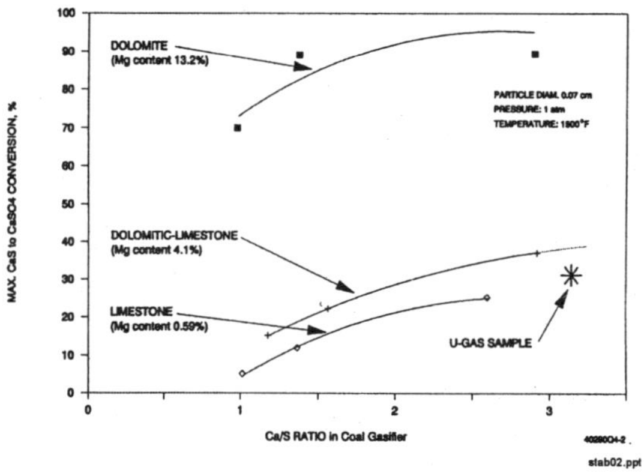

Figure 9. The Extent of Sulfation of Different Ca-Based Sorbents (Coarse Particles)

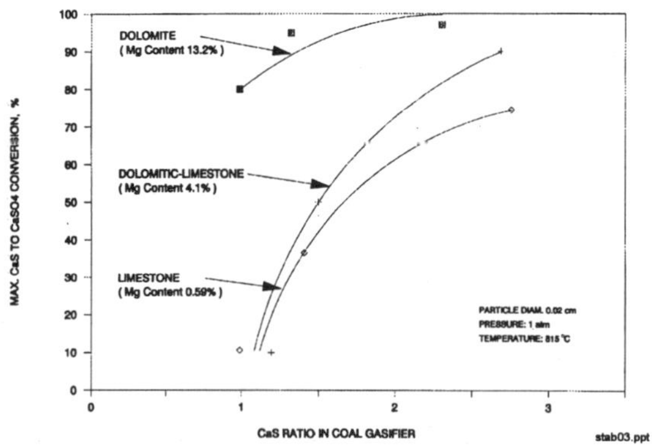

Figure 10. The Extent of Sulfation of Different Ca-Based Sorbents (Fine Particles)

The data in Figures 9 and 10 indicate that sulfided dolomite can be sulfated to levels approaching 100% when the Ca/S ratio in the gasifier is greater than 1.5. To achieve high levels of sulfation with sulfided limestone or dolomitic limestone, very high Ca/S ratio (Ca/S >5) and sorbent particles with diameters smaller than 0.008 inch (2 X 10-4 m) should be used.

The distribution of the reaction products as a function of temperature in the tests conducted in the fluidized-bed reactor are shown in Figures 11 through 13, indicating that the fraction of sulfur converted released on SO_2 significantly increases with increasing temperatures.

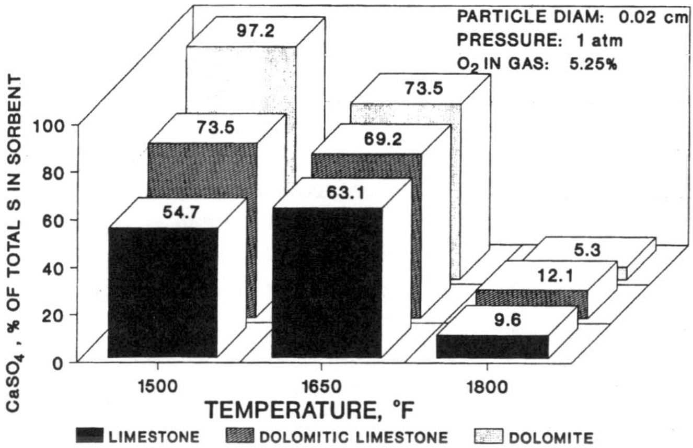

Figure 11. Maximum Conversion of CaS to $CaSO_4$ (Fine Partially Sulfided Sorbents)

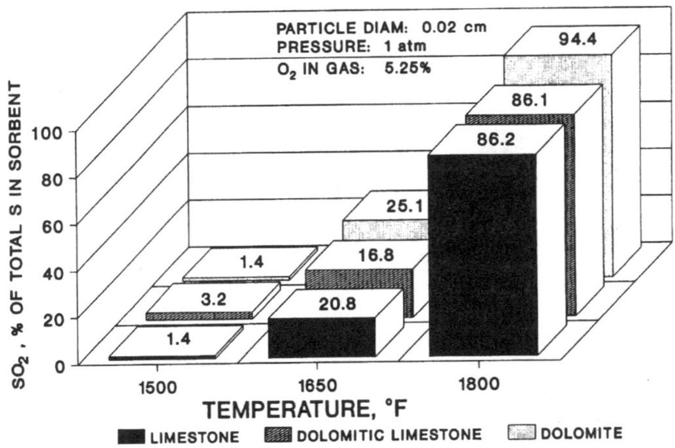

Figure 12. Maximum Conversion of CaS to SO_2 (Fine Partially Sulfided Sorbents)

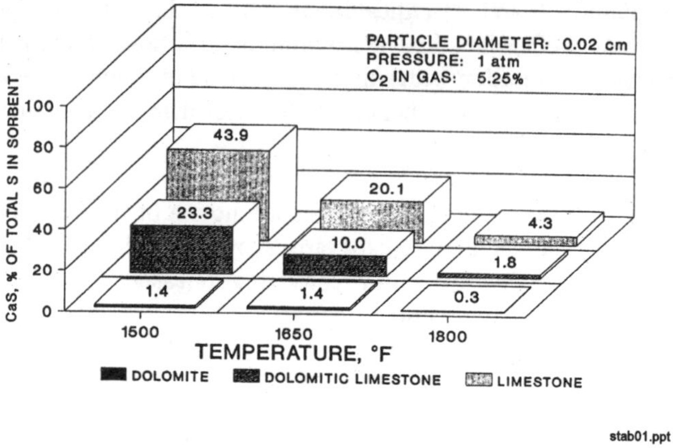

Figure 13. Unreacted CaS Remaining After Oxidation (Fine Partially Sulfided Sorbents)

Samples of sulfided dolomite were regenerated with water and carbon dioxide at 75 to 200 psig (0.6 to 1.6 MPa) in the temperature range of 160° to 260°F (70° to 125°C). Analyses of regenerated product indicated consistent sulfur removal levels above 97%. The optimum operating temperature and pressure appear to be at 160°F (70°C) and 75 psig (0.6 MPa), respectively.

A sample of sulfided dolomite was regenerated over four successive cycles with and four cycles without, adding binder during pelletization. The regenerated sorbents were analyzed after each test and the extent of regeneration was determined. The results of these analyses are given in Table 2, indicating that a high regeneration efficiency can be maintained in successive sulfidation/regeneration cycles.

Table 2. Regeneration Efficiency in Successive Cycles

Cycle No.	Binder	Regeneration Efficiency
1	--	97.7
2	--	98.7
3	--	98.6
4	--	9.7
1	--	97.7
2	Methyl Cellulose	98.7
3	Bentonite	97.3
4	Bentonite	97.8

The reactivities of the pelletized sorbents (with and without binder) were determined in the TGA before sulfidation tests were conducted in the following cycle. The sorbent pellets were crushed and screened to -80+100 mesh (dp = 0.01 cm) for the TGA test. A first order homogenous reaction model was applied and the initial rate constant (0% to 50% conversion) was calculated.

The sulfidation reaction rate constant in successive cycles with and without the binder is shown in Figure 14, indicating that the sorbent reactivity is not decreased in successive cycling in the SSRP process.

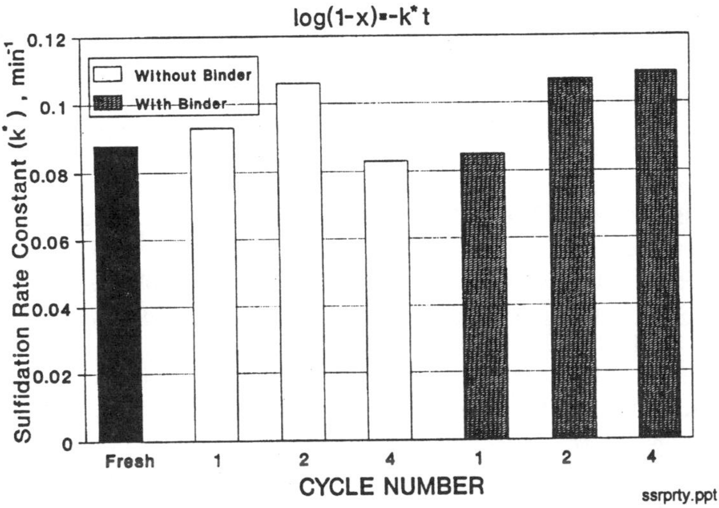

Figure 14. Reactivity of Spent Dolomite in Successive Regeneration Cycles

4 CONCLUSIONS

The results obtained in this study indicate that the highest levels of sulfation of the sulfided calcium-based sorbents can be achieved in the temperature range of 1500° to 1650°F (815° to 870°C).

The fraction of CaS converted to SO_2 during CaS oxidation significantly increases with increasing temperature in the range of 820° to 980°C, while the fraction of CaS remaining as unreacted decreases with increasing temperature.

The rate and the extent of sulfation reaction are higher for samples with lower levels of sulfidation. The maximum level of CaS, as well as selectivity toward conversion to $CaSO_4$ is higher for dolomite compared to limestone.

The results also indicate that sulfided calcium based sorbent can be regenerated with H_2O and CO_2 with high regeneration efficiencies without loss of sorbent reactivity toward H_2S.

5 REFERENCES CITED

1. Abbasian, J., Rehmat, A. and Banerjee, D. D., "Sulfation of Partially Sulfided Calcium-Based Sorbents," Ind. Eng. Chem. Res. 30, No. 8, 1991.
2. Abbasian, J., Rehmat, A., Leppin, D. and Banerjee, D. D., "An Advanced Coal Gasification Desulfurization Process." Paper presented at the 25th IECEC Conference, Reno, Nevada, August 1990.
3. Torres-Ordonez, R. J., Longwell, J. P. and Saroffin, A. F., "Physical Transformations During CaS(s) Oxidation," Energy & Fuel 3, 1989.
4. Torres-Ordonez, R. J., Longwell, J. P., and Saroffin, A. F., "Intrinsic Kinetics of CaS(s) Oxidation," Energy & Fuel 3, 1989.
5. Torres-Odronez, R., "The Oxidation of CaS(s) Crystals During Simulated Coal Combustion," Ph.D. Thesis, Massachusets Institute of Technology, June 1986.
6. Kamphius, B., Potma, A. W. and Spitsbergen, U., "Regenerative Sorbents for High Temperature Desulfurization of Coal Combustion Gases." Proceedings of the Sixth Annual International Pittsburgh Coal Conference 2, 994, September 10-14, 1990.
7. Yang, R. T. and Shen, M.-S., "Direct Evidence of the Existence of Gaseous Intermediates in the Calcium Sulfide-Calcium Sulfate Reaction," AICHE J. 25, No. 3, 548 (1979).
8. Yang, R. T. and Shen, M.-S., "Calcium Silicate: A New Class of Highly Regenerative Sorbents for Hot Gas Desulfurization," AICHE J. 25, No. 5, 812 (1979).
9. Pham, H. N., "An Oxidation Study of Fluidized-Bed Ashes Containing Calcium Sulfide." Paper presented at AICHE Annual Meeting — Clean Coal Technology II Session, Chicago, November 11, 1990.

Mixed Metal Oxide Sorbents

Ates Akyurtlu

Department of Chemical Engineering, Hampton University, Hampton, Virginia, USA

Abstract. High temperature desulfurization can be accomplished by using solid sorbents containing oxides of metals which form stable sulfides. These oxides by themselves cannot meet the requirements of Integrated Gasification Combined Cycle (IGCC) and Integrated Gasification Fuel Cell (IGFC) applications. When metal oxides are mixed together and properly calcined, they form solid solutions and/or new solid compounds with markedly different thermodynamics, kinetics and physical properties. Thus, by adding other oxides to the primary desulfurization species, it may be possible to improve its sulfur removal efficiency, prevent or slow the reduction to elemental metal, prevent sulfate formation, improve dispersion and modify porosity and pore size, and impart structural strength and stability.

Development of sorbents for high temperature operation (above 800 K) is still continuing while there is increased interest in the development of sorbents for operation in the medium temperature range (650-800 K). In the high temperature group various versions of zinc titanate sorbents and Z-Sorb are being tested under realistic conditions in pilot scale units. Work on copper-based sorbents are still at the laboratory level. In addition to these, there are reports of superior performance by several zinc-based proprietary sorbents.

1 Introduction

This paper consists of two parts. Part one (Sections 2 and 3) is a general overview of bulk mixed metal oxide sorbents for high temperature H_2S removal. Part two describes the recent developments in this field with an emphasis on the developments in the United States.

Considering the low cost of coal and the high efficiency of combined cycles the integration of coal gasifiers with combined cycles is expected to provide the lowest cost baseload electric power in the future. The high efficiency of the IGCC plants (42-44%) will be further improved by 1 to 2% if the low temperature sulfur removal systems are replaced by hot gas cleanup units.

High temperature desulfurization can be accomplished by using solid sorbents containing oxides of metals which form stable sulfides. Early applications have generally used oxides of calcium and iron. These oxides by themselves cannot meet the requirements of Integrated Gasification Combined Cycle (IGCC) and Integrated Gasification Fuel Cell (IGFC) applications

NATO ASI Series, Vol. G 42
Desulfurization of Hot Coal Gas
Edited by A. T. Atimtay and D. P. Harrison

1.1 Requirements for a High Temperature Desulfurization Sorbent

The requirements for high temperature desulfurization is illustrated by the following excerpt from a presentation by Daniel Cicero[1]:

- "Regenerability - the primary goal
- Withstand highly reductive gas atmospheres
- Temperatures 538 to 760°C (343 to 538 °C)
- Pressures 2 to 4 MPa
- Separate small volume percentage of H_2S from fuel gas (less than 0.5%)
- Regeneration temperatures: 580 to 780°C
- Recover reactivity and resist attrition
- Long life at low cost
- Sulfur removal efficiency - The sorbent must exhibit and consistently demonstrate less than 100 ppmv H_2S, with a goal of 10 ppmv in the outlet stream."

The high temperature interval for sulfidation was established considering the temperature of the gasifier gases, the operating temperature of the turbines, and the overall efficiency of the IGCC plant. There are several reasons for lowering this temperature interval. Recent economic analyses indicate that incremental gain in efficiency becomes small above 623 K. On the other hand the capital and operating costs increase significantly as the operating temperature is increased. These considerations have lead Department of Energy Morgantown Energy Technology Center (DOE-METC) to expand the temperature interval of interest down to 623 K.

The requirements for a high-temperature sorbent arise from economic and technical considerations. Long life and low cost requirements are important because of their favorable effect on the economics of the whole gas cleanup process. At the present there are sorbents which can effectively reduce the H_2S levels to the desired levels Current efforts are directed towards the improvement of the economics of the process. In both the moving-bed and the fluidized-bed processes, spent sorbent is continuously removed as material below a certain size and fresh sorbent is added to make up for this. The reduction in the sorbent attrition and deactivation rates will make the whole power generation process more economical. The sorbent costs are expected to come down from about $15/kg to about $7-8/kg due to the economy of scale when sorbents are produced in large quantities. However, it should not be expected that the cost of all sorbents will be reduced the similarly. The production method of some sorbents may involve the use of expensive materials and operations and may have inherently higher costs than the others.

1.2 Limitations of Single Metal Oxides

The requirement of high sulfur removal efficiency (very low levels of H_2S in the clean gas), combined with the limited temperature range for desulfurization, precludes a large number of metal oxides from being used as a high temperature desulfurization sorbent. Westmoreland and Harrison[2] in their 1976 paper screened 28 solids with respect to their high temperature desulfurization potentials. Two criteria were used: A minimum of 95% equilibrium sulfur removal efficiency and the stability of the unreacted solids. Only eleven metals were found to be possible candidates for desulfurization over at least a portion of the temperature range considered (673-1473 K). These are barium, calcium, cobalt, copper, iron, manganese, molybdenum, strontium, tungsten, vanadium, and zinc. Oxides of all these metals except barium and strontium whose behavior is similar to that of calcium but are inferior to calcium with respect to cost and temperature range, have been studied as high temperature desulfurization sorbents.

If vanadium is excluded because of its effect on the turbines and calcium is excluded because of the difficulty in regeneration, and if the lower sulfur efficiency requirement of below 10 ppmv H_2S is considered with the high temperature interval, the only possible metal oxides for use with a low Btu gas will be those of zinc and copper. For high Btu gas we can add cerium oxide to the list but it is still a very short list. Of course if the lower temperature interval is considered, the use of other metal oxides becomes thermodynamically possible but may be kinetically limited.

No single metal oxide can meet all the requirements stated above. For high efficiency you need to retard the reduction of copper and enhance the reduction of cerium oxide to a lower oxide. Highly reductive fuel gases reduce copper oxides to metallic copper and the sorbent can no longer meet the sulfur removal efficiency requirement thermodynamically. Reduction of zinc oxide leads to evaporative zinc loss at temperatures above 873 K. Because zinc sulfide has almost 50% larger volume than zinc oxide large pores with thick walls are needed to accommodate the volume increase and withstand the internal mechanical stresses. During regeneration under highly oxidative atmospheres zinc sulfate forms at temperatures below 973 K. Huge volume changes eventually result in the physical deterioration of the sorbent. Sintering leads to surface area loss and decline in sulfur capacity. At lower temperatures, although sintering and evaporation will no longer be a problem and thermodynamics will be more favorable, highly dispersed large surface area sorbents will be required to compensate for slower kinetics.

2 Mixed metal oxides

When metal oxides are mixed together and properly calcined they form solid solutions and/or new solid compounds with markedly different thermodynamics, kinetics and physical properties. Thus by adding other oxides to the primary desulfurization species, it may be possible to improve its sulfur removal efficiency, inhibit or slow the reduction to elemental metal, prevent sulfate formation, improve dispersion and modify porosity and pore size, and impart structural strength and stability. In some cases metal oxides do not form compounds and the oxides remain separate in a dispersed form such as the CuO-CeO system, but the existence of one oxide can still affect the properties of the other oxide. Generally one metal oxide is used as the primary sorbent to capture bulk of the sulfur. A different oxide is added to improve the thermodynamics and chemical properties of the primary metal oxide. A third oxide may than be added to provide support and modify the structurof the pellet.

The following examples will illustrate the improvements obtained by combining several metal oxides

2.1 Zinc-Based Sorbents

2.1. 1 Zinc Ferrite

Zinc ferrite ($ZnFe_2O_4$) is one of the first and most extensively tested mixed oxide sorbent. Upon contact with the reducing coal gas, zinc ferrite decomposes into ZnO and Fe_2O_3 which reduce and sulfide. One of the main shortcomings of zinc ferrite is the attrition loss upon cycling. Independently[3].

Swisher and Schwerdtfeger[4] have summarized the thermodynamics of the Zn-Fe-O-S system and discussed the implications for desulfurization with zinc ferrite. They indicate that $ZnFe_2O_4$ and Fe_3O_4 are mutually soluble and therefore Fe_3O_4 can be added to zinc ferrite without causing a phase transformation. Addition of iron oxide would reduce the vapor pressure of zinc over the compound, thus reducing zinc vaporization. The authors have calculated that at 1000 K the vapor pressure of zinc over zinc ferrite can be up to an order of magnitude lower than the vapor pressure over zinc oxide. In order to be able to take advantage of this low vapor pressure, zinc ferrite needs to be stable. Since as stated above, zinc ferrite is readily reduced to ZnO and either Fe_3O_4, FeO, or iron; when contacted with reducing gases, addition of iron oxide to zinc oxide cannot stabilize zinc oxide against reduction and reduce zinc loss.

Addition of iron would also expand the oxygen partial pressures over which the compound is stable. Swisher and Schwerdtfeger[4] have constructed a phase dominance area diagram for the Zn-Fe-O-S system at 1000K and for Fe/Zn = 2. The left side of this diagram where the oxygen partial pressure is about 10^{-20} atm or less, corresponds to coal gas conditions that exist during sulfidation. The right hand side corresponds to the regeneration conditions. Under the coal gas atmosphere the stable phases of iron are Fe_3O_4, FeO, or metallic iron. The phase dominance diagram also indicates that careful selection of regeneration atmosphere and temperature is needed to minimize the formation of iron and zinc sulfates.

Considering the large number of phase transformations the sorbent undergoes as it is cycled between sulfidation and regeneration conditions, the low mechanical strength of the sorbent is to be expected.

2.1.2 Zinc Titanate

Currently zinc titanate is the most widely tested and used high temperature desulfurization sorbent. Addition of TiO_2 to ZnO alters both the thermodynamics and structural properties if zinc oxide. In the Zn-Ti-O system the solid phases present depends on the Zn/Ti atomic ratio and the calcination temperature. With increasing temperature the following phase transformations are observed: $Zn_2Ti_3O_8 \rightarrow ZnTiO_3 \rightarrow Zn_2TiO_4$. Above 1273 K, Zn_2TiO_4 is the only stable phase for all $ZnO\text{-}TiO_2$ mixtures. It is also the stable phase for mixtures with $Zn/Ti \geq 2$ calcined at temperatures ≥ 973 K. At temperatures below 1073 K, all three phases may be present[5].

Swisher and Schwerdtfeger[4] have prepared a phase predominance area diagram for Zn-Ti compounds with Zn/Ti = 2 at 1000 K. It can be observed from this diagram that zinc titanate is stable down to an oxygen partial pressure of about 10^{-30} which means that zinc titanate will not be readily reduced in the coal gas atmosphere, thus evaporation losses of zinc will be less. The equilibrium constants for the reduction of zinc oxide and zinc titanate have been presented by Lew, et al.[6]. Their results show that the difference in reduction equilibrium constants becomes more significant as the temperature increases. The effect of TiO_2 addition on the initial rate of reduction was reported by Lew, et al.[6]. It can be seen from their results that the addition of up to about 25 mole % TiO_2 significantly decreases the reduction rate.

Another observation from the phase predominance area diagram of Swisher and Schwerdtfeger[4] is that zinc titanate does not undergo as many phase transitions as zinc ferrite as it is cycled between the coal gas and regeneration gas atmospheres. Thus from thermodynamic considerations it is expected that zinc titanate will stabilize the sorbent against zinc loss and improve its structural stability. These predictions have been confirmed by numerous test results.

Lew, et al.[5] report that the initial rate of sulfidation for ZnO at 873 and 973 K was approximately 1.5-2 times higher than for all Zn-Ti-O sorbents containing ≥ 25 mole % Ti and that the initial rate was similar for different zinc titanate phases. Figure 4 shows the effect of TiO_2 content on the initial sulfidation rate as reported by Lew, et al.[6]

Lew, et al.[5] also report that zinc titanate sorbents with less than 50 mole % TiO_2 had higher surface areas and pore volumes than the ZnO sorbent. They pointed out that TiO_2 inhibits the sintering of ZnS by acting as a physical barrier to the growth of ZnS particles.

Since zinc titanate reacts with H_2S without decomposing into constituent oxides, its desulfurization efficiency is expected to be different than that of zinc oxide. Flytzani-Stephanopoulos, et al.[7] calculated the equilibrium H_2S concentrations for various sorbents in a gas mixture with hydrogen to steam mole fraction ratio of 0.8. It is clear from their results that the sulfur removal efficiency of zinc titanate is lower than that of zinc oxide, but it can still bring the H_2S level in the exit gas stream to a level below 10 ppmv around 800 K. Experimental results indicate that for some zinc titanate sorbents H_2S removal efficiencies comparable to that of zinc oxide could be obtained[7]. Explanations offered for the observed sub-equilibrium H_2S levels include chemisorption, and formation of surface sulfides and solid solutions of different sulfidation thermodynamics than bulk sulfides.

As discussed above, zinc titanate sorbents with sulfur removal efficiencies comparable to that of zinc oxide can be produced and the higher operating temperature of zinc titanate sorbents compensates for the slower sulfidation kinetics. Therefore, the main trade-off is the loss of sulfur retention capacity since TiO_2 does not sulfide. A potential problem that might limit the number of sulfidation-regeneration cycles that a zinc titanate sorbent can be subjected to is the migration of zinc atoms to the surface upon repeated cycling. Siriwardane, et al.[8] investigated the Zn/Ti distributions in the fresh, sulfided, and regenerated zinc titanate samples. They observed an increase in the Zn/Ti ratio on the surface of the regenerated and sulfided samples compared to that on the fresh samples. They have concluded that the migration of zinc to the surface and elemental rearrangement during sulfidation and regeneration could lead to the presence of free zinc oxide. This in turn will cause zinc losses and sorbent deterioration.

The trade-offs involved in the selection of titanium oxide as an additive to zinc oxide can be summarized as follows:

Unfavorable effects

- H_2S removal efficiency of zinc titanate is less than that of ZnO.
- Initial sulfidation rate of zinc titanate is less than that of ZnO.
- Sulfur capacity of zinc titanate is less than that of ZnO.

Favorable effects

- Zinc titanate is more stable under reducing atmospheres.
- The initial rate of reduction of zinc titanate is significantly slower than that of ZnO.
- Zinc titanate has higher surface area and pore volume.
- Titanium oxide inhibits the sintering of ZnS.
- Zinc titanate has a higher structural strength.

2.1.3 Other Zinc-Based Sorbents

An example of ternary mixed oxides is the zinc titanate sorbents with MoO_3 additive[9,10]. Molybdenum was added as a structure modifier. It was found that molybdenum does not directly affect the desulfurization mechanism, except for the modification of the pore structure. Molybdenum-containing sorbents had higher median pore diameter and higher bulk density. These are desirable properties that enable the sorbent to accommodate volume increases during sulfidation. Crush strength and specific pore volume were found to depend mainly on the calcination temperature.

Another important zinc based sorbent is Z-Sorb which is a proprietary sorbent developed by Phillips Petroleum Company. It is based on zinc oxide supported in a porous matrix and contains a nickel oxide promoter. Originally it was developed for tail gas cleanup and modified for high temperature coal gas cleanup service. Since details of the composition and manufacture of this sorbent is not disclosed, this sorbent will be discussed in terms of its performance in the section on the recent developments.

2.2 Copper-Based Sorbents

The result of Flytzani-Stephanopoulos, et al.[7] indicate that while CuO has an excellent sulfur removal efficiency, elemental copper cannot reduce H_2S concentrations below 40 ppmv above 800 K. Unfortunately, when contacted with the reducing coal gases, copper oxide is rapidly reduced to elemental copper and thus, the superior thermodynamics of copper oxides cannot be utilized. If the copper oxides can be stabilized by the addition of other metal oxides, a sorbent with extremely high sulfidation efficiency may be possible.

Flytzani-Stephanopoulos, et al.[7] investigated the thermodynamics of the Cu-Fe-Al-O system and reported the following order of decreasing sulfidation efficiency:

$CuO, Cu_2O > CuAl_2O_4, CuAlO_2 > CuFe_2O_4, CuFeO_2 >> Cu >> Fe_3O_4 > FeAl_2O_4$

They have also investigated sorbents containing Cu-Fe-Al-O and Cu-Al-O solids. Their results indicate that the addition of Al_2O_3 into CuO enhances the dispersion of CuO and retards its reduction to copper metal. The copper aluminates, $CuAl_2O_4$, $CuAlO_2$ reduce more slowly than CuO and inhibit the reduction of copper to elemental form. Patric, et al.[11] report that the reduction rate of $CuAl_2O_4$ is an order of magnitude lower than that of CuO, and it proceeded through $CuAlO_2$ and Cu_2O intermediates. Thus the Cu-Al-O and Cu-Fe-Al-O mixed oxides have much higher sulfur removal efficiencies than that of metallic copper.

In Cu-Fe-Al-O ternary mixed oxide sorbents, aluminum oxide has a stabilizing effect on iron and prevents segregation of uncombined FeO. These sorbents also have a higher mechanical strength than Cu-Al-O sorbents. Another advantage of iron is the formation of high sulfur compounds such as Cu_5FeS_4 and $CuFe_2S_3$ which increase the sulfur loading of the sorbent (under sulfidation conditions the stable copper sulfide is Cu_2S)

Some Cu-Fe-Al-O sorbents were reported to remove H_2S to lower than 2-5 ppmv levels up to about 50% sorbent conversion[7,12]. Some CuO-Al_2O_3 sorbents produced prebreakthrough H_2S levels below 50 ppmv up to 100% conversion[11].

ElectroChem Inc. studied Cu_2O-Mn_2O_3 binary oxides as desulfurization sorbents[13,14]. They have shown that these sorbents contained two stable oxide phases, $CuMnO_2$ and $CuMn_2O_4$. During sulfidation these oxides easily sulfided to $CuMnS_2$ and $CuMn_2S_4$ which could easily be regenerated back to $CuMnO_2$ and $CuMn_2O_4$. Hydrated copper manganese sulfate was found to be unstable at 923 K. Sintering was found to be less than that observed with zinc ferrite. Sorbents with the stoichiometric composition gave the best performance. Effluent H_2S concentrations were less than 40 ppm up to a conversion of 50 %. Although these sorbents will have high sulfur capacity at high temperatures they may have instability problems in the presence of water vapor.

Abbasian, et al.[13] evaluated several copper-based mixed oxide sorbents for gas cleanup above 923 K. The reduction and sulfidation equilibrium constants they have calculated for various copper compounds show that $CuCr_2O_4$ and Cu_3MgO_4 have the highest stability against reduction and all cuprous oxides are more stable than cupric oxides. This observation suggest the involvement of intermediate cuprous oxides in sulfidation reactions in the reductive coal gas atmosphere. Cupric compounds were shown to have a better sulfidation equilibria than cuprous compounds and with respect to sulfidation equilibria copper aluminates were found to be the best mixed oxides. Experimental evaluation by Abbasian, et al.[13] confirmed the high reduction stability of $CuCr_2O_4$ and $CuCr_2O_4$ sorbents. In sulfidation tests the H_2S removal efficiency of copper chromite sorbent exceeded that of metallic copper up to high conversion levels.

Among the sorbents tested, $CuCrO_2$ was chosen for reduction stability, $CuCr_2O_4$ for H_2S removal efficiency and $CuMnO_2$ for high sulfur capacity.

2.3 Tin-Based Sorbents

Topsoe in Europe[14] has developed a SnO_2 - ZnO sandwich sorbent which consists of successive beds of a tin oxide sorbent and zinc a zinc oxide sorbent. Sulfur removal to a few ppm H_2S was achieved at 400 °C sulfidation temperature. Regeneration was done with steam-oxygen mixtures at 450-550°C. Sulfate formation was eliminated and sulfur capacity and mechanical strength were not affected by cycling through six sulfidation-regeneration cycles.

TDA Research[15] has developed a similar process but used the SO_2 from the regeneration of ZnO to regenerate SnO_2 and produce elemental sulfur. They have found that addition of 0.2 % NiO increased the sulfur capacity of tin oxide at 673 K by about an order of magnitude. They have concluded that the addition of NiO increases the sulfidation rate of tin and that NiO stabilizes SnO_2 against reduction by compound formation. Another advantage of NiO is that Ni is a good catalyst for NH_3 decomposition.

2.4 Vanadium Promoted Zinc Ferrite Sorbents

The possibility of the inhibition of zinc reduction during sulfidation and sulfate formation during regeneration by the addition of vanadium to zinc ferrite sorbent was investigated by Akyurtlu and Akyurtlu[16]. It was hoped that the high amount of chemisorption by vanadium will result in subequilibrium levels of H_2S and a relatively small decrease in capacity. Another advantage of vanadium would be the promotion of elemental sulfur formation. It was found that vanadium promotion imparted stability during cycling and resulted in the formation of larger amounts of elemental sulfur during regeneration. The unwanted effects of vanadium on the turbines may prevent the practical application of this sorbent.

3 Recent Developments

Development of sorbents for high temperature operation (above 800 K) is still continuing while there is increased interest in the development of sorbents for operation in the medium temperature range (650-800 K). In the high temperature group various versions of zinc titanate sorbents and Z-Sorb were tested under realistic conditions in pilot scale units. Work on copper-based sorbents are still at the laboratory level. In addition to these, there are reports of superior performance by several zinc-based proprietary sorbents.

Current focus in sorbent development is on the production of pellets or extrudates for moving-bed application and agglomerated or granulated particles for fluidized-bed or transport reactors. Each sorbent developed for a specific task is given a separate designation by the developers. These designations will be used when discussing each group of sorbents in the following sections.

3.1 Zinc-Based Sorbents for Moving-Bed and Fixed-Bed Applications

Side by side bench-scale testing of zinc titanate, L-4103B2 of United Catalysts Inc. (UCI) and Z-Sorb of Phillips Petroleum Company have been performed at General Electric (GE)[17]. Through eight sulfidation-regeneration cycles both sorbents achieved 99% H_2S removal prior to breakthrough with outlet H_2S levels of 10 ppm for Z-Sorb and 20 ppm for zinc titanate. Sulfur loadings on Z-Sorb were 10.6% at the outlet and 16.2 wt % at the inlet (maximum theoretical sulfur capacity of Z-Sorb is 20 kg/100 kg of fresh sorbent). The corresponding figures for zinc titanate were 6 and 8.9 wt % (the maximum theoretical capacity for zinc titanate is 25 kg/100 kg fresh sorbent). Although Z-Sorb has a higher sulfur capacity per unit mass, because of its lower bulk density sulfur loading per unit volume of reactor will be comparable for the two sorbents. When regenerated at 1033 K in a high SO_2/air atmosphere mechanical durability of zinc titanate deteriorated dramatically while Z-Sorb showed a higher mechanical durability at this elevated regeneration temperature indicating that the greater porosity of Z-Sorb could accommodate the large volume changes due to sulfate formation. Both zinc titanate and Z-Sorb were found to be suitable for operation in a moving-bed when regenerated at temperatures below 1033 K.

The results of further bench-testing of Z-Sorb III at GE[(10)] show a loss in bed capacity upon cycling. Analysis of samples taken from the inlet and outlet of the reactor indicated that the heavily sulfided samples from the inlet did not loose reactivity and capacity while the partially sulfided sorbent from the exit location suffered losses in reactivity and capacity. Capacity after two hours of sulfidation was about 60 to 65% of that of the fresh sorbent.

Z-Sorb III was also tested in the GE moving-bed pilot facility[(18)]. The sorbent lost more than 50% of its reactivity and sulfur capacity after 160 hours of testing in the pilot moving-bed facility. The effect of steam in the coal gas or the regenerator gas was suggested as a probable cause for this loss.

Z-Sorb was further tested in the GE moving-bed facility which contained a chloride guard bed before the reactor. The outlet H_2S concentrations were under 10 ppmv for 1000 ppmv inlet concentration. The outlet concentrations remained under 50 ppmv when inlet H_2S level increased to 3500 ppmv. 30% sulfur loading was achieved, but attempts to increase it to 50 % were not successful. Sulfur levels in the regenerated sorbent were around 0.5 % during steady operation with regenerator temperature controlled at 1005 K. Mechanical attrition of sorbent was less than 0.2 %.

A zinc titanate sorbent, T-2535M, formulated for moving-bed applications, and containing 2.5 % MoO_3 was tested in a bench-scale reactor[(10)]. The outlet prebreakthrough H_2S level was less than 10 ppmv. An average of 25 % sulfur loading was obtained at breakthrough. No significant effect of pressure was seen on this sorbent up to 2 MPa. ASTM attrition resistance of the pellets was 98.5 %. By the adjustment of the regeneration conditions pellet decrepitation could be prevented.

TDA Research[(19)] reported the development of a new sorbent structure for moving-bed applications. The sorbent is strong. It can expand and contract without affecting the pellet structure, it provides for fast diffusion of H_2S into the pellet, and it can have high loadings of the active material without spalling. After twenty sulfidation-regeneration cycles there was no change in the characteristics of the sorbent except for a slight improvement in the crush strength, porosity and surface area, even when regenerated with a gas stream containing 5% SO_2 at 933 K. The TDA sorbent TMZ-6 has a theoretical maximum sulfur loading of 19 kg S/100 kg sorbent, but the conversion was significantly higher(90%) than the conversion of L-4103 B2 sorbent (27-40%) and comparable to that of Z-Sorb (58-88%). Because of the higher bulk density of TMZ-6, the sulfur loading per unit reactor with TMZ-6 is about 1.5 times higher than those with the other sorbents. The future cost of the TDA sorbent was estimated to be about the same as the zinc titanate sorbent and lower than that of Z-Sorb.

One of the latest sorbents developed by TDA Research, ZVT12LN, could remove H_2S to less than 10 ppmv for several hours before breakthrough occurred when sulfided at 763 K with simulated Texaco gasifier gases containing 1% H_2S. Attrition resistance was better than 98%[20].

A sorbent, METC7, containing 50% zinc oxide is being developed at DOE-METC for fixed-bed operations[21]. METC7 was subjected to repeated sulfidation-regeneration cycles in a high-pressure fixed-bed. Sulfidations were at 260 kPa and 811 K with a feed gas containing 2000 ppmv H_2S. Regenerations were carried out at 272 kPa in three temperature stages (811, 866, and 922 K) with a gas containing 50% steam. Sorbent stabilized after the third cycle. Outlet H_2S concentration was zero for 15 hours. No physical deterioration of the sorbent was observed.

3.2 Zinc-Based Sorbents for Fluidized-Bed and Transport Reactors

The zinc titanate sorbent ZT-4 which was developed by the Research Triangle Institute(RTI) has been extensively tested. A life-cycle test (100 cycles) was performed in a bench-scale fluidized-bed unit using simulated Texaco oxygen-blown gasifier gas containing 11500 ppmv H_2S. The results of this tests can be summarized as follows[22]. The ZT-4 sorbent consistently reduced the H_2S content of coal gas from 11400 ppmv to < 20 ppmv at 1023 K and 1.5 MPa. The sorbent was found to be fully regenerable at 993 to 1033 K using 2-2.5% O_2 in N_2, with negligible residual sulfate. The sulfur capacity of the sorbent at the breakthrough of 500 ppmv H_2S ranged from 12.6 wt % in cycle 1 to 5.8 wt % in cycle 100. This decline in sulfur capacity was found to correlate with a decrease in the BET surface area, pore volume, and internal porosity. Attrition resistance of the sorbent after 100 cycles of testing was significantly higher (3% over 5-h loss) over that of the fresh sorbent (40 % 5-h loss). No significant sorbent loss (< 2 %) from the reactor was detected.

The sorbent ZT-4L prepared in a larger granulator by Contract Materials Processing for RTI was also tested in the same unit for 40 cycles with similar results. In this case the decrease in sulfur capacity correlated better with the decrease in the median pore diameter and pore volume which were attributed to the interaction of nickel with zinc titanate.

In a slipstream testing with actual coal gas over three cycles ZT-4L demonstrated 99 % removal of H_2S[23]. Attrition resistance of the sorbent improved significantly over three cycles.

ZT-4L was also tested in the Enviropower's pressurized fluidized-bed gasification pilot plant together with the zinc titanate sorbent UCI-5 manufactured by United Catalysts Incorporated (UCI)[24,25]. The reactivity and the sulfur capacity was found to be 1.5-2 times higher than that of UCI-5. UCI-5 was also regenerated with steam diluted air and steam was found to be as effective as nitrogen and did not have any detrimental effect on sorbent performance.. During steady operations, sorbent attrition and elutriation was found to be satisfactory. With use, the particle density increased and the porosity decreased for both sorbents. On the other hand the surface area of ZT-4L increased while that of UCI-5 decreased. The contact with real coal gas for 5-6 days was reported to have no detrimental effect on the sorbent physical and chemical properties.

Sulfate in the regenerated sorbent was 0.14 wt % for UCI-5 and 0.038 wt % for ZT-4L. Sulfided UCI-5 contained 9.4 wt % sulfur and ZT-41 contained 10.4 wt %.

A zinc titanate sorbent,CMP-5, was prepared by Contract Materials Processing (CMP) for RTI by spray drying[26]. This sorbent was tested in a bench-scale fluidized-bed reactor for several sulfidation-regeneration cycles. Sulfidations were performed at 873 K and 1.8 MPa with simulated coal gas containing 1% H_2S. For dry regeneration with 3-4% oxygen at 873-923 K there were no changes in the physical and chemical properties of the sorbent except for a decline in the attrition resistance. When 50 % steam was added to the regenerator gas degradation in sorbent performance was observed.

A sorbent developed at Department of Energy's Morgantown Energy Technology Center (DOE-METC) for fluidized-bed applications, METC6, was tested in a high-pressure fixed-bed[21]. The sorbent was chemically stable during the first 15 cycles of testing and was not adversely affected by steam regeneration. No sorbent decrepitation was observed.

Efforts at Phillips Petroleum[27] to produce a more steam resistant fluidized-bed sorbent resulted in formulation Z-Sorb IV. This sorbent did not loose sulfur capacity when regenerated in air diluted with 42 % steam at 922 K. Regeneration at 1033 K decreased the sulfur capacity from 24 wt % to 16-17 wt %, but this is significantly better than the performance of the earlier formulation Z-Sorb III.

Z-Sorb was also tested in a transport reactor[28]. There was a small loss in capacity over 15 cycles.

3.3 Novel High Temperature Mixed Oxide Sorbent Development

3.3.1 Nickel-Based Sorbents

Swisher, et al.[29] studied two nickel-based sorbents: 26 wt % Ni in alumina and 24 wt % Ni plus 7% Cu in alumina. Since Ni is a better sorbent than NiO, nickel is reduced to metallic state before sulfidation. Therefore these sorbents cannot be considered as true mixed oxide sorbents.

3.3.2 Manganese-Based Sorbents

Malcolm Hepworth's group in Minnesota have done extensive work on manganese oxide sorbents[30,31]. The sulfur removal efficiency of MnO is less than that of ZnO (57 ppmv H_2S at 1000 K) but manganese-based sorbents posses high chemical and physical stability and good mechanical strength. Although the manganese ore-based sorbents naturally contain other oxides, since they are not mixed intentionally they will not be considered as mixed metal oxides.

3.3.3 Copper-Based Sorbents

Copper-based sorbents appear to be the most promising novel mixed oxide sorbents. The research is mainly conducted at Institute of Gas Technology with cooperation from

General Electric Research and Tufts University[32,33,34]. Abbasian, et al.[33] studied copper oxide in combination of oxides of Cr, Ce, Al, Mg, Mn, Ti, Fe, and Si. The cyclic sulfidation-regeneration tests were carried out in a packed-bed reactor at ambient pressure, in the range of 1023-1123 K in sulfidation tests and at 1023 K in regeneration tests. A simulated fuel gas containing 5000 ppmv H_2S was used for sulfidation, while the regeneration gas contained 30% air and 70% nitrogen. Their initial results showed that the mixed binary oxides of copper and chromium posses the best properties for high temperature desulfurization applications. In a 25-cycle test of copper chromite sorbent at 1023 K the sorbent exhibited excellent desulfurization efficiency (H_2S<10 ppm) up to 50% conversion but the reactivity of the sorbent decreased with cycling. This was attributed to the agglomeration of metallic copper produced by reduction leading to a decrease in the surface area.

Li and Flytzani-Stephanopoulos[34] studied the effectiveness of copper-chromium and copper-cerium oxide sorbents in the desulfurization of fuel gas in the temperature range 923 to 1123 K. Their results showed that the reactivity of the sorbents with equimolar ratios of $CuO{:}Cr_2O_3$ and $CuO{:}CeO_2$ at temperatures between 923 and 1123 K were high enough to remove H_2S in the fuel gas to less than 1-10 ppm. Good regeneration was obtained with 30 % air in nitrogen above 1023 K. The CuO-Cr_2O_3 was found to be more difficult to reduce because of compound formation. The activation energy for the reduction of CuO-Cr_2O_3 was about twice that for CuO-CeO_2, but the activation energies for sulfidation were closer to each other. Water vapor was found to reduce the sulfur capacity of both sorbents at high conversions, but did not affect the reduction rate or the sulfidation rate at low conversions.

Ayala, et al.[32] report some work on developing copper-based sorbents for fluidized-bed applications by modifying the chromium-rich copper chromite sorbents by the addition of Fe, Mo, or Co to enhance their reactivities.

3.3.4 Medium Temperature Desulfurization Sorbents

After the recent economic analyses indicating that desulfurization operations at temperatures as low as 600 K may be economically feasible, work on the development of sorbents which can perform at these lower temperatures gained importance. Some existing sorbents were reevaluated at lower temperatures and efforts to develop novel sorbents for operation at the medium temperature range were initiated.

Grindley and Everitt[35] investigated the sulfidation of zinc ferrite in the temperature range 588-811 K. They have found that prereduction of zinc ferrite was necessary for good desulfurization. They concluded that zinc ferrite can be used below 800 K by contacting the regenerated sorbent with clean coal-derived gas. They point out that this step will also eliminate the residual sulfates that may be present .

Greenwood, et al.[36] and Khare, et al.[27] report the results of fixed, fluidized, and moving bed testing of Z-Sorb. Z-Sorb was found to have a high sulfidation reactivity at temperatures between 643 K and 723 K. When sulfided at 672 K and 2 MPa, Z-Sorb had a sulfur capacity of 20.3 wt % which declined to 17.6 % after eight cycles. After a following sulfidation at 622 K, its sulfur capacity was 15.4 wt %. The fluidized-bed tests were conducted at 20 atm with a gas stream containing 1 % H_2S, 5 % H_2O, 28.5 % H_2, 58.9 % CO,

In another report Khare, et al.[37] indicate that with Z-Sorb sulfur loadings of 6 to 22 % can be obtained in the temperature range 588-644 K with a coal gas containing 4.2 % H_2S in CO_2/N_2 mixture. Z-Sorb retained 40 to 50 % of its capacity at 811 K. A fixed-bed formulation, when subjected to 700 cycles at a sulfidation temperature of 700 K, retained a large portion of its sulfur capacity which fell to 8 % from 14 %. A fluidized-bed formulation of Z-Sorb was subjected to several 672 K sulfidation - 873 K regeneration cycles at 20 atm. with Shell gas containing 1 % H_2S. Over eight cycles its sulfur capacity remained around 18 %.

Using a thermodynamic analysis in the 623-823 K range at 2 MPa and for U-GAS and Texaco fuel gases, Ayala, et al.[32] selected Zn, Fe, Ce, and Cu as the primary desulfurization species. Molybdenum, Co, Cu, and Fe were selected as desulfurization enhancers. Titanium, Zr, and Ce were selected as supports and structure stabilizers.

RTI and Hampton University are cooperating in the development of medium temperature zinc-based sorbents[38,39]. Most recent sorbent MRCH-25 was tested in a laboratory-scale reactor for 100 sulfidation-regeneration cycles. Sulfidation was at 700 K with a simulated gas containing 1% H_2S, and regeneration was done with 10% air in nitrogen at 850 K. Conversion was reported to increase with cycling becoming 100% after the thirtieth cycle. The sorbent was also reported to retain 100 % of its sulfur capacity over 100 cycles.

4 Conclusion

Zinc titanate and Z-Sorb are currently the most advanced mixed oxide sorbents. Among the novel mixed oxide sorbents copper-based sorbents are the most promising. They need to undergo multi-cycle pilot scale testing to validate their superior performance under more realistic conditions.

Medium temperature mixed oxide sorbent development shows a big potential for research. and opens the door for the consideration of metal oxides which were found to be unsuitable for use at high temperatures

References

1. Cicero, Daniel C. : "METC Hot Gas Desulfurization Program Overview," *Proceedings of the Coal-Fired Power Systems 94 -- Advances in IGCC and PFBC Review Meeting,* Volm. I, pp. 219-221, 1994.

2. Westmoreland, P. R. And D. P. Harrison: "Evaluation of Candidate Solids for High-Temperature Desulfurization of Low-Btu Gases," *Environ. Sci. Technol. 10,* 659-661, 1976.

3. Pineda M., J. L. Fierro, J. M. Palacios, C. Cilleruelo, and J. V. Ibarra: "Kinetic

Behavior and Reactivity of Zinc Ferrites for Hot Gas Desulfurization," *J. Materials Sci. 30*, 6171-6178, 1995.

4. Swisher, J. H. And K. Schwerdtfeger: "Thermodynamic Analysis of Sorption Reactions for the Removal of Sulfur from Hot Gases," *J. Mater. Eng. Perform. 13,* 565-572,1992.

5. Lew, S., A. F. Sarofim, and M. Flytzani-Stephanopoulos: "Sulfidation of Zinc Titanate and Zinc Oxide Solids,"*Ind. Eng. Chem. Res. 31,* 1890-1899, 1992.

6. Lew, S., A. F. Sarofim, and M. Flytzani-Stephanopoulos: "Mechanistic and Kinetic Studies of High-Temperature Coal Gas Desulfurization Sorbents," final report on DE-FG22-88PC88927 to DOE-PETC, 1991.

7. Flytzani-Stephanopoulos, M., G. R. Gavalas, K. Jothimurugesan, S. Lew, P. K. Sharma, M. J. Baajewicz, and V. Patrick: "Detailed Studies of Novel Regenerable Sorbents for High-Temperature Coal-Gas Desulfurization," Final report for DE-FG21-85MC22193 to DOE-METC, 1987.

8. Siriwardane, Ranjani V., James A. Poston, and Steven D. Woodruff: "Sulfur Capture Mechanisms," *Proceedings of the Twelfth Annual Gasification and Gas Stream Cleanup Systems Contractors Review Meeting*, volume II, pp. 498-501, Morgantown, WV, September 1992.

9. Ayala, R. L., E. Gal, and R. P. Gupta: " Enhanced Durability of High-Temperature Desulfurization Sorbents for Moving-Bed Applications," *Proceedings of the Twelfth Annual Gasification and Gas Stream Cleanup Systems Contractors Review Meeting*, volume I, pp. 230-239, Morgantown, WV, September 1992.

10. Ayala, Raúl E. , Timothy Chuck, Raghubir P. Gupta: "Moving-bed Sorbents," *Proceedings of the Advanced Coal-Powered Systems '95 Review Meeting*, Volume II, pp. 591-600, 1995.

11. Patrick, V., G. R. Gavalas, M. Flytzani-Stephanopoulos, and K. Jothimurugesan: "High-Temperature Sulfidation-Regeneration of $CuO-Al_2O_3$ Sorbents," *Ind. Eng. Chem. Res. 28,* 931-940, 1989.

12. Tamhankar, S. S. , M. J. Baajewicz, G. R. Gavalas, P. K. Sharma, and Flytzani-Stephanopoulos, M.; "Mixed-Oxide Sorbents for High-Temperature Removal of Hydrogen sulfide,"*Ind. Eng. Chem. Des. Dev. 25,* 429-437, 1986.

13. Abbasian, J., A. Rehmat, M. Flytzani-Stephanopoulos, and Z. Hu: "Development of Novel Copper-Based Sorbent for Hot Gas Cleanup," paper presented at the 1992 AIChE Spring National Meeting, New Orleans, LA, 1992.

14. Hojllund, N. and I. G. Sigurdardottir: "Hot Gas Cleaning of Coal Gases by

Sequential Absorption," *Proceedings of Twelfth Annual International Pittsburg Coal Conference*, pp. 1074 -1079, 1995.

15. Copeland, Robert J., Dan Feinberg, D. Wickham, Brian Windecker, and J. Yu: "High Temperature Hydrogen Sulfide Removal with Stannic Oxide" *Proceedings of the Coal-Fired Power Systems 94-- Advances in IGCC and PFBC Review Meeting*, Volume I, pp. 444-457, 1994.

16. Akyurtlu, J. F. and A. Akyurtlu: "Hot Gas Desulfurization With Vanadium Promoted Zinc Ferrite Sorbents,"*Gas Sep. Purif. 9,* 17 - 25, 1995.

17. Ayala, R. E. , Timothy Chuck, Eli Gal, and Raghubir P. Gupta: "Development of Sorbents for High-Temperature Desulfurization in Moving-Bed Systems," *Proceedings of the Coal-Powered Systems 95 -- Advances in IGCC and PFBC Review Meeting*, Volume II, pp. 637-645, 1994.

18. Ayala, Raúl E., Alan Feitelberg, Anthony Furman: "Development of a High Temperature Moving-Bed Coal Gas Desulfurization System," *Proceedings of Twelfth Annual International Pittsburgh Coal Conference*, pp. 1053 -1058, 1995.

19. Copeland, Robert J. , Mike Cesario, Margarita Dubovic, Dan Feinberg, and Brian Windecker: "A Long Life ZnO-TiO_2 Sorbent," *Proceedings of the Advanced Coal-Powered Systems '95 Review Meeting*, Volume I, pp. 394-401, 1995.

20 Copeland, Robert J., Mike Cesario, Margarita Dubovic, Dan Feinberg, Brian Windecker, and J. Yang: "A Long Life ZnO-TiO_2 Sorbent for moving-bed reactors," to be presented at the 1996 International Pittsburgh Coal Conference, Pittsburgh, PA, July 1996.

21. Sirivardane, Ranjani V. , Ulrich Grimm, James Postsn, and Steven Monaco: "Desulfurization Sorbent Development at the Morgantown Energy Technology Center," *Proceedings of the Coal-Fired Power Systems 94-- Advances in IGCC and PFBC Review Meeting*, Volume I, pp. 662-667, 1994.

22. Gupta, Raghubir P. and Santosh K. Gangwal: "Bench-Scale Testing of Fluidized-Bed Sorbents - ZT-4," *Proceedings of the Advanced Coal-Powered Systems '95 Review Meeting*, Volume I, pp. 229-240, 1995.

23. Portzer, Jeffrey W. and Santosh K. Gangwal: "Slipstream testing of Hot-Gas Desulfurization with Sulfur Recovery," *Proceedings of the Advanced Coal-Powered Systems '95 Review Meeting*, Volume I, pp. 220-228, 1995.

24. Konttinen, Jukka, Reza Ghazanfari, Arto Lehtovaara, Wahab Mojtahedi: " Pilot Scale Experience on IGCC Hot Gas Cleanup," *Proceedings of the Advanced Coal-Powered Systems '95 Review Meeting*, Volume I, pp. 202-214, 1995.

25. Lehtovaara, Arto and Jukka Konttinen: “Pilot Scale experience on IGCC Hot Gas Cleanup,” *Proceedings of Twefth Annual International Pittsburgh Coal Conference*, pp. 1065 -1072, 1995.

26. Gupta, Raghubir P. : “Spray- Dried Fluid-bed Sorbents Tests - CMP-5,” *Proceedings of the Advanced Coal-Powered Systems ‘95 Review Meeting*, Volume II, pp. 601-608, 1995.

27. Khare, G. P., G. A. Delzer, D. H. Kubicek, and G. J. Greenwood: “Hot Gas Desulfurization with Phillips Z-Sorb Sorbent in Moving Bed and Fluidized bed Reactors,” *Environmental Progress, 14,* 146-150,1995.

28. Campbell, William M. and Gunnar B. Henningsen: “Hot gas Desulfurization Using Transport Reactors,” *Proceedings of Twelfth Annual International Pittsburgh Coal Conference*, pp. 1059-1064, 1995.

29. Swisher, J.H., M. Jhunjhunwala, L. D. Gasper-Galvin, T. H. Gardner, and K. Hammerbeck, “Properties of Sulfur Sorbents Containing Dispersed Nickel in anAl_2O_3 Matrix,”*J. Mater. Eng. Perform., Vol. 5(No. 2),* pp. 247- 255, 1996.

30. Hepworth, Malcolm T. and Rachid Ben-Slimane: “Hot Coal Gas Desulfurization With Manganese-Based Sorbents,” *Proceedings of the Advanced Coal-Powered Systems ‘95 Review Meeting*, Volume I, pp. 367-375, 1995.

31. Hepworth, Malcolm T. and Rachid Ben-Slimane: “Hot Coal Gas Desulfurization With Manganese-Based Sorbents,” *Proceedings of the Coal-Fired Power Systems 94-- Advances in IGCC and PFBC Review Meeting*, Volume I, pp. 337-355, 1994.

32. Ayala, Raúl E., Venkat S. Venkataramani, Javad Abbasian, Andy H, Hill: “Advanced Low-Temperature Sorbents,” *Proceedings of the Advanced Coal-Powered Systems ‘95 Review Meeting*, Volume I, pp. 407-416, 1995.

33. Abbasian, Javad, Andy H. Hill, Francis S. Lau, Maria Flytzani-Stephanopoulos, and Franklin I. Honea: “Fuel Gas Desulfurization at elevated Temperatures with Copper-Based Sobents,” *Proceedings of Twelfth Annual International Pittsburgh Coal Conference*, pp. 1189 - 1194,1995.

34. Li, Z. and M. Flytzani-Stephanopoulos: “Development of Novel Copper-Based Sorbents for Hot-Gas Cleanup,” final report prepared for Institute of Gas Technology, 1994.

35. Grindley, Thomas and C. Elaine Everitt: “Absorption of Hydrogen Sulfide by Zinc Ferrite in the Temperature Range 315 to 538°C,” *Proceedings of the Coal-Fired Power Systems 94-- Advances in IGCC and PFBC Review Meeting*, Volume I, pp. 431-443, 1994.

36. Greenwood, G. J., G. P. Khare, D. H. Kubicek, G. A. Delzer, and D. L. Kinsinger: "Pilot Plant Tests of Z-Sorb Sorbent," *Proceedings of the Coal-Fired Power Systems 95-- Advances in IGCC and PFBC Review Meeting*, Volume I, pp. 241-246, 1995.

37. Mangnus, P. J., A. Bos, and J. A. Moulijn: "Temperature-Programmed Reduction of Oxidic and Sulfidic Alumina-supported NiO, WO_3, and NiO-WO_3 Catalysts," *J. Catal.* 146, 437-448 (1994).

38. Khare, G. P., G. J. Greenwood, D. H. Kubicek, and G. A. Delzer: "Regenerable Sorbents for Low Temperature Desulfurization of Coal Gas," *Proceedings of Twelfth Annual International Pittsburgh Coal Conference*, pp. 1080 -1085, 1995.

39. Jothimurugesan, K., A. A. Adeyiga, and S. K. Gangwal: "Development of Advanced Hot-Gas Desulfurization Sorbents," *Proceedings of the Advanced Coal-Powered Systems '95 Review Meeting*, Volume I, pp. 402-406, 1995.

40. Jothimurugesan, K. and Santosh K. Gangwal: "Advanced Hot-gas Desulfurization Sorbents," to be presented at the *Advanced Coal-Powered Systems '96 Review Meeting*, Morgantown, WV, July 1996.

Development of Supported Sorbents for Hydrogen Sulfide Removal from Fuel Gas

A. T. Atimtay

Environmental Engineering Department, Middle East Technical University,
06531 Ankara, TURKEY

Abstract. A general overview of the supported sorbents for the removal of H_2S from coal gas at high temperatures is given. As support material γ-Alumina, zirconia, zeolites, etc. can be used. A single or a mixture of metal oxides can be dispersed on the surface and in the pores of the support material. Thus, it may be possible to improve dispersion of active metal oxides in the solid matrix, to increase the surface area for the reaction, and to impart structural strength and stability. As an example to supported sorbents, a sorbent developed by using SiO_{2-} rich zeolite as a support and CuO-MnO as active metal oxides is presented here. The zeolite support in the sorbent will make the sorbent price higher than bulk sorbents. However, the higher stability can compansate the higher cost when in continuous use in fluidized bed or moving bed reactors.

1 Introduction

Today in the world there is an ever increasing demand for electrical energy due to the increase in population, economic development, and technical advancements. Within the last 25 year period the total energy consumption has almost doubled. For the purpose of meeting this increasing demand, research is conducted both on new sources of energy and on the improvement of the present energy production technologies. The efficiency of the energy production technology is quite an important issue. In both, i.e. the new energy sources and the improvement of the present technologies, pollution prevention becomes indispensable for the sake of sustainable development.

The energy demand of the world is met by different sources of energy. Currently, only about 15% of the world energy use is from renewable sources such as wood, crop residues, hydroelectric, solar, wind, geothermal and biomass. Unfortunately, these are usually quite local and their exploitation is quite capital intensive as reported by Sage[1]. Therefore, immediate developments in this area are not very likely.

The World Energy Council study [2] and statistical review of world energy[1] suggest a massive increase in future global energy demand from 8 Gtoe in 1994 for a population of about 6 billion to 20 Gtoe in 2050 for a population estimate of 10.1 billion. At the present the alternatives for energy sources are mainly non-renewable

NATO ASI Series, Vol. G 42
Desulfurization of Hot Coal Gas
Edited by A.T. Atimtay and D. P. Harrison

ones such as coal, oil, and natural gas due to their abundance. The use of coal-gas and its related technology for energy generation is becoming quite an object of interest. The reason is the hope to generate electrical energy at higher efficiencies than today by using clean, coal-derived, pressurized gaseous fuel and utilize it in a combined steam/gas turbine cycle. This is why **"Integrated Gasification Combined Cycle (IGCC)"** system seems to be one of the most promising technologies. In an IGCC system energy conversion efficiencies of up to 43 % can be reached today as reported by Bissett and Strickland[3] and Harrison[4].

Although in IGCC systems about 30 % less SO_x and NO_x are being emitted per unit of energy generated as compared to pulverized coal plants, H_2S formed in the gasifier creates a problem as stated by Harrison [4]. Approximately 95 % of the sulfur present in the coal is converted to H_2S in the gasifier[5].

Having H_2S removed without cooling the coal gas is a very important task for having a high efficiency in the IGCC system. This fact emphasizes the importance of **"hot gas cleanup (HGCU)"**. For the hot gas desulfurization various sorbents are being developed and used. Various metal oxides in pure (single) and mixed forms have been tested in order to remove H_2S. Metal oxides can also be loaded on a support material, thus structurally more stable form of the sorbent is obtained. When a sorbent is developed the requirements for a good sorbent are: Performance stability, high H_2S removal efficiency and high sulfur capacity, good regeneration characteristic, high structural stability and attrition resistance.

The reason for the development of the supported sorbents lies also in the fact that there is a possibility of using supported sorbents in a fluidized bed reactor for H_2S removal. The advantages of the fluidized bed reactors over the fixed bed reactors will be :

- Excellent gas-solid contact due to vigorous agitation of small particles
- Temperature uniformity
- Minimum diffusional resistance
- Faster overall kinetics
- Possibility of pneumatic transport
- Sulfidation and regeneration can be conducted in separate fluid beds, therefore continuous sorbent exchange between the two beds can be possible
- Continuous steady-state operation
- Capability of continuously adding or removing sorbents
- Control of highly exothermic reactions e.g. during regeneration
- Prevention of solid segregation due to vigorous solid agitation
- Ability of withstanding greater particle loading in the process stream than fixed beds
- Heat removal system can be incorporated in the sorbent regeneration unit
- Absence of shut-off valves operating at HTHP and high corrosion conditions
- Lower sorbent capacity/utilization is acceptable.

All these advantages make the fluidized bed reactors superior to the fixed bed reactors. Therefore, if suitable sorbents which can withstand fluidized bed conditions can be developed, many drawbacks that fixed beds introduce can be overcome.

This lecture will review the developments in supported sorbents for hydrogen sulfide removal from coal gas.

2 Previous Studies

As indicated before about 95 % of the sulfur contained in the coal is turned into H_2S during the gasification process [5]. The produced H_2S should be lowered from the typical gasifier output concentration of 5000 ppm down to a tolerable limit of about 150 ppm as stated by Woods, et al.[6].

Research has been conducted to find the best metal oxide for desulfurization of coal gas. The history of supported sorbents dates back to 1936. The earliest work to evaluate copper sorbents was conducted at Johns Hopkins University by Huff [7]. Huff mainly used copper oxides and prepared sorbents by mixing metal oxides with binder components such as clay, pumice, alundum and lime. Bench scale desulfurization experiments were done in the temperature range of 350-820°C. The desulfurization capacity of the sorbent declined with the repeated cycles of sulfidation and regeneration, therefore sustainability of the sorbent failed. The next significant use of copper in hot gas desulfurization was that of Nachod, et al. [8]. Heavy metals like copper, zinc and lead were supported on aluminosilicates. This sorbent was used to remove sulfur from petroleum distillates and H_2S from gases in the temperature range of 90-540°C. Bench scale experiments were conducted with this sorbent, however, the work was terminated because of low removal efficiency [8]. The regeneration was done with air/steam mixture.

Another study of supported sorbent was carried out by Lewis [9] who used copper supported on silica to desulfurize H_2S/H_2 gas mixtures. The temperature range studied was 300 - 1000°C. The final application of the sorbent was intended for cleanup of coke oven gas. The desulfurization evaluation tests were done with concentrations of H_2S of 1-1.5 % in hydrogen with the sorbent in fluidized bed or stirred gas-solid reactors. Sulfided sorbents were regenerated by oxidation with air in the temperature range of 400-740°C. Sulfur removal efficiencies of 98.5% were obtained in bench scale tests and the optimum desulfurization temperature was found to be 350 - 600°C. Information about this process is only available from the patent literature and the reason for termination of the work is reported to be unknown in a report prepared for DOE by Jalan [10].

One of the other efforts to develop supported sorbents is by Wheelock [11, 12] to use La_2O_3 on alumina. The temperature range of the study for desulfurization and regeneration was 150-930°C. The regeneration was done with air/steam mixture. Bench scale experiments were conducted. The research was sponsored by Exxon, however not much information is available about the study because it was held proprietary. Another study conducted by Steiner [13] was supported by Foster-Wheeler. Ni was used on a support. The sorbent was tested in the temperature range of 540-790°C in bench scale experiments. The regeneration temperature was the same and air was used for regeneration. However, the results of the study was not published.

For the supported sorbent development studies where ZnO is used as the metal oxide on a support, ZnO process developed by Institut Francais du Petrole (IFP) in 1979 can be mentioned [14]. The sorbent was tested in a bench scale experiment. The sulfidation temperature range was 400-600°C and the regeneration temperature range was 600-900°C. The project was supported by IFP and the results of the study was reported by Onursal [14].

Sorbent development studies have also been carried out by Jalan[10] from the Giner, Inc. located at Waltham, MA. The studies were initially geared to high temperature desulfurization/regeneration reactions of metal oxides for fuel cell developments. Work has been conducted to identify and evaluate which metal/metal oxide systems are the most appropriate systems as hot gas regenerable sorbents based on the previous thermodynamic study of Westmoreland and Harrison [15]. Copper oxide shows a very favorable thermodynamic equilibrium with H_2S. In a reducing atmosphere the dominant specie is Cu metal because copper oxides are reduced to metallic copper or Cu(I) oxide at temperatures above 200°C [16]. The equilibrium H_2S concentrations are too low (around 1-5-ppm at 1000°K) with Cu based sorbents Additionally, Cu based sorbents do not suffer from metal volatility problems like Zn based sorbents (Zn volatilizes around 700°C, but the melting point of Cu is 1083°C). Therefore, Cu based sorbents can be used at temperatures around 800°C or higher.

Supported desulfurization sorbents evaluated by Jalan [10] have consisted of: ZnO on zeolite, CuO on zeolite, CuO on silica or alumina, mixed CuO/ZnO and V_2O_5 (unsupported and supported on alumina). Both fixed bed and fluidized bed reactor configurations have been used in these studies. The parameters evaluated were: the nature of the support, sorbent particle size, water partial pressure, desulfurization temperature, regeneration mode, desulfurization of inorganic and organic sulfur compounds as well as the interference from the reactor materials. The sorbent evaluation program for fuel cells at Giner, Inc. during late 1970's demonstrated that the desulfurization agents like CuO, V_2O_5 and mixed oxides CuO/ZnO are worth further investigation as hot regenerable sorbents. Additionally, it was shown that **desulfurization over supported sorbents offered significant improvements in the efficiency of sorbent utilization** [10]. Therefore, in a project supported by DOE the experiments conducted at Giner were concentrated on the above mentioned metal oxides and tests were carried out with a simulated coal gas mixture containing 200 ppm H_2S and about 6 % water vapor. The breakthrough concentration of H_2S was 10 ppm. The temperature of the sulfidation and regeneration was 650°C. Air/steam mixture was used during regeneration.

As a result of the studies conducted at Giner with different metal oxides, ZnO supported on zeolite (17.8% by wt. ZnO) showed 20 mole % conversion at 10 ppm breakthrough concentration. However, the initial performance was not maintained at subsequent sulfidation/ regeneration cycles. This lower performance was attributed to sintering and reduction of small particles of ZnO to Zn and subsequent volatilization of Zn at 700°C [10]. On the other hand, the desulfurization tests conducted at 650°C with CuO on prestabilized alumina (17.8% by wt. CuO) provided sulfur uptake capacity of 6-12 mole % when prereduced in hydrogen at 650°C. The regeneration of the sorbent was carried out either in air at 650°C followed by the sulfate decomposition at 750°C under nitrogen atmosphere or in air/steam mixture at 650°C. Similar desulfurization tests were also conducted with CuO/SiO_2 sorbent (a commercial sorbent) at 500 to 600°C. The desulfurization performance of CuO/Al_2O_3 sorbent at 650°C was compared with that of CuO/SiO_2. The desulfurization performance of CuO/Al_2O_3 sorbent at 650°C was found to be superior (the sulfur loading of about 12 mole %) in terms of stoichiometric conversion over CuO/SiO_2 tested at 500 to 600°C (the sulfur loading of about 1.3 to 4.5 mole %) [10]. On the other hand, sulfur loading capacity of CuO with a zeolite (faujasite-Y) support at 500°C ranged from about 16 to 23 mole % as compared to 0.7 mole % for unsupported CuO sorbent as reported by Jalan. As a result of these studies, among the different sorbents tested by Jalan, CuO/ZnO mixed sorbent and the CuO/zeolite supported sorbent have demonstrated the capability of maintaining the H_2S concentrations below 1 ppm at significant loadings.

The results of Jalan[10] indicated that the highest mole % conversion were realized with zeolite supported CuO and ZnO. However, the sulfur loading of these two systems decreased during repeated desulfurization/ regeneration cycles. The initial high capacity of the supported sorbents were believed to be occurring due to very high dispersion of loaded metal oxides on the support (zeolite). However, the decrease in capacity in the later cycles were thought to be caused by sintering or in case of ZnO due to volatilization. At the end of the study Jalan reached the conclusion that the mixed sorbent CuO/ZnO provided the larger sulfur loadings than the supported sorbents after several cycles. Therefore, further studies were done on CuO/ZnO sorbent in a fluidized bed reactor.

Very detailed and systematic studies on **Novel Sorbents** for high temperature regenerative H_2S removal were carried out by Flytzani-Stephanopoulos and her coworkers [17]. In a project carried out for DOE between 1983 and 1985 at JPL, detailed studies were conducted. The aim of the project was to reduce the H_2S level in coal gas from 0.5% - 1.0 mole % down to 1 ppm in a single stage process. The objective of the work was defined as to synthesize, characterize and test two classes of sorbents. One class was consisting of eutectic mixtures of metal oxides in the form of a melt coating the pore surface of a high surface area solid support. In this context $ZnO-V_2O_5$ mixture and $CuMoO_4-MoO_3$ mixture was tested. The second class of sorbents included unsupported mixed oxides in the form of solid solutions or solid compounds prepared as highly dispersed microcrystalline solids. Since this chapter deals with the supported sorbents, the second class of sorbents will not be talked about here.

In the work carried out by Flytzani-Stephanopoulos [17] and her co-workers experiments were performed at 650-700°C and 538-650°C with Zn-V-O and Cu-Mo-O, respectively. Simulated gas mixture was used for sulfidation, and air-nitrogen or air-steam-nitrogen mixture was used for regeneration tests. Both sorbents exhibited very high H_2S removal efficiency and the pre-breakthrough H_2S concentrations were always less than 1 ppm. This was attributed to the presence of eutectic phases with improved sulfidation equilibria over the constituent pure oxides and/or additional capacity for H_2S retention. Steep breakthrough curves were obtained which were indications of fast reactions and diffusion rates [17].

A stable conversion of about 65% over 5 cycles was obtained for Zn-V-O system. Surface area loss was not observed during reaction. It was noticed that the zinc was stabilized by vanadium in the alumina support, thus the evaporation loss of zinc was supressed. Vanadium was not forming any bulk vanadium sulfide, however, up to 0.2 mole H_2S per mole of V_2O_5 was captured through chemisorption. This value was substantially higher than the chemisorption occurring with TiO_2 [17]. Among the supported sorbents tested by Flytzani-Stephanopoulos, similar results were obtained for CuO-V_2O_5 supported on alumina.

For the sorbent of Cu-Mo-O, it was found that [18] lower Mo contents enhanced both the stability and the overall sorbent conversion. The surface area of the sorbent was high (32.3 m^2/g). In sulfidation at 650°C pre-breakthrough H_2S levels lower than 10 ppm were obtained and the evaporative Mo losses were supressed in the range of 538-650° C. However, if the temperature was above 650°C and the Mo/Cu ratios were high, it was found that sorbent lost Mo compounds by evaporation [18].

Kyotani, et al.[19] studied five kinds of Cu containing solids, pure CuO, a physical mixture of CuO and SiO_2, CuO supported on SiO_2, CuO dispersed in a SiO_2 matrix, and CuO supported on zeolite (mordenite and clinoptilolite) for the removal of H_2S from coal gas at high temperatures. After the sorbents were prepared they were subjected to sulfidation and regeneration tests (sulfidation at 400-800°C and regeneration at 700°C). The reaction behaviors of these sorbents were investigated by TGA analysis. Fresh, sulfided and regenerated sorbents were characterized physically and chemically. At the end of the study the reactivity of pure CuO was found to be lowest due to the formation of a dense sulfide layer during the sulfidation step. The supported CuO sorbents showed higher activities in sulfidation and regeneration stages. No decline in reactivity was observed through repeated sulfidation and regeneration cycles. It was found interesting to note that the physically mixed sorbent showed almost the same reaction behavior as the supported sorbents. The results of the study suggested that physical mixing of the CuO with inert solids was good enough to increase the reactivity and lifetime of the sorbent. Presence of water vapor was found to have no dominant retardation effect on reactivity of Cu-containing sorbents. The sulfate formed during regeneration has further decomposed.

Wakker and Gerritsen [20] have developed a steam regenerative process using MnO or FeO on γ-alumina acceptors to remove H_2S and COS. The temperature range was 400° to 800°C. The regeneration was carried out at the same temperature as sulfidation. The regeneration gas contained steam. No oxidation or noticable heat effect was observed due to steam regeneration. Sulfidation-regeneration behavior of the sorbent was extensively investigated in a packed bed microreactor for cyclic test. The results showed that H_2S can be removed to a level of 0-50 ppm and little deactivation of the sorbent was observed after 400 cycles. The sorbents were also tested in a 200 KW thermal gasifier for four weeks of continuous sulfidation and regeneration conditions[21]. After these tests no deactivation of the sorbents were observed and the fuel gas was cleaned to a level of 0-20 ppm H_2S. The breakthrough capacity of the sorbents was found to be quite low. The researchers have concluded that the iron based sorbents seemed to be less sensitive to deactivation caused by the use of a realistic fuel gas than the Mn based sorbents. Hydrogen sulfide concentration of the regeneration off-gas was found to be high enough to send to a Claus plant.

As a continuation of the previous work, a recent study was carried out by Atakul, Wakker and Gerritsen [22] in which MnO supported on γ-Al_2O_3 (MnO/γ-Al_2O_3) was used as a regenerable sorbent for high temperature removal of H_2S from gases. The sorbent contained 8% by wt. MnO and was tested for successive sulfidation-regeneration cycles. Sulfidation was carried out at 600°C. The regeneration was carried out at the same temperature with both N_2/H_2 and N_2/H_2/steam mixtures. The optimum temperature for both sulfidation and regeneration was determined to be 600°C. The breakthrough capacity of the sorbent decreased slightly as the gas flow rate increased. The conversion of the active component of the sorbent corresponding to the total capacity (34-35%) was not affected significantly. Consumption of steam for regeneration increased slowly until 60-70% of the regeneration was completed, and then found to rise rapidly. The manganese conversion ranged from 15-19% at the breakthrough point, to 32-35% at the end of sulfidation.

While several formulations of metal oxides mentioned above are rugged enough, when pelletized, to endure many successive sulfidation/ regeneration cycles, there is a need to improve the durability of these materials and reduce the costs for replacing the sorbents. Also, more durable sorbents are needed for fluid-bed and moving bed reactors.

In a recent study, van der Ham [23] investigated use of supported sorbents for H_2S and COS removal in a **Circulating Fluidized Bed (CFB)** reactor. Firstly, a sorbent which consisted of iron oxide deposited on an inert carrier (SiO_2) up to a concentration levels of 50% by wt. was used. It was shown experimentally that at 400°C the reactions proceed very fast and the sulfur capacity of the sorbent is favorably high. Unfortunately, the sorbent was very brittle to withstand the conditions in a CFB. A sorbent which can withstand the attrition forces during transport from the absorber to the regenerator or vice versa was needed. Therefore, FCC particles with a mean particle diameter of 0.1 mm was decided to be used. About 20% by wt. Fe_2O_3 was deposited on the particles to absorb H_2S. The experiments were conducted at atmospheric pressure at various temperatures by using small scale fixed and fluidized bed set-ups. H_2S removal efficiencies of 99% were achieved for a gas-solid contact time of 0.04 sec. The observed sulfur capacity was 4% by wt. on solid basis. The particle conversion rate was successfully described with Grain Model which required the input of kinetic data obtained by analysis of breakthrough curves from the small fixed bed reactor.

The regeneration reaction was also studied in a laboratory scale fluidized bed reactor at 500°C by using various O_2/N_2 mixtures. The results showed that stable operation was possible with gas mixtures containing less than 15% by vol. O_2. Careful control of the sorbent temperature was necessary to prevent sintering of the sorbent particles.

The desulfurization potential of Mn, according to Westmoreland and Harrison [24] is more than Fe and less than Zn. It can withstand high temperatures (melting point of all Mn oxides are above 830°C). Therefore, they are suitable to be used at high temperature conditions. In a gasifier atmosphere MnO is the active reagent with H_2S. Mn metal is not formed in a reducing (producer gas) atmosphere since the equilibrium constant of the reducing reaction is small ($K_{1000° K} = 5.6x10^{-7}$).

The combination of Cu oxide and Mn oxide seems to be a promising sorbent that can be used at high temperatures. Moreover, there are good indications in the literature that CuO and MnO can form a **spinel** structure which might help to increase the H_2S sorption capacity of the sorbent.

Based on these thermodynamic considerations this combination was used by Desai et al.[25] to prepare copper manganese sorbents and they studied the copper based mixed oxide sorbent (Cu and Mn oxides) by running sulfidaion and regeneration experiments. The reactor temperature was held constant in the range of 500-800°C during the experiment. The primary conclusion from these experiments was that a copper manganese oxide sorbent with Cu/Mn =1:1 atomic ratio would perform the best in the temperature range of 550-600°C. Sorbents with Cu/Mn=1:2 had the highest sulfur capacity at higher temperature. The outlet H_2S concentrations were reduced from 1% down to 20 ppm. However, these sorbents were prepared as bulk sorbents. Since the main topic of this chapter is supported sorbents, the detailed information on the bulk sorbents are not included in here.

In a study by Atimtay et al.[26] Cu oxide and MnO combination was used to develop a sorbent suitable for high temperature fluidized bed applications. This sorbent was tested by using a simulated coal gas mixture at temperatures of 540°-870°C. Metal oxides were supported on a silica rich zeolite to provide a good structural stability for the sorbent. The sorbent was shown through sulfidation and regeneration experiments to be capable of reducing hydrogen sulfide level in coal gas from 2000 ppmv down to 200 ppmv for about 60-70 min before breakthrough point. Regeneration of the sorbent with a 50 mole % air -50 mole % steam mixture is preferred over air regeneration due to decreased sulfate formation in the presence of steam. The metal coating on the zeolite extrudates especially the Mn appears to promote the increased resistance to crushing. Detailed results about this sorbent will be given in the Results and Discussion Section of this paper.

Bakker, et al. [27] have recently investigated the structure and properties of supported MnO and manganese aluminate as a high temperature sorbent for desulfurization of coal gas. The paper is given in this book. A sorbent having bulk $MnAl_2O_4$ as the bulk phase and MnO as the dispersed phase gave the best result for desulfurization of dry coal gas. The capacity of sulfur absorption was found to be up to 20% by wt. sulfur and the sorbent survived 110 cycles of sulfidation and regeneration. It was interesting to find out by the researchers that if the sorbent was regenerated with SO_2, elemental sulfur was formed. The operating temperature range was reported as 400-1000°C, at 850°C giving the optimum performance.

The following section will give the summary of the studies carried out by the author of this chapter and her coworkers in developing Cu-based supported mixed oxide sorbents.

3 Experimental

3.1 Sorbent Preparation

The sorbents investigated in this study[26] were prepared by wet impregnation method. SP-115 zeolite was used as the support material. SP-115 had the SiO_2 > 99% by weight, surface area = 482 m^2/g, and crush strength = 6.6 kg / cm.

For the wet impregnation method solutions of cupric acetate and manganese acetate were used. Three samples were prepared with this process (SP1C5M5, SP2C5M5, SP2C9M1). The nomenclature of these sorbents can be interpreted as follows: SP denotes SP-115 support, 2 denotes the number of impregnation, C and M denote the copper and manganese metals loaded on the support, respectively. The numbers after each letter show the relative molar concentrations of each of the previously listed metals in solutions. The atomic absorption spectrophotometer was used for the analysis of the metal contents. Cu and Mn loadings on the sorbent SP1C5M5 were 1.82% and 1.64% by wt., respectively. SP2C5M5 had 4.39% by wt. Cu and 6.11% by wt. Mn. The corresponding values for SP2C9M1 were 5.15% and 0.57% by wt., respectively.

3.2 Experimental Procedure

A schematic diagram of the experimental system is shown in Fig. 1. The system consisted of a gas mixing system, a fixed bed reactor, a condenser, a water knock-out pot, a gas sampling system, and a tail gas cleanup system. The gas mixture fed to the reactor simulated the partially quenched exit gas of an air-blown KRW fluidized bed coal gasifier which had the following molar composition:
42.5% N_2, 11% CO_2, 12.5% CO, 13.8% H_2, 1% CH_4, 19% H_2O and 2000 ppmv H_2S. Gases were supplied from gas cylinders and the flow rates were controlled and monitored through mass flow controllers. After the gases were mixed in a manifold (except the H_2S gas), they were fed into the fixed bed reactor. H_2S gas was introduced into the reactor separately just before the reactor inlet. Steam was provided by adding a predetermined amount of water into the gas stream before it entered the reactor via a high pressure pump.

The reactor consisted of a zirconia tube with a 2.22 cm inside diameter and 25.4 cm length. It was placed vertically in a tubular furnace. The temperatures of the gas at the inlet and the center of the reactor were measured by K-type thermocouples located along the central axis of the reactor. The reactor was designed to withstand a pressure of 15 psig and a temperature of 1600°C. The sorbent sample was placed in the center of the reactor and the bed height was 5 cm. The gas lines were heavily insulated to prevent the condensation of steam in the gas lines. The exit gas from the reactor was filtered, cooled in the condenser and sampled for gas analysis.

The experiments were conducted at 538°, 677°, 871° C and consisted of a sequence of sulfidation and regeneration runs. The reactor temperature was held constant during the sulfidation and regeneration experiments. Regeneration was carried out at the reaction temperature. The space velocities were 2000 hr^{-1} for the

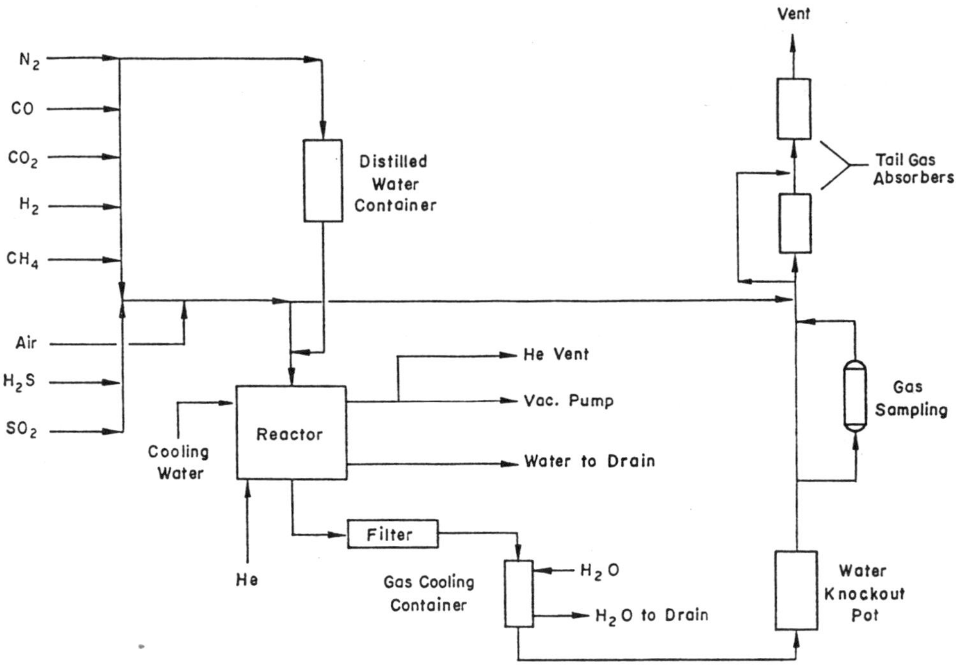

Fig. 1. Schematic diagram of the fixed-bed reactor system [26]

sulfidation and 600 hr^{-1} for the regeneration runs. For the sulfidation runs the breakthrough hydrogen sulfide concentration in the exit gas was defined as 200 ppm. After the sulfidation runs the reactor was purged with nitrogen. Regeneration runs were carried out until the SO_2 concentration of the effluent gas was about 500 ppm. The sorbents were subjected to five sulfidation and regeneration cycles to establish sulfur sorption capacity and regenerability.

Regeneration of the sulfided sorbent was done with 50 % air and 50 % steam mixture. The sorbents were examined and analyzed for total weight, metal and sulfur content, crush strength, surface area, average pore diameter, pore volume, and pore size distribution before and after the reaction. Gas grab samples obtained during experimentation were analyzed by gas chromatography.

4 Results of the Experiments

The H_2S breakthrough curves showed that the breakthrough times were 38, 49 and 17 min at 538, 677 and 871°C, respectively. The sorbent had a greater capacity at higher temperatures than at 538°C. However, the pre-breakthrough H_2S level was lower at 538°C. The reproducibility of the experimental method was checked by running the same experiment twice with different samples.

The sorbent regenerability was first tested by conducting the two and one-half cycle experiments at 677°C with air as the regenerating gas. Then five cycle-tests were conducted by using a 50 mole % steam/ 50 mole % air mixture. Fig. 2 shows the H_2S breakthrough curves after the sorbent was regenerated with steam/air. The results showed that when the sorbent was regenerated with steam/air mixture less sulfate formation occurred as compared to regeneration with air. The SO_2 levels measured during the sulfidation reaction after regeneration reached to 500-600 ppm when the regeneration was done with air. Less sulfate formation with steam/air regeneration was expected because the presence of steam reduces the overall oxidative properties of the regenerant stream. The formation of sulfates during

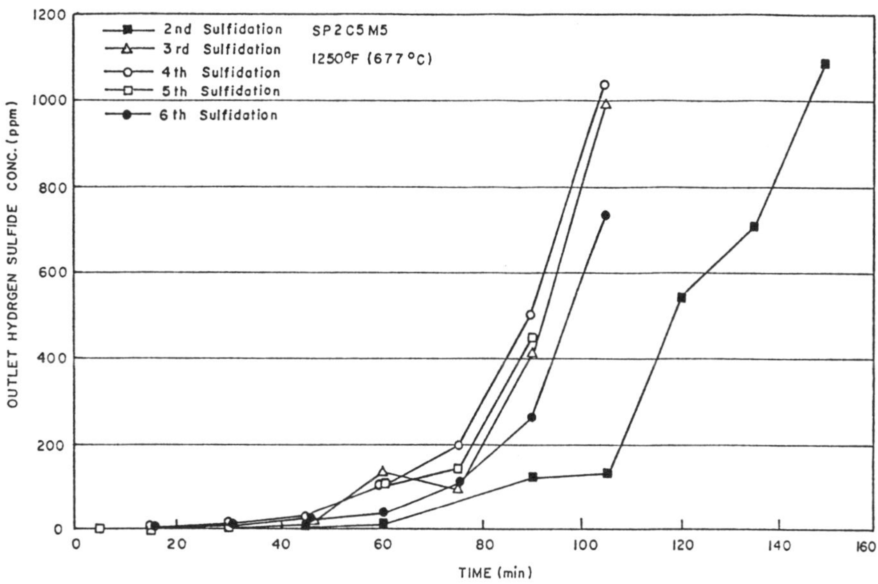

Fig.2. Hydrogen sulfide breakthrough curves for multiple sulfidation of SP2C5M5 sorbent at 677°C with 50 mole % steam/50 mole % air used for regeneration [26]

regeneration and the decomposition of these sulfates during sulfidation should be avoided because this will require a reductive regeneration of the sorbent after oxidative regeneration and prior to sulfidation. However, the economics of adding a reductive regeneration to the system is poor because of the capital cost and operational difficulties of an additional reactor.

The effects of combining two metal oxides, namely, Cu oxide and Mn oxide was compared by conducting separate experiments with the sorbents prepared by loading CuO and MnO separately on the SP-115 zeolite with the same sorbent preparation technique. The results of these experiments showed that the sulfur absorption capacity for Cu oxide is much greater than that of MnO at 871°C. By combining Cu oxide and Mn oxide, the efficiency and the capacity of the sorbent was increased over that obtainable by using the zeolite-supported Cu oxide alone.

Appreciable concentrations of carbonyl sulfide (COS) were detected in the outlet gas during sulfidation, although no COS was added to the reactor. The COS concentrations measured were usually greater than 50 ppm. It went up to 100-250 ppm after a certain minimum was passed during the experiment as can be seen in Fig. 3. It was thought that COS was formed due to the reaction of H_2S with CO/or CO_2 and zeolite might have catalyzed the reaction [26]. At temperatures between 230-730°C, the thermodynamics of the reaction is more favorable for CO-H_2S combination. The formation of COS during the reaction passed through a minimum and then increased later. This initial decrease in COS concentration was thought to be due to reduced availability of CO, because of the competing reactions in which CO was used in reducing copper oxide. The results also showed that COS production was more at 677°C than the other temperatures, and the production increased with the number of cycles on the sorbent. More experimental data is needed in order to explain these results.

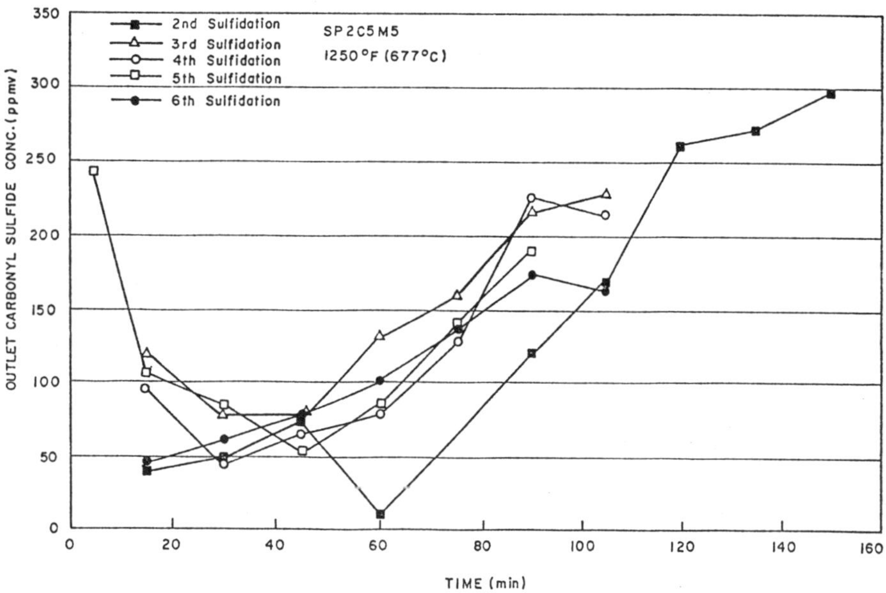

Fig.3. Outlet COS concentration for multiple sulfidations of SP2C5M5 sorbent at 677°C with 50 mole % steam/50 mole % air used for regeneration [26]

The change in the surface area of the sorbent was also investigated by determining the area before and after the reaction by N_2 adsorption method. The unreacted

SP1C5M5 had an average surface area of 434 m^2/g as compared to 482 m^2/g for fresh zeolite. This indicated some pore blockage due to metal oxide coating. After single sulfidation at 538, 677 and 871°C, the surface area decreased to 379, 370 and 237 m^2/g, respectively. The loss of surface area after sulfidation reactions was thought to be due to coalescence of pores in the zeolite, or to a certain extent, silicic acid formation by reaction of the zeolite with high-temperature steam.

Cyclic tests were conducted on the sorbent SP2C5M5. The ratio of the weight of metal reacted to the total weight of the metal in the sorbent at breakthrough points were found as 20%, 7%, 7%, 13% and 12%, for each cycle (first through fifth), respectively. The percent utilizations were not very consistent. Firstly, some assumptions were made in calculating the % utilizations. Low utilizations might be due to ineffective coating of the zeolite with metal oxides, or even the pores were effectively coated, there could be a significant diffusional resistance for gas transport for H_2S into the pores.

The loss of Cu or Mn was checked for different samples by analyzing the metal contents before and after the reactions. No evidence of Cu or Mn loss was observed. The crush strength of the sorbent extrudates were also determined. The crush strength of the fresh zeolite was 6.58 kg/cm. After the metal loading the crush strengths increased to 11.1 to 29.6 kg/cm depending on the number of metal coatings. The average crush strengths of 6 samples tested after 5 cycles decreased by only 6 % and that of 3 samples tested after 3 cycles increased by 3 %. Therefore, it was seen that the crush strength of the zeolite was increased by metal loading and changed little by reactions.

5 Conclusion

A zeolite-supported Cu/Mn oxides sorbent was developed for the removal of H_2S from coal gas at high temperatures (around 850°C). The sorbent is regenerable and more durable for fluidized-bed and moving bed reactors. The sorbent was shown to be capable of reducing the H_2S concentration in coal gas from 2000 ppm to 200 ppm level for about 60-70 min before the breakthrough point. Decreased sulfate formation was obtained with a 50 mole% air - 50 mole % steam mixture over air regeneration. The metal coating on the zeolite support, especially the Mn, appeared to increase crush resistance of the sorbent. The zeolite support in the sorbent is expected to make the sorbent price higher than bulk sorbents. However, the higher stability can compansate the higher cost when in continuous use in fluidized bed or moving bed reactors. In order to make the economical feasibility studies, more data needs to be collected for many cyclic sulfidation and regeneration runs.

Acknowledgment

The financial support provided by the Oak Ridge Associated Universities and the Department of Energy, Morgantown Energy Research Center are greatly appreciated.

References

1. Sage, P.W. and Mills, S.J., "Overview of Clean Coal Technologies and Current Status of the Air Blown Gasification Cycle", NATO ASI Series on Desulfurization of Hot Coal Gas With Regenerable Metal Oxide Sorbents, 7-19 July, 1996, Kusadasi, Turkey.
2. World Energy Council, "Global Energy Perspectives to 2050 and Beyond", WEC/III ASA Report, 1995.
3. Bissett, L.A. and Strickland, L.D. "Analysis of a Fixed Bed Gasifier IGCC Configuration", *Ind. Eng. Chem. Res.,* **30** (1), pp.170-176 (1991).
4. Harrison, D.G., "Control of Gaseous Contaminants in IGCC Power Systems, An Overview", Presented at the 12th Annual International Pittsburgh Coal Conference, pp. 1047-1052, Pittsburgh, Sept. 1995.
5. Watkinson, A.P., Lucas, J.P., Lim, C.Y., "A Prediction of Performance of Commercial Coal Gasifier", *Fuel*, **70**, April, pp.519-609 (1991).
6. Woods, M.C., Gangwal, S.K., Harrison, D.P., Jothimurugesan, K., "Reaction Between Hydrogen Sulfide and Zinc Oxide-Titanium Oxide Sorbents: I. Single Pellet Kinetic Studies", *Ind. Eng. Chem. Res.*, **29**, pp.1160-1166 (1990).
7. Huff, W. J, Logan, L., AGA Proceedings, **18**, 724 (1936).
8. Nachod, F.C., "Desulfurization of Hydrocarbons", *U.S. Patent*, No. 2,442,982 (1948).
9. Lewis, W. K., *U.S. Patent,* No. 3,079,223 (1963).
10. Jalan, V., "Studies Involving High-Temperature Desulfurization Regeneration Reactions of Metal Oxides for Fuel Cell Development", *Final Report*, DOE/MC/16021-1486, Oct. 1983.
11. Wheelock, K. S., Aldridge, C.L., *U.S. Patent*, No.3,974,256 (1976).
12. Wheelock, K. S., Say, R.S., *U. S. Patent* , No. 4,002,720 (1977).
13. Steiner, P., *U.S. Patent* , No. 4,039,619 (1977).
14. Onursal, A. B., EPA-700/7-79-169, July 1979.
15. Westmoreland, P.R., Harrison, D.P., "Evaluation of Candidate Solids for High-Temperature Desulfurization of Low-Btu Gases," *Environ. Sci. Technol.*, **10**, 659 (1976).
16. Chemistry of Hot Gas Cleanup in Coal Gasification and Combustion, DOE/Morgantown Energy Research Center, MERC/SP-78/2, 1978.
17. Flytzani-Stephanopoulos, M, Gavalas, G.R., Tamhankar, S.S., Sharma, P.K., "Novel Sorbents for High Temperature Regenerative H_2S Removal", Final Report for the project DOE/MC/20417-1898 (DE86004261), Jet Propulsion Laboratory, October 1985.
18. Flytzani-Stephanopoulos, M., *Environ. Sci. and Technol.*, **11**, 488-491 (1987)
19. Kyotani, T., Kawashima, H., Tomita, A., "High-Temperature Desulfurizing Reaction With Cu-Containing Sorbents", *Environ. Sci. Technol.,* **23** (2), pp. 218-223 (1989).
20. Wakker, J.P., Gerritsen, A.W., "High Temperature H_2S Removal from Process Gases in a Steam Regenerative Process Using MnO or FeO on Gamma-Alumina Acceptors", *Prep. Pap. Am. Chem. Soc., Div. Fuel Chem.*, **35** (1), pp. 170 (1990).

21. Wakker, J.P., Gerritsen, A.W., "Coal Gasification:High Temperature H_2S Removal in a Steam Regenerative Process Under Realistic Conditions", *Prep. Pap. Am. Chem. Soc., Div. Fuel Chem.*, **35** (1), pp. 179-187 (1990).
22. Atakul, H., Wakker, J.P., Gerritsen, A.W., van den Berg, P.J., "Removal of H_2S From Fuel Gases at High Temperatures Using $MnO/\gamma\text{-}Al_2O_3$", *Fuel*, **74** (2), pp.187-191 (1995).
23. van der Ham, A.G.J., "Regenerative High Temperature Desulfurization of Coal Gas in a Circulating Fluidized Bed ", Ph.D. Thesis, University of Twente, Enschede, The Netherlands, 1994.
24. Westmoreland, P.R., Gibson, J.B., Harrison, D.P., "Comparative Kinetics of High Temperature Reactions Between H_2S and Selected Metal Oxides," *Environ. Sci. and Technol.*, **11**, 488-491 (1977).
25. Desai, M., Brown, F., Chamberland, B., Jalan, V., "Copper Based Sorbents for Hot Gas Cleanup", *Prep. Pap. Am. Chem. Soc., Div. Fuel Chem.*, **35** (1), pp. 87-94 (1990).
26. Atimtay, A.T., Gasper-Galvin, L.D., Poston, J.A., "Novel Supported Sorbent for Hot Gas Desulfurization", *Environ. Sci. Technol.*, **27** (7), pp. 1295-1303 (1993).
27. Bakker, W.J.W, van Rossen, J.C.P., Mugge, J.M., Kapteijn, F., Moulijn, J.A., "Performance of Manganese-Based Sorbents in High Temperature Coal Gas Desulfurization", this book.

Regeneration of Sulfided Sorbents and Direct Production of Elemental Sulfur

Douglas P. Harrison

Department of Chemical Engineering, Louisiana State University, Baton Rouge, LA 70803, U.S.A.

Abstract. Although equally important, the regeneration phase of the hot gas desulfurization process has received much less attention than the sulfidation phase. The standard method of sorbent regeneration, which involves the oxidation of metal sulfide to metal oxide and SO_2, creates several problems. The reaction is highly exothermic and temperature excursions can cause the porous, high surface area sorbent to sinter. Addition of a diluent to the regeneration feed gas is effective for temperature control, but complicates the ultimate SO_2 control problem. In addition, the O_2 and SO_2 partial pressures in the regeneration reactor often favor the formation of metal sulfate, currently thought to be the major cause of sorbent deterioration

The most severe instances of deterioration are associated with fixed-bed reactors where sorbent near the sulfidation gas inlet may be almost completely sulfided while little sulfidation occurs near the gas exit. The durability problem appears to be less severe in moving-bed and fluidized-bed reactors where the maximum level of sulfidation may be controlled and all sorbent is more or less equally sulfided. Semi-batch fluidized-bed tests of as many as 100 sulfidation-regeneration cycles, and totally integrated tests using dual circulating fluidized-bed sulfidation-regeneration reactors coupled to a coal gasifier have been completed with only moderate loss of sulfidation activity with time.

Controlling the emission of sulfur compounds produced during sorbent regeneration is also of major concern. One option involves the use of fluidized-bed gasifiers with in-bed injection of a calcium-based sorbent for bulk sulfur capture. The regenerable metal oxide is used as a sulfur polishing step with SO_2 produced during regeneration recycled to the gasifier. Effectively all sulfur is captured as $CaSO_4$ and disposed of as landfill. Regeneration gas recycle in the GE moving-bed process results in an SO_2 content of about 12%, which is ideal as feed to a H_2SO_4 plant. However, the SO_2 content from other regeneration processes is too small to serve as H_2SO_4 feed. The Research Triangle Institute is developing the direct sulfur recovery process (DSRP) which has been shown to be effective in converting almost 100% of the SO_2 in dilute regeneration off-gas to elemental sulfur.

Recently, research has been initiated on the direct production of elemental sulfur during sorbent regeneration. This approach will reduce the heat released during regeneration, minimize the metal sulfate formation problem, and avoid the formation of a dilute SO_2 off-gas and the associated requirement of an additional chemical processing step to produce elemental sulfur.

NATO ASI Series, Vol. G 42
Desulfurization of Hot Coal Gas
Edited by A.T. Atimtay and D. P. Harrison

1 Introduction

In order to be economical, high temperature metal oxide desulfurization sorbents must maintain strength and reactivity through a large number of sulfidation-regeneration cycles. The regeneration portion of the cycle is, therefore, equally as important as the sulfidation phase, and, in many respects, more difficult to carry out properly. In spite of the importance and difficulty associated with regeneration, this portion of the cycle has received much less attention than has sulfidation.

In this paper, the regeneration problems are classified in terms of sorbent multicycle stability and recovery/disposal of sulfur in the regenerator off-gas. The stability problems are caused by the highly exothermic nature of the regeneration reactions, and the complex set of reactions which occur as the temperature, pressure, and composition of the regeneration gas are varied. Metal sulfate formation is favored at appropriate temperatures and partial pressures of O_2 and SO_2. Zinc sulfate is thought to be a major cause of zinc-based sorbent deterioration. Increasing the regeneration temperature can minimize sulfate formation, but sintering of the porous, relatively high surface area sorbent may result. Decreasing the partial pressure of reactant gases may be used to control the reaction exotherm and also to minimize sulfate formation. However, this approach increases the time required for regeneration and reduces the concentration of sulfur species in the regenerator off-gas, thereby complicating the ultimate recovery or disposal of the sulfur.

In most regeneration processes, sulfur is liberated as SO_2 and further processing is required to prevent the SO_2 from being discharged to the environment. In fluidized-bed gasification processes which use limestone or dolomite injection for bulk sulfur capture, regeneration off-gases from the desulfurization polishing step may be recycled to the gasifier. There the SO_2 is reduced to H_2S, H_2S reacts with CaO, and all the sulfur is eventually removed as $CaSO_4$ in the bottom ash. However, this concept is not applicable to all gasifier types, and, even when applicable, a sizeable solid waste disposal problem is created.

Recovery of the sulfur in the regenerator off-gas is preferable to land disposal. The production of sulfuric acid is a logical option although SO_2 concentrations are generally well below those associated with H_2SO_4 manufacture. The General Electric moving-bed process (Smith et al., 1987), which is designed to produce a regenerator off-gas containing about 12% SO_2, is an exception. However, power plant sites may be far removed from sulfuric acid markets and long distance transportation is both expensive and potentially dangerous. Recovery of elemental sulfur is the preferred option. A market for elemental sulfur exists, and it can be stored and transported safely at relatively low cost. Researchers at the Research Triangle Institute (Dorchak et al., 1991; Gangwal and Portzer, 1995) have developed the direct sulfur recovery process (DSRP) aimed specifically at producing elemental sulfur from low SO_2 content gases. The drawbacks of DSRP include the use of a slipstream of the clean coal gas to reduce SO_2 to elemental sulfur, and the liability of adding another chemical process to the electric power plant. Research on alternate approaches to sorbent regeneration in which elemental sulfur is produced directly in the regeneration step has recently been initiated at the Research Triangle Institute (Gangwal et al., 1995) and at Louisiana State University (Harrison et al., 1995).

2 Standard Regeneration

The standard regeneration process involves the reaction between sulfided sorbentand oxygen to reform the metal oxide and liberate sulfur as SO_2. In the following the sorbent is assumed to be ZnO, and the regeneration reaction is

$$ZnS(s) + \frac{3}{2} O_2(g) \rightarrow ZnO(s) + SO_2(g)$$

Although the quantitative aspects vary with different sorbents, the nature of the problems and the qualitative analysis are similar. The reaction is highly exothermic; the standard heat of reaction (298K) for ZnS oxidation is $\Delta H_R = -106$ kcal/mol ,and is almost independent of temperature.

2.1 Temperature Control

Temperature control in the regeneration reactor can be quite difficult as illustrated by Grindley (1986). A fixed-bed of zinc ferrite ($ZnFe_2O_4$) sorbent was first sulfided using simulated coal gas containing 2.7% H_2S. The sulfidation cycle was terminated when the H_2S concentration in the product gas reached 500 ppmv. As a consequence, sorbent near the sulfidation entrance was almost totally converted to ZnS and FeS, while only small conversions were achieved near the sulfidation exit. Regeneration gas containing 10% O_2, 40% N_2, and 50% steam at 2.5 atm and 204°C was fed in the opposite direction. Temperature profiles as a function of time, measured by four thermocouples placed axially within the bed are shown in Figure 1. Temperature increases on the order of 550°C were measured as the

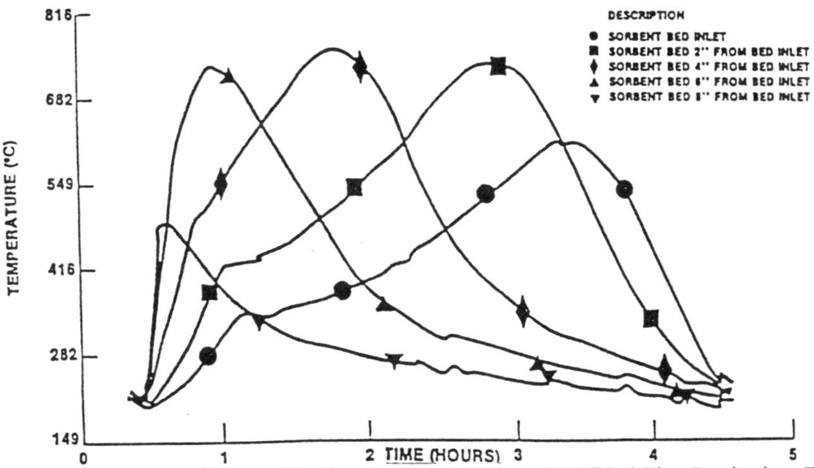

Figure 1. Temperature Increase During the Regeneration of Sulfided Zinc Ferrite in a Fixed-Bed Reactor (Grindley, 1980)

reaction front moved past the thermocouple positions. Regeneration was continued until the SO_2 concentration dropped to 500 ppmv. The sorbent was then removed from the reactor and the chemical composition and structural properties of each section were determined. Residual ZnS and $ZnSO_4$ were found in sorbent located near the regeneration outlet (sulfidation inlet). No FeS was found, and the majority of the Fe_2O_3 and ZnO had recombined to form $ZnFe_2O_4$. Surface area and porosity were lower than in the fresh sorbent, presumably due to sintering and the presence of ZnS and $ZnSO_4$. The maximum decrease in surface area occurred at the regeneration inlet (sulfidation outlet). The crush strength of the pellets increased, which was attributed to the decrease in pellet porosity. The residual sulfate decomposed rapidly when exposed to simulated coal gas in a subsequent sulfidation cycle.

The results illustrate quite clearly the nature of the regeneration problems. Even with the low initial temperature of 204°C, the rate of oxidation was sufficiently fast to permit almost complete regeneration in less than five hours, and to produce a 550°C temperature increase. The fact that $ZnSO_4$ was found in the regenerated sorbent was a preview of future problems associated with pellet durability.

Reducing the maximum temperature in the sorbent bed to control sintering is most easily accomplished by reducing the oxygen content (increasing the diluent concentration) of the regeneration feed gas, which provides the added benefit of reducing the tendency for sulfate formation. This is counterbalanced by the increased stability of the sulfate at low temperature, by the increased duration of the regeneration cycle, and by reduced SO_2 concentrations in the regenerator product gas.

2.2 Zinc Sulfate

The pervasive nature of the zinc sulfate problem may be understood from the following thermodynamic analysis. Figure 2 shows the equilibrium mol fraction of SO_2 as a function of temperature and pressure for the reaction

$$ZnO(s) + SO_2(g) + ‰\, O_2(g) \rightarrow ZnSO_4(s)$$

By assuming that the gas phase is ideal and that each solid specie forms a separate phase having unit activity, standard methods of analysis show that

$$y_{SO_2} = \left[K\, y_{O_2}^{1/2}\, P^{3/2}\right]^{-1}.$$

K is the equilibrium constant, P is the reaction pressure (atm), and y_{SO2} and y_{o2} are equilibrium mol fractions of SO_2 and O_2, respectively, in the gas phase. In Figure 2, the value of y_{O2} is fixed at 0.01, which is representative of current operating practice. $ZnSO_4$ formation is favored for all values of y_{SO2} above and to the left of the appropriate total pressure curve while $ZnSO_4$ is not thermodynamically stable for values of y_{SO2} below and to the right of the appropriate curve. For example, at 25 atm and 1000K, $ZnSO_4$ formation is favored for values of $y_{SO2} > 10^{-4}$; at 1 atm and 1000Kit would be necessary to maintain $y_{SO2} < 0.01$ to prevent $ZnSO_4$

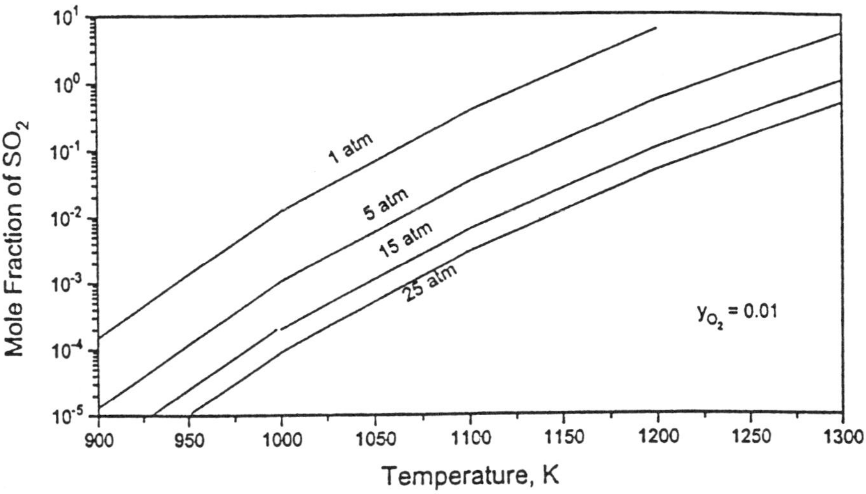

Figure 2. Thermodynamic Analysis of $ZnSO_4$ Formation During Sorbent Regeneration

Therefore, reducing the O_2 concentration as a temperature control measure does not eliminate the potential formation of $ZnSO_4$.

2.2.1 Fixed-Bed Test Results

Initially, the problem caused by $ZnSO_4$ was believed limited to the release of relatively large concentrations of sulfur gases (both SO_2 and H_2S) in the early stages of a subsequent sulfidation cycle. This led to the concept of two-stage regeneration, with oxidative regeneration followed by reductive regeneration using either sulfur-free coal gas or partially combusted natural gas to achieve sulfate decomposition. Results from a number of reductive regeneration tests are presented by Grindley (1988).

However, it soon became apparent that sulfate formation was the cause of more severe problems associated with multicycle sorbent durability. Gangwal et al. (1988) performed fifteen cycle sulfidation-regeneration tests with zinc ferrite in a fixed-bed reactor using simulated coal gas in the sulfidation phase and a mixture of air and steam during regeneration. The initial regenerator temperature was 540°C and the composition of the feed gas was increased in stages from 8% air - 92% steam initially to 26% air - 74% steam near the end. The air flow was controlled so that the temperature throughout the bed was in the range of 730 to 790°C to minimize $ZnSO_4$ formation. Beginning with the fourth cycle, the sorbent was removed from the reactor following each regeneration and screened. The amount of sorbent fragments gradually increased from about 0.5% of the sorbent following cycle 5 to roughly 1.5% of the sorbent following each of cycles 10 through 15. Cumulatively, approximately 18% by weight of the original charge was fragmented to the extent that the particles passed through a 7 mesh (2.83 mm) screen.

Leininger et al. (1992) reported test results using zinc titanate sorbent L-3140 (1.5 $ZnO{\cdot}TiO_2$) in a fixed-bed reactor, with sulfidation gas from the Texaco process development gasifier. Sorbent utilization at breakthrough decreased from 73% in cycle 1 to 53% in cycle 2, with the decrease attributed to incomplete cycle 1 regeneration. The oxygen content of the regeneration gas was 1.6% in cycle 1 and 1.4% in cycle 2 while the regeneration temperature ranged from 540 to 750°C in cycle 1 and from 590 to 700°C in cycle 2. Following the second regeneration, the sorbent was removed in eight axial sections for chemical analysis and structural property characterization. The sulfur content ranged from 4.3% in that section nearest the sulfidation feed to 1.9% at the sulfidation outlet. Surface area ranged from 3.3 to 4.1 m^2/g, compared to 3.0 m^2/g for the unused sorbent. Pore volume ranged from 0.31 to 0.37 cm^3/g, which represented a small to moderate decrease from the 0.38 cm^3/g pore volume of the unused sorbent. Most significantly, the crush strength of the pellets subjected to two cycles decreased by from 11% to 46%. In addition, the zinc to titanium molar ratio decreased from 1.5 for the fresh sorbent to an average of 1.38, which represented a zinc loss of about 8%. Severe cracking of the pellets was apparent except near the sulfidation outlet (regeneration inlet) where relatively little reaction occurred. The sorbent had fragmented to the extent that 3% (by weight) passed through a 14-mesh (1.19 mm) screen at the end of five cycles. Subsequently, four different zinc titanate formulations having zinc to titanium ratios from 0.8 to 2.0 were tested through 4½ cycles (five sulfidations and four regenerations) at the same reaction conditions. All four sorbents cracked and spalled with a loss of as much as 14% by weight of one sample through a 14-mesh screen. The crush strength was reduced to the point that pellets could be crushed between two fingers.

The probable reason for sorbent deterioration when $ZnSO_4$ is formed is the large increase in molar volume when either ZnO (14.5 cm^3/mol) or ZnS (24.1 cm^3/mol) is converted to $ZnSO_4$ (45.6 cm^3/mol). The additional solid fills the pores and creates mechanical stresses sufficient, in some cases, to cause the pellet to spall or fragment into smaller pieces. The severity of the problem depends on the amount of sulfate formed, the porosity and strength of the pellet, reaction temperature and pressure, and even on the radial distribution of sulfate within the pellet.

2.2.2 Moving-Bed Test Results

The regeneration concept in the General Electric moving-bed desulfurization process is unique in several respects. Moving-bed sorbents are specially formulated for greater pellet strength and the pseudo-steady-state operating mode results in more even sulfur loading in all pellets. For example, in a recent integrated test involving the GE gasifier and moving-bed desulfurization and regeneration reactors, the sorbent circulation rate was fixed to produce a sulfur loading in the sorbent entering the regenerator of 25% of theoretical (Bevan et al., 1995).

Partially sulfided sorbent is discharged from the desulfurization reactor located above the regeneration reactor. This freshly sulfided sorbent, which includes tar carried over from the gasifier, is quite reactive. Regeneration is carried out in stages with the oxygen content to the first stage limited to control the temperature increase resulting from the exothermic sorbent regeneration and tar oxidation reactions. Additional oxygen is then added to complete the regeneration. Recycle gas in the regeneration loop is designed to operate with about 12% SO_2 so that the regenerator product gas is suitable for feed to a sulfuric acid plant. Nitrogen entering with the air provides the required dilution, eliminating the need for external diluent such as pure nitrogen or steam.

Numerous changes have been made in the regenerator design to improve operation. The original design called for three-stage regeneration. Single-stage operation was tested for a time before settling on the current two-stage regeneration. The internals of the regeneration reactor have also changed from the original design, which called for the entire vessel to be filled with sorbent, to the current annular flow design (Bevan et al. 1995) which was adopted to improve solid flow and temperature control. Regenerated sorbent exits from the bottom of the reactor into a lock hopper containing a vibrating screen for fines removal, and is transported by bucket elevator to a second lock hopper above the desulfurization reactor. The ability to remove fines, which may be caused by ash carryover from the gasifier or by attrition or fragmentation of sorbent pellets, is an important capability which is not possible in fixed-bed operation.

Ayala et al. (1994) compared the durability of zinc titanate and Z-Sorb pellets designed for moving-bed applications. The eight-cycle test was carried out in a fixed-bed which was operated to simulate the moving-bed system. Sorbent properties and operating conditions are presented in Table 1. The zinc titanate pellets had a zinc to titanium ratio of 2 and contained bentonite binder and MoO_3 as a pore modifier. The composition of Z-Sorb is proprietary, but it is known to consist of ZnO on a porous matrix with a nickel oxide promoter. The theoretical sulfur capacity of the zinc titanate was somewhat larger. Z-Sorb pellets were considerably smaller, less dense, and had significantly higher attrition resistance. The sulfidation gas simulated the product from an oxygen-blown gasifier similar to that proposed for the Tampa Electric clean coal demonstration project (U.S. Department of Energy, 1995). The regeneration gas in cycles 1 through 5 contained no SO_2, and the O_2 content was increased from 0.5% initially to 21% at the

Table 1. Sorbent Properties and Operating Conditions for the Eight-Cycle Comparative Test of the Performance of Zinc Titanate and Z-Sorb (Ayala et al., 1994)

Sorbent Properties		
	Zinc Titanate L-4103B	Z-Sorb B
Composition	Zn_2TiO_4 with ~2% MoO_3 and 3% bentonite	Proprietary
Pellet Length, mm	5.0	3.6 - 3.9
Pellet Diameter, mm	4.0	2.6 - 2.9
Pellet Mass, mg	200	27 - 32
Crush Strength, lb_f	23.2	7.1 - 7.6
Attrition Resistance, %	95 - 96.6	99.1 - 99.6
Bulk Density, lb_m/ft^3	107	~60
Sulfur Capacity, % S	25	20

Operating Conditions	
Sulfidation	
Pressure, atm	1
Temperature, °C	540
Gas Comp., % CO	39
CO_2	10
H_2	30
H_2O	20
H_2S	1
Regeneration	
Pressure, atm	1
Temperature, °C	540 - 700
Gas Composition	
Cycles 1 through 5	0.5 to 21% O_2 balance N_2
Cycles 6 through 8	0.5 to 21% O_2 5% SO_2 balance N_2
Extended Cycle 8	
Pressure, atm	1
Temperature, °C	760
Gas Composition	20% O_2, 5% SO_2

conclusion to simulate the two-stage operation of the moving-bed process. 5% SO_2 was added to the regenerator feed in cycles 6 through 8 to simulate the effect of recycled SO_2. Finally, after regeneration cycle 8, the sorbent was exposed for an extended period in an atmosphere of 20% O_2, 5% SO_2 at 760°C. Prebreakthrough H_2S concentrations of less than 10 ppmv were measured with Z-Sorb, and less than 20 ppmv with the zinc titanate. Both provided greater than 99% H_2S removal. The sulfur content of the Z-Sorb was higher on a mass basis, but the sulfur content per unit reactor volume was approximately equal because of the lower bulk density of Z-Sorb.

The sorbents were removed from the reactor following the fifth, eighth, and extended eighth cycles, examined visually for cracks, and small samples were removed for chemical analysis, and crush strength and attrition resistance measurements. No obvious cracks were observed in either the zinc titanate or Z-Sorb pellets following the fifth and eighth cycles. However, serious cracking was found in the zinc titanate following the extended exposure to harsh conditions with little cracking observed in the Z-Sorb. Figure 3 provides a visual comparison of changes in crush strength, attrition resistance, and sulfate content of the two sorbents at various stages of the test program. The crush strength of zinc titanate (Fig. 3a) increased after the first five cycles in which the feed gas contained no SO_2. Addition of SO_2 in cycles 6 through 8 caused a slight decrease in crush strength, and the harsh final treatment reduced the final crush strength to only about half that of the fresh zinc titanate pellets. The attrition resistance shown in Figure 3b did not change appreciably during the eight normal cycles, but decreased dramatically after the final treatment. Little zinc sulfate was present in the regenerated zinc titanate pellets (Fig. 3c) following the first five cycles. Addition of 5% SO_2 resulted in about 2% $ZnSO_4$ following cycle 8, and the further harsh treatment produced an increase in the amount of $ZnSO_4$ to about 4%.

The crush strength of the Z-Sorb pellets increased at each stage of the test to the point that the crush strengths of the final Z-Sorb and zinc titanate pellets were approximately equal. Increases in Z-Sorb attrition resistance were measured after five and eight cycles, followed by a slight decrease in attrition resistance following the final harsh treatment. However, the Z-Sorb attrition resistance was significantly better than that of the zinc titanate at all stages of the test. Sulfate levels of about 6% were found in the Z-Sorb pellets at the end of the eight regeneration cycles and following the final harsh treatment. Despite the higher sulfate levels, the mechanical properties of the Z-Sorb did not deteriorate in the same manner as the zinc titanate. One can speculate that the Z-Sorb pellets were able to accommodate more zinc sulfate because of the lower initial bulk density.

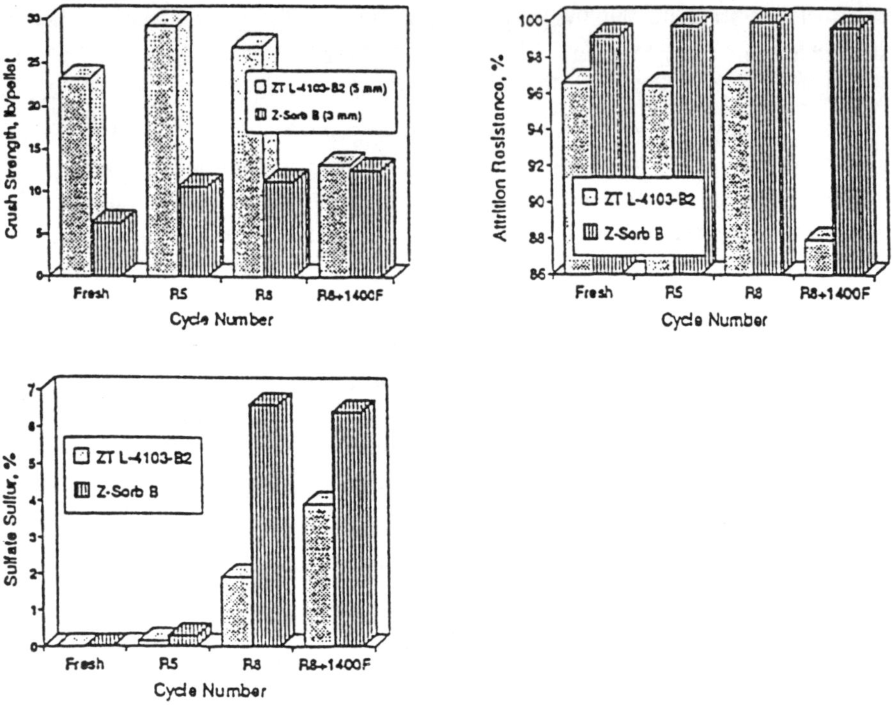

Figure 3. Comparison of the Mechanical Properties of Zinc Titanate and Z-Sorb During an Eight Cycle Test (Ayala et al., 1994)

2.2.3 Fluidized-Bed Test Results

Much of the research on fluidized-bed sorbents has been carried out at the Research Triangle Institute. Zinc ferrite, zinc titanate, and Z-Sorb have been tested, with the primary objective being the development of fluidizable sorbent particles in the size range of 50 to 300 μm having an acceptable combination of reactivity, regenerability, and attrition resistance.

The patented zinc titanate ZT-4 sorbent prepared by a granulation technique has been tested extensively. Properties of ZT-4 are summarized in Table 2 (Gangwal and Gupta, 1994). Sorbent labelled ZT-4 was prepared on a small-scale for laboratory testing while ZT-4L represented a scale-up in the manufacturing process by a factor of 17.5. This sorbent was subjected to 100 sulfidation-regeneration cycles in the laboratory (Gangwal and Gupta, 1995a). Sulfidation used a simulated Texaco gas containing 1.14% H_2S with reaction conditions of 750°C and 15 atm. Regeneration was performed at 15 atm over a temperature range of 720 to 760°C with the regeneration gas containing 2.0 to 2.5% O_2 in N_2.

Table 2. Physical Properties of ZT-4 and ZT-4L (Gangwal and Gupta, 1994)

	ZT-4	ZT-4L
ZnO-to-TiO_2, molar ratio	1.5	1.5
Binder content (bentonite)	5 wt%	5 wt%
Particle diameter range, μm	100 to 300	100 to 300
Mean particle diameter, μm	174.6	179.9
Surface area, m^2/g	3.53	3.4
Mercury pore volume, cm^3/g	0.2229	0.1968
Median pore diameter, Å	2,175	1,734
Attrition resistance		
5-hr loss	39.7%	17.0%
20-hr loss	89.0%	71.0%

Highlights from this life-cycle test are summarized below:

- H_2S content of the sulfidation product gas was below 20 ppmv (>99.8% H_2S removal) during the prebreakthrough period of each of the 100 cycles.
- The sulfur capacity at breakthrough (500 ppmv H_2S) decreased from 12.6% S sorbent in cycle 1 to 5.8% S in cycle 100. The average capacity was 9% S sorbent, and the majority of the capacity loss occurred during the first 50 cycles.
- The sorbent was fully regenerable with negligible sulfate in the regenerated sorbent. Reactor temperature control due to the exothermic regeneration reaction was not a problem.
- The attrition resistance of the sorbent after 100 cycles was significantly higher than the freshly prepared sorbent.

- Less than 2% of the sorbent was lost due to carryover of fines from attrition and particle fragmentation. Zinc loss was negligible.

- Loss of sulfur capacity correlated with decreases in the surface area and pore volume. X-ray diffraction indicated small amounts of zinc silicate, which was attributed to reaction with the ceramic gas distributor, and was believed to be responsible for some of the reactivity loss.

The fluidizable ZT-4 zinc titanate has also been tested on actual coal gas using a slip-stream from the METC fluid-bed gasifier (Gangwal and Portzer, 1995). A single semi-batch reactor which cycled between the sulfidation and regeneration phases was used. Operating conditions and a comparison osf sorbent properties before and after the three-cycle test are shown in Table 3. Inlet and outlet H_2S concentrations during the third sulfidation cycle are shown in Figure 4. Greater than 99% of the H_2S was removed during the prebreakthrough periods of the sulfidation cycles with breakthrough capacity as large as 20% S sorbent. The fact that exit H_2S concentration was relatively consistant while the inlet concentrations varied widely illustrates the rapid response of the fluidized-bed reactor. The SO_2 content of the regeneration product gas approached 1.5%, which is the stoichiometric value associated with 2.25%

Table 3. Operating Conditions and Properties of Fresh and Sulfided ZT-4 Sorbent in Fluidized-Bed Test Using Actual Coal Gas (Gangwal and Portzer, 1995).

Reaction Conditions	Sulfidation	Regeneration
Temperature, K	600	730
Pressure, atm	17.7	17.7
Feed Gas	Coal Gas	2.25% O_2 in N_2
Properties of ZT-4 Sorbent		
	Fresh	3 Sulfidation Cycles
Surface Area, m^2/g	3.2	7.6
Pore Volume, cm^3/g	0.23	0.11
Mean Pore Diameter, Å	2500	1800
Mean Particle Size, μm	112	95
Attrition Loss, %		
5-hr	16	1.4
20-hr	36	6.2
Zn/Ti Ratio	1.35	1.42

Table 4. The Build-up of Trace Contaminants in ZT-4 Sorbent (Gangwal and Gupta, 1995a)

	Concentration, μg/g	
Contaminant	Fresh Sorbent	After Three Sulfidations
As	0.7	8.4
Se	<0.57	0.72
Pb	<30	50
Cl	NM	38

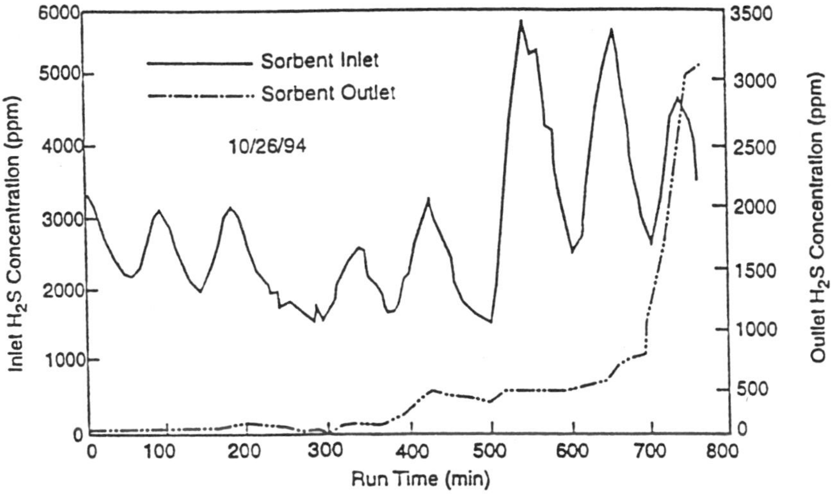

Figure 4. Inlet and Outlet H_2S Concentrations in the Third Cycle of a Fluidized-Bed Test Using Zinc Titanate ZT-4 (Gangwal and Gupta, 1995a)

O_2 in the feed gas. Therefore, essentially all of the oxygen reacted with little or no formation of $ZnSO_4$. Trace contaminants present in the coal gas accumulated on the sorbent as shown in Table 4. No mercury was detected in either the coal gas or the sorbent.

Results of completely integrated tests in which dual circulating fluid-bed sulfidation-regeneration reactors were coupled to a U-gas gasifier have been reported by Salo et al. (1994). Two zinc titanate formulations, UCI-5 from United Catalysts and ZT-4L from Research Triangle Institute, were used in tests providing 5 to 6 days of continuous operation. High sulfur removal efficiency was achieved as well as continuous production of SO_2 in the regenerator off-gas. Little, if any, $ZnSO_4$ was formed using either steam or N_2 as the regeneration diluent gas. The changes in sorbent properties were comparable to those observed in laboratory tests using simulated coal gas with cyclic sulfidation and regeneration.

3 Control of SO_2 in the Regenerator Off-Gas

Three options for the control of SO_2 in the regenerator off-gas are being considered -- conversion to sulfuric acid, reduction to elemental sulfur, and land disposal as $CaSO_4$.

3.1 Sulfuric Acid Production

The recycle gas loop of the General Electric moving-bed process is specifically designed to produce about 12% SO_2 so that the regenerator product gas can be fed to a sulfuric acid plant. This approach is limited only by the fact that the coal-fired power plant may be located far from existing H_2SO_4 plants and markets. While construction of a new H_2SO_4 plant adjacent to the power plant may be feasible, the cost and hazards of transporting H_2SO_4 to markets must still be faced. Other regeneration approaches in which the SO_2 content is no more than 1 to 2% do not integrate well with conventional sulfuric acid plants.

3.2 Reduction to Elemental Sulfur

The direct sulfur recovery process (DSRP) utilizes a slip stream of cleaned coal gas to reduce SO_2 produced during sorbent regeneration to elemental sulfur. This process, shown schematically in Figure 5 (Gangwal and Portzer, 1995), is suitable for SO_2 contents in the 1 to 2% range, and DSRP operating conditions integrate well with desulfurization/ regeneration conditions. The following reactions occur in a fixed-bed catalytic reactor

$$SO_2 + 2H_2 \text{ (or } 2CO) \rightarrow \frac{1}{n} S_n + 2H_2O \text{ (or } 2CO_2)$$

$$H_2 + \frac{1}{n} S_n \rightarrow H_2S$$

$$2H_2S + SO_2 \rightarrow \frac{3}{n} S_n + 2H_2O$$

DSRP was originally developed as a two-stage process with sulfur to be removed by condensation following each stage. Approximately 95% of the sulfur was to be recovered in stage 1 followed by 80% recovery in stage 2 to produce an overall sulfur recovery of 99%. However, recent tests have shown that 99% recovery can be achieved in a single stage when reaction conditions are carefully controlled.

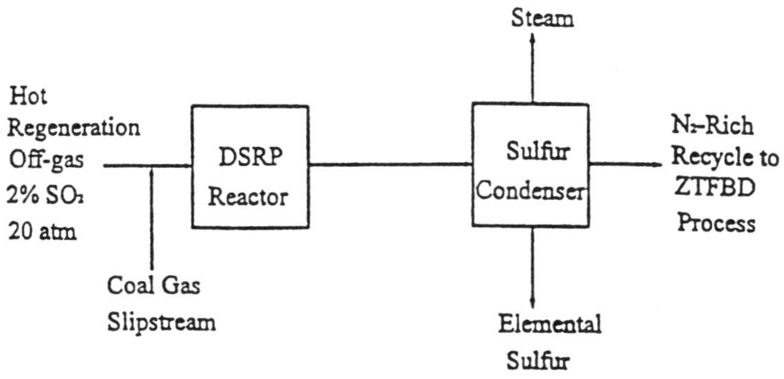

Figure 5. The Single-Stage Direct Sulfur Recovery Process (Gangwal and Portzer, 1995)

The single-stage DSRP has been tested in a semi-integrated fashion where a slip stream from the METC gasifier was fed to a single fluidized-bed reactor which alternated between the desulfurization and regeneration phases. The duration of the regeneration phase was insufficient for the DSRP to reach steady-state conditions using actual regenerator product gas. However, the DSRP reactor was fed with a simulated regenerator off-gas and actual coal gas from the zinc titanate desulfurization reactor to provide semi-integrated operation. DSRP reaction conditions as well as selected results using the simulated regenerator off-gas are presented in Table 5. Run 1 used fresh catalyst and produced SO_2 conversions equal to or greater than 98% with 100% selectivity to elemental sulfur. The SO_2 conversion decreased in the latter part of the test, but no H_2S was produced. In runs 3 and 4, the SO_2 conversion was always near 100% but the selectivity to elemental sulfur was generally in the range of 95 to 96%. The decreased selectivity was attributed to a non-optimum stoichiometric ratio of reducing gas to SO_2.

3.3 Land Disposal as $CaSO_4$

Figure 6, from Virr (1992), is a schematic flow diagram showing a U-gas fluidized-bed gasifier with in-bed calcium-based sorbent injection used for bulk sulfur removal and dual circulating fluidized-bed sulfidation-regeneration reactors for sulfur polishing. The reactions involved in in-bed sulfur capture are calcination followed by calcium sulfide formation

$$CaCO_3 \rightarrow CaO + CO_2$$
$$CaO + H_2S \rightarrow CaS + H_2O$$

Calcium-based sorbent should be capable of removing 90 to 95% of the H_2S, depending on operating conditions, with the remainder removed using an appropriate zinc-based sorbent in the polishing step. The off-gas from regeneration of the polishing sorbent,containing about 1½% SO_2 in N_2/steam, is recycled to the gasifier. There, theSO_2 will be reduced to H_2S where it will have another opportunity to react with CaO to form CaS. The oxidizing zone at the bottom of the gasifier is used to convert CaS to environmentally acceptable $CaSO_4$ for land disposal. Effectivelycomplete sulfide conversion is required before the combined limestone and coal ash (LASH) or dolomite and coal ash (DASH) can meet stability requirements for landfilling.

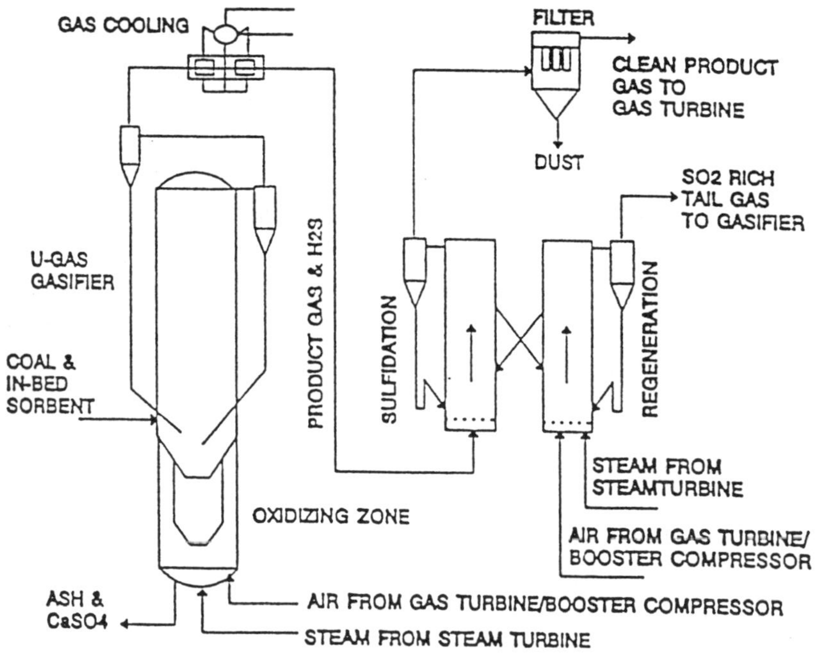

Figure 6. An Integrated Process Consisting of a U-Gas Gasifier with In-Bed Sorbent Injection and Circulating Fluidized-Beds for Sulfidation and Regeneration (Virr, 1992)

Table 5. DSRP Reaction Conditions and Test Results (Gangwal and Portzer, 1995)

Temperature (°C)	550-610
Pressure (atm)	17.7
Space velocity (std $cm^3/cm^3 \cdot h$)	4,560
Inlet SO_2 (%)	1.8
Stoichiometric Ratio (mol reducing gas per mol SO_2)	2.0

Run Time (min)	SO_2 Conversion, %	SO_2 Conversion H_2S (%)	SO_2 Conversion to Sulfur (%)
Run			
No. 1	99.4	0.0	99.4
32	99.4	0.0	99.4
36	99.5	0.0	99.5
40	99.7	0.0	99.7
44	99.5	0.0	99.5
48	98.5	0.0	98.5
52	98.0	0.0	98.0
56			
Run			
No. 3	99.6	1.2	98.4
117	100.0	4.1	95.9
121	100.0	4.9	95.1
125	100.0	2.4	97.6
129			
Run			
No. 4	100.0	4.9	95.1
175	100.0	4.1	95.9
179	100.0	4.9	95.1
183	100.0	4.7	95.3
199			

Complete oxidation of CaS is not necessarily a trivial task because of the large increase in solid molar volume associated with conversion to $CaSO_4$. Shires et al. (1994) have studied the reactions at 925°C and 1 atm in air, and found that sulfide conversions after 1 hour ranged from 22 to 80% using limestones, and from 93 to 100% using dolomites. The improved performance of dolomite may be attributed to the more open pore structure resulting from $MgCO_3$ calcination, and the fact that MgO is inert to H_2S at typical sulfidation conditions.

4 The Direct Production of Elemental Sulfur

Research on the feasibility of the direct production of elemental sulfur during sorbent regeneration has recently been initiated. Although direct production of elemental sulfur has never before been a primary research objective, the literature contains a number of reports in which significant amounts of elemental sulfur have been found. Three concepts for elemental sulfur production identified from the literature are described below.

4.1 Regeneration Concepts

Reaction between SO_2 and metal sulfides, which is represented by the following generic reaction, are known to occur

$$2\,MeS(s) + SO_2(g) \leftrightarrow 2\,MeO(s) + {}_{2}\,S_2(g)$$

Schrodt and Best (1978) studied this reaction using sulfided fly ash sorbent (active specie Fe_2O_3). The same reaction was investigated in more detail by Tseng et al. (1981). Anderson and Berry (1987) studied SO_2 regeneration of a sulfided cobalt titanate sorbent. More recently, Patrick and Gavalas (1993) report complete regeneration of a sulfided Fe_2O_3-Al_2O_3 sorbent using SO_2. Copeland (1993) has received a patent on an SnO_2 sorbent in which the sulfided form, SnS, can be regenerated with SO_2 to liberate elemental sulfur. Nielsen and Rudbeck (1993) have recently proposed a modification in which an SO_2-H_2O mixture is used for SnS regeneration. FeS regeneration with SO_2 has received the most attention, and the limiting factor seems to be the fact that SO_2 conversion is limited to a few percent. The regeneration product gas would have to be cooled to condense elemental sulfur, then reheated and recycled. The energy requirements associated with the large recycle stream are reported to render this option unattractive.

Reversing the desulfurization reaction using steam forms another regeneration concept.

$$MeS(s) + H_2O(s) \rightarrow MeO(s) + H_2S(g)$$

Although elemental sulfur is not a direct product of regeneration, high concentrations of H_2S can be readily converted to elemental sulfur using the Claus process. Tamhankar et al. (1985) have studied the steam regeneration of FeS, while Nielsen et al. (1991) have examined the regeneration of SnS with steam. Since this reaction is the reverse of the desulfurization reaction, it is obvious that sorbents having the greatest affinity for H_2S in the sulfidation cycle such as ZnO will be least amenable to steam regeneration.

A third concept involving regeneration in a steam-oxygen mixture operated under O_2-starved conditions has been reported. This concept, referred to as partial oxidation, would involve a number of simultaneous gas-solid and gas-phase reactions. In order to form elemental sulfur it is necessary to operate with substoichiometric O_2 as thermodynamics favors total oxidation to SO_2 when excess O_2 is present. Joshi et al. (1979) reported that 75% of the sulfur was liberated in elemental form when FeS was regenerated in a mixture of 95% steam-5% air at 538°C and 1 atm using a gas hourly space velocity of 600. Grindley and Steinfeld (1981) reported similar results at the same reaction conditions and found that smaller steam-to-air ratios resulted in less elemental sulfur and increased amounts of SO_2. No attempt was made to identify conditions which would maximize the elemental sulfur yield. Van der Waal (1987) performed extensive tests on the regeneration of iron-based sorbents and reported that low temperature, low O_2 content, and large space velocity increased the yield of elemental sulfur. Kay and Wilson (1989) describe a process in which elemental sulfur is formed during the regeneration of Ce_2O_2S, the product of the reaction between CeO_2 and H_2S. In a follow-up paper, Kay et al. (1993) state that "O_2-starved" conditions are required for the formation of elemental sulfur.

4.2 Thermodynamic Analysis

The free-energy minimization program CHEMQ (Kirkpatrick and Pike, 1994) was used to evaluate sorbents for their ability to remove H_2S from Texaco and KRW coal gases, and the feasibility of producing elemental sulfur using the three regeneration concepts previously described. Results of this study were presented by Harrison et al. (1995), and a summary is presented below.

4.2.1 Reduction/Sulfidation Analysis

The desulfurization potential of the eight metal oxides listed in the first two columns of Table 6 was examined over a temperature range of 650 to 1150 K and pressure range of 1 to 25 atm. The oxidation state of the excess sorbent in the reducing coal gas, the nature of the sulfided product, the possible formation of molten or volatile species, possible carbon deposition, and the equilibrium fractional H_2S removal was examined. Column 3 lists the nature of the sulfided product which, except for cerium, is the metal sulfide. Ce_2O_2S is the favored product from the sulfidation of CeO_2. As shown in column 4, the analysis confirmed prior knowledge that cobalt and copper oxides would be reduced to the elemental form, and that iron, manganese, and molybdenum would be subject to partial reduction. SnO_2 was stable over a portion of the temperature and pressure range, but reduction to liquid metallic tin, Sn(l), was favored at the more severe conditions. In addition, the vapor pressure of the sulfidation product, SnS, is sufficient that formation of SnS(g) can become a problem at higher temperatures. Similarly, the zinc oxide analysis showed the well-known tendency for reduction and subsequent formation of Zn(g). Molten and volatile species which may be formed are listed in column 5. The last column shows the maximum fractional H_2S removal at conditions where carbon deposition and the formation of volatile or molten species are not thermodynamically favored. These values range from 0.88 for CeO_2 to 1.00 for Co(s), MoO_2, and ZnO(s). Three types of temperature effect on H_2S removal were found. With most sorbents, the equilibrium H_2S removal decreased monotonically with increasing temperature. However, with SnO_2 the maximum H_2S removal potential occurred at an intermediate temperature, and the equilibrium H_2S removal with CeO_2 increased with increasing temperature. These effects are illustrated in Figure 7 for desulfurization of a Texaco gas at 5 atm.

4.2.2 Regeneration Analysis

Questions addressed in the thermodynamic analysis of regeneration included the extent of regeneration, i.e., the fraction of sulfur transferred from the solid to the gas phase, the quantity of gas-phase elemental sulfur which could be formed, the oxidation state of the regenerated oxide, and the possible formation of undesired products such as metal sulfate and molten or volatile species. Copper and cobalt were not included in the regeneration analysis because of their tendency to be reduced to the elemental form in the sulfidation cycle.

4.2.2.1. Reaction With SO_2

The reaction between metal sulfide and SO_2 should proceed essentially as written in eqn (9). Therefore, complete regeneration to the metal oxide and elemental sulfur is theoretically possible if sufficient excess SO_2 is provided. Reactions between

Table 6. Reduction/Desulfurization Analysis Summary

Sorbent System	Highest Oxidation State	Sulfided Product	Excess Metal	Volatile/ Molten Species	Maximum Fractional H_2S Removal*
Cerium	CeO_2	Ce_2O_2S	CeO_2	None	0.88
Cobalt	Co_3O_4	Co_3S_4	Co(s)	None	1.00
Copper	CuO	Cu_2S	Cu(s)	None	0.99
Iron	Fe_2O_3	FeS	FeO, Fe_3O_4	None	0.96
Manganese	MnO_2	MnS	MnO	None	0.99
Molybdenum	MoO_3	MoS_2	MoO_2	None	1.00
Tin	SnO_2	SnS	SnO_2	Sn(l), SnS(g)	0.98
Zinc	ZnO	ZnS	ZnO	Zn(g)	1.00

*at conditions where carbon deposition and volatile/molten species are not favored

MnS, MoS_2, and ZnS and SO_2 were, for practical purposes, thermodynamically impossible at temperature and pressure conditions of interest. FeS was somewhat reactive, although a molar ratio of SO_2 to FeS of about 50 to 1 would be required for complete regeneration. Reaction with SnS was somewhat more favorable, although care must be taken to avoid $SnSO_4$ formation at high pressure, and to avoid SnS(g) at low pressure. The thermodynamics of the Ce_2O_2S-SO_2 reaction are quite favorable.

An example of the thermodynamic results obtained when 1 mol of sulfided sorbent (SnS, Ce_2O_2S or MnS) was "mixed" with three mols of SO_2 is shown in Table 7. Excess SO_2 is available so that complete regeneration is possible if the thermodynamics are favorable. Ideal performance would correspond to completeregeneration with the gas phase containing equal quantities of sulfur in the oxidized (SO_2) and elemental forms (S_n). Although these conditions would not match actual regeneration conditions, they provide a valid comparison of sorbent

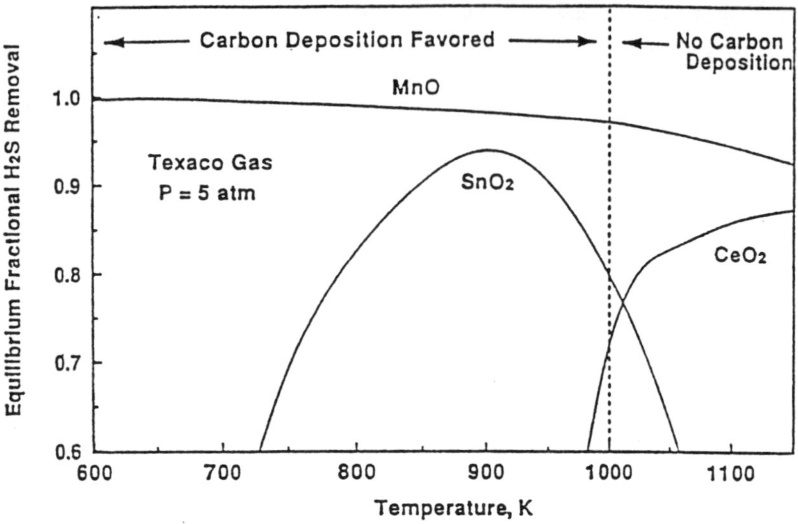

Figure 7. The Effect of Temperature on Equilibrium H_2S Removal Using Manganese, Tin, and Cerium Sorbents (Harrison et al., 1995)

performance. With SnS at 900K and 25 atm, Table 7 shows that sulfur is added to the condensed (as $SnSO_4$(s) and S(l)). Fractional regeneration is actually negative at this condition. At 1050K and 25 atm, the $SnSO_4$ should decompose and S(l) vaporize. Essentially complete regeneration is possible as indicated by the fact that only trace quantities of SnS would remain at equilibrium. The sulfur distribution in the gas phase of 47% elemental and 53% oxide is reasonably close to the ideal 50:50 split between the two forms. Complete regeneration of Ce_2O_2S to CeO_2 is possible at both 900 and 1050K. The gas phase sulfur distribution is close to the ideal 50:50 split between elemental and oxide forms, with the distribution being somewhat improved at the lower temperature. The practical impossibility of regenerating MnS with SO_2 is shown in the last two columns of Table 7. Only trace amounts of MnO can be formed, and essentially all sulfur in the equilibrium gas remains in the oxide form.

The thermodynamic attractiveness of Ce_2O_2S-SO_2 regeneration at 1000K and 25 atm is shown in greater detail in Figure 8 where the equilibrium fraction of the gas phase sulfur in elemental form is shown as a function of the molar ratio of SO_2 to Ce_2O_2S. Complete regeneration, i.e., complete transfer of all sulfur to the gas phase is favored for SO_2 to Ce_2O_2S ratios as small as 1.3. The fraction of the gas phase sulfur in elemental form increases from about 0.3 at a ratio of 5:1

to about 0.85 at the minimum ratio of 1.3:1. Further reduction in the SO_2 to Ce_2O_2S ratio is not limited by incomplete regeneration, but by the condensation of S(l) at the 25 atm pressure. Consequently, still smaller SO_2 to Ce_2O_2S ratios are thermodynamically possible at a lower regeneration pressure.

Table 7. Results of Equilibrium Analysis for the Regeneration of SnS, Ce_2O_2S, and MnS With SO_2 (1 mol metal sulfide : 3 mols SO_2): P = 25 atm

System	SnS		Ce_2O_2S		MnS	
Temperature, K	900	1050	900	1050	900	1050
Condensed Phase Composition, mol fraction						
Oxide	0.00	1.00	1.00	1.00	trace	trace
Sulfide	0.00	trace	0.00	0.00	1.00	1.00
Sulfate	0.77	0.00	0.00	0.00	0.00	0.00
Sulfur (liq)	0.23	0.00	0.00	0.00	0.00	0.00
Sulfur Distribution (as S), fraction						
Condensed	0.33	trace	0.00	0.00	0.25	0.25
Gas	0.67	1.00	1.00	1.00	0.75	0.75
Sulfur Distribution (as S) Within Gas Phase, fraction						
Elemental (S_n)	0.62	0.47	0.49	0.47	trace	trace
Oxide (Sox)	0.38	0.53	0,51	0.53	1.00	1.00
Sulfide (H_2S)	0.00	0.00	0.00	0.00	0.00	0.00

4.2.2.2 Steam Regeneration

The desired product from steam regeneration is not elemental sulfur, but H_2S in sufficiently high concentrations to permit the production of sulfur via the Claus process. Since this concept is a direct reversal of the desulfurization reaction, it is obvious that sorbents most efficient for H_2S removal will be least amenable to steam regeneration.

SnS and Ce_2O_2S are thermodynamically capable of reacting with steam. Both reactions should proceed according to stoichiometric eqn (10) with negligible quantities of SO_2 or $S_x(g)$ formed. However, both reactions produce H_2 and H_2S in equimolar quantities, so that a H_2 separation step will be required. In order to achieve complete regeneration, it should only be necessary to supply sufficient excess steam to satisfy the thermodynamics. Figure 9 shows the minimum steam to sorbent ratio required for complete regeneration of SnS and Ce_2O_2S as a function of temperature at 15 atm. For SnS, the ratio is in the range of 20 to 25 and is a relatively weak function of temperature. However, as the temperature approaches 1000K, small amounts of SnS(g) may be formed in the equilibrium product. Ce_2O_2S is even more amenable to steam regeneration. As shown in Figure 9, the minimum ratio required for complete regeneration is in the range of 5 to 10 mol H_2O per mol Ce_2O_2S. The ratio increases marginally with temperature, and is effectively independent of pressure since there is no change in the number of gas phase mols.

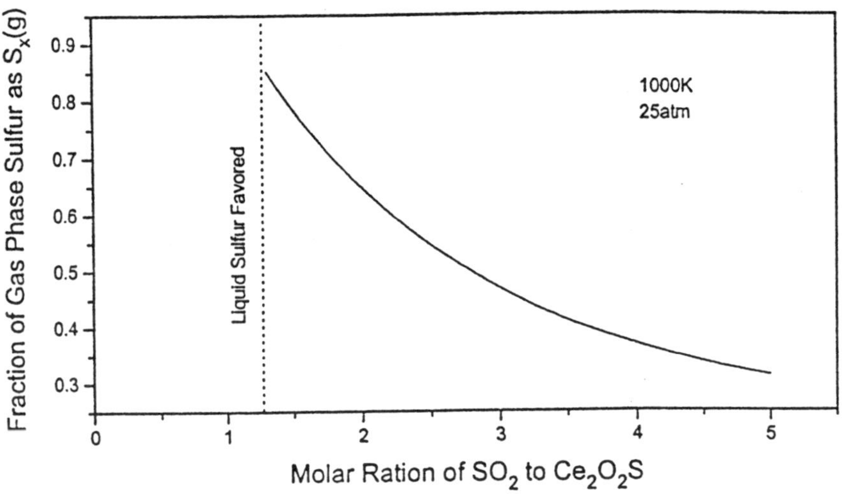

Figure 8. Equilibrium Yield of Elemental Sulfur in the Product Gas From the Regeneration of Ce_2O_2S With SO_2

4.2.2.3 Partial Oxidation

Duplication of the regeneration conditions in a flow reactor operated in an O_2-starved mode with large excess steam is impossible because of the inherent batch nature of the thermodynamic analysis. However, legitimate comparisons of the potential for elemental sulfur formation have been obtained by equilibrating the sulfided sorbents with varying amounts of substoichiometric oxygen and excess steam. This approach is illustrated in Figure 10 at 1000K and 15 atm for an initial mixture containing 1 mol of sulfided sorbent, 50% of the stoichiometric oxygen required for complete regeneration of the metal sulfide, and a molar steam-to-oxygen ratio of 5 to 1. Partial regeneration of all sorbents is thermodynamically possible in the presence of oxygen. Almost complete regeneration of SnS and Ce_2O_2S is possible and appreciable quantities of elemental sulfur may be produced.

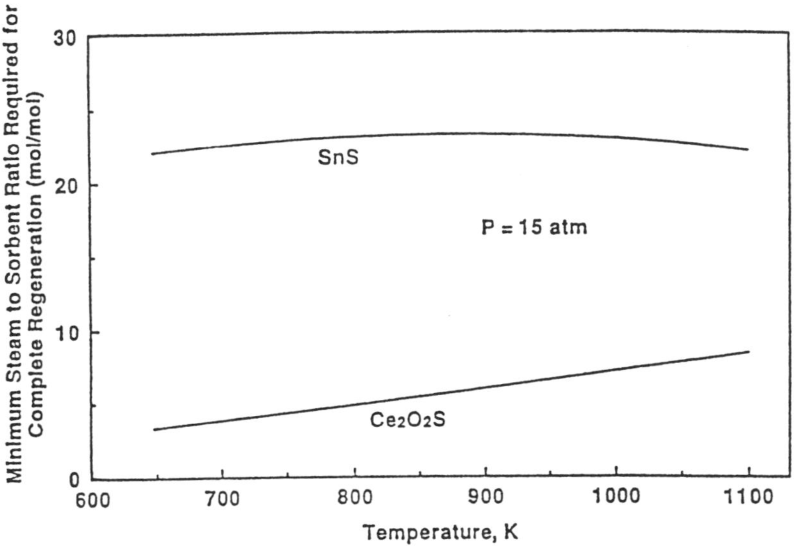

Figure 9. Minimum Steam to Sorbent Ratio for the Complete Regeneration of Ce_2O_2S and SnS_2 Sorbents (Harrison et al., 1995)

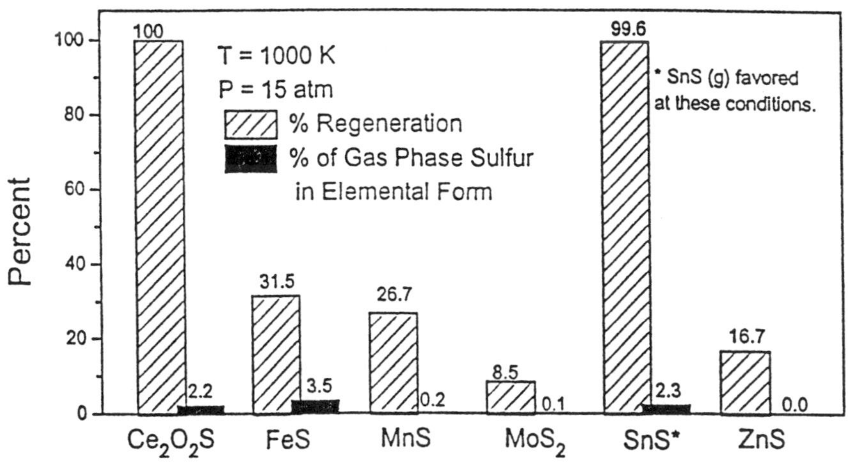

Figure 10. Thermodynamic Limits on Percent Regeneration and Elemental Sulfur Production During Sorbent Regeneration by Partial Oxidation

FeS can only be partially regenerated but the quantity of elemental sulfur which may be produced exceeds the amounts from SnS and Ce_2O_2S regeneration. Essentially no elemental sulfur can be produced from the other sorbents.

4.2.2.4 Process Implications

The thermodynamic analysis showed that sorbents having the greatest affinity for H_2S in the reduction/desulfurization cycle would be the most difficult to regenerate with elemental sulfur (or H_2S) as a direct product. Ce_2O_2S, SnS and, to a lesser extent, FeS show promise for elemental sulfur production.

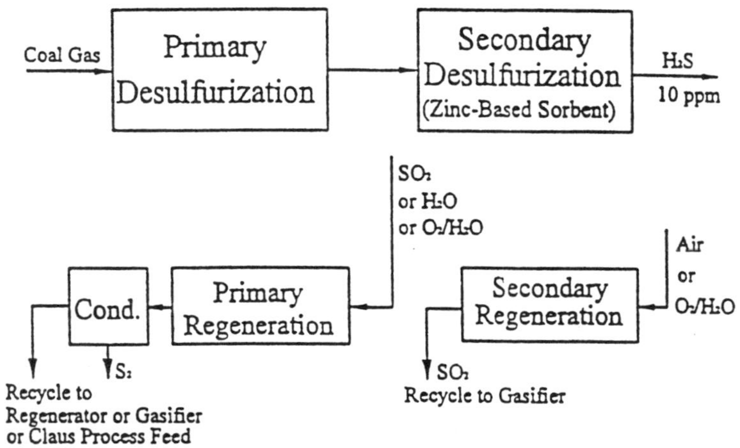

Figure 11. A Two-Step Desulfurization-Regeneration Process (Harrison et al., 1995)

Because of the reduced affinity of these sorbents for H_2S, a two-step desulfurization process such as shown in Figure 11 would be required. A primary desulfurization reactor containing an oxide such as iron, tin, or cerium would be followed by a secondary desulfurization reactor containing a highly efficient zinc-based sorbent. Regeneration of ZnS would be accomplished in the traditional manner and the product gas containing SO_2 along with $O_2/N_2/H_2O$ would be recycled to the gasifier. Removal of the majority of the H_2S in the primary desulfurization step would permit the secondary reactor to be smaller and/or require less frequent regeneration. Regeneration of the primary sorbent would be accomplished using one of the three concepts previously described. In the case of regeneration with SO_2, elemental sulfur would be condensed and excess SO_2 recycled. For regeneration with steam, the regeneration product would be fed to a Claus unit. Finally, for partial oxidation regeneration, elemental sulfur would be condensed and the remaining gases recycled to the entrance of the regeneration reactor and/or to the gasifier.

5 LSU Experimental Program

As a result of the literature search and thermodynamic analysis, experimental studies on elemental sulfur production from the regeneration of FeS using the partial oxidation concept and the regeneration of Ce_2O_2S using al l three concepts are being carried out.

The first phases of the FeS regeneration study used electrobalance reactors to investigate the kinetics of the reaction of FeS in atmospheres of O_2/N_2, H_2O/N_2, and $O_2/H_2O/N_2$. The rates were found to be first-order in gas concentration and only weakly dependent on temperature. FeS reacts much faster with O_2 than with H_2O, which explains why large ratios of H_2O to O_2 are required for the production of significant amounts of elemental sulfur. Product gas from the electrobalance reactor was not analyzed, and no direct information on elemental sulfur formation was obtained. Sulfidation and regeneration reactions of cerium oxide are not amenable to electrobalance studies since there is no mass change associated with the conversion of $2CeO_2$ to Ce_2O_2S.

A laboratory-scale fixed-bed reactor which incorporates product gas analysis has been constructed to study the reactions in more detail. Real time analysis of the regeneration product gas, which may contain $S_n(g)$, SO_2, and H_2S, has proven to be a major problem because of the difficulty in handling elemental sulfur. The reactor system is shown schematically in Figure 12. N_2, air, SO_2, and H_2S are obtained from high purity cylinders with flow rates controlled by high pressure mass flow controllers. Liquid water is fed using a high pressure syringe pump. Lines downstream of the syringe pump are heat traced to induce vaporization. Combined gases flow downward over the sorbent contained inside a reactor insert tube, and exit from the bottom of the reactor. The aluminized stainless steel pressure vessel is capable of operation to approximately 800°C and 15 atm.

The temperature and pressure of the regeneration product, which may contain SO_2, H_2S, S_n, N_2, O_2, H_2, and H_2O, must be such that all components (particularly elemental sulfur) remain in the gas phase. A portion of the product gas (see Figure 13) flows through a capillary flow restrictor to an oxidation tube where O_2 is added and all sulfur compounds are oxidized to SO_2 at about 1000°C. Gases exiting the oxidation tube pass through a membrane separator for H_2O removal and into a UV-

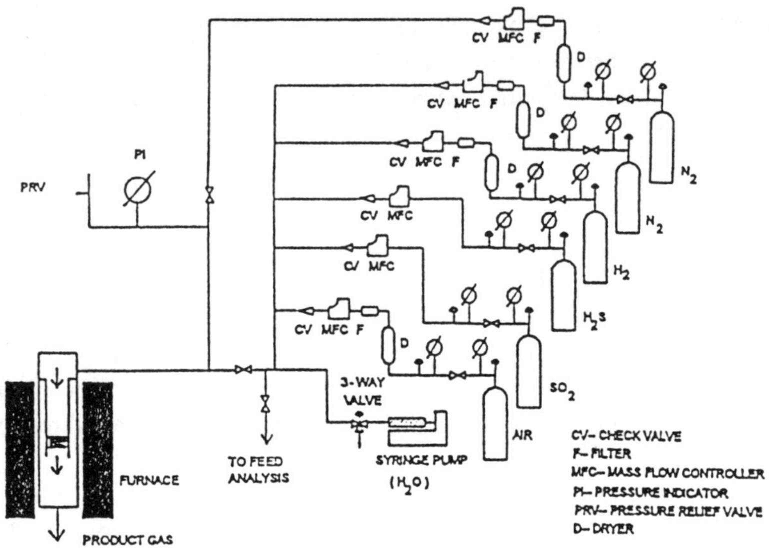

Figure 12. Schematic Diagram of the Reactor System Used in the Elemental Sulfur Production Project at LSU

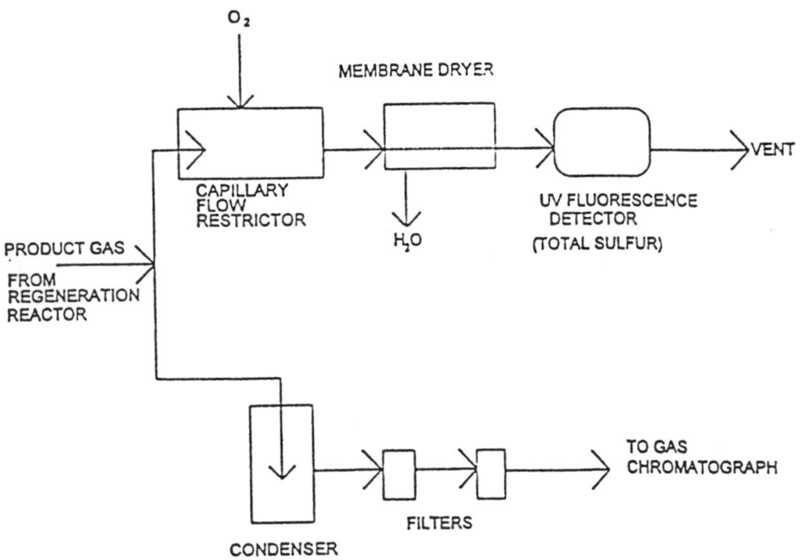

Figure 13. Schematic Diagram of the Gas Analysis Method Used in the Elemental Sulfur Production Project at LSU

fluorescence detector where the SO_2 concentration is determined. The flow rate of reactor product through this arm of the analytical train is determined solely by the reactor pressure and resistance of the flow restrictor. A change in the resistance due to sulfur condensation or carryover of particulate matter will alter the flow rate and the response of the UV detector. Frequent recalibration of both the flow rate and the response of the UV detector has been necessary.

The remainder of the regenerator product gas passes through a condenser and a series of filters, where water and elemental sulfur are removed, and through a backpressure regulator where the reactor pressure is controlled. SO_2 and H_2S concentrations are determined using gas chromatography. Experimental problems with this portion of the analytical system have been experienced due to incomplete removal of elemental sulfur in the condenser and filters and subsequent plugging of downstream lines.

Results from one FeS partial oxidation run are shown in Figure 14 where total sulfur, SO_2, and H_2S concentrations in the product gas are plotted versus time. Regeneration conditions were 600°C, 4.4 atm, and 5700 hr^{-1} (stp) space velocity with the regeneration feed gas containing 0.25% O_2 and 20% H_2O in N_2. After a brief time delay associated with the reactive gases flowing through the reactor to the detectors, both the H_2S and total sulfur concentrations began to increase. SO_2 did not appear in the product gas for approximately one hour. The mol fraction of total sulfur reached a maximum of 0.0025 about 0.9 hours after the reaction began, and declined steadily until the run was terminated after almost 8 hours. The H_2S mol fraction reached a maximum of 0.0006 after about 1.6 hours, and declined steadily

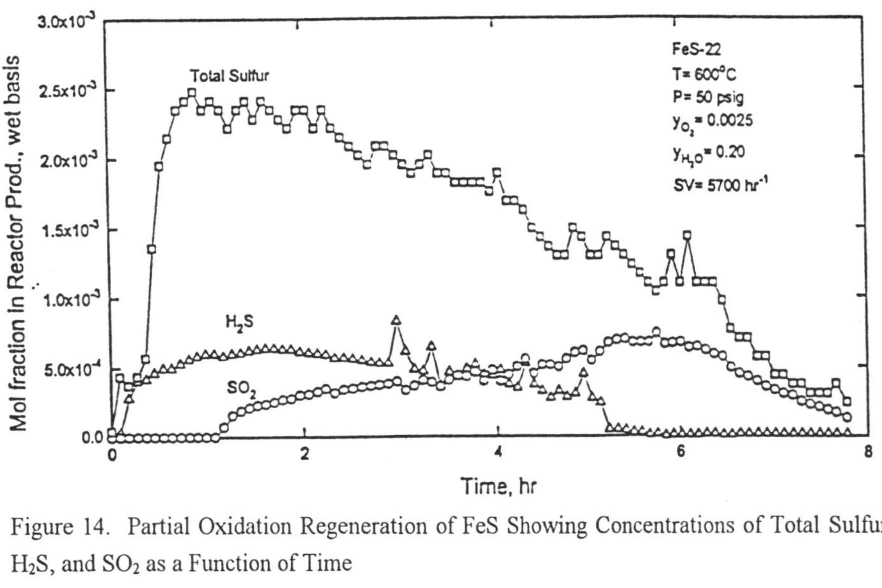

Figure 14. Partial Oxidation Regeneration of FeS Showing Concentrations of Total Sulfur, H_2S, and SO_2 as a Function of Time

from that time to approach zero after about 5 hours. The initial appearanch of SO_2 in the product gas occurred as the H_2S mol fraction was approaching the maximum value. SO_2 mol fraction increased steadily for about five hours to a maximum of 0.0007, and then decreased for the remainder of the run.

The total quantity of SO_2, H_2S, and total sulfur produced during the run was obtained by integrating the respective concentrations as a function of time. The total quantities of each product, expressed as a percent of the sulfur contained in the initial charge of FeS, amounted to 24% as SO_2, 22% as H_2S, and 99% total sulfur. Thus, approximately 50% of the original sulfur in FeS was released during regeneration as elemental sulfur. Approximately 75% of the sulfur released during the early stages of the reaction was in elemental form. This distribution gradually decreased to the point that about 25% of the sulfur released during the latter stages of the run was in elemental form.

The sulfidation phase of the CeO_2 tests has been plagued with "over-sulfidation." That is, the quantity of H_2S removed during the sulfidation cycle exceeds the stoichiometric amount required for the complete conversion to Ce_2O_2S. This problem is attributed to reaction of H_2S with the aluminized stainless steel reactor walls. A quartz insert tube should eliminate reaction with the vessel walls and allow the cerium tests to proceed. Ce_2O_2S regeneration results, although very limited, are encouraging. Figure 15 shows that only elemental sulfur was produced in a regeneration test at 750°C, 5 atm, and 5000 hr^{-1} space velocity with the feed gas containing 1.0% O_2 in N_2. The test had to be terminated after about 1.1 hours because elemental sulfur plugged the chromatograph feed lines. The shape of the sulfur response curve and the absence of SO_2 in the product suggests that SO_2 formed at the entrance of the reactor subsequently reacted downstream with Ce_2O_2S to liberate elemental sulfur. Improvements in the condenser and filters for removing elemental sulfur have subsequently been made.

6 Research Triangle Institute Experimental Program

Gangwal et al. (1995) at the Research Triangle Institute are also examining the feasibility of the direct production of elemental sulfur during sorbent regeneration. Concepts for elemental sulfur production considered by RTI included (1) SO_2 regeneration, (2) substoichiometric (partial) oxidation, (3) steam regeneration followed by H_2S conversion to elemental sulfur, and (4) steam-air regeneration. Their experimental effort is concentrating on the SO_2 regeneration concept since elemental sulfur should be the only sulfur product, and reactant SO_2 can be obtained by burning a portion of the elemental sulfur produced.

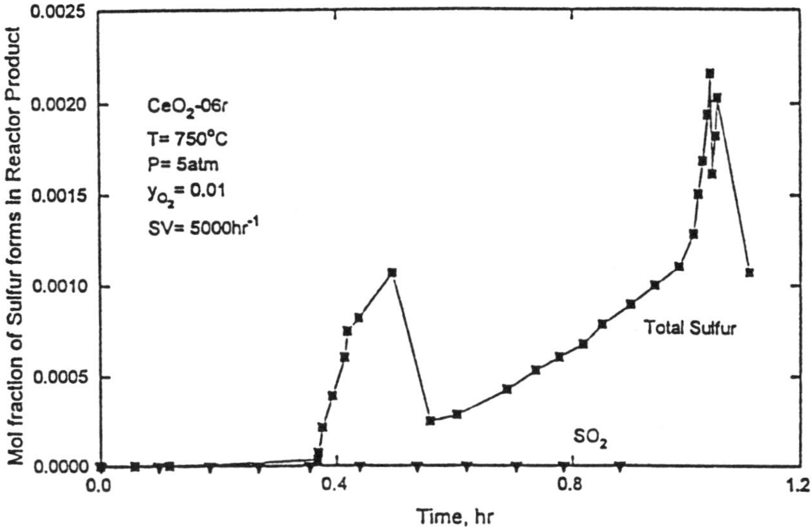

Figure 15. Elemental Sulfur Production During the Regeneration of Ce_2O_2S

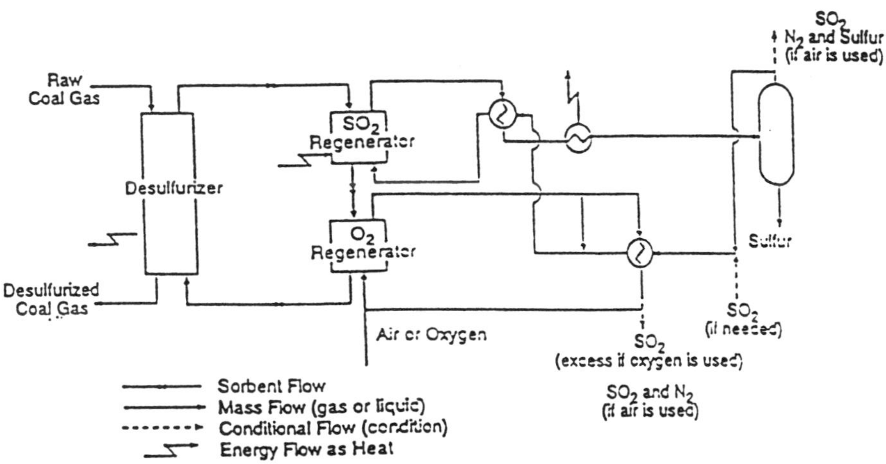

Figure 16. Research Triangle Institute Process for the Production of Elemental Sulfur (Gangwal et al., 1995)

The RTI process concept, shown in Figure 16, involves single-stagedesulfurization and two-stage regeneration. Sorbent composed of a mixture of iron- and zinc-oxides would be used to reduce the H_2S concentration in the treated coal gas to the required level in a single-stage desulfurization reactor. Iron sulfide would be regenerated in the first reactor using SO_2 while zinc sulfide would be regenerated in the second reactor in the traditional manner using dilute oxygen. SO_2 produced in the second regeneration stage could be used for the first regeneration stage or recycled to the gasifier.

7 References

Anderson, G.L., and Berry, F.O., 1987, Development of a Hot Gas CleanupSystem, Proceedings of the Seventh Annual Gasification and Gas Stream Cleanup Systems Review Meeting, DOE/METC-87/6079, p. 642.

Ayala, R.E., et al., 1994, Development of Sorbents for High-TemperatureDesulfurization in Moving-Bed Systems, Proceedings of the Coal-Fired Power Systems 94 -- Advances in IGCC and PFBC Review Meeting,DOE/METC-94/1008, Vol. 2, p. 637.

Bevan, S., et al., 1995, Integrated Testing of a Pressurized Fixed-Bed Gasifier, Hot Gas Desulfurization System and Turbine Simulator, Proceedings of the Advanced Coal- Fired Power Systems '95 Review Meeting, DOE/METC-95/1018, Vol.1, p. 187.

Copeland, R.A., 1993, High Temperature Regenerable H_2S Removal Agents, U.S. Patent 5,271,907.

Dorchak, T.P., et al., 1991, The Direct Sulfur Recovery Process, Environmental Progress, 10, 68.

Gangwal, S.K., et al., 1988, Multicycle Testing of Zinc Ferrite, Final Report, DOE/MC/23126-2644.

Gangwal, S.K., and Gupta, R.P., 1994, Fluidized-Bed Sorbents, Proceedings of the Coal-Fired Power Systems 94--Advances in IGCC and PFBC Review Meeting, DOE/METC-94/1008, Vol. 2, p. 646.

Gangwal, S.K., and Portzer, J.W., 1995, Slipstream Testing of Hot-Gas Desulfurization With Sulfur Recovery, Proceedings of the Advanced Coal-Fired Power Systems '95 Review Meeting, DOE/METC-95/1018, Vol.1, p. 220.

Gangwal, S.K., and Gupta, R.P., 1995, Bench-Scale Testing of Fluidized-BedSorbent -- ZT-4, Proceedings of the Advanced Coal-Fired Power Systems'95 Review Meeting, DOE/METC-95/1018, Vol. 1, p. 229.

Gangwal, S.K., et al., 1995, Advanced Sulfur Control Concepts, Proceedings of the Advanced Coal-Fired Power Systems '95 Review Meeting, DOE/METC-95/1018, Vol. 2, p. 622.

Grindley, T., and Steinfeld, G., 1981, Development and Testing of Regenerable Desulfurization Sorbents, Final Report, DOE/MC/16545-1125.

Grindley, T., 1988, Sidestream Zinc Ferrite Regeneration Tests, Proceedings of the Eight Annual Gasification and Gas Stream Cleanup Systems Review Meeting, DOE/METC- 88/6092, Vol. 1, p. 58.

Grindley, T., 1986, Analysis of Fuel Gas Cleanup for Molten Carbonate Fuel Cells, Proceedings of the Sixth Annual Meeting on Contaminant Control in Coal-Derived Gas Streams, DOE/METC-86/6042, p. 190.

Harrison, D.P., et al., 1995, Advanced Sulfur Control Concepts, Proceedings of the Advanced Coal-Fired Power Systems '95 Review Meeting, DOE/METC-95/1018, Vol. 2, p. 610.

Joshi, D.K., et al., 1979, Hot Low-Btu Producer Gas Desulfurization in Fixed-Bed of Iron Oxide-Fly Ash, Final Report, DOE/FE-2257-3.

Kay, D.A.R., and Wilson, W.G., 1989, Method for the Regeneration of Sulfided Cerium Oxide Back to a Form That is Again Capable of Removing Sulfur From Fluid Materials, U.S. Patent 4,857,280.

Kay, D.A.R., et al., 1993, High Temperature Thermodynamics and Applications of Rare Earth Compounds Containing Oxygen and Sulfur in Fuel Gas Desulfurization and SO_x and NO_x Removal, J. of Alloys and Compounds, 192, 11.

Kirkpatrick, M., and Pike, R.W., 1994, Linking Database Management Systems and Chemical Engineering Application Programs -- Prediction of Chemical Equilibrium, AIChE Symposium Series, Vol. 90, No. 298, p. 174.

Leininger, T.F., et al., 1992, Integration and Testing of Hot Desulfurization and Entrained Flow Gasification for Power Generation Systems, Proceedings of the Twelfth Annual Gasification and Gas Stream Cleanup Systems Review Meeting, DOE/METC-92/6128, Vol. 1, p. 95.

Nielsen, P.E.H., et al., 1991, Steam Regenerable Sulfur Absorption Masses andTheir Application in IGCC Plants, paper presented at EPRI Coal Gasification Conference, San Francisco.

Nielsen, P.E.H., and Rudbeck, P., 1993, Hot Gas Cleaning in IGCC Power Plants, Power Generation Technology, ___________.

Patrick, V., and Gavalas, G.R., 1993, Reduction, Sulfidation, and Regeneration of Mixed Iron-Aluminum Oxide Sorbent, Industrial and Engineering Chemistry Research, 32, 519.

Salo, K., et al., 1994, Enviropower Hot Gas Desulfurization Pilot, Proceedings of the Coal-Fired Power Systems 94--Advances in IGCC and PFBC Review Meeting, DOE/METC-94/1008, Vol. 1, p. 236.

Schrodt, J.T., and Best, J.E., 1978, Sulfur Recovery From Flue Gas Desulfurization Sorbent, AIChE Symposium Series, 74, No. 175, 189.

Shires, P.J., et al., Stabilization of Spent Calcium-Based Sorbent, Proceedings of the Coal-Fired Power Systems 94 -- Advances in IGCC and PFBC Review

Meeting, DOE/METC-94/1008, Vol. 2, p. 668.

Smith, D.P., et al., 1987, Design Studies for Hot Gas Cleanup System in Load Following Mode, Proceedings of the Seventh Annual Gasification and GasStream Cleanup Systems Review Meeting, DOE/METC-88/6092, Vol. 1, p. 32.

Tamhankar, S.S., et al., 1985, Kinetic Study on the Reactions Involved in Hot Gas Desulfurization Using a Regenerable Iron Oxide Sorbent. III. Reactions of the Sulfided Sorbent With Steam and Steam-Air Mixtures,Chemical Engineering Science, 40, 1019.

Tseng, S.C., et al., 1981, Kinetic Studies on the Reactions Involved in Hot Gas Desul furization Using a Regenerable Iron Oxide Sorbent.II. Reactions of Iron Sulfide With Oxygen and Sulfur Dioxide, Chemical Engineering Science, 36, 1287.U.S. Department of Energy, 1995, Clean Coal Technology Demonstration Program- Program Update 1994, DOE/FE-0330.

Van der Waal, W.J.J., 1987, Desulfurization of Process Gas by Means of Iron Oxide on Silica Sorbents, Ph.D. Dissertation, University of Utrecht.

Virr, M.J., 1992, Toms Creek IGCC Demonstration Project, Proceedings of the Twelfth Annual Gasification and Gas Stream Cleanup Systems Review Meeting, DOE/METC-92/6128, Vol. 1, p. 42.

Elemental Sulfur Recovery from SO_2-Rich Streams

M. Flytzani-Stephanopoulos, W. Liu[#], T. Zhu and L. Kundakovic

Department of Chemical Engineering, Tufts University, Medford, MA 02155, U.S.A.
[#]Department of Chemical Engineering, Massachusetts Institute of Technology; presently at Amoco Research Center, Naperville, IL 60563, U.S.A.

Abstract. Single-stage catalytic reduction of SO_2 to elemental sulfur is a promising new technology better suited to treat SO_2-rich streams in power plants than the conventional multi-stage Claus process. SO_2-containing streams are produced in IGCC power plants as well as in conventional coal-fired boilers employing regenerative flue gas desulfurization. In this paper we describe the performance of a new highly active and selective catalyst for dilute- or concentrated- SO_2 reduction by CO, CH_4 or $CO+H_2$ reductants. The catalyst comprises a transition metal oxide as a minor component in a fluorite-type oxide matrix. The properties of this type of composite oxide are illustrated here by the system of copper or nickel metal oxide in cerium oxide or doped cerium oxide. The reaction of SO_2 with CO is discussed within the redox framework. Strong synergism for this reaction is displayed by the association of the transition metal with the fluorite-type oxide.

1 Introduction

Hot gas desulfurization (HGD) by regenerable sorbents is integrally related with sulfur recovery from the regeneration off-gas. The ZnO-based sorbents presently under development by the U.S. DOE and its contractors for IGCC application will be regenerated by diluent air so that an off-gas stream with about 1-3 mol. % SO_2 in N_2 will be produced. This needs to be processed further in order to obtain a saleable byproduct, such as elemental sulfur, sulfuric acid or liquid SO_2.

A similar situation is envisioned for coal-fired power plants employing flue gas desulfurization (FGD) by regenerable sorbents. FGD is presently accomplished by non-catalytic processes such as wet or dry scrubbing. Wet scrubbers are the most common worldwide and typically produce sulfated lime or limestone sorbents that must be disposed of as a waste in landfills. In the U.S., a total of 19 million tons of disposable FGD byproduct are produced, transported and disposed of in landfills annually [1].

The use of regenerable sorbent technologies has the potential to reduce or eliminate solid waste production, transportation and disposal. In a typical dry regenerative FGD process, sulfur dioxide in the flue gas steam is removed by a solid sorbent in the sorber unit. The saturated sorbent is subsequently regenerated by a reducing gas in the regenerator, while SO_2 is released in much higher concentration than in the dilute flue gas. Sulfur recovery from the regenerator off-gas is a major issue when determining the commercial viability of regenerative FGD. At the present time the most advanced regenerative FGD under development is the CuO-process of the U.S. DOE/PETC and

NATO ASI Series, Vol. G 42
Desulfurization of Hot Coal Gas
Edited by A. T. Atimtay and D. P. Harrison

UOP Inc. [2]. The concentration of SO_2 in the regenerator off-gas can be low or high depending on the reducing gas composition (e.g.,H_2, syngas or CH_4).

Direct reduction of sulfur dioxide in the regenerator off-gas to elemental sulfur is an attractive approach, because it involves a simplified flowsheet and produces a saleable product. Even when there is no market for the sulfur, elemental sulfur is only one-third of the volume of the equivalent $CaSO_4$ and is totally innocuous. Therefore, no secondary pollution issues ensue from this approach. Development of a highly efficient, single-stage catalytic converter for this purpose will expedite the commercialization of both the regenerative HGD and the new generation FGD process. Such a process is clearly much more attractive and economical than the multi-stage complex and expensive Claus plants.

Most previous studies of reduction of sulfur dioxide to elemental sulfur examined carbon monoxide as the reductant. A variety of catalyst materials have been studied, including transition metals and mixed oxides. The perovskite-type mixed oxides have received a lot of attention. A class of ABO_3-type mixed oxides consisting of alkaline earth, rare earth, and transition metal elements were disclosed in a patent by Whelan [3]. Happel et al. [4,5] proposed a redox reaction mechanism involving surface oxygen/oxygen vacancy participation for La-Ti-O perovskite oxide catalysts. The SO_2 reaction with CO over $Sr_xLa_{1-x}CoO_3$ perovskite catalysts was also discussed by Hibbert and Campbell within the redox framework [6,7]. Perovskite oxides, such as $LaCoO_3$, have both excellent electronic conductivity and oxygen ion mobility [8]. The previous studies suggest that the oxygen vacancy or mobility may be an important catalyst property for SO_2 reduction by CO. However, perovskite oxides are not stable under the reaction conditions and decompose into metal sulfide and oxysulfide. The COS reaction intermediate mechanism was proposed for this type of catalysts [9], that is, CO reacts with metal sulfides to form COS and then, COS reduces SO_2 to elemental sulfur. A single-stage catalytic process for the conversion of SO_2 to sulfur is under development at Lawrence Berkeley Laboratory using mixed metal oxide catalysts [10]. Sulfur yields as high as 98% have been reported at 500°C with dry or wet syngas as the reductant at atmospheric pressure and low space velocity ($2100h^{-1}$). Another process under development is the Direct Sulfur Recovery Process (DSRP) at the Research Triangle Institute [11]. This reports high sulfur yields at 627°C at relatively low space velocities ($3750h^{-1}$) and high pressure (21.4 atm).

In recent work, we have found that cerium oxide undoped or doped with rare earth oxides is a highly active and stable catalyst for SO_2 reduction by CO [12,13] and zirconium oxide also has comparable activity to cerium oxide [14,15]. Cerium and doped zirconium oxide are fluorite-type oxides well known for their high oxygen vacancy and mobility properties [16]. These refractory materials keep their crystal structure in the highly reactive sulfur atmosphere. In addition, the catalytic activity and resistance to impurities such as water vapor of the fluorite oxide catalysts can be significantly enhanced by addition of small amounts of transition metals, such as Cu, Ni, Co, etc [14,15]. The resulting transition metal/fluorite oxide composite catalysts have superior performance to the other catalysts disclosed in the literature. With dry CO as the reductant, the new catalysts yield 97% sulfur at 470°C, atmospheric pressure and high space velocity ($>45,000h^{-1}$). At the stoichiometric CO/SO_2 molar ratio, the yield of elemental sulfur is maximized. This is true both for dilute as well as concentrated SO_2 streams [12-15].

In this paper we discuss the salient features of SO_2 reduction by CO over the fluorite oxides and transition metal/fluorite oxide catalysts. Further details may be found in recent publications [15,17,18]. Among the various catalyst compositions we examined, we chose the Cu-Ce(D)O_x (D=dopant) as a model system, while the other catalysts are briefly discussed. We also present here new information on the reactivity of copper or nickel-modified cerium oxide for the reduction of SO_2 by methane and hydrogen-rich syngas [19].

2 Experimental

2.1 Catalyst Preparation and Characterization

Low purity cerium nitrate containing about 1.5 wt. % La was typically used for catalyst synthesis. The cerium from this precursor is designated as Ce(La) in the catalyst formula throughout the paper. High purity cerium nitrate (99.9%, Aldrich) was used to prepare the catalysts containing different dopant ions (Sr, Sc, La, Gd). Pure cerium oxide of medium high surface area (ca. 26 m^2/g) and mesopore size (ca. 20nm) was made from direct thermal decomposition of cerium acetate (Aldrich) for 4 h at 750°C. γ-Al_2O_3 support was supplied by LaRoche. Cerium oxide doped with alkaline earth or rare earth oxide, Ce(D)O_2, was prepared by complexation with citric acid [13, 17]. This method produces a solid foam with high pore volume and large fraction of macropores (>1μm). However, the resulting catalyst was bulky and had a low packing density of typically 0.1g/cc so that a large reactor volume would be required. The transition metal/fluorite oxide composite catalysts were mainly synthesized from nitrate salts by a conventional coprecipitation method using ammonium carbonate and were typically calcined in air for 4 h at 650°C. The low metal-containing catalysts, Cu-Ce(La)-O and Ni-Ce(La)-O, studied for the reduction of SO_2 by CH_4 and syngas were prepared by urea precipitation of the metal carbonates from nitrate salts at 100°C [20].This method provides more homogeneous, well-dispersed mixed oxides or oxide compounds. The supported catalysts were prepared by conventional wet impregnation using aqueous salt solutions of the metals. The impregnated samples were dried for a few days at room temperature and then heated in air for 4 h at 650°C. Further details of the catalyst preparation procedure are given in [15].

For elemental analysis, the as-prepared catalyst prior to calcination was dissolved in concentrated hydrofluoric acid solution and after dilution with de-ionized water. The resulting solution was analyzed by the Inductively Coupled Plasma (ICP) atomic emission spectrometry (Perkin Elmer Plasma 40). The catalyst surface composition was analyzed by X-ray Photoelectron Spectroscopy (XPS). The total BET surface area was routinely measured by single-point N_2 adsorption and desorption on a Micromeritics 2000 instrument. Selected catalysts were analyzed on a Micromeritics ASAP 2000 apparatus by multipoint BET surface area and pore size distribution measurements. Powder X-ray diffraction (XRD) analyses were performed on a Rigaku 300 X-ray Diffractometer. Copper Kα1 radiation was used with power setting of 50kV and 200mA. XPS analyses were performed on a Perkin Elmer 5100 system with 2 mm spatial resolution. The measurements were carried out at room temperature without any sample pretreatment. A Mg Kα X-ray source was used with incident energy of 1253.6eV. The X-ray generator power was typically set at 15kV and 20mA. C1s peak was found in all measurements resulting from ambient hydrocarbons in the sample, and was used as an internal standard. Therefore, all binding energies were adjusted relative to C1s at 284.6eV.

2.2 Apparatus

Catalyst activity was tested in a laboratory-scale, quartz tube packed bed reactor with a porous quartz frit placed at the middle for supporting the catalyst. A 0.6 cm I.D x 50 cm long reactor was used in regular catalyst tests, while a 1.0 cm I.D. reactor was used for high volume catalyst loading. The reactor tube was heated by a Lindberg furnace. The reaction temperature was monitored by a quartz tube-sheathed K-type thermocouple placed at the top of the packed bed and controlled by a temperature controller. The reacting gases, all certified calibration gas mixtures with helium, were measured with mass flow controllers and mixed prior to the reactor inlet. The resulting gas mixture flowed downward through the packed bed. Water vapor was introduced with helium bubbling through a heated water bath. The experiments were carried out at nearly atmospheric pressure. Elemental sulfur produced by the SO_2 reduction was condensed out in a cold trap installed at the outlet of the reactor. A filter was installed at the entrance of GC gas sample loop to protect it against entrained particles. The product gas was analyzed by a HP 5880A Gas Chromatograph (GC) with a Thermal Conductivity Detector (TCD). A 1/4" O.D. x 6' long packed column of Chromosil 310 (from SUPELCO) operated at 60°C provided baseline separation of CO, CO_2, COS, H_2S, CS_2, and SO_2.

Some quantities referred in the text, namely; CO/SO_2 molar ratio (R_{CO}), CH_4/SO_2 molar ratio (R_{CH4}), SO_2 conversion (X-SO_2), and sulfur yield (Y-sulfur), are defined as follows:

$$R_{CO} = [CO]_i/[SO_2]_i$$

$$R_{CH4} = [CH4]_i/[SO_2]_i$$

$$X\text{-}SO_2 = ([SO_2]_i-[SO_2]_e)/[SO_2]_i$$

$$Y\text{-sulfur} = [Sulfur]_e/[SO_2]_i$$

where $[Sulfur]_e$ is calculated from the difference: $[Sulfur]_e = [SO_2]_i - [H_2S]_e - 2[CS_2]_e - [COS]_e - [SO_2]_e$; $[CO]_i$, $[SO_2]_i$, and $[CH_4]_i$ are molar fractions of CO, SO_2, and CH_4 in the feed gas; $[Sulfur]_e$, $[SO_2]_e$, $[H_2S]_e$, $[COS]_e$, and $[CS_2]_e$ are molar fractions of elemental sulfur, SO_2, H_2S, COS, and CS_2 in the exit gas.

3 SO_2 Reduction by CO

The overall reaction may be written as:

$$SO_2 + 2CO \rightarrow S_2 + 2CO_2$$

The objective of this study was to find a highly active catalyst system that forms no COS byproduct and is resistant to poisoning by the impurities in the feed gases. For example, coke gasification is one likely industrial CO source where the existence of H_2O and CO_2 impurities is inevitable.

3.1 Catalytic Properties of Fluorite Oxide Catalysts

Figure 1 shows the crystal structure of fluorite oxides. The fluorite oxide has a face-centered cubic lattice in which the large tetravalent (+4) ions, such as Ce^{+4}, Zr^{+4}, Th^{+4}, etc., have an eight-coordination of oxygen ions. A unit cell contains four cations occupying four opposite vertices. Oxygen vacancies are created when the tetravalent lattice cations are replaced by divalent or trivalent dopant (impurity) ions. Alkaline earth and rare earth oxides, such as MgO, La_2O_3, Y_2O_3, etc., are common dopant oxides. These oxides generally have substantial solubility in fluorite oxides ranging from 2 to 100% [21]. In contrast, transition metal oxides have little solubility in fluorite oxides. High oxygen vacancy concentration and oxygen mobility are well-known properties of fluorite oxides. These materials have been extensively studied as oxygen ion conductors [16]. In the catalysis field, they have been used as promoters in the automotive three-way catalyst [22], but usually they are not considered as active catalyst components.

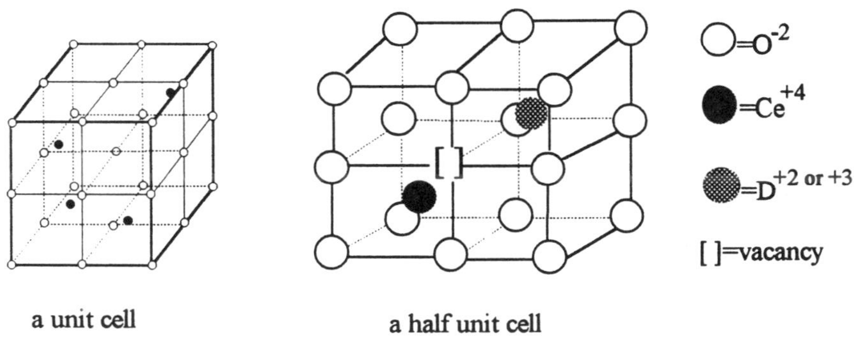

Figure 1. Lattice structure of a fluorite-type oxide.

Catalytic behavior of fluorite oxides for SO_2 reduction by CO is illustrated in Figure 2. Only sulfur yield is presented in the Figure, because COS as the only byproduct over these catalysts was produced in a small amount (less than 5%) and sulfur yield basically represents the SO_2 conversion. The branch I in Figure 2 is the light-off curve showing the temperature at which reaction starts. For these tests, we started with the fresh catalyst without any pretreatment by a reducing agent, raised the reaction temperature from 500 to 700°C in 50°C-steps, and held each temperature constant for about half an hour until complete reaction occurred. We found that the reaction was lighted off in the neighborhood of 600°C over all the La-doped ceria catalysts, independent of dopant content and the amount of catalyst used. The yttrium-doped zirconia has a similar light-off temperature. However, a higher light-off temperature (~690°C) was measured for Mg and Sc-doped ceria catalysts. The branch II in Figure 2 represents the fall-off behavior. This curve was obtained by lowering the reaction temperature in 50°C-steps until the reaction was quenched. The fall-off temperatures for the fluorite oxide catalysts were in the neighborhood of 500°C. The light-off and fall-off temperatures could not be lowered by increasing the contact time above the value of approximately 0.02 g·s/cc(STP). Also, the light-off temperature could not be lowered by increasing the dopant level. These temperatures appear to be associated with the intrinsic properties of the fluorite oxide.

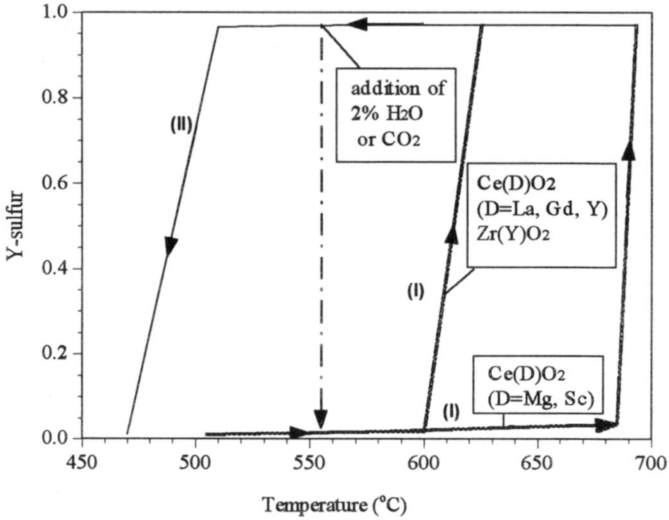

Figure 2. Catalytic performance of fluorite oxides for SO_2 reduction by CO (1% SO_2, 2% CO, 0.030 g·s/cc (STP); (I) ascending temperature; (II) descending temperature).

The fluorite oxides showed high stability and selectivity to elemental sulfur. Only the fluorite structure was identified by XRD for both fresh and used catalysts. However, the fluorite oxide catalysts were very sensitive to impurity gases. The catalyst completely lost its activity at 550°C when 2% H_2O or CO_2 was introduced into the reacting gas stream. Resistance to water vapor poisoning could not be improved by using alkaline earth and rare earth oxide dopants. The catalyst could be activated by either pre-reduction or by reaction at high temperatures. Hence, a partially reduced surface comprised the working catalyst. The surface state of cerium oxide was easily observed during the experiment, because bulk ceria appeared pale yellow, while a reduced surface had a dark blue color.

The catalytic behavior of fluorite oxides is explained within a redox reaction framework consisting of reduction of the catalyst surface by CO and oxidation of the reduced catalyst surface by SO_2.

$$\text{cat-O} + CO \rightarrow \text{cat-[ ]} + CO_2 \tag{1}$$

$$\text{cat-[ ]} + SO_2 \rightarrow \text{cat-O} + SO \tag{2}$$

$$\text{cat-[ ]} + SO \rightarrow \text{cat-O} + S \tag{3}$$

An oxygen vacancy is created as a surface capping oxygen is removed by CO. Then, SO_2 donates its oxygen to the vacancy to form SO. The SO is mobile on the surface until it finds another vacancy to donate its oxygen or a vacancy may migrate to a neighboring site to accept its oxygen. High oxygen ion mobility in the catalyst will facilitate the oxygen transfer from one site to another on the surface or from the bulk to the surface. However, the oxygen vacancy can be taken up by other oxidizing agents such as H_2O. The stronger these impurity oxygen atoms attach to the vacancy, the more severe their inhibition effect. The activation tests proved that the creation of oxygen vacancies on the catalyst surface is a key step to start the reaction.

The specific catalytic activity of cerium oxide can be improved by using small amounts (a few atomic percent) of Y or La dopant [13]. But, alkaline earth and rare earth dopants are not effective in lowering the reduction temperature of the surface capping oxygen and changing the surface properties of the fluorite oxide. Water vapor not only takes up the oxygen vacancy, but also inhibits CO adsorption on cerium oxide according to recent FTIR studies [23]. Therefore, the oxygen vacancy property of fluorite oxides is relevant for catalytic reduction of SO_2 by CO, but not sufficient for an active and stable catalyst that is resistant to water vapor poisoning.

3.2 Transition Metal/Fluorite Oxide Catalysts

Significant improvement of fluorite oxide catalysts in both activity and tolerance to water vapor was achieved by addition of small amounts of transition metals, such as Cu, Ni, Co, Pt [15,17]. Typical results of the transition metal/fluorite oxide composite catalysts are illustrated in Figures 3 and 4. Figure 3 shows the light-off curves of SO_2

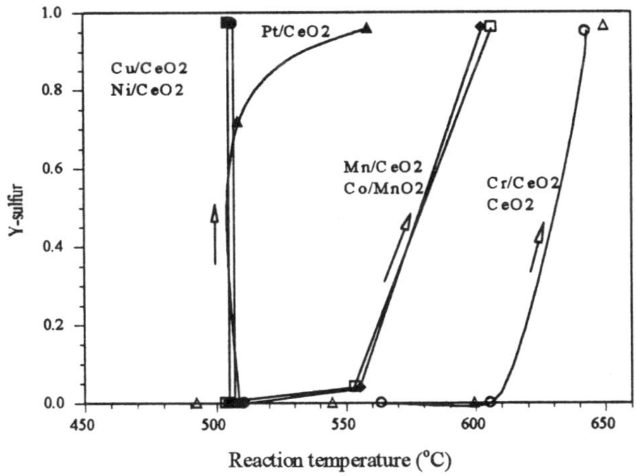

Figure 3. Light-off behavior over several transition metal-impregnated ceria catalysts (1% SO_2, 2% CO, 97% He).

reduction by CO over transition metal-impregnated ceria catalysts. The 90% conversion temperatures are 500°C for Cu/CeO_2 and Ni/CeO_2, 550°C for Pt/CeO_2, and 600°C for Co/CeO_2 and Mn/CeO_2. Thus, addition of transition metals significantly lowered the light-off temperature of ceria. Figure 4 shows sulfur yields and light-off temperatures for coprecipitated Cu-Ce(La)-O and Cu-Zr(Y)-O catalysts. The light-off temperatures over the Cu-Ce(La)-O and Cu-Zr(Y)-O catalysts were lowered to ca. 500°C from ca. 600°C for Ce(La)O_2, while a conventional alumina-supported Cu catalyst showed a light-off temperature of about 550°C. A synergistic effect was clearly demonstrated for the Cu-fluorite oxide systems. It is interesting that the copper content in the Cu-Ce(La)-O catalysts did not affect the light-off temperature and sulfur yield under the same reaction conditions.

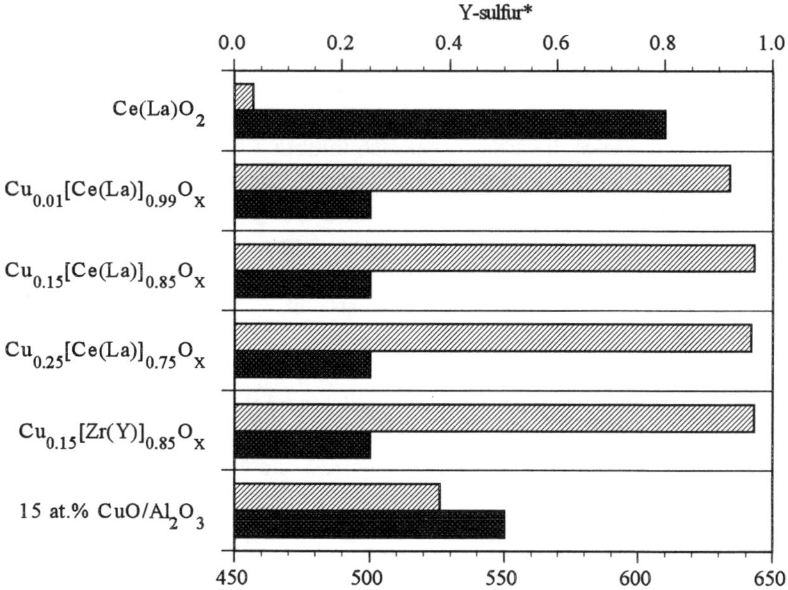

Figure 4. Light-off temperature and sulfur yield of SO_2 reduction by CO over various oxide composite catalysts (1% SO_2, 2% CO, 0.09 g·s/cc (STP)).
sulfur yield (Y-sulfur); Light-off temperature.
*Sulfur yield at 470°C after catalyst was activated.
**Light-off temperature at which 90% conversion occurred.

Figure 5 shows the variations of SO_2 conversion and sulfur yield with R_{CO} over a Cu-Ce(La)-O catalyst at a constant contact time. The selectivity to sulfur at all reaction temperatures tested is greater than 95% for R_{CO}=1.5 and R_{CO}=2.0. For R_{CO}=2.5, complete SO_2 conversion occurred over the test temperature range from 440 to 610°C, but only 50 to 90% sulfur yield was obtained. The sulfur yield from stoichiometric conversion is 50% for a feed gas mixture of R_{CO}=2.5. Thus, the catalyst exhibited some selectivity to elemental sulfur in the presence of excess CO. There is a trend that sulfur yield increases with temperature. 90% sulfur yield was obtained at 610°C for R_{CO}=2.5. Extensive parametric studies showed that as long as stoichiometrically excessive amount of CO is not used, elemental sulfur comprises almost all converted SO_2 over this catalyst system under various process conditions.

Separate studies showed that water vapor did not poison the catalyst but the sulfur yield decreased a little as a result of promotion of the water-gas-shift reaction and production of H_2S [15,17]. For example, no catalyst deactivation was observed during a 34-h run with the $Cu_{0.15}[Ce(La)]_{0.85}O_x$ catalyst at 470°C in the presence of 2 % H_2O. Similarly, no deactivation of this catalyst took place with 3% CO_2 in the feed gas [15].

3.3 Reaction Mechanism

Transition metals themselves are active catalysts for SO_2 reduction by CO [24-26]. However, transition metals are typically sulfided at the reaction conditions. A reaction mechanism involving COS intermediate was proposed for such catalysts [25,26].

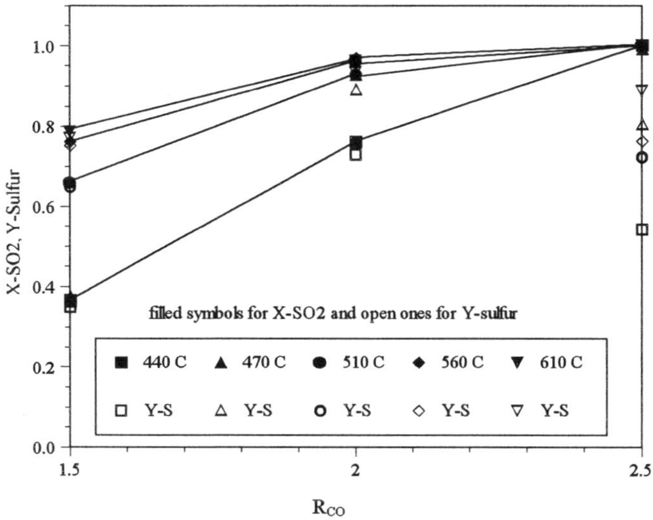

Figure 5. Effects of R_{CO} and reaction temperature on product distribution over the $Cu_{0.15}[Ce(La)]_{0.85}O_x$ catalyst under a constant contact time of 0.09s·g/cc.

Based on this hypothesis, COS is formed by reaction of CO with metal sulfide as expressed by reaction (4) and then COS reduces SO_2 to produce elemental sulfur through reaction (5).

$$MS_x + CO \rightarrow MS_{x-1} + COS \tag{4}$$

$$2COS + SO_2 \rightarrow 3/x\ S_x + 2CO_2 \tag{5}$$

The above COS intermediate hypothesis anticipates high COS level at short contact times as observed in [11]. However, in our parametric studies in the fixed bed microreactor and mechanistic studies in a thermogravimetric analyzer (TGA) coupled with on-line mass spectrometry [17], we have not detected COS as a reaction intermediate. The plots of typical kinetics results obtained in the fixed bed reactor with the $Cu_{0.15}[Ce(La)]_{0.85}O_x$ catalyst are shown in Figure 6. In these measurements, we varied inlet SO_2 concentration from 0.1 to 2% while keeping R_{CO} close to 2. The contact time was controlled by changing both catalyst loading and flow rate. With all inlet SO_2 concentrations and contact times tested, elemental sulfur was the dominant product. Conversion and sulfur yield were a little higher with 0.1% inlet SO_2 than those with 2% inlet SO_2. The higher activity and sulfur yield observed with lower inlet SO_2 concentration are explained by lower SO_2 inhibition on the catalyst activity and less partial pressure of elemental sulfur for the COS reaction, respectively. The SO_2 inhibition effect will be further discussed below. The kinetics data suggest that the shorter the contact time, the higher the sulfur yield or the less the COS production obtained, a trend that contradicts what is expected from the COS intermediate hypothesis. Thus, we believe that instead of an intermediate, COS is a byproduct formed from reaction of produced sulfur with excess CO.

The kinetics data in Figure 6 further confirm that the Cu-Ce-O catalyst is effective for SO_2 reduction by CO over a wide range of process conditions. The catalyst was so active that more than 30% SO_2 conversion was obtained at 510°C at a very short contact time, 0.0045 s·g/cc(STP), which corresponds to a space velocity of about 1,000,000 v/v/h.

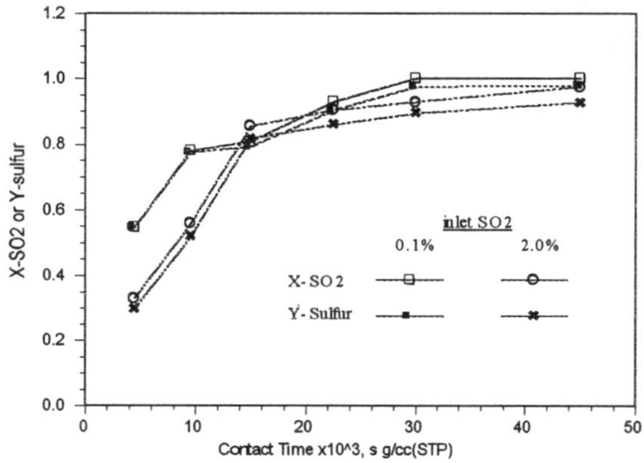

(a) T=510°C

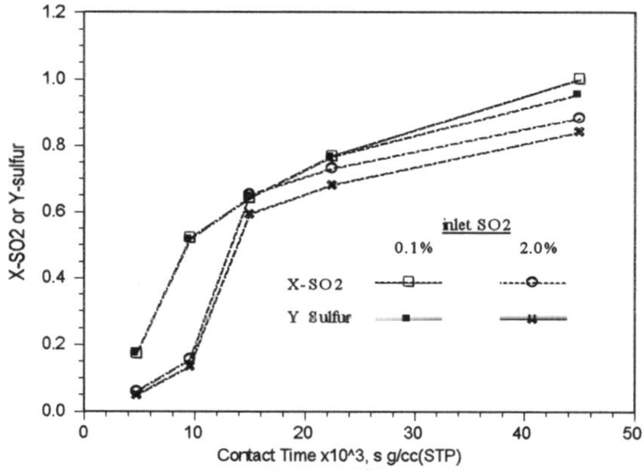

(b) T=470°C

Figure 6. Variations of SO_2 conversion and sulfur yield over $Cu_{0.15}[Ce(La)]_{0.85}O_x$ catalyst with contact time (R_{CO}=2).

Taking into account the kinetics results and extensive catalyst structural and composition analyses by XRD, XPS, and STEM, we propose the redox mechanism in reactions (1)-(3) for the composite catalyst.

The fluorite oxide structure is stable under the present reaction conditions and serves as the backbone of the composite catalyst structure where the redox process takes place. Transition metals or oxides are dispersed in the fluorite oxide as additives and in turn modify the properties of the fluorite oxide. Therefore, cerium oxide in the Cu-Ce-O composite is not simply a support material for copper dispersion. There is a strong interaction between these two kinds of materials. Such an interaction or synergism results in higher activity than each of the individual components.

Extensive characterization results of the fresh Cu-Ce-O catalyst by XPS, XRD, and STEM/EDX were recently reported elsewhere [27]. For this class of composite catalysts prepared by 650°C air-calcination, no CuO phase was detected at copper contents lower than 15 at.%. Copper was distributed in the cerium oxide in the form of isolated ions, clusters, and bulk CuO particles. The copper clusters were found in all the Cu-Ce-O catalysts, while the CuO particle content increased with the bulk copper content. Cu^{+1} species were observed by XPS on the Cu-Ce-O catalyst and are considered the result of the interaction of copper oxide clusters and cerium oxide lattice. However, sulfur compounds are so reactive that the catalyst composition often changed after reaction. XRD and STEM analyses of the spent catalysts [17] identified that the cerium oxide matrix was not attacked by sulfur compounds and still held its fluorite crystal structure, while bulk CuO was completely converted into copper sulfide crystals, and copper and sulfur were extensively distributed on the cerium oxide matrix.

The surface compositions of two Cu-Ce(La)-O catalysts after reaction are shown in Table 1. More copper was measured by XPS than the nominal bulk value, indicating surface enrichment in copper. Significant amount of sulfur was found in all the catalysts. The oxygen content is much higher than the nominal values calculated by assuming Cu as CuO and Ce as CeO_2. These large amounts of excess oxygen suggest that sulfur mainly exists as sulfates. Generally, more oxygen and sulfur were found with the catalysts in the inactive state than in the active state. The surface states of these elements were determined by the binding energies of Ce3d, Cu2p, S2p and O1s core electrons [15,17]. Cerium existed as cerium sulfate and cerium oxide. Copper existed as copper oxide, sulfide, and sulfate. Oxygen existed as metal oxide and sulfate. Sulfur existed as sulfate and sulfide. The major conclusion from the XPS analyses is that a working Cu-Ce(La)-O catalyst comprised partially sulfided copper and partially sulfated cerium, while the cerium oxide in a deactivated catalyst was fully capped by sulfate.

The catalytic properties of the Cu-Ce(La)-O catalyst can be understood on the basis of the proposed redox reaction scheme. Surface reduction by CO is still the key step to initiate the reaction. However, under the reaction conditions, SO_2 and CO would compete for the surface oxygen. Reaction of CO with the surface capping oxygen produces CO_2 and an oxygen vacancy (reaction (1)), while reaction of SO_2 with the surface oxygen forms sulfate:

$$SO_2 + 2O\text{-cat} \longrightarrow SO_4 - \text{cat} \qquad (6)$$

The sulfate is a strongly bound surface species and its formation tends to terminate the redox reaction. Therefore, the rate of the surface oxygen reduction by CO has to be greater than that of sulfation by SO_2 if the fresh catalyst is activated in the reacting gas mixture. After the surface oxygen and sulfate are removed, the redox reaction is initiated and the reaction can be operated at lower temperatures. We propose that copper and cerium oxide play different roles in the redox reaction. Cerium oxide comprises the matrix of the catalyst and provides the oxygen/oxygen vacancy sources, while copper as an additive promotes the reducibility of cerium oxide and provides the surface sites for CO adsorption.

Table 1. Surface composition of used Cu-Ce(La)-O catalysts[a]

element	$Cu_{0.02}[Ce(La)]_{0.98}O_x$		$Cu_{0.15}[Ce(La)]_{0.85}O_x$	
	6 h at 440°C[d]	two days[e]	2 h at 433°C[f]	2 h at 510°C[g]
Ce	23.2	25.2	9.9	23.1
Cu	2.6	5.3	7.3	9.2
S	5.9	5.6	13.7	6.8
O	68.2	63.9	69.1	60.8
[b]$O_{theoretical}$	49.0	55.7	27.1	55.4
[c]$O_{difference}$	19.2	8.2	42	5.4

a. reaction gas: 1% SO_2, 2% CO , 97%He. Contact time: W/F=0.09 g·s/cc (STP).
b. oxygen content calculated on the basis of CuO and CeO_2.
c. difference in oxygen content between the measured and calculated values.
d. catalyst was completely deactivated at this temperature.
e. catalyst was tested in the active state for two days.
f. catalyst was completely deactivated.
g. catalyst was fully activated at this temperature and complete SO_2 conversion occurred.

The catalyst characterization results have shown that Cu^{+1} species tend to be stabilized by interaction of the copper clusters and the cerium oxide lattice [26]. It is known that Cu^{+1} sites are strong sites for CO chemisorption. The Cu^{+1} species were observed by XPS on the fresh catalyst. However, the catalyst surface became complex in the CO+SO_2 reacting atmosphere. CuO particles formed copper sulfide crystals that have low surface area and contributed little to the catalytic activity. It is not known at the present time how copper ions and clusters in the cerium oxide matrix are affected by sulfur compounds. According to a recent study by Badley et al. [28] of SO_2 and CO adsorption on Cu/Al_2O_3, SO_2 can block CO adsorption on Cu^{+2} sites but not on Cu^{+1} sites, while the pre-adsorption of SO_2 does not prevent CO adsorption. Therefore, copper in the Cu-Ce(La)-O composite catalyst may still provide strong sites such as Cu^{+1} for CO adsorption even in the presence of sulfur compounds and water vapor. However, the cerium oxide surface can be completely capped by sulfate when the composite catalyst is exposed to SO_2 gas at temperatures below or around 500°C. The sulfated cerium oxide surface would have no oxygen vacancy sites to catalyze the reaction. The proposed active catalyst configuration suggests that only small amounts of copper are needed to achieve the synergism. Given the fact that copper oxide is not miscible with cerium oxide, use of excess copper only generates more bulk CuO particles that are converted to copper sulfide at the reaction conditions. This explains why the catalyst light-off temperature and activity were not much affected by the copper content.

4 SO_2 Reduction by Natural Gas and Syngas

The work described above has shown that the catalytic activity of fluorite-type oxides, such as ceria and zirconia, for the reduction of SO_2 by CO to elemental sulfur can be significantly promoted by active transition metals, such as copper. The performance of this catalyst was consistently good over a wide range of SO_2 inlet concentration (0.1 to 10%) indicating its flexibility in treating SO_2 tail gases as well as high concentration streams.

In work presently underway at Tufts University and Arthur D. Little, Inc. [19], we have been examining these types of catalysts as robust and flexible systems with regard to choice of reducing gas (methane, CO, or syngas). In addition to COS, H_2S and CS_2 are likely byproducts from these reactions. The objective is to find a catalyst of high selectivity to elemental sulfur with a variety of reducing gases.

$$SO_2 + (CH_4 \text{ or } H_2 + CO) \rightarrow S + H_2S + COS + CS_2 + H_2O + CO_2$$

4.1 SO_2 Reduction by CH_4

The early industrial practice of using natural gas to reduce SO_2 into elemental sulfur dates back to 1970 for a sulfide ore roasting facility [29], where bauxite catalyst was used. The reduction kinetics of SO_2 by CH_4 were investigated by Sarlis and Berk [30]. The maximum elemental sulfur yield achieved in these studies in a single reaction step was about 50%. Therefore, the residual gas comprising of SO_2, H_2S, COS, and small amounts of CS_2 had to go through a downstream Claus plant in order to accomplish sulfur recovery. But, this technology was reported only suitable for concentrated SO_2 streams. A supported catalyst [31] containing thorium oxide in combination with various other refractory metal oxides or chromites also required high reaction temperature and showed low selectivity to elemental sulfur. It is generally recognized that CH_4 activation is difficult over metal oxide catalysts and even more difficult in the presence of sulfur compounds.

Since the transition metal/fluorite oxide composite was identified as an active catalyst system for SO_2 reduction by CO and for CH_4 oxidation [27], we have evaluated this catalyst formulation for SO_2 reduction by CH_4. Screening tests were performed with stoichiometric gas mixture (R_{CH4}=0.5) at a contact time of 0.09 s·g/cc(STP) (or 80,000 v/v/h). The catalysts tested were CeO_2, $Ce(La)O_2$, Ni-Ce(La)-O, CeO_2-ZrO_2, Mn-Ce(La)-O, Co-Ce(La)-O, and Fe-Ce(La)-O. The 5 at% Ni-containing Ni-Ce(La)-O catalyst was identified to be the most active. Figure 7 shows the effect of molar ratio R_{CH4} on SO_2 conversion and sulfur yield over the $Ni_{0.05}[Ce(La)]_{0.95}O_x$ catalyst. SO_2 conversion monotonically increased with R_{CH4}. At low SO_2 conversions (<0.5), elemental sulfur, as a major product, increased with SO_2 conversion. But, a further increase in SO_2 conversion resulted in significant production of byproducts, mostly H_2S and hence, sulfur yield was not enhanced. As a result, about 50% sulfur yield was obtained with either partial or complete SO_2 conversion.

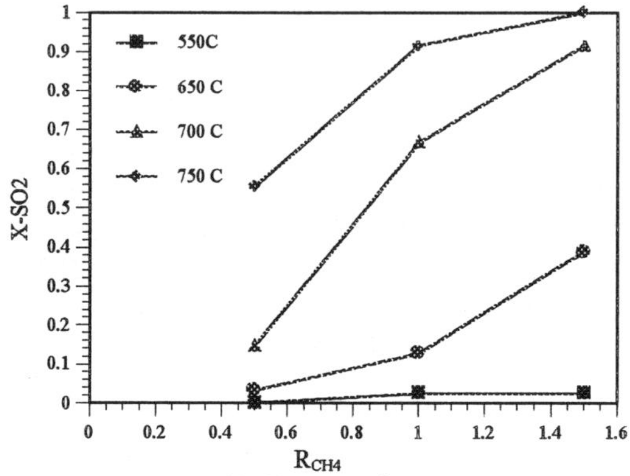

(a) SO_2 conversion

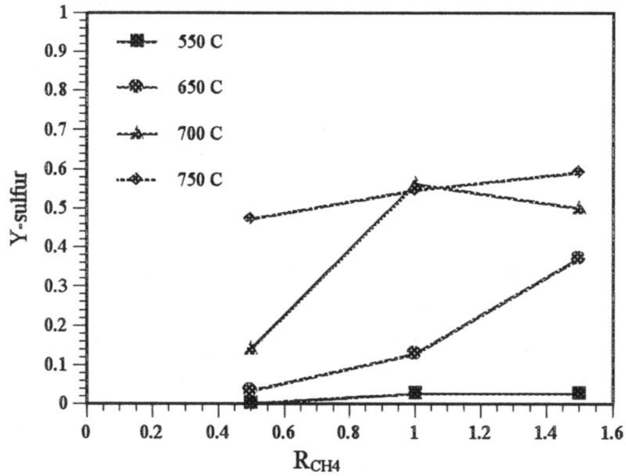

(b) sulfur yield

Figure 7. Effect of molar ratio of CH_4 to SO_2 (R_{CH4}) on the activity and selectivity of $Ni_{0.05}[Ce(La)]_{0.95}O_x$ catalyst (1% SO_2, 0.18 s·g/cc(STP) or 40,000 v/v/h).

Figure 8 shows the effect of contact time on the activity and selectivity of this catalyst at different temperatures with a feed gas mixture of R_{CH4}=0.5 (stoichiometric). SO_2 conversion increased with contact time. Generally, sulfur yield also increased with contact time as a result of increased conversion. But, high conversion did not always represent a high sulfur yield. For example, at a contact time of 0.48s·g/cc(STP), SO_2 conversion at 750°C and 700°C was 0.89 and 0.81, respectively, while the corresponding sulfur yields were 0.61 and 0.71, respectively, that is, a higher sulfur yield was achieved at a lower SO_2 conversion. The preliminary parametric studies suggest that sulfur yield probably can be maximized by optimizing reaction temperature, contact time, and R_{CH4}. However, the 0.71 sulfur yield obtained in the present work at a space velocity of ca. 15,000 v/v/h is already higher than that over the conventional catalyst [29,30].

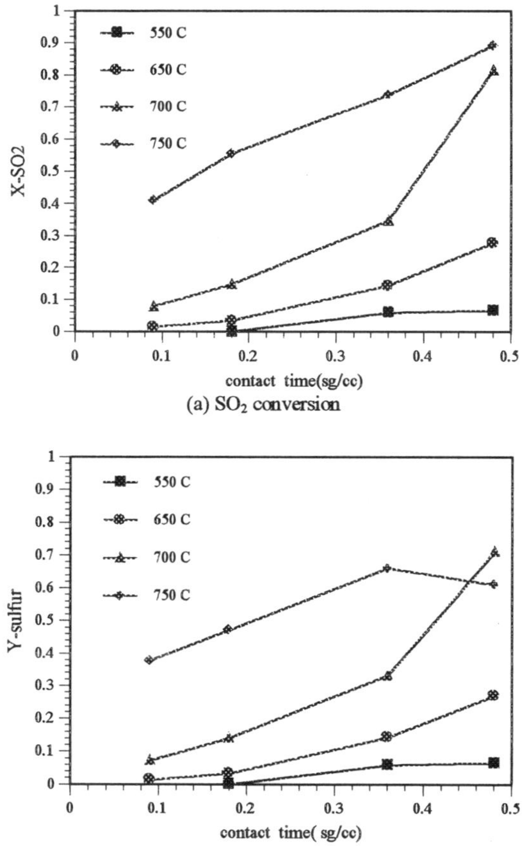

(a) SO_2 conversion

(b) sulfur yield

Figure 8. Effect of contact time on the activity and selectivity of $Ni_{0.05}[Ce(La)]_{0.95}O_x$ catalyst (1% SO_2, 0.5% CH_4).

The activity of Ni-Ce(La)-O catalyst can be further improved by changing the lanthanum dopant content and/or by the use of other dopants. The Cu-Ce(La)-O catalyst was also tested for SO_2 reduction by CH_4. The results are shown in Figure 9 and are compared to the $Ni_{0.05}[Ce(La)]_{0.95}O_x$ catalyst. Interestingly, the 5 at% Cu-Ce(La)-O catalyst had a much higher activity than the 15 at% Cu-containing catalyst and was comparable to or slightly better than the $Ni_{0.05}[Ce(La)]_{0.95}O_x$ catalyst at high temperatures. The effect of the Cu content is explained by the synergistic model discussed above for SO_2 reduction by CO. Small amounts of transition metals are helpful to promote the fluorite oxide activity and also preferable to avoid agglomeration and sintering at high reaction temperatures. Overall, much higher reaction temperatures are required for the SO_2 reduction by CH_4 than by CO, which reflects the refractory nature of CH_4. The reaction mechanism of the SO_2 reduction by CH_4 appears to be more complex than a redox model.

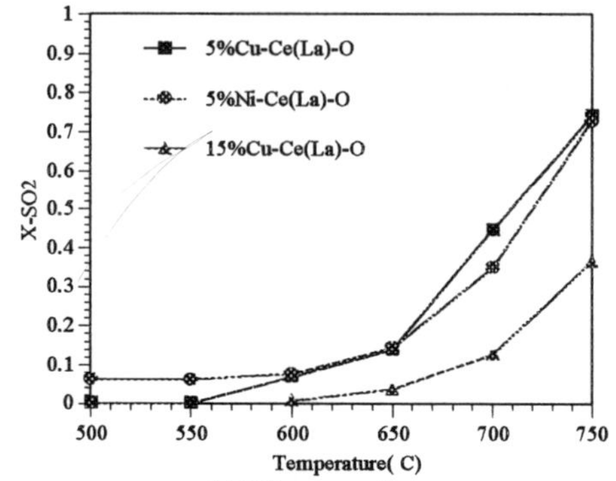

(a) SO_2 conversion

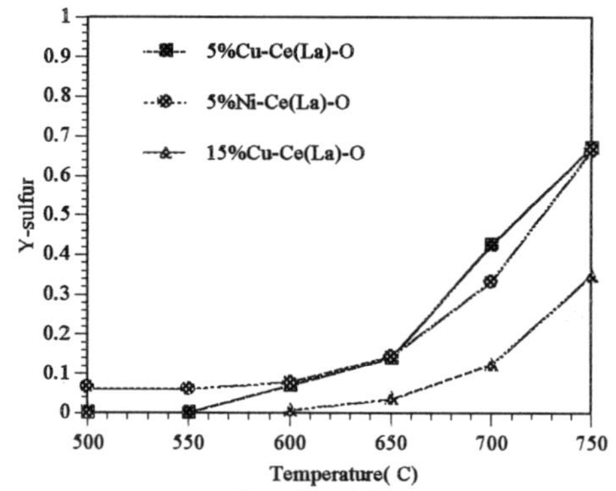

(b) sulfur yield

Figure 9. Activity and selectivity of Cu-Ce(La)-O catalyst for SO_2 reduction by CH_4 (1% SO_2, 0.5% CH_4, 0.36s·g/cc (STP) or 20,000 v/v/h).

4.2 SO_2 Reduction by Syngas

To assess the flexibility with respect to choice of reducing gas, the $Ni_{0.05}[Ce(La)]_{0.95}O_x$ and $Cu_{0.05}[Ce(La)]_{0.95}O_x$ catalysts were tested using a simulated syngas of $H_2/CO=2$ as the reductant. Figure 10 shows the activity of these two catalysts. The $Ni_{0.05}[Ce(La)]_{0.95}O_x$ catalyst had higher activity but lower selectivity to elemental sulfur than the $Cu_{0.05}[Ce(La)]_{0.95}O_x$ catalyst. The major byproduct from the reaction was H_2S and its production became significant above 450°C while the COS level was nearly constant from 350 to 750°C. Complete SO_2 conversion was not achieved because the H_2S production consumed more reductant. The overall catalyst performance looked similar to the SO_2 reduction by CO. Although low sulfur yield was obtained in the present tests with H_2-rich syngas, it is expected that higher sulfur yield is achievable with a syngas of higher CO content as is typical in many coal gasifier exit gas streams. In addition, as long as the stoichiometric $SO_2/(H_2+CO)$ ratio is used, complete conversion of the reducing gas over the catalyst results in a stoichiometric gas mixture of H_2S and SO_2 so that high sulfur recovery can be realized by a downstream Claus plant. Alternatively, recycle designs may be used after removal of elemental sulfur.

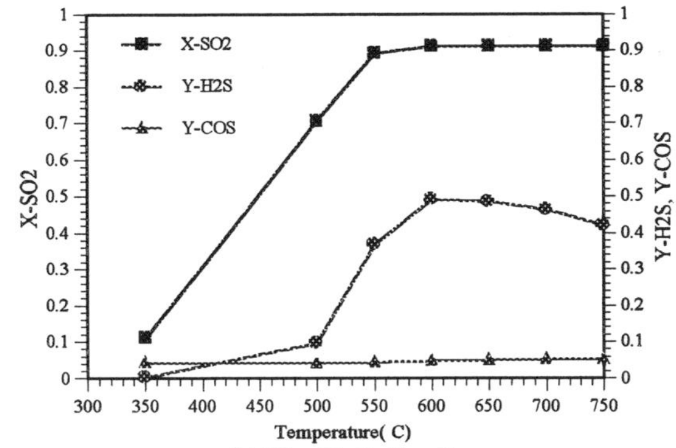

(a) $Ni_{0.05}[Ce(La)]_{0.95}O_x$

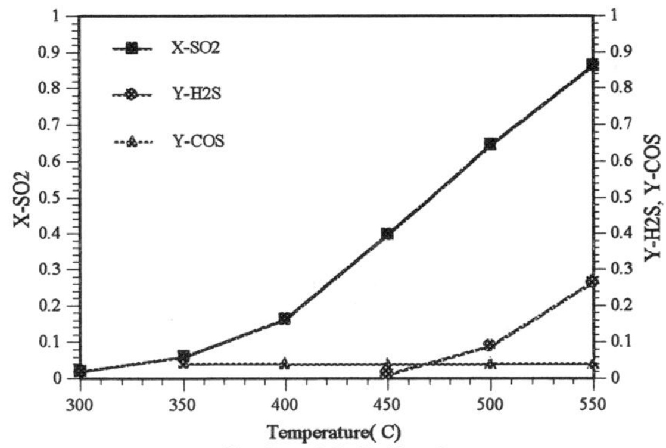

(b) $Cu_{0.05}[Ce(La)]_{0.95}O_x$

Figure 10. Activity and selectivity of Ni-Ce(La)-O and Cu-Ce(La)-O catalysts for SO_2 reduction by syngas (1% SO_2, $(H_2+CO)/SO_2 \approx 2$, $H_2/CO=2$, 0.18 s·g/cc(STP) or 40,000 v/v/h).

5 Conclusions

Fluorite-type oxides, such as CeO_2 and ZrO_2, are active catalysts for the reduction of SO_2 by CO. More than 95% sulfur yield is achieved at reaction temperatures about 500°C or higher with the feed gas of stoichiometric composition. Reaction of SO_2 and CO over these catalysts demonstrated a positive correlation of catalytic activity with the catalyst oxygen mobility. The catalyst activity and tolerance to H_2O and CO_2 poisons are significantly enhanced by addition of small amounts of active transition metals, such as Cu, Ni, etc. The model Cu-Ce(La)-O catalyst showed extremely high activity. Significant SO_2 conversion at 510°C was obtained at a space velocity as high as 1,000,000 v/v/h. The catalyst is also effective over a wide range of SO_2 concentrations.

The catalytic performance of transition metal/fluorite oxide composite catalysts is explained by the redox reaction mechanism and a synergistic catalyst model was derived from comprehensive kinetics and structural studies of the Cu-Ce(La)-O model catalyst. Reaction takes place over surface oxygen vacancy sites which are created by reduction by CO. In the Cu-Ce(La)-O system, the stable cerium oxide serves as the backbone of the catalyst structure and as the oxygen vacancy source, while copper promotes the reducibility of the cerium oxide and provides surface sites for CO adsorption. SO_2 has a strong affinity to cerium oxide, such that a working Cu-Ce(La)-O catalyst comprises both partially reduced and partially sulfated cerium oxide.

The transition metal-fluorite oxide composite catalysts were also evaluated for SO_2 reduction by CH_4. The Ni-Ce(La)-O and Cu-Ce(La)-O catalysts of low transition metal contents, about 5 at%, were identified as the most active catalysts. Compared to SO_2 reduction by CO, more byproducts were formed because of the complex reaction scheme and about 200°C higher reaction temperature was required due to the refractory nature of CH_4. A 71 % sulfur yield was achieved over the $Ni_{0.05}[Ce(La)]_{0.95}O_x$ catalyst at 700°C and a space velocity of ca. 15,000 v/v/h with a stoichiometric mixture of SO_2 and CH_4. The present studies indicate that the catalyst activity and sulfur yield can be further improved by optimizing the catalyst formulation, surface area and process conditions.

Studies of SO_2 reduction by syngas showed that the catalyst activity is similar to that for the SO_2 reduction by CO. However, the presence of H_2 promotes production of H_2S and lowers the sulfur yield. A syngas of high CO/H_2 ratio is preferable to achieve high sulfur yield in a single-step reaction.

In summary, the Ni-Ce(La)-O and Cu-Ce(La)-O oxide composites of low transition metal content have been identified as promising new catalysts for application to single-stage, highly efficient catalytic converter to achieve elemental sulfur recovery from SO_2 streams with a variety of industrially possible reducing agents.

6 Acknowledgments

This work was supported by the U.S. Department of Energy, University Coal Research Program, under Grant No. DE-FG22-92PC92534. Also, support by the U.S.DOE/Arthur D.Little, Contract No. DE-AC-95PC95252, under subcontract to Tufts University is acknowledged. The authors would like to thank Mr. Eric Mullins, a visiting student at MIT from Cumberland College under DOE/ORISE program, for his assistance in some of the catalyst activity tests.

7 References

1. Tyson, S., “Guidance for coal ash use: codes and standards activities for high volume applications” in Proceedings of the 9th Annual International Pittsburgh Coal Conference, October, 1992.
2. Yeh, J.T., Demski, R. J., Strakey, J. P. and Joubert, J. I., Environ. Prog. **4**(1985), 223.

3. Whelan, J.M., U.S Patent 4 081 519 (1978).
4. Happel, J., Hnatow, M.A., Bajars, L. and Kundrath, M., Ind. Eng. Chem., Prod. Res. Dev., **14**, No.3 (1975) 154.
5. Happel, J., Leon, A.L., Hnatow, M.A. and Bajars, L., Ind. Eng. Chem., Prod. Res. Dev., **16**, No.2 (1977)150.
6. Hibbert, D.B. and Campbell, R.H., Appl. Catal., **41** (1988) 273.
7. Hibbert, D.B. and Campbell, R.H., Appl. Catal., **41**(1988) 289.
8. Minh, N.Q., J. Am. Ceram. Soc., **76**(3) (1993) 563.
9. Baglio, J.A., Ind. Eng. Chem. Prod. Res. Dev., **21** (1982) 38.
10. Jin, Y., Yu, Q., and Chang, S.G., U.S. Patent 5 494 879 (1996).
12. Flytzani-Stephanopoulos, M. and Hu, Z., U.S. Patent No. 5 242 673 (1993).
13. Liu, W. and Flytzani-Stephanopoulos, M., in Armor, J. N.(Editor), Environmental Catalysis (ACS Symposium Series 552), American Chemical Society, Washington, DC, 1994, p.375.
14. Flytzani-Stephanopoulos, M. and Liu, W., U.S. Patent No. 5 384 301 (1995).
15. Liu, W., Sarofim, A.F. and Flytzani-Stephanopoulos, M., Appl. Catal. B: Environmental, **4** (1994) 167.
16. Hagenmuller, P., and Van Gool, W., Solid Electrolytes, Academic Press, New York, 1978.
17. Liu, W., Sc.D. Thesis, Massachusetts Institute of Technology, May 1995.
18. Liu, W., Wadia, C. and Flytzani-Stephanopoulos, M., Catal. Today, **28**(4), 1996.
19. Arthur D. Little, Inc. "Advanced Byproduct Recovery: Direct Catalytic Reduction of Sulfur Dioxide to Elemental Sulfur", Quarterly Technical Report to DOE/PETC, Contract No. DE-AC-95PC95252, July 1996.
20. Aiken, B., Hsu, W.P. and Matijevic, E., J. Am. Ceram. Soc., **71**, 845(1988).
21. Kim, D.J., J. Am. Ceram. Soc., **72**(8) (1989) 1415.
22. Crucq, A.(Editor), Catalysis and Automotive Pollution Control II, Elsevier Science Publishers B.V., Amsterdam, 1991.
23. Li, C., Sakata, Y., Arai, T., Domen, K., Maruya, K.,and Onishi, T., J. Chem. Soc., Faraday Trans. 1, **85** (1989) 929; ibid, 1451-1461.
24. Ryason, P.R. and Harkins, J., J. Air Pollut. Contr. Ass., **17** (1967) 796.
25. Haas, L.A. and Khalafalla, S.E., J. Catal., **30** (1973) 451.
26. Bazes, J.G.L., Careto, L.S. and Noble, K., Ind. Eng. Chem., Prod. Res. Develop., **14** (1975) 264.
27. Liu, W. and Flytzani-Stephanopoulos, M., J. Catal., **153** (1995) 304; ibid, 317-332.
28. Badley, M.B., Rochester, C.H., Hutchings, G.J. and King, F., J. Catal., **148** (1994) 438.
29. Rinckhoff, J.B. and Pfeiffer, B.(Eds.), "Sulfur Removal and Recovery from Industrial Processes", Advances in Chemistry Series, No. 139, American Chemical Society, 1975.
30. Sarlis, J. and Berk, D., Ind. Eng. Chem. Res. **27** (1988) 1951-1954.
31. Stiles, A.B., U.S. Patent 3 856 459 (1975).

Experience of Regenerable Sulphur Sorbents in Real Fuel Gas systems

Ian R Fantom and Scott Radford

Coal Technology Development Division, British Coal Corporation, PO Box 199, Stoke Orchard, Cheltenham, Glos, GL52 4ZG, England

Abstract. A coal gasification based combined cycle power generation system known as the Air Blown Gasification Cycle is being developed jointly by an industry led group comprising GEC Alsthom, PowerGen, Mitsui Babcock Energy and British Coal Corporation towards a demonstration of the technology. Sulphur control to meet current EU legislation is inherent in the process, however tighter emissions limits are anticipated for new power plant in the near future. This paper describes the approach adopted to meeting lower sulphur emissions limits using hot gas cleaning techniques and results of recent testing of metal oxide sorbents.

1 Introduction

Within the European Union, demonstration size coal gasification combined cycle power generation systems are either being built, designed or operated, eg, a 250 MW_e IGCC plant at Buggenum, The Netherlands (mid 1993 start-up), a 300 MW_e IGCC plant by RWE, Germany and a 340 MW_e plant at Puertollano, Spain. In the UK, proposals for an 87 MW_e Prototype Integrated Plant based on the Air Blown Gasification Cycle (ABGC) have been made.

The Coal Technology Development Division of British Coal (CTDD) is developing components of a coal gasification based combined cycle power generation system known as the Air Blown Gasification Cycle (ABGC) [1]. The ABGC is being developed jointly by an industry led group comprising GEC Alsthom, PowerGen, Mitsui Babcock Energy and British Coal Corporation, towards a demonstration of the technology.

The cycle offers significant increase in power generation efficiency compared to conventional coal fired systems with the added advantage that this can be achieved at reduced capital cost. A number of work programmes are being undertaken at CTDD to develop the component parts of the cycle ie, high pressure gasifier, char

NATO ASI Series, Vol. G 42
Desulfurization of Hot Coal Gas
Edited by A. T. Atimtay and D. P. Harrison

combustor, gas cleaning and gas combustion systems. The information from the various work programmes will be used to design a Prototype Integrated Plant (PIP).

For such advanced power generation systems there is the need to remove dust, sulphides, ammonia, cyanides and halides from the fuel gas. Existing technologies to remove these pollutants are based on wet gas cleaning systems which have an adverse effect on cycle efficiency, can have high capital cost and can produce an aqueous effluent requiring expensive treatment before disposal. The development and introduction of hot gas cleaning systems offers the potential of a lower cost approach to pollutant control with associated cycle efficiency advantages. The hot gas systems which will replace the conventional wet gas technologies will need to operate at 300-600°C and at pressures of 10-25 bar_a. In order to realise fully the cost and environmental advantages, it is essential that the systems developed give not only efficient removal of the respective pollutant but also have the reliability and availability to match the respective wet gas system.

Sulphur control to meet current UK legislation is inherent in the ABGC. However, tighter emissions limits are anticipated in the next decade. This paper describes briefly existing methods for the reduction of sulphur emissions from the ABGC. A more detailed description of recent developments in ex-bed sulphur removal systems to reduce emissions further is also given. Fig.1 shows a schematic of the ABGC.

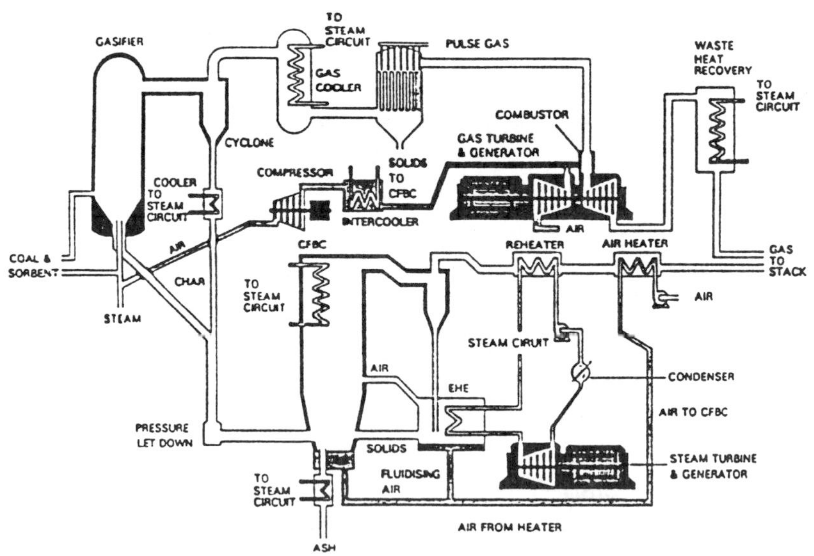

Fig. 1 Schematic of the ABGC

2 Sulphur Retention in the Gasifier

Sulphur is present in coal in two primary forms, in the organic structure as compounds such as thiophenes and aromatic sulphides and also as pyrite (FeS_2) inclusions. During gasification much of the organic sulphur is converted to gaseous hydrogen sulphide (H_2S) and carbonyl sulphide (COS), only a small amount is retained as organic sulphur in the residual char [2]. The pyrite decomposes into pyrrhotite (FeS) and H_2S. Calcium compounds such as limestone are added to the gasifier and together with calcium compounds present in the coal ash, retain the hydrogen sulphide as calcium sulphide. At the operating conditions of the gasifier of around 1000°C and 25 bar the limestone decomposes to form calcium oxide which reacts with the hydrogen sulphide via the following reaction.

$$CaO + H_2S \rightleftharpoons CaS + H_2O$$

This reaction has been shown to be in equilibrium at the gasifier operating conditions providing that there is an excess of sorbent of approximately 100 %. Sorbent of an appropriate size and porosity are also required.

Test work at CTDD has been undertaken on 0.3 m diameter atmospheric and high pressure gasifiers to test the effectiveness of various sorbents over a range of operating conditions. The tests have identified the optimal operating conditions ie sorbent type, feed rate, size, fluidising velocity etc. To date work has focused on carboniferous limestones, future work will look at dolomitic limestones.

Laboratory and theoretical studies are being conducted in support of this work in collaboration with a number of universities. The object of this work is to evaluate the kinetics and hydrodynamics of the calcium carbonate decomposition and subsequent reaction with hydrogen sulphide. The results will be used to develop a mathematical model of the desulphurisation process to be used as part of an overall dynamic model of the fluidised bed gasification process.

In addition, tests are being undertaken, on a small atmospheric pressure gasifier, to look at the effects of operating conditions, in particular various sulphur sorbents, on bed agglomeration.

3 Sulphur Retention in the CFBC

The char material passed from the gasifier, cyclone and filter contains organic sulphur, iron sulphide and calcium sulphide. During combustion of the char, calcium sulphide is converted to calcium sulphate. The organic sulphur and iron sulphide are oxidised to sulphur dioxide (SO_2) which is subsequently retained as calcium sulphate by unreacted calcium oxide.

Tests have been undertaken on a small scale CFBC (200 KW), at the University of British Columbia, using char produced during a pressurised gasifier test. The sulphur sorbent used in the gasifier test was limestone. 95 % sulphur retention was demonstrated during combustion of the char and around 75 % of the calcium sulphide was converted to sulphate. Sulphur retention can in part be increased by addition of more limestone. If this is not possible then an alkali scrubber will have to be used down stream of the bag filter. Further test work will be undertaken to look at different sulphur sorbents such as dolomite. This material is more porous and should improve the sulphide to sulphate conversion. Waste residues containing calcium sulphide are costly to dispose of if the sulphide is leachable. Early evidence, however, shows the ash should not be classified as a hazardous waste.

Larger scale test work will be undertaken at CTDD using a 1.8 MW CFBC. The residence time in this plant is more representative of a commercial plant. This should improve the sulphur retention and provide a higher sulphide to sulphate conversion.

4 Ex-bed Sulphur Removal

The combined emissions of H_2S and COS in the fuel gas, predicted to be around 300 vppm for a commercial plant, should meet the emissions limits for new coal fired power plants. The emission limit is 400 mg m_n^{-3} (6% O_2, dry) [3] and is equivalent to around 340 vppm in the fuel gas. This figure assumes that a high sulphur retention is achieved in the CFBC. However, in anticipation of more stringent emissions limits and to match emission levels achieved by other developing IGCC systems ex-bed sulphur removal systems are being developed.

Existing technologies to remove these compounds are based on wet gas cleaning systems which have an adverse effect on cycle efficiency, can have high capital cost and can produce an aqueous effluent requiring expensive treatment before disposal. The development and introduction of hot gas cleaning systems offer the potential of a lower cost approach to pollutant control with associated cycle efficiency advantages. The hot gas systems which will replace the conventional wet gas technologies will need to operate at 400-600°C and at pressure up to 25 bar.

The basis of the technology is to contact the hydrogen sulphide with a metal oxide producing a metal sulphide. The sulphide is then regenerated using an appropriate oxidant to produce a concentrated sulphur dioxide stream from which sulphur can be recovered using conventional techniques. The sorption process requires two separate vessels one for sorption the other for regeneration. A variety of contacting methods are being considered for the process including fixed bed, moving bed and fluidised bed reactors. It is unclear as yet which system will be used for the ABGC. The process will form part of an integrated gas cleaning system to reduce levels of sulphur, chlorine and nitrogen compounds in the fuel gas. Each of these processes will have an influence on the type of reactor required.

In the ABGC, the fuel gas from the gasifier is cooled and filtered at between 400°C and 600°C. Thermodynamic studies by many workers have shown that H_2S levels can be reduced further at these temperatures by contacting with metal oxides such as zinc and iron. At the appropriate temperature these sorbents can reduce H_2S levels to 50 vppm or less [4]. Many organisations are undertaking development of regenerable sorbents and gas contacting systems. Much of the work is done using laboratory scale reactors. The purpose of the study reported here was to test available sorbents using real fuel gas supplied from a pressurised gasifier at CTDD. Data on the performance of the materials under laboratory conditions will be obtained from the sorbent suppliers. The effects of the fuel gas contaminants such as tars, hydrogen chloride and trace metals could then be derived.

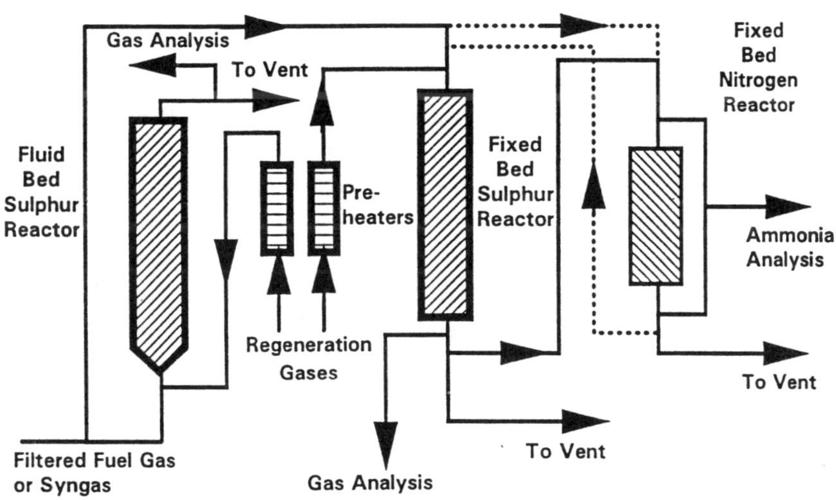

Fig.2 Sulphur sorbent test facility

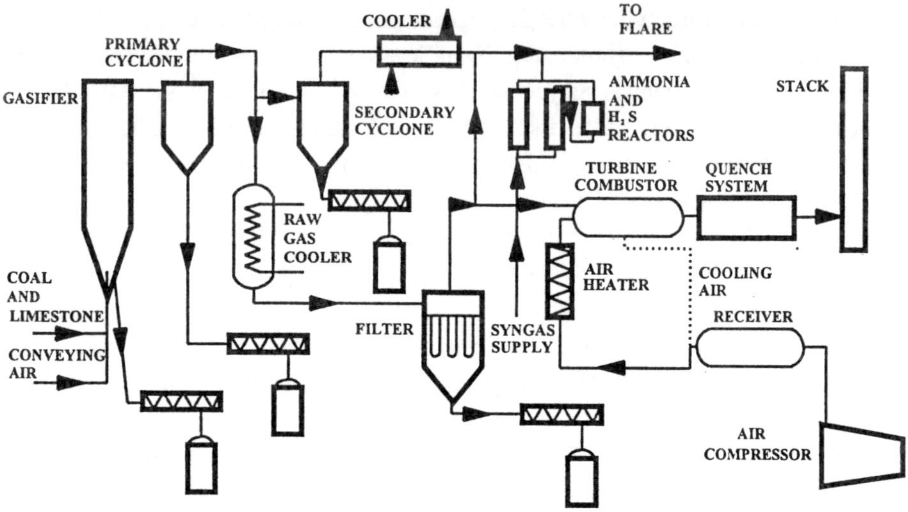

Fig.3 Gas utilisation facility

4.1 Description of Test Facilities

A fixed bed and a fluidised bed reactor skid was installed as a sidestream on a high pressure gasifier. A schematic of the system is given as Fig.2. The rig was designed to allow the two reactor systems to operate independently either on fuel gas or on a synthetic gas supplied from a trailer. Therefore, each system has independent temperature, flow and pressure control. Regeneration gases are supplied via a preheater system which is also used to preheat and purge the reactor vessels prior to the introduction of hot fuel gas. The fixed bed system also includes a nitrogen catalyst bed. This has been used to test ammonia decomposition catalysts. Each vessel is 0.15 m in diameter, manufactured in 310 SS and designed to operate at up to 750°C and 20 bar. The reactor system is attached to the Gas Utilisation Facility (GUF) at CTDD. A schematic is shown as Fig.3.

4.2 Fixed Bed Sorbent Performance

A zinc based sorbent Z-Sorb™ supplied by Phillips Petroleum was tested over four sorption regeneration cycles. The bed was filled with sorbent to a depth of 0.9 m. Sorption was carried out at around 540°C, 11 bar g and at a space velocity of just over 3000 h^{-1}. The operating conditions are shown in Table 1.

The inlet H_2S (COS) concentrations varied widely for each test due to the gasifier operating conditions. For cycle 1 the average inlet H_2S concentration was 1700 vppm with average values of around 500 vppm for cycles 2, 3 and 4. COS levels were around 5 - 10 % of the H_2S values. For each cycle outlet H_2S concentrations were below 10 vppm in the initial part of sorption rising slowly to around 30 vppm then rapidly at breakthrough. Less well defined profiles were observed in cycles 3 and 4.

The sulphur uptake at breakthrough approximated to around 10 % wt S for cycles 1 and 2, 5 - 6 % wt S for cycle 3 and less than 4 % wt S for cycle 4. Values from literature suggest that 15 to 20 % wt S could be obtained [5]. These data were obtained in a laboratory under ideal conditions using a small scale reactor and bottled fuel gas. The decrease from cycle 3 onward is almost certainly due to reduced reactivity of the sorbent. After cycles 2 and 4 sorption the bed was sectioned carefully and analysed for sulphide content. In each case shallow profiles (Fig.4) were seen indicating that the reaction was rate controlled (either reaction rate or mass transfer). Note that cycle 2 was not taken to breakthrough because of operational problems. In cycles 3 & 4 sorption the breakthrough curve was not as sharp as previous cycles again indicating reduced reactivity.

In the first cycle the fuel gas flow was turned up slowly to prevent the bed temperature exceeding the design temperature of the vessel (720°C). After approximately 50 minutes full flow was obtained and the peak bed temperature stabilised at around 650°C, 140°C above the inlet gas temperature. The exotherm continued through the bed for around 6 hours. In the second cycle the exotherm lasted for around 2 hours and in the third and fourth cycles for only 40 minutes. The exotherm was too large to account for the water gas shift reaction alone. Substantial methanation must have occurred.

One theory for this phenomena is that the fresh sorbent contains active catalyst for the methanation reaction. This is slowly poisoned as the run proceeds reducing its activity. The catalyst is only partly regenerated and in subsequent cycles the activity is reduced.

In cycle 1 COS concentrations were measured at the outlet. The values were around 1-2 vppm, well below the inlet levels of 30 - 60 vppm. The zinc sorbent reacts only with H_2S. Therefore, significant conversion of COS to H_2S must have occurred in the bed. At equilibrium, assuming a H_2S concentration of 5 vppm, less than 1 vppm COS would be predicted. This indicates that the conversion approached equilibrium.

Table 1 Fixed bed reactor tests

Sorption		1	2	3	4
Inlet T	°C	535	535	545	550
Bed T max	°C	690	730	640	635
Pressure	barg	10.5	10.5	10.5	11
Inlet H_2S (COS)	vpm	1700-1900 (30 - 60)	300 - 700 (15-30)	600 - 800 (N/D)	300 - 400 (N/D)
Steady State Outlet H_2S (COS)	vpm	< 10 (<5)	N/D	0 - 30 (<2)	0 - 10 (<2)
Initial Bed wt	g	13750	13750	13650	13750
S Capacity	% wt S	10.5	4.9 (10.5[a])	5.5	3
Regeneration		**1**	**2**	**3**	
Inlet T	°C	580	560	590	
Bed T range	°C	580 - 730	540 - 750	580 - 720	
Pressure	barg	7	8	8	
Space Velocity	h^{-1}	1100	1100	1100	
Oxygen Conc,	% v	1.4	1	1	

a.Calculated by extrapolating final bed S profile to breakthrough.

Regeneration was carried out at a lower pressure and space velocity using an air nitrogen mixture to give a gas containing around 1 % vol O_2 at around 600°C. The bed temperature varied markedly with peak temperatures of around 740°C. The peak temperature travelled down the bed as regeneration proceeded. Residual sulphide was observed after regeneration together with some sulphate formation.

No degradation or loss of pellet strength was observed. However, around 3 % of the bed was lost. One explanation is that the sorbent gained moisture during storage. A second is that the chlorine content of the gas assisted volatilisation of the sorbent.

4.3 Fluidised bed Sorbent Performance

Two zinc based fluidised bed sorbents Z-Sorb™ from Phillips Petroleum and ZT-4 from RTI were tested under nominally the same conditions. The reactor was initially charged with sorbent to give a static bed height of 0.15 m.

Table 2 Fluidised bed reactor tests

Parameter		**ZT4**			**Z-SORB™**			
Sorption cycle		1	2	3	1	2	3	4
Inlet T	°C	490	525	500	540	540	520	520
Pressure	barg	10.5	9	9	9	9.5	11	10.5
Fluid.vel.	cms^{-1}	8.4	9.9	9.7	10.7	10.2	9	9.5
Inlet H_2S (COS)	vpm	300-400 (15-50)	1700-1900 (30-60)	1500 (30-60)	300-400 (15-30)	350-450 (15-30)	300-400	300-400
Outlet H_2S (COS)	vpm	10 - 40	0 - 10 (5 - 10)	30 - 50 (4 - 6)	<10 (<2)	<10 (<2)	<10	<10
Initial Bed wt	g	3450	3550	<3550	2270	<2270	2275	<2275
Final bed Wt	g	3420	<3550	3510	<2270	1820	<2275	2125
Uptake	%wtS	7	14	10	13	13	9.5	7.5
Regeneration		1	2		1	2	3	4
Inlet T	°C	600	580		590	600	575	585
Bed T max.	°C	620	670		685	690	710	717
Pressure	barg	10	8		8	8	8	8
Fluid Vel.	cms^{-1}	9	10.5		12.7	12.7	12	12
O_2	% v	N/D	1.9		1.9	1.9	2.7	2.8

4.3.1 ZT-4

ZT-4, a zinc titanate formulation from Research Triangle Institute (RTI) USA was tested at 10 bar over three cycles at 490, 500 and 525°C. The sorption capacity was very sensitive to temperature with capacities of 7,10 and 14 % wt S respectively at the above temperatures. Outlet H_2S levels were initially very low at less than 30 vppm for each cycle although in the lowest temperature test the levels rose slowly to around 50 vppm before breakthrough. Regeneration was complete when undertaken using 1.9 2% vol O_2 in nitrogen at an inlet temperature of around 600°C. No sulphate was formed during regeneration. A downstream filter was used to capture the elutriate from the bed. Around 1 % wt of the initial bed was captured indicating no significant sorbent degradation.

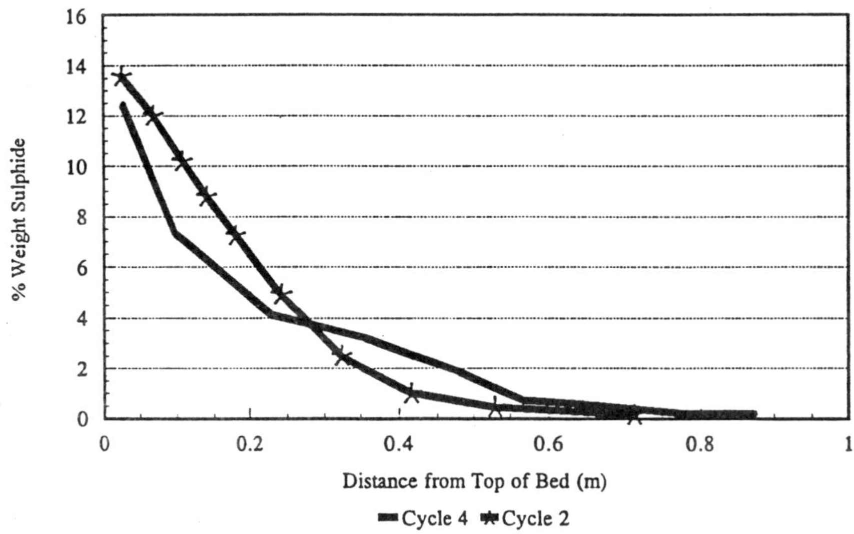

Fig.4 Fixed bed sulphide profiles

The sorbent did not catalyse fuel gas reactions such as the shift reaction, methanation and carbon deposition. However, some chloride was formed by reaction with the HCl in the fuel gas. This could lead to the formation of volatile zinc compounds and possible down stream fouling and corrosion problems.

4.3.2 Z-Sorb™

Z-Sorb™ was tested for four cycles at between 9 and 11 bar g with an inlet temperature of between 520 and 540 °C. The inlet H_2S concentrations varied between 300 and 450 vppm with COS levels at between 15 and 50 vppm. The fluidising velocity was between 9 and 10.7 cm s^{-1}. The outlet hydrogen sulphide profiles for each cycle are shown in Fig.5. H_2S concentrations were below 10 vppm for most of the tests rising rapidly at breakthrough. No exotherms were observed and the bed temperature was constant throughout indicating no methanation or shift activity. The calculated uptake at breakthrough based on the final bed weights was around 13 % wt S for the first two cycles decreasing to 10 and 8 % in cycles 3 & 4 respectively. Theoretical and laboratory scale tests suggest that values of up to 20 % wt S are attainable [6]. The above results show much lower values on real fuel gas and also that the activity decreases markedly even over 4 cycles. A commercial plant would require a life of several hundred cycles.

Regeneration for all cycles was carried out at 8 bar g. The temperature of the inlet nitrogen air mixture was controlled at around 580 - 600 °C. A fluidising velocity of 12.7 cm s^{-1} and an oxygen concentration initially of 1.9 % vol was used giving a peak temperature of 690°C in cycles 1 & 2. After cycle 2 the bed was analysed and 0.4 % sulphate observed with some residual sulphide (0.5 % wt) indicating incomplete regeneration. In cycles 3 and 4 the O_2 concentration was increased to 2.8 % vol with a corresponding temperature increase to 720°C. The bed after cycle 4 contained around 0.1 % wt sulphide indicating more complete regeneration. However, sulphate was still observed at 0.55 % wt. In addition a trace amount of chloride, (0.03 % wt) was observed. Whilst this was small it could still cause formation of volatile chlorides leading to slow depletion of the zinc bed.

After cycles 2 and 4 the bed was extracted and weighed together with elutriate that was captured in a down stream filter. During cycles 1 & 2 the bed weight was depleted by around 20 % of which 3 % was captured in a metal filter. This suggests that around 17 % wt was gaseous loss from the initial bed. In cycles 3 & 4 the same bed was used with additional material from a short syngas test to ensure that the whole bed was fully calcined. In cycles 3 & 4 around 7 % wt of the bed weight was lost. Again 3 % wt was solid elutriate captured in a new ceramic filter (see below) and 4 % unaccounted for and probably gaseous loss. The sorbent manufacturer suggests that the initial loss was due to poor calcination of the material supplied and loss of adsorbed moisture. The size distribution of the fresh sorbent, the final bed and the filter capture was measured. 99.5 % of the filter capture was less than 125 μm. This was the largest particle that could be elutriated at the maximum fluidising velocity used. Over 80 % was less than 20 μm. This suggested that there was no mal-operation of the rig.

A ceramic candle filter system was installed between cycles 2 and 3. During cycles 3 & 4 the temperature, pressure drop and flow in the filter were monitored and the permeance calculated. This value is the pressure drop divided by the filtration velocity and gives an indication of the rate of filter cake thickness growth with time ie, the elutriation rate from the fluidised bed. During purge periods the rate of increase in permeance was low. However, during both sorption and regeneration the rate was around 15 times higher. This suggests that the prime degradation mechanism in the fluidised bed is chemical reaction and not simply physical attrition. The ceramic candle was cleaned periodically using a nitrogen pulse jet to remove the filter cake.

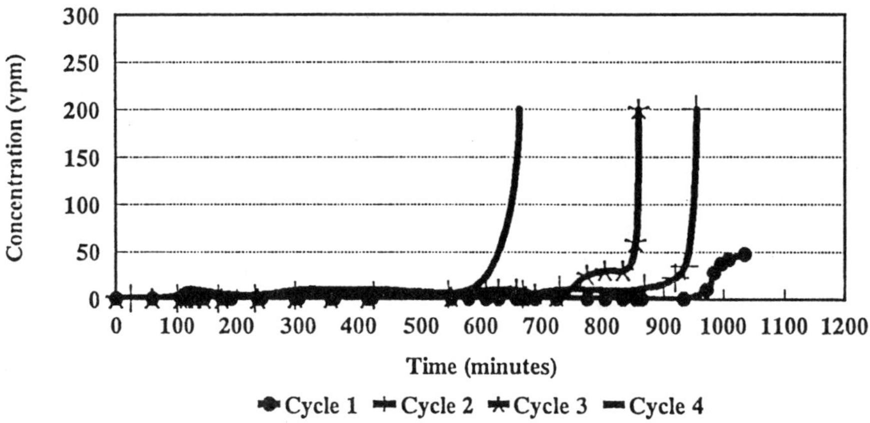

Fig.5 Fluidised bed outlet H_2S profiles

5 Conclusions

- The zinc based sorbents were capable of reducing H_2S and COS levels to below 10 vppm in real fuel gases. These emissions levels are equal to conventional wet gas cleaning techniques.
- The fluidised bed sorbents performed better than the fixed bed sorbent. A fluidised bed also offers the most practical contacting system for a continuous process.
- The activity of the sorbents tested reduced over the cycles performed. The activity was also lower on real fuel gas than clean synthetic laboratory gases.
- The Z-Sorb™ fluidised bed sorbent showed some degradation during sorption and regeneration. A more attrition resistant material would be required for a commercial application.

References

1. Minchener A J, Gale J J, Davison J E, Smith M A, Development of the CRE Gasifier For Use In Integral Partial Gasification Combined Cycle Systems, paper presented at 12th EPRI Gasification Power Plant Conference, San Francisco, USA, October 1993.
2. Davies N H, Hayhurst A N, Laughlin K M, Reed G P, The British Coal Topping Cycle Approach To Reducing Sulphur Emissions, Desulphurisation III, Sheffield, 1993.
3. Vernon J, Emissions Standards for Coal Fired Boilers, /IEA Coal Research IEACR/11 1988.
4. Fantom I R, Sage P W, Smith M A, Sulphur Control in the ABGC - An Integrated Approach, IChemE Symposium No.138, Desulphurisation IV, Sheffield 1995.
5. Khare G P, Greenwood G, Kubicek D H, Delzer G A , Regenerable Sorbents For Low Temperature Desulphurisation of Coal Gas, Phillips Petroleum Co, Proceedings of the 12th Pittsburgh Coal conference university of Pittsburgh, 1995.
6. Gupta R, McMichael, Gangwal S K, Development of A Hot Gas Desulphurisation System For IGCC Applications, AIChE Annual Meeting Miami, Clean Coal Technology, Paper No.2440f, RTI, November 1992.

Note:- The majority of the paper is also to be presented at the "Gas Cleaning At High Temperatures" conference in Karlsruhe, Sept 1996.

SUMMARY OF LECTURES AND DISCUSSION ON FUTURE RESEARCH

Rapporteurs:

Dr. Isabel CABRITA
INETI
Department of Combustion and End-Use Technologies
Lisbon, PORTUGAL

Dr. Sibel OZDOGAN
Marmara University
Department of Chemical Eng.
Istanbul, TURKEY

The lecturers presented the general picture as well as an overview of coal gasification systems. Technical issues are close to being solved in cold gas clean-up systems. However, further improvement is needed in terms of hot gas clean-up systems. Within this frame hot gas clean-up is the main barrier for coal gasification and IGCC systems. The 2-3% increase of efficiency of IGCC systems to be achieved with hot gas clean-up systems is justified not only economically but also environmentally.

Reactor and sorbent selection is of utmost importance. Fixed bed is out due to zoning problems. Fluidized beds are perhaps the best option, however, different fluidization schemes have different features. Hence, sorbents should be tested in fluidized bed systems under various operating conditions. Even the sorbents for fluidized beds need good reactivity. Synergetic effects, different chemistries and structural issues have to be addressed, as well as durability.

Modeling is an important tool in the prediction of the sorbent behavior. On the other hand, longer testing and higher number of cycles are needed to determine the real performance of the sorbents. More detailed studies are needed in areas of sorbent regeneration process, its kinetics, sulfate formation and final removal. Among these options elemental sulfur is the best option in terms of the environment, but further development is needed to have a one step economic conversion process. Hence, further research on applied catalysis is recommended.

In the discussion, questions were put forward by participants, and these were answered by a panel composed of all the lecturers.

The questions which emerged are grouped into three different sections, presented below:

1. Sorbents: properties, advantages and disadvantages of the various types,
2. Procedures for sample preparation and reactor selection,

NATO ASI Series, Vol. G 42
Desulfurization of Hot Coal Gas
Edited by A. T. Atimtay and D. P. Harrison

3. Applications: demonstration and deployment aspects, economics and practical aspects of plant operation,

The conclusions of the debate and the answers given by the panel are presented below, together with noted questions.

1. Sorbents

• *Questions*

With respect to sorbent properties, most of the questions were focused on the use of metals and metal oxides, the effect of sorbent loading, dispersion and type of support, operating temperature (high versus low temperature - advantages and disadvantages), structural stability and its influence over the sulfidation/ regeneration capability, as well as the effect of the sorbent structure and the potential use of ores.

• *Answers and Conclusions*

The desired sorbent properties were summarized by the lecturers. However, it was underlined that there exists no sorbent which comprises all the desired properties. Hence, there is much room for future improvement in terms of properties such as absorption, regeneration, attrition resistance, etc. Within this frame, the effect of metal oxide/metal, support, as well as the weight loading and the degree of dispersion, on sorbent properties were thoroughly discussed. Lecturers also pointed out that mixed metal oxide sorbents may show distinguished properties which cannot be attributed only to the relative amounts of the particular oxides.

For the selection of a particular sorbent, one should consider thermodynamic limitations first, followed by regenerative capacity and, finally, the kinetics of sulfidation. Once a sorbent has been selected, its structure and performance as a sorbent can be improved by adding dopers, binders, etc. In terms of comparison, it was reported that with zinc titanate 3000 cycles have been achieved in a fluidized bed, and iron oxide on FCC has shown no loss of activity up to 50 cycles.

2. Procedures for Sample Preparation and Types of Absorbers

• *Questions*

Part of the second group of questions referred to the techniques for sample preparation, sorbent fixation on a monolithic support, characteristics of a sample (size, shape, grain size, porosity, mechanical strength, etc.). Participants were also interested in coal gas desulfurization test installations where one can refer to when developing a sorbent and, finally, in the advantages and disadvantages of the fluidized bed technology over the fixed bed technology.

- *Answers and Conclusions*

It was referred that there has been extensive work on sample preparation, and techniques to characterize particles and pellets are well documented. Also there are laboratory and test facilities available, namely in the case of the Institute of Gas Technology, in the Unites States, it is possible to conduct such a study on contract basis.

The use of fluidized and fixed bed technology was debated both for absorption and regeneration purposes. In terms of vigorous mixing, uniformity of temperature distribution and the use of the heat liberated by exothermic reactions, the fluidized bed offers advantages to the process. However, attrition occurs to a higher degree in a fluidized bed, so that structural stability and sorbent loading have to be well controlled. As an example, alumina supports are quite resistant, but zeolites offer problems at high temperatures and high pressures with regard to mechanical strength and stability.

3. Plant Applications

- *Questions*

With regard to the aspects related to the final use of sorbent in plants, a lot of interest was shown on comparative evaluation of different sorbents in various desulfurization processes. Many questions focused on the hot/ warm gas removal and turbine efficiencies in terms of operation temperatures and technological limitations. Capital and operating costs, as well as deployment aspects related to factors affecting site, system, and capacity selection in countries at various stages of development were also referred to. Some questions addressed the final disposal and/or utilization options of spent sorbents, as well as sulfurous compounds.

- *Answers and Conclusions*

The participants from the industry emphasized the fact that industrial application of a sorbent requires acceptable durability of the sorbent during successive regenerative cycles without undergoing too much of a loss of its properties. It was agreed upon that the sorbent feed should be changed at most 3-4 times annually assuming about 8000 hours of IGCC operation. However, it was also reported that, at present, naturally occurring cheap ores such as limestone, dolomite, and iron ores are used as sorbents, mainly in fluidized bed applications, in countries such as Japan and USA in spite of the fact that they do require frequent replacement of the sorbent due to relatively fast loss of sorbent properties during the usage.

The IGCC technology will achieve about 42-43% efficiency with available gas clean-up systems. Hot gas clean-up systems, properly optimized in terms of sorbent properties and system temperature, are expected to improve efficiency to about 45% by the year 2000. Experts also emphasize the fact that the economic advantage of efficiency increase disappears when gas clean-up temperatures are above 600°C , given the current market prices of special alloys and equipment. The targeted efficiencies around 50% require development of new generation turbines and may occur in about 20-25 years. Some studies are available in which efficiency gains/losses are correlated with gas clean-up temperatures (*e.g.*, Electrical Power Research Institute (EPRI) in the USA, COMPRIMO-study in Europe, and NOVEM in the Netherlands).

With respect to comparative economic evaluation of IGCC and FBC systems with conventional coal combustion systems such as PFC + FGD, it was emphasized that IGCC should not be directly compared with fully commercialized systems. Electrical Power Research Institute (EPRI) has already published economical data on PFC + FGD systems. Data on IGCC systems will be provided during the upcoming Pittsburgh Coal Conference in fall 1996. For developing countries, the high capital costs of IGCC systems appear to be the main barrier. The physical coal cleaning option prior to coal utilization cannot achieve the extremely low levels of desulfurization even under current conditions. Hence, its application alone to the power plant scheme is out of question from the environmental compliance point of view, considering the organic sulfur content of coals.

Studies are going on with respect to disposal and/or reuse of spent sorbents. In the USA, zinc and copper based sorbents may be reused by the metal industries if the metal content is above a certain limit. Calcium carbonate based sorbents converted to calcium sulfate may be either utilized in the road construction industry (*e.g.* in China) or disposed as landfill material.

Conclusions (Research Topics and Future Trends)

Interest was shown by the participants in identifying areas where further research is needed, as well as in new areas for development and plant constructions. Participants of the 16 countries present reported on their energy and environmental policies as well as trends with regard to research and development.

The recommended areas of research are presented below:

- Product layer diffusion
- Mechanical strength
- Potential of low temperature supported sorbents
- Warm desulfurization as an alternative for hot gas clean-up systems

- Regeneration process, as well as subsequent steps for sulfur removal
- Catalysis for regeneration process and sulfur removal, as well as the final recovery process
- Off-gas utilization
- Removal of contaminants such as HCl, ammonia, etc., and coupling the process with the basic clean-up process

The size of IGCC plants is governed primarily by the requirements of available large industrial gas turbines. The largest current turbines typically produce 200-250 MWe, although improvements in efficiency, mainly by use of higher inlet temperatures, will raise the output of the same machines by up to 20-30%. In addition there is also the steam turbine which produces between 30-50% of the power depending of the IGCC plant of interest. This suggests plants of the size 500-600 MWe based on the use of a single gas turbine. To increase the output, plants can be built in modules as foreseen in countries like the United States and the Netherlands. With regard to developing countries, namely Asian markets, as advanced technologies lead to very high investments, smaller size turbines may have better applicability, even if they are not so efficient because steam in supercritical conditions cannot be used.

Subject Index

NATO ASI Series G

Vol. 1: **Numerical Taxonomy.** Edited by J. Felsenstein. 644 pages. 1983. (out of print)

Vol. 2: **Immunotoxicology.** Edited by P. W. Mullen. 161 pages. 1984.

Vol. 3: **In Vitro Effects of Mineral Dusts.** Edited by E. G. Beck and J. Bignon. 548 pages. 1985.

Vol. 4: **Environmental Impact Assessment, Technology Assessment, and Risk Analysis.** Edited by V. T. Covello, J. L. Mumpower, P. J. M. Stallen, and V. R. R. Uppuluri. 1068 pages.1985.

Vol. 5: **Genetic Differentiation and Dispersal in Plants.** Edited by P. Jacquard, G. Heim, and J. Antonovics. 452 pages. 1985.

Vol. 6: **Chemistry of Multiphase Atmospheric Systems.** Edited by W. Jaeschke. 773 pages. 1986.

Vol. 7: **The Role of Freshwater Outflow in Coastal Marine Ecosystems.** Edited by S. Skreslet. 453 pages. 1986.

Vol. 8: **Stratospheric Ozone Reduction, Solar Ultraviolet Radiation and Plant Life.** Edited by R. C. Worrest and M. M. Caldwell. 374 pages. 1986.

Vol. 9: **Strategies and Advanced Techniques for Marine Pollution Studies: Mediterranean Sea.** Edited by C. S. Giam and H. J.-M. Dou. 475 pages. 1986.

Vol. 10: **Urban Runoff Pollution.** Edited by H. C. Torno, J. Marsalek, and M. Desbordes. 893 pages. 1986.

Vol. 11: **Pest Control: Operations and Systems Analysis in Fruit Fly Management.** Edited by M. Mangel, J. R. Carey, and R. E. Plant. 465 pages. 1986.

Vol. 12: **Mediterranean Marine Avifauna: Population Studies and Conservation.** Edited by MEDMARAVIS and X. Monbailliu. 535 pages. 1986.

Vol. 13: **Taxonomy of Porifera from the N. E. Atlantic and Mediterranean Sea.** Edited by J. Vacelet and N. Boury-Esnault. 332 pages. 1987.

Vol. 14: **Developments in Numerical Ecology.** Edited by P. Legendre and L. Legendre. 585 pages. 1987.

Vol. 15: **Plant Response to Stress. Functional Analysis in Mediterranean Ecosystems.** Edited by J. D. Tenhunen, F. M. Catarino, O. L. Lange, and W. C. Oechel. 668 pages. 1987.

Vol. 16: **Effects of Atmospheric Pollutants on Forests, Wetlands and Agricultural Ecosystems.** Edited by T. C. Hutchinson and K. M. Meema. 652 pages. 1987.

Vol. 17: **Intelligence and Evolutionary Biology.** Edited by H. J. Jerison and I. Jerison. 481 pages. 1988.

Vol. 18: **Safety Assurance for Environmental Introductions of Genetically-Engineered Organisms.** Edited by J. Fiksel and V.T. Covello. 282 pages. 1988.

Vol. 19: **Environmental Stress in Plants. Biochemical and Physiological Mechanisms.** Edited by J. H. Cherry. 369 pages. 1989.

NATO ASI Series G

Vol. 20: **Behavioural Mechanisms of Food Selection.**
Edited by R. N. Hughes. 886 pages. 1990.

Vol. 21: **Health Related Effects of Phyllosilicates.**
Edited by J. Bignon. 462 pages.1990.

Vol. 22: **Evolutionary Biogeography of the Marine Algae of the North Atlantic.**
Edited by D. J. Garbary and G. R. South. 439 pages. 1990.

Vol. 23: **Metal Speciation in the Environment.**
Edited by J. A. C . Broekaert, Ş. Güçer, and F. Adams. 655 pages. 1990.

Vol. 24: **Population Biology of Passerine Birds. An Integrated Approach.**
Edited by J. Blondel, A. Gosler, J.-D. Lebreton, and R . McCleery.
513 pages. 1990.

Vol. 25: **Protozoa and Their Role in Marine Processes.**
Edited by P. C. Reid, C. M . Turley, and P. H. Burkill. 516 pages. 1991.

Vol. 26: **Decision Support Systems.**
Edited by D. P Loucks and J. R. da Costa. 592 pages. 1991.

Vol. 27: **Particle Analysis in Oceanography.** Edited by S. Demers. 428 pages. 1991.

Vol. 28: **Seasonal Snowpacks. Processes of Compositional Change.**
Edited by T. D. Davies, M . Tranter, and H . G. Jones. 484 pages. 1991.

Vol. 29: **Water Resources Engineering Risk Assessment.**
Edited by. J. Ganoulis. 551 pages. 1991.

Vol. 30: **Nitrate Contamination. Exposure, Consequence, and Control.**
Edited by I. Bogárdi and R. D. Kuzelka. 532 pages. 1991.

Vol. 31: **Industrial Air Pollution. Assessment and Control.**
Edited by A. Müezzinoğlu and M. L. Williams. 245 pages. 1992.

Vol. 32: **Migration and Fate of Pollutants in Soils and Subsoils.**
Edited by D. Petruzzelli and F. G. Helfferich. 527 pages. 1993.

Vol. 33: **Bivalve Filter Feeders in Estuarine and Coastal Ecosystem Processes.**
Edited by R. F. Dame. 584 pages. 1993.

Vol. 34: **Non-Thermal Plasma Techniques for Pollution Control.**
Edited by B. M. Penetrante and S. E. Schultheis.
Part A: Overview, Fundamentals and Supporting Technologies. 429 pages. 1993.
Part B: Electron Beam and Electrical Discharge Processing. 433 pages. 1993.

Vol. 35: **Microbial Mats. Structure, Development and Environmental Significance.**
Edited by L. J. Stal and P. Caumette. 481 pages. 1994.

Vol. 36: **Air Pollutants and the Leaf Cuticle.**
Edited by K. E. Percy, J. N. Cape, R. Jagels and C. J. Simpson. 405 pages. 1994

Vol. 37: **Azospirillum VI and Related Microorganisms.**
Edited by I. Fendrik, M. del Gallo, J. Vanderleyden, and M. Zamaroczy.
588 pages. 1995.

Vol. 38: **Molecular Ecology of Aquatic Microbes.**
Edited by I. Joint. 423 pages. 1995.

Vol. 39: **Biological Fixation of Nitrogen for Ecology and Sustainable Agriculture.**
Edited by A. Legocki, H. Bothe, and A. Pühler. 339 pages. 1997.

NATO ASI Series G

Vol. 40: National Parks and Protected Areas. Keystones to Conservation and Sustainable Development.
Edited by J. G. Nelson and R. Serafin. 1997.

Vol. 41: Physiological Ecology of Harmful Algal Blooms.
Edited by D. M. Anderson, A. D. Cembella and G. M. Hallegraeff. 1998.

Vol. 42: Desulfurization of Hot Coal Gas
Edited by A. T. Atimtay and D. P. Harrison. 1998.

Springer and the environment

At Springer we firmly believe that an international science publisher has a special obligation to the environment, and our corporate policies consistently reflect this conviction.

We also expect our business partners – paper mills, printers, packaging manufacturers, etc. – to commit themselves to using materials and production processes that do not harm the environment. The paper in this book is made from low- or no-chlorine pulp and is acid free, in conformance with international standards for paper permanency.

Springer

Druck: Strauss Offsetdruck, Mörlenbach
Verarbeitung: Schäffer, Grünstadt